Introduction to Microdisplays

Wiley-SID Series in Display Technology

Series Editor:
Anthony C. Lowe

Consultant Editor:
Michael A. Kriss

Display Systems:
Design and Applications
Lindsay W. Macdonald and **Anthony C. Lowe (Eds)**

Electronic Display Measurement:
Concepts, Techniques and Instrumentation
Peter A. Keller

Projection Displays
Edward H. Stupp and **Matthew S. Brennesholz**

Liquid Crystal Displays:
Addressing Schemes and Electro-Optical Effects
Ernst Lueder

Reflective Liquid Crystal Displays
Shin-Tson Wu and **Deng-Ke Yang**

Colour Engineering:
Achieving Device Independent Colour
Phil Green and **Lindsay MacDonald (Eds)**

Display Interfaces:
Fundamentals and Standards
Robert L. Myers

Digital Image Display:
Algorithms and Implementation
Gheorghe Berbecel

Flexible Flat Panel Displays
Gregory Crawford (Ed.)

Polarization Engineering for LCD Projection
Michel G. Robinson, Jianmin Chen, and **Gary D. Sharp**

Fundamentals of Liquid Crystal Devices
Deng-Ke Yang and **Shin-Tson Wu**

Introduction to Microsdisplays
David Armitage, Ian Underwood and **Shin-Tson Wu**

Introduction to Microdisplays

David Armitage
Consultant, Los Altos, California, USA

Ian Underwood
MicroEmissive Displays Ltd, Edinburgh, UK

Shin-Tson Wu
University of Central Florida, Florida, USA

John Wiley & Sons, Ltd

Telephone (+44) 1243 779777

Email (for orders and customer service enquiries): cs-books@wiley.co.uk
Visit our Home Page on www.wiley.com

Other Wiley Editorial Offices

John Wiley & Sons Inc., 111 River Street, Hoboken, NJ 07030, USA

Jossey-Bass, 989 Market Street, San Francisco, CA 94103-1741, USA

Wiley-VCH Verlag GmbH, Boschstr. 12, D-69469 Weinheim, Germany

John Wiley & Sons Australia Ltd, 42 McDougall Street, Milton, Queensland 4064, Australia

John Wiley & Sons (Asia) Pte Ltd, 2 Clementi Loop #02-01, Jin Xing Distripark, Singapore 129809

John Wiley & Sons Canada Ltd, 6045 Freemont Blvd, Mississauga, ONT, L5R 4J3

Wiley also publishes its books in a variety of electronic formats. Some content that appears in print may not be available in electronic books.

British Library Cataloguing in Publication Data

A catalogue record for this book is available from the British Library

ISBN-13 978-0-470-85281-1 (HB)
ISBN-10 0-470-85281-X (HB)

Typeset in 9/11 pt Times by Thomson Digital
Printed and bound in Great Britain by Antony Rowe Ltd, Chippenham, Wiltshire
This book is printed on acid-free paper responsibly manufactured from sustainable forestry in which at least two trees are planted for each one used for paper production.

Series Editor's Foreword

There is an old saying that good things are worth waiting for and that is certainly true of this latest addition to the Wiley-SID Series. I only wonder at its title; the authors have been modest; this is much more than an Introduction. It is an in-depth treatment of the subject written by three experts in the field with extensive academic backgrounds, but with very differing experience outside academia. David Armitage has been a consultant to the industry for more than 20 years, Ian Underwood has founded and successfully floated his company on the London Stock Exchange and Shin-Tson Wu, after many years in industry, has returned to academia. Between them the authors bring a range of different and complementary knowledge and backgrounds to this work.

In the late 1970s, when David Armitage and I were working in IBM Research Division on liquid crystal effects for what would become known as LCOS displays, I do not believe that either of us anticipated the impact microdisplay technology would have or the economic importance it would achieve in the display industry. It is the basis of digital cinema, projection TV, computer projectors, near-to eye displays and some aspects of digital signage. The projection technologies that have become commercially successful on a large scale have been transmissive or reflective liquid crystal and micromechanical devices; the first based on polysilicon or transferred crystalline silicon and the other two on CMOS technology. Slower growth has occurred in head mounted and virtual reality displays. However, emissive OLED microdisplay technology is now appearing on the market and will surely drive increased growth of near-to-eye displays. It is a given that a dynamic and growing market will encourage the development of improvements to the existing technologies and the invention of new ones.

The book covers this commercially important and technologically diverse subject to a depth which will benefit engineers at many levels from practitioners in the field to potential users of microdisplays. The introduction provides a useful review of human factors, specifications, display technologies and applications. The second chapter covers all aspects of analog and digital addressing schemes for LCDs, DMDs and OLEDs and the next two discuss reflective and transmissive backplane technologies. There follow three chapters which deal respectively with transmissive LC microdisplays, reflective LC microdisplays and LCD assembly and testing. The next two chapters discuss the two important microdisplay technologies other than LCD; micromechanical and emissive (predominantly OLED) devices. The book concludes with two chapters on optical and psychophysical aspects of projection and near-to-eye systems.

Coverage of the subject is comprehensive not just in the sense that all the technical issues are addressed but because significant practical information relating to the economics of manufacturing and assurance of product reliability are also included. General bibliographies are provided where appropriate and each chapter contains comprehensive references.

This twelfth book in the series is the first to deal exclusively with microdisplays and creates a welcome extension of our coverage into this important area.

Anthony Lowe

Series Editor
Braishfield, UK

Contents

Preface

Devices that are now recognized as "microdisplays" have been around for many years, although the name itself is more recent. Initially, helmet-mounted miniature CRTs with magnifying optics delivered an image to an individual viewer, such as an aircraft pilot confined to a small cockpit. Personal display development demanded smaller, lighter-weight, lower-voltage alternatives to the CRT. In recent years, display designers have been presented with a range of microdisplay options satisfying many of their demands.

The flat panel dominance of direct view displays continues at small scale with advances in microdisplay projectors. Conference rooms quickly adopted the new lightweight projectors, replacing 35 mm slide shows and overhead transparencies with electronic images. Home TV projection followed, and today cinema projection is converting to microdisplays. Cinema is the most demanding application and shows electronic projection at its best, but at considerable expense. Film is long gone from home projection, which provides a huge potential market to support development and manufacturing of microdisplay projectors. The blossoming of high-definition television (HDTV), requiring higher resolution and larger screens, enhances the quest for market share amongst competing TV systems. Microdisplays are basic components of projection, having great influence on performance and price.

Display technology brings together a number of disciplines – chemistry, physics, materials science, electronics, mechanical engineering, optics, and others – in addition to which it involves other specialized areas such as visual perception and psychology. In 2002, Shin-Tson Wu, recognizing the need for a book bringing together various aspects of microdisplay design, fabrication, and application, invited David Armitage to co-author the work. The series editor, Tony Lowe, suggested that a third author, Ian Underwood, be added. The team would thus bring together a range of knowledge and experience that allows them to explain performance and clarify issues in microdisplay technologies and applications. In total, the authors of this book have injected more than 60 person-years of experience in microdisplay technology and applications into writing it. Responsibility for the structure and content of each individual chapter has resided with the author having the most relevant experience in the given topic area. General coordination of material and structure fell to David Armitage. As a guideline, we have erred a little on the side of repetition from chapter to chapter in order to allow each chapter to be read without too much cross-referencing to other chapters.

As part of the SID series, the book serves the display community, but also accommodates readers unfamiliar with display details. Many engaged in the display industry concentrate on a particular discipline such as electronics or optics; others are limited to segments such as direct-view LCD, or plasma display. The book offers easy access to microdisplay details for readers seeking to broaden their background.

We examine the dominant microdisplay forms in detail, revealing their strengths and weaknesses. Microdisplays that once attracted attention are discussed briefly, noting the reasons for their decline. Others that are in the early stages of development are mentioned in relation to their promise. A range of basic projection system designs are discussed in relation to different microdisplay types. Other important projection components, such as light sources and polarizing beam-splitters, relate to microdisplay aspects. The need to provide compactness and viewing comfort as well as magnification complicates the optics of personal displays. A review of several successful designs includes output-pupil expansion techniques to enhance viewing comfort.

The introduction places microdisplays in context and discusses human factors such as visual acuity and flicker that influence the design of all display systems. The second chapter is concerned with electronic addressing methods, particularly the digital techniques supported by silicon backplane devices. A chapter devoted to silicon technology charts development of the silicon backplane structure with evolution of the silicon fabrication industry in general. The chapter on transmission microdisplays highlights the challenge of maintaining a high aperture ratio under pressure of shrinking pixel size. Two chapters on liquid crystals cover the design options and address the field fringing issues of concern in microdisplays. The chapter devoted to the burgeoning field of electromechanical systems emphasizes the development of the DMD microdisplay. An emissive microdisplays chapter is dominated by OLED development to complete the coverage of microdisplay devices. The final chapters are devoted to applications areas – projection then near-to-eye.

The authors thank Tony Lowe (Editor) and Mike Jin (SpatiaLight) for their initial comments on the content and structure of the book. We thank members of the display community for supplying pictures and diagrams that enliven and clarify the text. We are particularly indebted to: Larry Hornbeck (Texas Instruments); Bob Melcher (Syntax-Brillian Corporation); Professor Ifor Samuel (St Andrews University); Professor Bill Crossland (Cambridge University); Dr Euan Smith and Terry Nicklin (Cambridge Display Technology Ltd) for supply of, and permission to reproduce, information and diagrams relating to polymer OLED materials and devices; MicroEmissive Displays Ltd for supply of, and permission to reproduce, information and diagrams relating to polymer OLED microdisplays; the University of Edinburgh for supply of, and permission to reproduce, information and diagrams relating to LCOS backplanes. Ian Underwood offers his grateful thanks to colleagues including Dr Georg Bodammer, Dr Alastair Buckley, Dr Dwayne Burns, Dr Christophe Miremont and Graeme Kelly for their helpful feedback on his drafts. Shin-Tson Wu is indebted to his post doctors and students for technical assistance, especially Dr. Xinyu Zhu and Dr. Simon Fan-Chiang for providing the simulation results.

About the Authors

David Armitage pursued academic interests in semiconductor and liquid crystal physics, leading to microdisplay activity at IBM and Lockheed Corporation. For many years, he has consulted in microdisplay development and applications. He has numerous publications on semiconductors, liquid crystals, and displays. Academic qualifications include PhD (Physics), Bath University, UK; MS (Physics), University of Newcastle-upon-Tyne, UK; BS (EE), Durham University, UK.

Ian Underwood entered the field of LCOS as a postgraduate student in 1983. For the next 16 years he was part of a research group that pioneered advances in LCOS devices and applications. In 1999 he shifted his attention to P-OLED microdisplays when he co-founded MicroEmissive Displays which he continues to serve today as Director of Strategic Marketing. Personal distinctions include a Fulbright Fellowship held at the University of Colorado (1991), the Ben Sturgeon Award of the UK chapter of SID (2000), seminar presenter at the SID Annual Meeting (2000 to 2004), the Alfred Woodhead best paper award of the UK chapter of SID (2002 and 2004), Ernst & Young Emerging Entrepreneur of the Year (2003), Fellow of the Royal Society of Edinburgh (2004), the Gannochy Award for Innovation of the Royal Society of Edinburgh (2004), and a keynote address on Microdisplays at Eurodisplay IDRC in 2005. In 2004 he was named Professor of Electronic Displays at the University of Edinburgh.

Shin-Tson Wu is a PREP professor at the College of Optics and Photonics, University of Central Florida (UCF). Prior to joining UCF in 2001, Dr Wu worked at Hughes Research Laboratories (Malibu, California) for 18 years. He received his PhD from the University of Southern California and his BS in physics from National Taiwan University. His studies at UCF concentrate on liquid crystal displays, liquid crystal materials, foveated imaging, tunable-focus liquid crystal/liquid lenses, bio-photonics, and laser beam steering. Dr Wu is a Fellow of the IEEE, SID, and OSA. He has co-authored **three** books: *Fundamentals of Liquid Crystal Devices* (Wiley-SID, 2006), *Reflective Liquid Crystal Displays* (Wiley-SID, 2001), and *Optics and Nonlinear Optics of Liquid Crystals* (World Scientific, 1993), **five** book chapters, and over 300 papers.

1

Introduction

1.1 Microdisplays

Advances in several technologies related to projection displays and near-to-eye (NTE) displays have intensified interest in these areas. Development of the microdisplay is a key technology. The term "miniature display" does not distinguish small displays, such as watch displays, from displays designed with magnification in mind as indicated in Figure 1.1.[1] Microdisplays are a natural extension of the familiar microfilm, where magnification is essential to readout. A coarse image is recognizable on a microdisplay, but the full resolution is only discernable with the aid of magnification. The term "microdisplay" now in general use means a compact display designed for use with a magnification system. Microdisplay projectors have taken over the consumer market from CRT projection displays that have long dominated large-screen television. NTE products such as camera viewfinders and head-mounted displays evolve with microdisplay developments.

The direct-view CRT display has grown bigger and better over the years. Competing flat panel technologies such as liquid crystal and plasma displays have followed a similar path. Conversely, microdisplays have become smaller, in step with the shrinkage in microelectronics. NTE displays have progressed from miniature CRTs to lightweight microdisplays with superior characteristics. Attractive image quality and improvements in viewing comfort have opened up a commercial market for NTE displays. Figure 1.2 shows an early head-mounted display (HMD) designed for the consumer market, employing dual liquid crystal microdisplays. The unit weighs 150 gm and simulates a 52-inch diagonal color image viewed from a distance of 7 ft with resolution $260 \times 346 \times 3$ RGB pixels.

The vacuum tube nature of the CRT has made it difficult to expand in scale over the years, and now appears to be saturating at a cumbersome 40-inch diagonal. Projection displays powered by special-purpose compact CRTs overcame the size limitation. In comparison, the simple slide projector provided a much better image, and obviously with the invention of a suitable "electronic slide" (microdisplay) would be the basis of a new video projection technology. A wide variety of electro-optic devices has struggled to fill the microdisplay role in projectors. The current choices,

Introduction to Microdisplays D. Armitage, I. Underwood and S-T. Wu

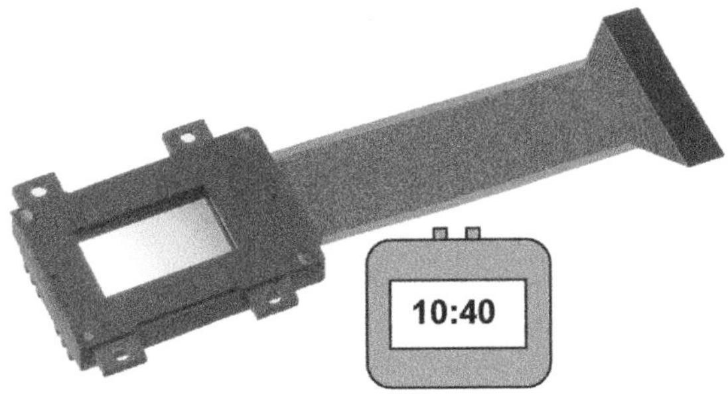

Figure 1.1 Miniature watch display compared with microdisplay having 9 μm pixel pitch, 1920 × 1080 resolution giving active area 0.78-inch diameter. Reprinted courtesy of Sony Corporation

liquid crystal and micromechanical devices, have proved commercially viable and will be difficult to displace. Price continues to exert a downward pressure on microdisplay area, due to the device cost and the influence of area on system cost. Figure 1.3 shows a portable business projector weighing 2.4 lb, with dimensions 2.6 × 6.1 × 7.8 inch, employing a single micromechanical microdisplay of 1024 × 768 resolution, diagonal 0.7 inches (DMD™), projecting full-color 1500 ANSI lumens at contrast ratio 1100:1.

Rear projection TV is a huge consumer market that drives the development of microdisplays. The latest products employ folded optic systems that reduce the unit depth to about 8 inches eroding the

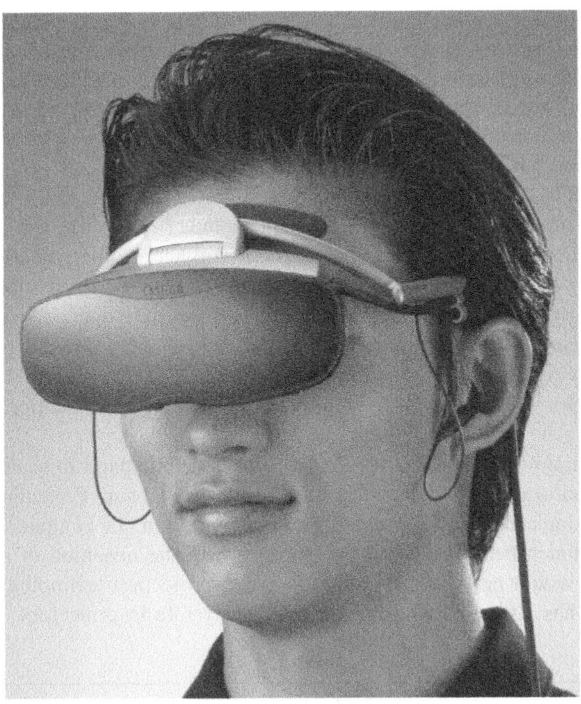

Figure 1.2 Head-mounted microdisplay GT270. Reprinted courtesy of Canon Inc.

Figure 1.3 Microdisplay portable business projector LP70+. Reprinted courtesy of InFocus Corp.

space advantage of flat panel displays. The highest quality products produce the best images of any display technology. Figure 1.4 shows a 65-inch diagonal rear projection TV, employing three XGA LCOS microdisplays: luminance 450 cd/m^2, contrast ratio 2000, weight 95 lb.

1.2 Human Factors

The display engineer needs some knowledge of human vision to understand display performance and comfort for a given application. An engineering specification of human vision provides details of resolution, sensitivity, response time, and wavelength dependence. Luminance measured in candela/m^2 (lumen/steradian/m^2) is a measure related to our sensation of brightness. The lumen is a photometric unit of light flux derived from the product of the eye's wavelength sensitivity and radiant power,

Figure 1.4 Microdisplay rear-projection TV model BR768HC. Reprinted courtesy of Syntax-Brillian Corp.

giving the visual response to different wavelengths directly. At peak wavelength sensitivity 555 nm, flux in lumens = 683(radiant flux in watts); the same radiant flux at any other wavelength is worth fewer lumens, falling to zero lumens outside the visible wavelength range. Ambient light, contrast, and color saturation also influence our sense of brightness.

1.2.1 Color

Perception of color depends on wavelength in a complicated way that allows three primary wavelengths to represent a wide color gamut according to the luminance values of the primaries. Red, green, and blue primaries (RGB) optimize the color gamut. An even wider gamut of colors follows from the addition of more primaries. The narrower the spectral range of each primary color, the more saturated the primary color becomes.

Reduction to three primaries is of great utility in electronic displays, where picture elements (pixels) grouped in RGB triads below visual resolution merge to provide a full-color display. Color pixels are standard for direct-view CRT, flat panel displays, and some microdisplays. One of the options in projector design is to superimpose primary color images on the screen, to achieve full color. Each primary has a dedicated microdisplay, requiring three microdisplays for this parallel color system. Alternatively, color sequential display systems exploit the eye's response time by presenting the primary colors in rapid succession to give the perception of a single color represented by the primaries. One microdisplay can handle the sequence of colors, but it must operate at high frame rate. Eye movement during the color sequence will cause some separation of primary colors on the retina, perceived as color breakup. Eye movement sets a lower limit on the color field rate needed to suppress color breakup.

Color temperature is a measure of the spectral distribution of a light source, by comparison with the temperature of equivalent black body radiation. A color temperature of 6504 K represents average daylight. An arc-lamp source can be adjusted to a given color temperature by spectral filtering, with some sacrifice in output. Primary color separation requires filtering, where strong color saturation incurs further loss. Some compromise between color temperature, color saturation, and luminance is required. Color vision is lost at low light level $<10^{-4}\,\text{cd/m}^2$; display engineering is essentially concerned with substantially higher luminance levels to provide good color vision.

1.2.2 Resolution

Display costs generally increase with resolution, making it wasteful to provide display resolution beyond viewer requirements. Light level, modulation depth, and wavelength of the signal influence eye resolution, peaking at maximum sensitivity 555 nm. The response to one-dimensional sinusoidal luminance patterns characterizes eye resolution as a function of spatial frequency (u), luminance (L), and contrast (C).

$$\textit{Contrast ratio} \quad CR = \frac{L_{max}}{L_{min}} \tag{1.1}$$

$$\textit{Contrast} \quad C = \frac{L_{max} - L_{min}}{L_{max} + L_{min}} = \frac{CR - 1}{CR + 1} \rightarrow 1 - \frac{1}{CR} \tag{1.2}$$

$$\textit{Perception boundary} \quad L = L_{av}\left[1 + C_{min} \sin u\right] \quad L_{av} = \frac{L_{max} + L_{min}}{2} \tag{1.3}$$

The contrast sensitivity function $S(u, L_{av}) = 1/C_{min}$, where C_{min} is the minimum contrast required to perceive the sinusoidal fringes at spatial frequency u and average luminance L_{av}. Contrast sensitivity plots reveal the most sensitive spatial frequency and upper frequency cutoff. Figure 1.5 shows contrast

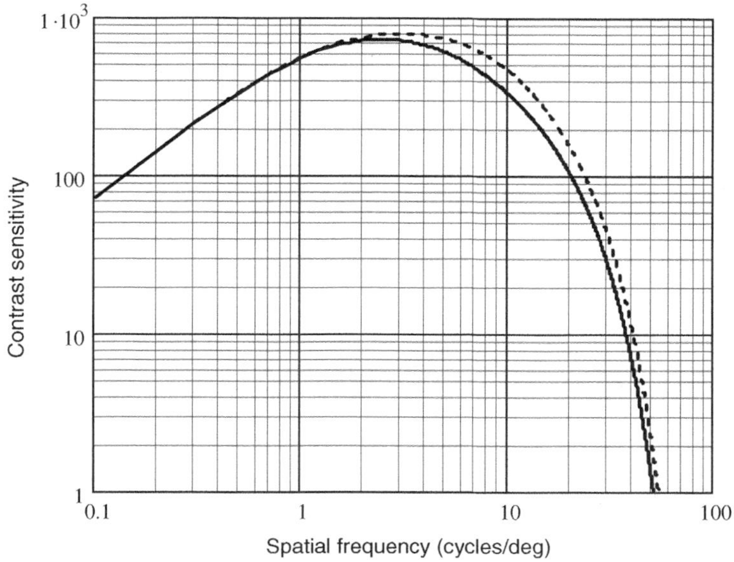

Figure 1.5 Contrast sensitivity for luminance: solid line 100 cd/m^2; dotted line 900 cd/m^2

sensitivity plots appropriate to displays. The plots are generated from formulae derived from a signal processing description of human vision, with parameters fitted to reported data.[2] Peak contrast sensitivity is about 4 cycles/deg, and cutoff about 60 cycles/deg, consistent with a limiting angular resolution of 0.5 minute of arc applicable to edge detection, and comparable with an accepted average value of 1 minute for human visual acuity. Increase in luminance provides a small increase in resolution at display luminance levels. The frequency cutoff corresponds to foveal vision, which is limited to about 2 degrees field of view. Eye and head movement compensate the limited field of view of the eye at high resolution, creating the impression of high resolution over a wide angle. Restriction of natural eye and head movement has a disturbing effect on vision that can result in eyestrain and discomfort.

The lower sensitivity of non-foveal vision contributes to the decline in sensitivity at low spatial frequency, where the spatial wavelength extends beyond the foveal range. Further low-frequency attenuation occurs at the neural processing level. Gradual variation in luminance of order 50% from the center to boundary of a display may not be noticeable, while shifts less than 1% are obvious at sensitive spatial frequencies. Color uniformity is similarly sensitive to spatial frequency. Moreover, we are more sensitive to color change than luminance change at low spatial frequency, making color uniformity more difficult to achieve in display systems.

Correlation in the visual process, known as hyperacuity, allows perception of an object's positional accuracy well beyond visual acuity. Vernier acuity is the ability to detect misalignment in lines and objects, which can be an order of magnitude higher than visual acuity. Spatial aliasing associated with pixelation gives rise to jaggy diagonal lines, detected by vernier acuity before perception of pixelation in general. Reducing pixel size below visual acuity may not suppress all the effects of pixelation.[3]

1.2.3 Flicker

Fluctuations in display luminance give rise to an annoying flicker. Frame-by-frame writing of electronic displays requires sufficient frame rate to avoid flicker. The early movies picked up the name flicks due to inadequate frame rate. With increase of flicker frequency, flicker attenuates, approaching

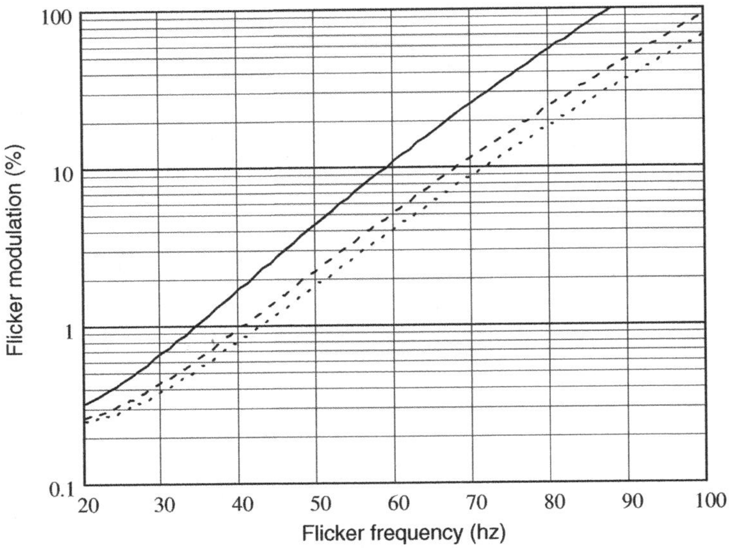

Figure 1.6 Threshold modulation for 50% probability of seeing flicker on a full white screen. Average luminance: solid line 100 cd/m^2; broken line 500 cd/m^2; dotted line 900 cd/m^2

threshold perception at a critical flicker frequency (CFF). Threshold perception is the level at which 50% of the population will detect flicker. Experimental studies of the dependence of flicker perception on luminance, modulation level, and frequency enable the elimination of display flicker.

The model for human vision referred to earlier generates Figure 1.6, on inserting typical parameters for threshold perception.[2] The flicker modulation m is of sinusoidal form $[1 + m \cdot \cos(2\pi wt)]$, at frequency w hertz. For non-sinusoidal flicker, m is the value of the fundamental Fourier component. The flicker parameters generating Figure 1.6 assume the worst case of a completely white screen at full brightness, a severe test of flicker compared with typical video projection. Perceptible flicker modulation decreases with increase of luminance, demanding a higher flicker frequency for suppression. An adjustment of the model parameters will give plots for a lower probability of flicker perception, predicting display conditions to eliminate flicker for essentially the entire population.

1.2.4 Contrast Ratio

Contrast is an important characteristic of image quality; low-contrast images have a washed-out appearance. Contrast sensitivity plots such as Figure 1.5 show that high contrast is required to achieve limiting eye resolution. High contrast ratio in all primary color channels promotes strong color saturation, and eliminates low-level color distortion, e.g. true black rather than dark purple.

The eye adapts to changes in average luminance, enhancing the visibility of dark scenes in a movie. To maintain image quality in dark scenes requires adequate contrast ratio at low luminance, implying high contrast at full luminance. Cinema-quality imaging requires $CR > 1000$ to reveal detail in dark scenes.

Ambient light reflecting from the display surface sets a minimum luminance level, requiring higher display luminance for a given CR. Screen luminance <60 cd/m^2 is accommodated by low lighting in a cinema supporting $CR > 1000$. Monitor displays designed for typical office lighting favor luminance of 150 cd/m^2 or higher, and light reflected from the screen may degrade $CR < 50$. Outdoor displays exposed to sunlight demand extreme luminance, prompting the development of reflective displays that exploit the high luminance environment. Shielded viewing inherent to NTE displays makes them insensitive to ambient light.

1.2.5 Grayscale

Analog displays such as CRT and LCD are continuously variable, making grayscale accuracy an issue of stability and noise. The advantages of digital systems result in a digital video signal converted to analog form to drive an analog display. Digital look-up tables (LUTs) assure the correct analog drive level, but digital bit-depth limits the number of gray levels. Making any display part of a digital system imposes grayscale quantization. Displays with only on/off luminance capability are inherently digital and achieve grayscale by pulse width modulation (PWM), averaging to the desired gray level over a frame period. Binary pulse code modulation generates equally spaced luminance levels over the luminance range; a poor match to the eye with an approximately logarithmic response to luminance, requiring progressively increasing steps in luminance for a perceptually uniform grayscale.

An inadequate number of gray levels gives rise to a contouring artifact in regions of near uniform luminance (e.g. image of sky), where the minimum quantized step is visible as an edge depicting the luminance contour.[4] A similar effect in color quantization appears as color contour boundaries, described as posterizing. At a given light level, fewer than 100 gray levels are required to avoid luminance contouring, provided the gray level steps are perceptually uniform, i.e. logarithmic. To provide adequate grayscale in dark scenes requires enhanced gray level count; cinema quality demands 1000 perceptually uniform gray levels.

Video image L_{in} transmitted with video signal voltage encoded $V_s \propto L_{\text{in}}{}^{0.455}$, to match typical receiving CRT luminance characteristic $L_{\text{out}} \propto V_s^{2.64}$, gives overall $L_{\text{out}} \propto L_{\text{in}}{}^{1.2}$. The display gamma ($\gamma = 2.64$) is adjusted to the viewer's preference of image gamma. The image gamma determines the distribution of grey levels in an image, where adjustment of gamma allows the viewer to optimize the appearance according to taste and viewing conditions. Video image gamma is typically set to 1.2, consistent with dimly lit viewing conditions; however, choice varies with image as well as individual.[2] It is interesting that fidelity in display engineering is usually overridden by viewer preference in choice of gamma. A similar distortion appears in choice of color temperature. Our vision evolved in response to daylight, making that the natural choice. However, given an adjustment in color temperature we will generally make color temperature higher than daylight, enhancing the blue region of the image spectrum.

1.2.6 Viewing Comfort

Adequate luminance and resolution are basic requirements for any display. The viewer's eyesight determines the upper limit on resolution. Tolerance to lower resolution depends on the information content of the display. The eye adapts to a wide range of luminance, making acceptable luminance dependent on ambient lighting and spurious display reflectance. Obviously, the display should be well engineered and in good working order, free from flicker and noticeable distortion. Direct-view displays should be set slightly below eye level, at a distance about two or three times the screen diagonal. Image quality is sensitive to viewing angle in some displays, and generally favors on-axis viewing.

Viewing comfort in head-mounted displays is much more sophisticated than in direct-view displays. Weight, size and balance have no counterpart in direct view, but are critical to HMD tolerance. Total-emersion HMDs concentrate vision on the display, shutting out all extraneous light to provide an artificial reality experience. It is very difficult to simulate normal vision effectively over a realistic field of view, including eye and head movement. Inadequate simulation gives rise to motion sickness and other discomforts known collectively as simulator sickness.

Partial-emersion HMDs allow some vision of the outside world, which preserves the viewer's orientation, and are tolerable for much longer periods, even when the outside view is restricted to peripheral vision. Eliminating the influence of head movement on the displayed image makes eye movement do all the work in scanning the image. A field of view in excess of about 35 degrees induces intolerable eye fatigue due to the scanning effort. Viewing discomfort is a difficult barrier to overcome for the HMD to gain general acceptance. The interest in wearable computers and portable internet displays provides an incentive for further development and innovation in HMDs.

1.3 Display Specifications

A display designer emphasizes resolution, luminance, etc., according to the targeted application.[5] In marketing the display, a list of specifications identifies its value in various applications.

1.3.1 Resolution and Size

Modulation transfer function (MTF) has the same functional form as Equation (1.1) for contrast, and describes the decline in contrast with spatial frequency in cycles/mm = (line pair)/mm = lp/mm. MTF is the standard method of describing the resolution of optical components such as a projection lens, and the product of component MTFs gives the overall MTF. Display resolution determined by raster scan or pixel count cannot be expressed in MTF form without loss of mathematical rigor. The number of TV lines or pixel array size characterizes display resolution. The video graphic adapter (VGA) notation, listed in Table 1.1, identifies standard pixel array formats, extended by inclusion of the high-definition television notation.

Screen diagonal characterizes the display size, a legacy of the early circular screen CRT. The cost of a direct-view display increases more rapidly than the area, since the defect probability and assembly problems increase. Plasma displays have demonstrated the largest diagonal in excess of 100 inches, followed closely by LCDs. The size and performance of flat panel displays competes with projection TV. When manufacturing costs have settled down, price will determine the target market for projection, now set at >40 inches diagonal. TV displays viewed at a distance two or three times the screen diameter barely resolve (1 arc min acuity) pixels at WXGA resolution, while computer monitors viewed at a distance comparable to screen diameter just resolve WUXGA. Higher visual acuity of lines and edges favors extended display resolution. HDTV at 1080 × 1920 resolution may pull the viewer closer to the screen, intensifying the experience, particularly for sporting events.

1.3.2 Luminance and Color Saturation

In CRT projectors and plasma flat panels, the peak luminance cannot be maintained over the entire screen due to limited power dissipation. A peak luminance and an average luminance should be specified. Consumer CRT projectors deliver an average of about 200 lumens to the screen and require high screen gain for adequate luminance of about 350 cd/m^2. Projectors in general can take advantage of screen gain to increase the screen luminance in a preferred direction at the expense of lower luminance in other directions. The focusing and scattering properties designed into the screen determine its gain.

Table 1.1 Video graphic adapter designation and array size

VGA	640 × 480	SVGA	800 × 600	XGA	1024 × 768
SXGA	1280 × 1024	UXGA	1600 × 1200	QVGA	320 × 240
WXGA	1365 × 768	WUXGA	1920 × 1200	GXGA	2560 × 2048
SDTV	729 × 480	HDTV(720P)	1280 × 720	SHDTV(1080P)	1920 × 1080

Increased screen gain has the downside of enhanced speckle as well as reduced viewing angle. Front projectors specify lumens delivered to the screen, since the luminance depends on screen area and gain. Rear-projection units have built-in screens and luminance quoted over a range of viewing directions. Acceptable luminance depends on ambient lighting. Cinema projectors illuminate a large area screen to give about $60\,cd/m^2$ screen luminance, and require a dark ambient environment to appreciate image quality. Microdisplay projectors aim for $500\,cd/m^2$ or higher to accommodate the higher ambient light favored by the business and consumer markets.

Color saturation is strongest in laser or LED driven displays, with sharply defined wavelength. Dichroic filters in arc-lamp projection displays determine the color, where stronger color saturation implies lower throughput lumens. Specification of projector output luminance is sometimes inflated by quoting the value obtained before color correction. A similar tradeoff applies to LCDs using dyed pixels for color, where optical absorption introduces severe throughput loss. Phosphor characteristics limit the color saturation of plasma and CRT displays; however, color saturation is sometimes enhanced by addition of dichroic filtering in CRT projectors.

Backlighting power and throughput efficiency determine the luminance of LCDs. Improvements in throughput efficiency should improve the already high luminance of $500\,cd/m^2$. Plasma displays are marketed with small-area luminance beyond $500\,cd/m^2$. Microdisplays reduce cost by shrinking the diagonal to 0.7 inches or less. The development of small-arc projection lamps has kept pace with microdisplay contraction, maintaining the optical collimation necessary for efficient lumen throughput. We are entering an intense stage of competition for large-screen home theatre, where several technologies vie for consumer attention.

1.3.3 Contrast Ratio and Grayscale

Contrast ratio is an important indication of image quality. It is quoted for dark ambient; room lighting always reduces the CR. The largest degradation in CR is associated with diffuse reflecting surfaces, such as the powder phosphors in CRTs and plasma displays. Projectors quote the serial all-on/all-off CR, along with the ANSI CR for a white/black checkerboard pattern; ANSI CR is always lower than the serial CR, due to internal light scattering, and provides a better indication of image quality. LCD CR is limited by the extinction ratio of polarizing optics and off-axis retardation dependence; however, optical compensation achieves $CR > 500$. Projection systems achieve $CR > 1000$, due to higher quality polarization optics and better compensation over the limited field angle of the projection lens. A recent development expands the effective CR by modulating the light source according to the image's average light level, to maintain excellent CR and grayscale in dark scenes.[6]

The standard 8-bit grayscale is adequate for most purposes, and applied to each primary color channel gives 24-bit color. More demanding applications such as cinema projection require 12-bit grayscale/channel. Look-up tables create appropriate luminance grayscale steps, taking into account the device characteristic. Pulse width modulation grayscale, necessary in plasma displays and other digital display devices, achieves high precision in grayscale; however, image contouring and motion artifacts associated with PWM requires an effective expansion in addressing bit depth.

1.3.4 Response Speed and Flicker

Video displays need to respond at a fast enough rate to avoid motional blurring and flicker. We are somewhat forgiving of motion blur, since our vision is less acute in observing movement. Movies shot at 24 frames/sec are acceptable, which is not very challenging for an electronic display response time. However, cinematographers structure the scenes and camera angles to minimize artifacts such as false wheel rotation; a higher frame rate is desirable. Theatres shutter the film projector to raise the flicker rate to 48 Hz or higher to suppress flicker perception. Motion picture frame time is 42 ms, but the frame changes abruptly, implying equivalent display response $\ll$42 ms to duplicate film. A video data rate of 60 frames/sec attenuates flicker, but requires response $\ll$17 ms to maintain gray level integrity and avoid trailing on fast-moving images. The response time is severely challenged in color field sequential displays, where color is formed by a rapid succession of primary color frames above the color fusion frequency of the eye. Color frame rates as high as 540 frames/sec are required to avoid color breakup caused by eye movement.

The image refresh rate introduces an intensity modulation at the refresh frequency, and half that frequency if the display mechanism is susceptible to odd/even asymmetry as in the LCD. Flicker issues are resolved in display design by minimizing the flicker modulation and raising the refresh frequency. The displayed frame rate may be doubled to reduce flicker if there is significant modulation at half frame rate.

A static image displayed for a substantial period may store some aspects of the image that persists for some time as a 'ghost image' superimposed on the newly addressed image. The effect is described as ghosting or image sticking. Liquid crystal displays are susceptible to image sticking due to ionic charging effects. Plasma displays are also prone to ghost image effects.

1.4 Displays in General

To place microdisplays and applications in context we discuss electronic displays in general and the shrinkage of direct-view displays into the micro domain. We restrict our attention to high-information-content displays, since there is little value in magnifying low information content capable of direct display.

1.4.1 Cathode Ray Tube

The CRT has been the dominant technology for many years. Reliability and low cost have outweighed its shortcomings, until recent developments in flat panels. The simplicity of electron-beam addressing remains appealing, but carries vacuum-tube baggage that occupies valuable space, and has become unfashionable. The shape of the tube has improved over the years, including attempts at a flat structure, but does not compare with a lightweight flat panel display. The tradeoff in luminance verses resolution is a technical limitation that is proving difficult to surmount. Diffuse reflection from the phosphor powder penalizes CR under modest room light. The direct-view CRT will play a prominent role in displays for many years and gradual improvement in performance will continue. However, flat panel performance is improving more rapidly and will continue to win market share from CRTs.

The luminance/resolution issue is more marked in projection CRTs, where inadequate luminance is the biggest drawback. Improved phosphors and the application of thin film phosphors should enhance performance. However, competitive pricing of high-quality microdisplay rear-projection HDTV is pushing the CRT units off the showroom floor. The development of large-area flat panel displays competes with projectors in general.

Miniature CRTs were developed for NTE displays such as military helmet displays. As microdisplays, they played an important role in the development of head-mounted displays, and are still in use. Microdisplays with superior performance and advantages in weight and volume have made the miniature CRT obsolete. Detailed discussion of microdisplays in later chapters excludes CRTs, because they do not provide much insight into current microdisplay development.

1.4.2 Matrix Addressed Displays

Electrode-addressed displays, such as liquid crystal displays, must use multiplex addressing to reduce the wiring complexity. Matrix addressing uses a rectangular network of electrodes similar to (x,y)

Cartesian coordinates, to address a pixel located at the (x,y) intersection of the electrodes. A rectangular array of $M \times N$ pixels can be addressed by $(N+M)$ electrode lines, plus the common return line. Detailed analysis of matrix addressing shows that the electro-optic response of the pixel must be highly nonlinear to address a large number of pixels in an independent manner. Moreover, line-by-line addressing implies average luminance decreases with increase in addressed lines, unless the pixel is bistable. Devices such as plasma displays have nonlinearity favoring high-resolution addressing and bistability, accounting for their early success as flat panels.

Early LCDs adopted cell designs that optimized the performance for passive matrix addressing, at the expense of response speed and CR. Active matrix addressing developed to eliminate the need for nonlinearity in the electro-optic response. A field effect transistor at each pixel activates the pixel according to the (x,y) addressing signal, and isolates the pixel between refresh cycles. The development of amorphous-silicon (α-Si) thin film transistor (TFT) arrays enabled large-area active-matrix liquid crystal displays (AMLCDs), with nematic liquid crystal (NLC) cell design optimized for display performance.

A large-area display has plenty of peripheral space to make connection to the matrix electrodes, or accommodate chip-on-glass methods to reduce the external connections. However, the small electrode pitch ($\sim$10 μm) of a microdisplay discourages external connections and peripheral space is minimal. The best solution is to incorporate the addressing electronics with the electrode matrix. α-Si transistors are inadequate for the addressing electronics, where polysilicon or single-crystal silicon (c-Si) devices are favored. High-temperature polysilicon circuitry on quartz provided the first AMLCD microdisplays used in projectors. Silicon wafer technology is cost-effective in microdisplays, but favors reflective rather than transmission optics, e.g. the liquid-crystal-on-silicon (LCOS) microdisplay.

1.4.3 Field Emission Displays

Field emission displays (FEDs) employ cathodoluminescence similar to the CRT, but the electrons are field-emitted from multiple cathodes, rather than thermo-emitted. A vacuum tube environment is still required, but multiple cathodes support a flat panel structure. Each source pixel contains many cathodes, with emission modulated by control gates, and focused to activate a similar sized phosphor pixel. The pixels are matrix addressed by driving the voltage between control gates and cathodes. A revival of interest in FEDs is making slow progress in competing with established flat panel displays, and rapid advances in OLEDs. Early development of an FED microdisplay yielded little published data, and transformed into a successful OLED program. FED remains an interesting technology, but is unlikely to compete in the microdisplay arena. In keeping with miniature CRTs, the need for a high-voltage and vacuum environment is an overwhelming handicap in NTE applications.

1.4.4 Plasma Displays

Light emitted from a gas discharge, similar to neon lighting, is the basis of plasma display panels (PDPs). Designing the gas discharges to emit UV light that excites primary-color phosphors provides a full-color display. The outstanding feature of the PDP is large screen area, with diagonal beyond 100 inches, and still growing. It provides peak luminance greater than 500 cd/m^2, and high CR > 1000 in a dark ambient. In ordinary room light, the CR may drop considerably due to optical scattering from the phosphors and plasma screening filter. Pulse width modulated grayscale and susceptibility to image burn-in are handicaps. The PDP competes in the business and consumer home-theater markets. The structural complexity to address and control the discharge has prevented the plasma display from shrinking to microdisplay dimensions.

1.4.5 Liquid Crystal Displays

Displays utilize the birefringence property of liquid crystals to modulate polarized light, or create strong optical scattering. High-resolution video displays favor polarization modulation, requiring the additional complication of polarizing optics. There are a number of nematic liquid crystal (NLC) cell configurations, such as twisted nematic (TN), and vertical aligned nematic (VAN), each having some advantage in a given application. A large electro-optic effect at a low voltage is the overwhelming advantage of liquid crystals, together with video response speed.

Liquid crystals modulate light, enabling operation in sunlight using a reflective configuration. Backlighting gives the best performance in ordinary room light, and pixel dyes provide primary color pixels to form full-color images. Pixel dye losses sacrifice optical throughput efficiency, which would benefit from improved color filtration methods. The gray levels are dependent on viewing angle, but various compensation schemes have evolved to minimize the off-axis image degradation. AMLCD covers a wide range of display sizes, from hand-held devices to large scale (>100-inch diagonal). Luminance $>500\,cd/m^2$ is achieved, with $CR > 500$.

AMLCD adapts to microdisplay dimensions by incorporating the matrix addressing circuits on the matrix periphery. High-temperature polysilicon on quartz provides circuitry for AMLCD projectors and head-mounted displays. The application of silicon chip technology to fabricate LCOS microdisplays achieves pixel dimensions $<10\,\mu m$. LCOS promises lower-cost higher-performance projectors and NTE displays.

Ferroelectric liquid crystals (FLCs) have a faster response, but require PWM grayscale. The manufacturing process of FLC displays is more difficult to control over large areas, compared with NLC, and FLC has a narrower temperature range. The advantages of FLC in microdisplays are faster switching enabling superior color sequential performance, and a thinner cell promoting higher pixel resolution.

1.4.6 Electroluminescent Displays

A thin film of electroluminescent material supports a large electric field in response to a modest applied voltage. The high field releases and accelerates electrons to several electron volts, exciting the activator atoms hosted by the film, which emit light on decaying to their ground state. Alternating-current thin film electroluminescent displays (ACTFELs) have been available for some time and have an advantage in rugged applications. Improvement in materials promotes competitiveness in large-screen consumer applications. PWM grayscale is a disadvantage.

The solid-state thin film structure of ACTFEL is readily adapted to microdisplay dimensions. A monochrome microdisplay designed for head-mounted displays employed c-Si-on-insulator addressing. The addition of a liquid crystal color shutter provided field-sequential color at the expense of lower luminance.[7] The excessive voltage requirement handicapped further development. Progress in OLED microdisplays has prompted the termination of the ACTFEL microdisplay program. The decline of interest in the ACTFEL microdisplay precludes detailed discussion.

Light emitting diodes (LEDs) exploit electron/hole recombination radiation, making them the most efficient electroluminescent elements. Inorganic LED arrays have been demonstrated, but the performance and cost is not competitive with established high-resolution displays. Recent developments in organic light emitting diodes (OLEDs) have produced full-color direct-view displays and microdisplays. The acceleration in OLED development challenges the established display technologies across the board.

1.4.7 Electromechanical Displays

Electromechanical displays have a long history, from simple clocks to motion picture film projection, and laser scanning. Development of electron-beam scanning provided a means of addressing a micromechanical structure designed to project an image by optical deflection or diffraction. Finally, microelectronic matrix addressing controls an array of micromechanical pixels. The revival of interest in micromechanics gave rise to the terms micro-electromechanical (MEM) devices and systems (MEMS), which may have non-optical functions such as pressure sensing, compared to micro-optical-electromechanical systems (MOEMS).

Scanning laser displays use mechanical scanning systems in conjunction with intensity modulation such as acousto-optic modulation. A very low power version scans the image directly onto the viewer's retina, and can take advantage of low-power LEDs. Voltage modulation of the LEDs and MEMS deflection makes a compact NTE system with full-color imaging. We have omitted further discussion of simple optical scanning systems of either high power or low power, on grounds that they do not have a physical microdisplay and magnification. However, MEM devices that started life as two-dimensional arrays, but which have found application as one-dimensional arrays plus scanning, are included.

The digital mirror device (DMD) is the first MEMS device to achieve large volume production. The DMD is designed for projection use, where the tilting mirror pixels deflect readout light out of the projection lens aperture, modulating the light on the projection screen. MEM devices are often designed in analog form, but maximum projector throughput and gray level precision are achieved by on/off binary operation. The downside is PWM gray levels. The DMD is fast enough for field-sequential color, does not require polarized light, and has high throughput and CR.

DMD-type angular modulation requires collimated light and is not suited to direct-view displays. However, other MEMS structures may have more general application. A prototype MEMS modulator based on optical interference is being developed in direct-view form.

1.5 Microdisplay Evolution

Early television investment favored CRT displays that could satisfy the huge demand for household receivers. CRT projection could also provide larger images with some sacrifice in luminance. Alternative projection methods were of interest, but did not justify sufficient investment for near term success. Work continued at a modest pace to develop alternatives to CRT technology. The emerging optical signal processing and computing community needed a two-dimensional optical modulator.[8] They used the term "spatial light modulator" (SLM) to distinguish it from shutters and acousto-optic modulators. The SLM impresses two-dimensional information onto an optical beam, where further processing with other optical components and SLMs produces the desired output signal. The resolution, response speed, and wavelength of the SLM are system issues that can extend well beyond the limits of human vision.

SLMs provide the two-dimensional parallelism essential to optical processing or computing, and experiments proceeded with whatever devices came to hand. Military funding developed SLMs for specific applications and general-purpose SLMs, where performance rather than cost was critical. If a missile is heading your way, you will pay a lot more to compute its trajectory a little faster. SLM programs had the advantage of steady funding geared to small production runs. Commercial display device programs needed to consider progression to large-scale production and competitive pricing.

Early SLMs employing magneto-optical materials were unsuited to display development. SLMs based on electron-beam scanned addressing could take advantage of the CRT manufacturing base. Electro-optic crystals were obvious candidates for the optical modulating element. Crystals such as lithium niobate require high voltage ($\sim$1000 V) while lower voltage crystals under development required more electrical charge. The product of charge and voltage is approximately constant for all electro-optic crystals. Fabrication, testing, and analysis of beam-addressed crystals showed that the resolution is limited by electric field fringing in the crystal, implying crystal thickness $<$100 µm. The device did not win favor for large-scale manufacturing, and languished as a potential high-power, high-cost projection component.[9] Electron-beam-addressed micromechanical devices showed more promise for consumer applications, but development could not surmount the manufacturing hurdle and lifetime problems.

The arrival of liquid crystals solved a number of problems. Low voltage and low charge simplified the addressing, while layers less than 10 μm thick improved resolution. They were not compatible with vacuum tubes, and photo-addressing became the favorite method.[10] A photoactive layer, such as cadmium sulfide, controlled the voltage applied to the liquid crystal according to the local addressing light intensity. Photo-addressing has intrinsic value in optical processing systems, but requires an optical input device such as a CRT to activate the photo address in display applications. A CRT coupled to a photo-addressed SLM with arc-lamp readout forms a light amplifying system for the CRT image. Projectors based on such light amplification schemes were marketed for many years, and gave rise to the term "light valve".[11] CRT photo-addressed liquid-crystal light-valve projectors have given way to electronic-addressed devices with similar performance at lower cost.

After the flat panel display community recognized the promise of liquid crystal displays, development of materials and device configurations quickly followed. However, the need for active matrix addressing slowed the progress towards displays with high information content. c-Si MOS technology could easily handle the electronic complexity of AM addressing on small-area devices. All the pieces were now in place for liquid crystal microdisplays. The first demonstration of MOS active matrix addressing employed the dynamic scattering effect in a nematic liquid crystal,[12] as did the later demonstration with integrated drivers.[13] Devices based on polarized light modulation replaced dynamic scattering, which soon become obsolete. The early displays demonstrated live TV images in 2-inch diagonal direct-view form, but the ability to shrink the scale was obvious. Addressing technology progressed to CMOS and the microdisplay became known as liquid-crystal-on-silicon (LCOS). LCOS microdisplays preceded the development of direct-view AMLCD displays, which awaited advances in α-Si thin film transistors.

The early LCOS devices suffered severe light leakage from the gaps between pixels, making them only of value in low-light-level applications such as camera viewfinders. An aluminum light shield under the pixels solves the problem, but the necessary multilayer metallization was not available in the early days. Chemical–mechanical polishing is a further advance in silicon technology that enhances the flatness and uniformity of LCOS microdisplays. LCOS will continue to benefit from advances in silicon wafer technology.

An alternative approach to electronic addressing employed a charge coupled device (CCD) structure to create a two-dimensional charge pattern, driven to the liquid crystal interface, in the same manner as photo-addressing. A dielectric mirror on the readout side shielded the silicon from output light, similar to the photo-addressed device. It had useful optical processing and projection applications, but could not compete in price with the developing AM addressed devices.[14]

While LCOS waited for advances in wafer silicon methods, flat panel demands pushed the rapid development of α-Si TFT addressing. The same companies were also interested in projection and soon produced the high-temperature polysilicon (HTPS) on quartz TFT for transmission liquid crystal microdisplays.[15] HTPS enabled the first AMLC microdisplay projector, and continues to support the LC projector market. Transmission devices enjoy simpler optical systems, but restriction in the pixel aperture sacrifices throughput compared with reflective displays.

The arrival of ferroelectric liquid crystals (FLCs) with enhanced response speed renewed interest in color sequential systems.[16] FLC microdisplays achieve color sequential with the help of pulsed LED readout in NTE displays. Further developments in FLC technology may provide sufficient speed for projector or flat panel color sequential applications.

Micro-optical-electromechanical systems (MOEMS) developed alongside liquid crystals, also benefiting from optical signal processing support. General interest in MEMS for a variety of applications other than displays generated considerable research activity and fabrication methods. Standard CMOS production methods can accommodate the integration of MOEMS with CMOS active matrix addressing. A complete microdisplay fabricated on wafer scale requires only testing and packaging to complete the production. Recognition of the advantages of digital mechanics led to the development of the digital mirror device (DMD), making this the only MEMS technology in projection display production.[17]

1.6 Microdisplay Applications

1.6.1 Projection Displays

The development of light valves governed the development of optical projection displays. The Eidphor oil-film projector and its descendants served the high-powered segment, while CRT/liquid-crystal systems had wider appeal. The arrival of AMLCD microdisplays captured the low-powered lower-resolution market. Microdisplay projectors gradually extended their range to higher resolution and higher power, essentially eliminating the earlier technologies. Liquid crystal and MEMS microdisplays now compete for projector market share.

The projection market should grow rapidly in the near future. The business projector market is well established and still expanding with a variety of projectors. Inexpensive, lightweight projectors with >1000 lumen output at XGA resolution suitable for conference rooms are available from many sources. Performance and price extend upwards according to auditorium demands. Top-of-the-line electronic cinema projection competes with established film projection, and must match or exceed film quality. Electronic cameras are beginning to displace film, making the entire movie system electronic, despite the wealth of artistic experience invested in film. The minimum specifications for electronic cinema projection are: 7000 lumen output, 1000 line resolution, 1000:1 CR, color equal to film, and faster response speed.

Rear-projection TV is the largest market. Microdisplay rear projection, already established, should take advantage of the move to HDTV. Acceptable performance requires about 500 lumens output at WXGA resolution, with color saturation and gamut comparable to CRT values. The consumer is sensitive to price, and the replacement cost of the arc lamp powering the display. The cost of the microdisplay is roughly proportional to area, but the cost of supporting optical components has a stronger dependence on area. The area should be minimized, consistent with delivering 500 lumens to the screen. Optical collimation and lifetime dictate a high-pressure mercury arc lamp. Such lamps (100 W power) have maintained screen lumens above 50% of initial value for more than 20 000 hours. The price of components will bottom with standardization of dimensions and performance, promoting extreme volume production.

High-power LEDs supplanting the arc lamp as the projection light source is an interesting development. Miniature projectors powered by LEDs are marketed, and home rear-projection TV has been demonstrated. LED illumination may dominate consumer TV production within the next five years, bringing enhanced saturation and color gamut, with lifetime beyond the product replacement cycle.

The tradeoff in performance and price to achieve consumer best buy status has yet to be established. Field-sequential color systems employ only one microdisplay, shedding cost and color convergence issues, compared with the three microdisplays of parallel color systems. However, the cost/performance tradeoff sacrifices throughput efficiency and image quality. Some of the loss and quality are restored by color band scrolling in a single microdisplay system. High-speed liquid crystal configurations exist for the single microdisplay option, but some increase in device area is favored to enhance throughput. The DMD projector employs a single microdisplay system for consumer markets and three microdisplays for high-performance markets.

1.6.2 Near-to-Eye Displays

Near-to-eye displays create an image for a single eye or pair of eyes, greatly reducing the output light compared with direct-view or projection displays. The light output is concentrated in an output pupil set by the magnification system. A large output pupil, or eye box, makes for easy viewing by relaxing critical eye positioning. Ambient light is not critical when masking the eye to receive little more than the desired image light.

NTE applications are invariably portable, such as camera viewfinders, mobile phones, and portable computers. Total power consumption is an important consideration in extending battery life. However, low light level relegates optical efficiency to a secondary consideration in video displays, where power lost in active matrix addressing is a primary consideration. In non-video applications such as text, diagrams, and charts, image storage (e.g. bistable nematic) has an advantage.

Compact lightweight systems are essential for head-mounted displays in particular, and a general requirement for portable equipment. Single microdisplay systems have an overwhelming advantage over multiple microdisplay options. LEDs in primary colors provide simple compact field-sequential operation, moreover their rapid switching compensates the marginal response time of the LC microdisplay. Price and field-of-view handicap the DMD in NTE display applications.

OLEDs are an attractive option for NTE displays. Lambertian emission fills the aperture of the magnifying optics, and power requirements are low. RGB pixels are the natural imaging method for OLEDs, which is a drawback at high magnification. However, color sequential filtering is possible with white light emitters. Several features of microdisplays favor early application of OLEDs. They are current rather than voltage controlled, and current control is easier to implement in c-Si. OLEDs are sensitive to contamination and stringent protection favors the compact structure of a microdisplay. The emergence of high-quality, lightweight, head-mounted displays should have a strong influence on expanding market opportunities in that area.

1.6.3 Other Applications

In principle, the DMD can operate over a wide range of wavelength determined by aluminum reflectivity at short wave, and diffraction at long wave; however, it is limited to near-UV at present. The DMD finds employment as a programmable mask in photolithography and related UV imaging applications.[18]

Liquid crystal microdisplays are handicapped in the UV range by photo-induced degradation, but material selection provides reasonable life in the near-UV range at modest intensity. Infrared operation is limited by absorption bands, and the birefringence required to achieve half-wave retardation. Military applications include infrared projection, creating an infrared scene to test detector arrays.[19]

Light modulating microdisplays favor incoherent light in display applications. The same devices can act as SLMs for coherent optical systems. Phase coherence exploited in such systems requires low phase distortion in the optical components, which will need some selection among available microdisplays. SLMs find application in programmable gratings and holograms found in optical systems; diffraction efficiency is much higher for phase-modulated structures, conferring advantage to liquid crystal devices. The following examples suggest the scope of optical signal processing, but further detailed accounts are beyond the scope of the book.

In holographic storage, a laser beam splits into signal and reference beams, and an SLM imposes a pattern of amplitude modulation on the signal beam. The modulated signal beam combines with the reference beam to record a volume hologram in a photosensitive material. Readout of the hologram by a reference beam retrieves the information in the form of the original modulated signal beam. A photodetector array transforms the two-dimension information back into the electronic domain, as required.[20]

Coherent optical correlation achieves pattern recognition by matched spatial filtering. An SLM writes the input image on a laser beam. A passive lens performs a two-dimensional Fourier transform of the input image, conferring translation invariance, while phase encodes location. A sequence of Fourier plane filters written to a second SLM interrogates the transformed image. A second lens transforms the output back to image space, where a photodetector array measures the output of each filter, allowing electronic assessment of the best match, along with location identification in the input plane. The system exploits the Fourier transform ability of a lens, and the product of Fourier transforms gives the correlation function.[8]

In telecommunications, optical signals routed over fiberoptic networks require optical switching and compensation devices. Microdisplays find application in array switching, and amplitude and phase control. Two-dimensional modulator arrays form programmable holograms providing wavelength discrimination and beam deflection.[21,22]

1.7 References

1 Ishino, H. and Inoue, T. (2004) QUALIA-004 full high-definition home theater projector using silicon crystal reflective display (SXRD) technology. *IDW04*, 1687–8.

2 Barten, P. J. (1999) *Contrast Sensitivity of the Human Eye and its Effects on Image Quality*. SPIE, Bellingham.

3 Curry, D. G., Martinsen, G. and Hopper, D. G. (2003) Capability of the human visual system. *Proc. SPIE*, **5080**, 58–69.

4 Brennesholtz, M. S. (1998) Luminance contouring at the display in digital television. *Proc. SPIE*, **3296**, 62–70.

5 MacDonald, L. W. and Low, A. C. (1997) *Display Systems*. John Wiley & Sons, Chichester.

6 Toyooka, T., Yoshida, S. and Lisaka, H. (2004) Illumination control system for adaptive dynamic range control. *SID Digest* **35**, 174–7.

7 Tuenge, R. *et al.* (1997) A field-sequential color VGA AMEL display. *SID Digest* **28**, 862–5.

8 Casasent, D. (1979) Coherent light valves. In R. Kingslake (ed.) *Applied Optics and Optical Engineering*, Vol. 6, 143–201. Academic Press, New York.

9 Farrayre, A. *et al.* (1985) Geometric resolution improvement of Sodern visualization system. *SID Tech. Digest* **16**, 266–9.

10 Grinberg, J. *et al.* (1975) A new real-time non-coherent light image converter the hybrid field effect liquid crystal light valve. *Opt. Eng.* **14**, 217–25.

11 Sterling, R. D. and Bleha, W. P. (1999) Electronic cinema using ILA® projector technology, *SID Digest* **30**, 216–19.

12 Ernstoff, M. N. *et al.* (1973) Liquid crystal pictorial display. *Tech. Digest IEEE Intl Electron Devices Mtg*, 548.

13 Lipton, L. T. *et al.* (1977) A fully integrated MOS liquid crystal video-rate display panel with drivers, *SID Digest* **8**, 64–5.

14 Welkowsky, M. S. *et al.* (1987) Status of the CCD addressed liquid crystal light valve, *Proc. SPIE*, **825**, 41–8.

15 Morozumi, S. (1984) A 25-in and 1.5-in B/W and full-color LC video displays addressed by poly-Si TFTs. *SID Digest* **15**, 316–17.

16 Handschy, M. A. *et al.* (1990) Fast ferroelectric-liquid-crystal spatial light modulator with silicon-integrated-circuit active back plane. *Proc. SPIE*, **1291**, 158–64.

17 Hornbeck, L. J. (1989) Deformable mirror spatial light modulators. *SPIE Critical Review Series*, **1150**, 86–102.

18 Dudley, D., Duncan, W. and Slaughter, J. (2003) Emerging digital micromirror device (DMD) applications, *Proc. SPIE*, **4985**, 14–25.

19 Wu, S-T. *et al.* (1985) Infrared liquid crystal light valve. *Proc. SPIE* **572**, 94–101.

20 Nesbitt, R. S. *et al.* (1999) Holographic recording using a digital micromirror device. *Proc. SPIE* **3637**, 12–20.

21 Duncan, W. *et al.* (2003) DLP™ switched blaze grating; the heart of optical signal processing. *Proc. SPIE*, **4983**, 242–9.

22 Crossland, W. *et al.* (2004) Liquid crystals in telecommunications systems. *Mol. Cryst. Liq. Cryst.* **413**, 363–83.

2

Electronic Addressing

2.1 Introduction

2.1.1 General Introduction

This chapter is concerned with the electronic circuits, integrated onto a microdisplay backplane, that are involved in reception of the electronic information (that represents the desired image), its processing, delivery to, and storage within, each pixel and with the method of conversion of the electronic information into "display" information within each pixel. The objective of the backplane circuitry then is the faithful and timely implementation of all those functions.

The primary focus is on CMOS active matrix backplane technology for LCDs (CMOS for LCOS) as it is the most widespread technology. Reference is made to TFT active matrix backplane technology. The nature of the CMOS electronics used for OLED and DMD microdisplays is also described.

2.1.2 Addressing Methods

In flat panel displays in general, a number of classes of electronic addressing exist[1] according to the number of electrical connections required in order to address effectively a number of pixels independently.

Direct addressing of a display involves an individual controlled electrical connection to, and driver circuit for, each pixel of the display on the substrate electrode and is therefore primarily suitable for displays with low pixel counts such as seven-segment displays as illustrated in Figure 2.1. In the case of displays that require a counter electrode, such as LCD and OLED, an advantage of direct addressing is that the counter electrode can be continuous or global – it does not require to be patterned at the pixel level. A display of $M \times N$ pixels driven by direct addressing requires $(M \times N) + 1$ drivers, one for each pixel and one for the counter electrode (that may be a static bias signal).

Introduction to Microdisplays D. Armitage, I. Underwood and S-T. Wu

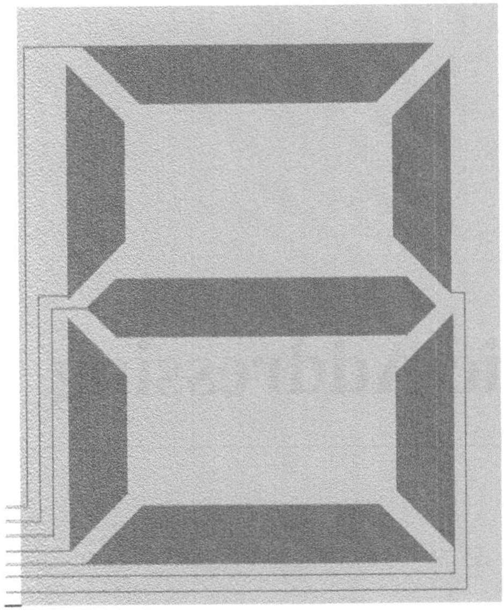

Figure 2.1 Electrode pattern for seven-segment direct-drive display

Passive matrix addressing is generally used for dot matrix displays in which the pixel array has a regular rectangular repeat pattern, and requires both the upper and lower electrodes to be patterned. Each electrode is patterned with conducting lines, one set running east to west, the other running north to south as shown in Figure 2.2. Thus a display of M columns by N rows, that is to say MN pixels, requires $M + N$ drivers of which typically M are on one (say the lower) substrate and N are on the other substrate. A significant limitation of straightforward passive matrix addressing is that each row can be actively and intentionally driven or addressed for no more than $1/M$ of the frame time – the upper limit of the active duty cycle is $1/M$. (Dual scan addressing in which the top half of the display is driven from above and the bottom half is driven from below doubles this to $2/M$.) The field address time, τ_{field}, is given by

$$\tau_{\text{field}} = N\,\tau_{\text{dwell}} \tag{2.1}$$

where τ_{dwell} is the row dwell time – the time required to fully switch the state of the liquid crystal (LC) in the row of pixels being addressed (10 ms to 100 ms or more for nematic LC).

This limits the field update rate for the display:

$$f_{\text{field,max}} = 1/N\,\tau_{\text{dwell}}\,. \tag{2.2}$$

Furthermore, capacitive crosstalk between the bus lines driving the pixels being addressed and other rows of pixels is difficult to control. A consideration of LCDs for notebook computers suggests that passive matrix addressing was used up to about SVGA definition; beyond that, active matrix addressing became necessary. In a bistable device such as plasma or FLC, the limit is dwell time. But devices that respond to an RMS signal, such as nematic LCs, have a different limit.

The corner region (Figure 2.3) of the matrix addressed array illustrated in Figure 2.2 shows pixels identified by p_{yx} addressed by y and x lines connecting to the applied signal voltage or current. A voltage (V_{23}) applied between y_2 and x_3, addressing p_{23}, also applies the voltage across the series combinations ($p_{13} + p_{14} + p_{24}$) and ($p_{33} + p_{32} + p_{22}$), etc., known as pixel crosstalk. Any crosstalk current always

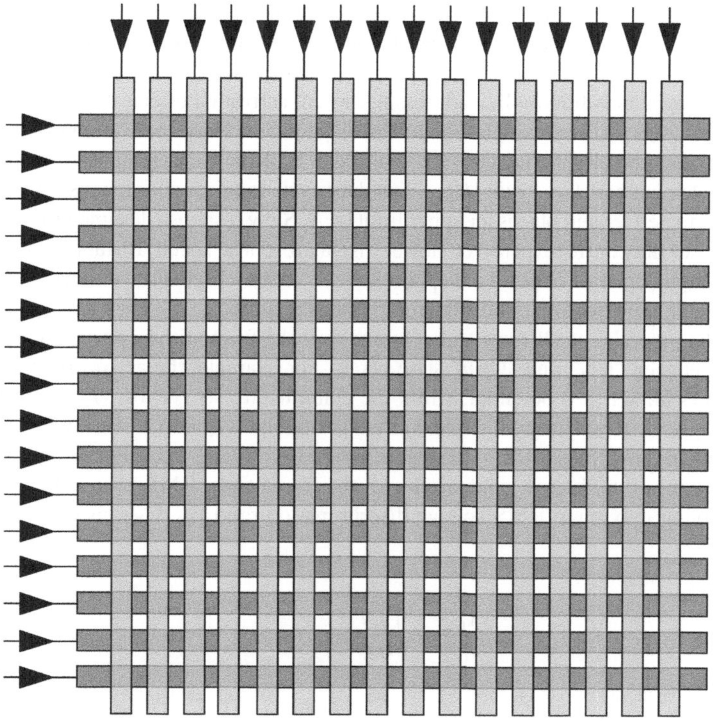

Figure 2.2 Plan view schematic of electrode pattern for passive matrix display

flows in the reverse direction of at least one of the pixels p_{yx}; therefore, a light-emitting-diode pixel that blocks reverse current flow eliminates such crosstalk. Nematic liquid crystal pixels present a symmetrical capacitive load to the applied voltage, making them susceptible to crosstalk. Passive matrix addressing requires sufficient nonlinearity in the pixel element to attenuate crosstalk to an acceptable

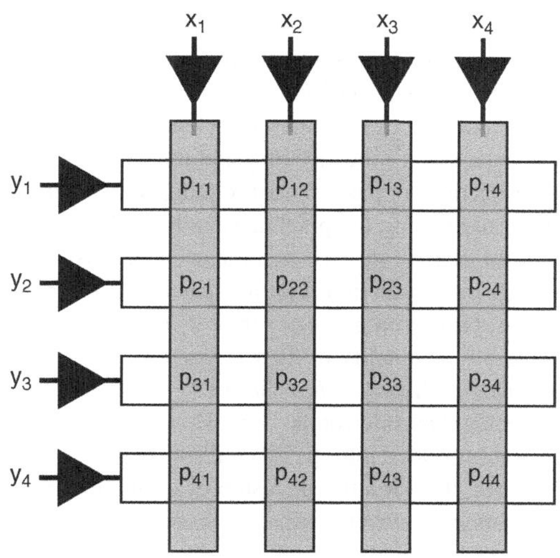

Figure 2.3 Corner of Figure 2.3 blown up

level. Liquid crystal configurations such as the super-twist cell evolved to provide appropriate threshold voltage levels and steep electro-optic transfer characteristics, optimizing the performance of passive matrix addressing.

Matrix addressing adopts line-by-line addressing, where activation of one y line (row) together with all the x lines (columns) writes a single line of the display. Successive activation of all rows writes a complete frame. Each pixel is activated for a limiting $1/N$ of the time, in an N-line display, where a fast responding element has *duty cycle* $1/N$. A low duty cycle requires a large pulse current or voltage, which is difficult to deliver and places an upper limit on N. Slow response designed into a liquid crystal cell provides higher duty at the expense of slower display response speed. Bistable switching provided by ferroelectric liquid crystals or plasma display elements enables high duty without sacrifice in speed, but raises pulse width modulation issues regarding image quality.

Controlling the potential of all rows and columns over the frame-time cycle optimizes the voltage discrimination between "ON" and "OFF" state pixels. For example, the active row potential set $+S$, with inactive rows all set to 0 potential, while the columns are set $-F$ or $+F$ according to desired "ON" or "OFF" pixel states. During the line address period the $V_{ON} = (S + F)$, while $V_{OFF} = (S - F)$; however, for the remaining $(N - 1)/N$ of the frame period $V_{ON} = V_{OFF} = \pm F$. For nematic liquid crystals in which the response time is slow compared with the frame time, RMS voltage becomes an appropriate metric, giving maximum RMS voltage discrimination ratio V_{ON}/V_{OFF}, as described by Alt and Pleschko:[2]

$$\frac{V_{ON}}{V_{OFF}} = \sqrt{\frac{\sqrt{N}+1}{\sqrt{N}-1}}. \tag{2.3}$$

Clearly, the discrimination between ON and OFF deteriorates with increasing N. Moreover, designing the liquid-crystal response long compared with refresh time cripples video imaging. Further improvement requires adapting the addressing voltages to each displayed frame (active addressing), implying a frame store and digital processing to evaluate the addressing pulse voltages. The method is known as multiple-line addressing and accommodates nematic response comparable to the frame rate, making video imaging possible. The additional costs of multiple-line addressing detract from the simplicity of passive addressing, while the superior performance of active matrix addressing remains the dominant factor. Some areas of display technology, such as plastic displays, remain passive addressed owing to the difficulty in fabricating active matrix devices.

Active matrix addressing is also generally used for dot matrix displays in which the pixel array has a regular rectangular repeat pattern, and requires only the lower electrodes to be patterned. Active matrix addressing involves, as a minimum, the inclusion of a nonlinear or switching element and a storage element within each pixel. The purpose of the nonlinear element is to provide controlled and switchable connection/isolation between addressing circuitry and the storage element within the pixel. This effectively removes the two restrictions of passive matrix addressing mentioned above, minimizing the detrimental effect of crosstalk and raising the pixel drive duty cycle close to unity. All microdisplays use active matrix addressing. In many cases the nonlinear element is a MOS transistor or thin film transistor operating as a switch in conjunction with a capacitor as the storage element, as shown in Figure 2.4. LCD microdisplays, built on a glass substrate using TFT active matrix technology, often reflect the architecture and pixel functionality of their larger cousins – conventional TFT LCDs. Like the equivalent passive matrix array, a display of M columns by N rows, that is to say MN pixels, requires $M + N$ line drivers (plus 1 counter electrode connection). However, the same capability that allows active switching elements to be incorporated into the pixel allows the line drivers to be incorporated onto the active substrate. Furthermore, it allows the inclusion of de-multiplexing circuitry that allows the number of external electrical connections to the substrate to be substantially reduced.

Microdisplays that use CMOS backplane technology, the primary focus of this chapter, typically leverage the complexity of the technology to integrate a great deal of interface capability into the

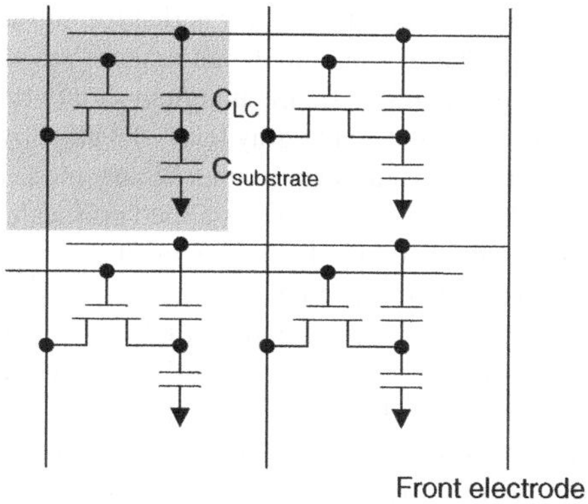

Figure 2.4 Active matrix

backplane; some even incorporate much greater functionality into the pixel itself in order to improve its performance.

2.1.3 Grayscale

The method chosen to produce the appearance of grayscale in a microdisplay system determines, to a significant extent, the design of the underlying architecture and pixel drive circuit.

The appearance of shades of gray (or levels of luminance) may be achieved on a display by one, or a combination, of a number of different means. A LC configuration with an inherently analog electro-optical response may produce a continuous range of gray levels or a set number of stepped gray levels depending upon the precision of control of the amplitude of an applied electronic drive signal of a continuously variable or stepped nature. Gray levels can also be perceived in an inherently binary system capable of only an ON/OFF electro-optical response by means of spatial segmentation of the pixel area into sub-pixels (or sub-pixel areas) of equal size, or sub-pixels whose relative size is binary weighted. In the case of the former, 2^n equal sub-pixels lead to 2^n gray levels or n-bit grayscale. In the case of the latter, n binary-weighted sub-pixels is sufficient to give 2^n gray levels or n-bit grayscale. Figure 2.5 shows the floor plan of a pixel capable of displaying four gray levels even when using an electro-optic

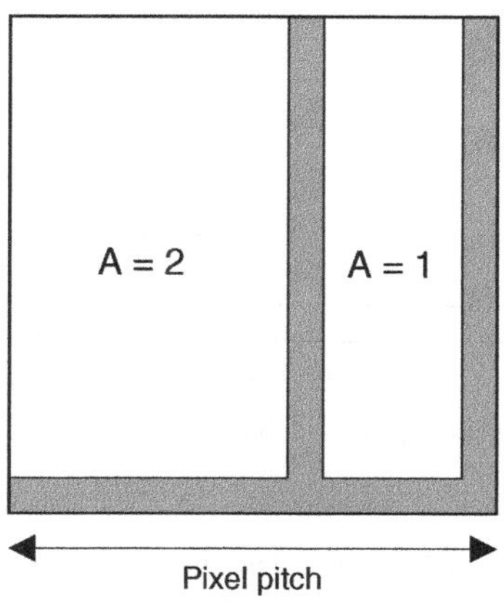

Figure 2.5 Schematic floor plan of pixel with 2-bit binary-weighted spatial sub-pixelation

effect capable of only binary output. An important factor is that the sub-areas should be too small to be visually resolved by the observer. (This should not be an issue in practice as, in any well-designed display system, the pixels themselves should be small enough to satisfy that condition.)

The time domain may also be used to control the gray level of a pixel by means of the duty cycle of the drive signal – the ratio of ON-time to overall frame-time. Binary-mode and bistable devices such as DMD and SSFLC are capable of achieving continuous or stepped grayscale by controlling (a) the width of a single pulse of light of fixed amplitude output from each pixel during a frame period (Single Pulse Width Modulation or S-PWM), (b) the ON/OFF state of a sequence of identical individual pulses from each pixel during a frame period (Pulse Count Modulation (PCM) or Count-based Pulse Width Modulation (C-PWM)), or (c) the ON/OFF state of a sequence of binary-weighted-in-time individual light pulses from each pixel during a frame period (Binary-Weighted Pulse Coded Modulation (B-PCM) or Binary-coded Pulse Width Modulation (B-PWM)). B-PWM[i] reduces the overhead of addressing of the pixel array in comparison to C-PWM as it requires only n addressing cycles (bit-planes) to achieve 2^n gray levels; C-PWM, on the other hand, requires 2^n addressing cycles in order to achieve 2^n gray levels. Consequently B-PWM is in widespread use whilst C-PWM is not. In externally illuminated systems, the binary weighting of the bit-planes can be achieved by temporal modulation of the illumination pulses, intensity modulation of the illumination pulses, or a combination of both.

A key issue here is that the refresh rate must be sufficiently fast for individual time slots to be irresolvable in the time domain by an observer, and the electro-optical switching must be fast enough to avoid distortion in the grayscale due to rise and decay of output during the switch ON and switch OFF periods respectively. (The former is analogous *in the time domain* to saying that the pixels or sub-pixels in a conventional display should be *spatially* unresolvable by the viewer whilst the latter is analogous to the fringing fields that cause output variation at the edges of each pixel.) As an illustrative example of combining spatial and temporal dithering, the pixel layout shown in Figure 2.6 is capable of displaying 6-bit grayscale provided the pulse width illumination is controlled to 3-bit precision.[3]

The above use of the time domain assumes that the electro-optical response time is fast enough to fully respond between consecutive changes in the drive signal. If the electro-optical response is significantly slower than that (as can be the case with, for example, nematic LC) the LC may respond to,

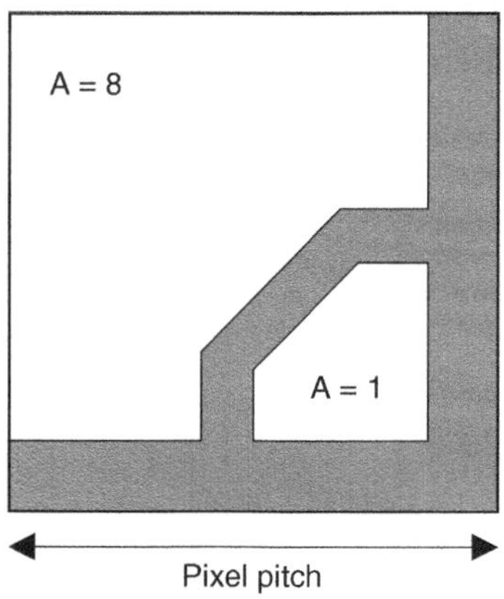

Figure 2.6 Schematic floor plan of pixel with binary-weighted spatial sub-pixelation in ratio 8:1

[i]Whilst the use of the term "pulse width modulation" to describe (b) or (c) above seems inaccurate, it is nevertheless in common use and, like other authors, we will use the general term "PWM" to cover all of the above and the more specific terms when appropriate.

for example, the RMS drive signal, allowing an analog electro-optical response to be controlled by the duty cycle of a square wave.

The interest in binary or digital electronic drive schemes for microdisplays derives from one or more of several possible sources.

- The design of a digital active-matrix backplane (or any digital IC) is generally considered to be a less difficult and more automated, and therefore a faster and less error-prone, process than the design of an equivalent analog IC. This may be a significant issue in terms of reducing the cycle time and risk associated with new product design. Jepsen[4] takes a position on the comparative benefits of the analog and digital approaches to microdisplay backplane design, arguing that analog is superior. But the 2005 publication of her study is neither comprehensive nor definitive, leaving the door open for further reports or rebuttals.
- Some electro-optical responses, such as the DMD and Surface Stabilized FLC (SSFLC), are inherently digital.
- Looking to the future, the storage of still and moving pictures is more likely to be in digital format (for example, the move from analog to digital broadcasting of TV and the move of camcorder storage to digital video cassette (DVC) and DVD.
- Today, and into the future, the digital domain in general is growing at the expense of the analog domain.

2.1.4 Color

The method chosen to produce the appearance of color in a microdisplay system determines, to a significant extent, the design of the underlying backplane architecture. Color can be produced in a number of ways.

Spatially segmented (or sub-pixelated) RGB system

In this case, a color backplane architecture is used and a single microdisplay suffices. Each pixel is split into separate red, green and blue sub-pixels[ii] as illustrated in Figure 2.7. The standard way of doing this

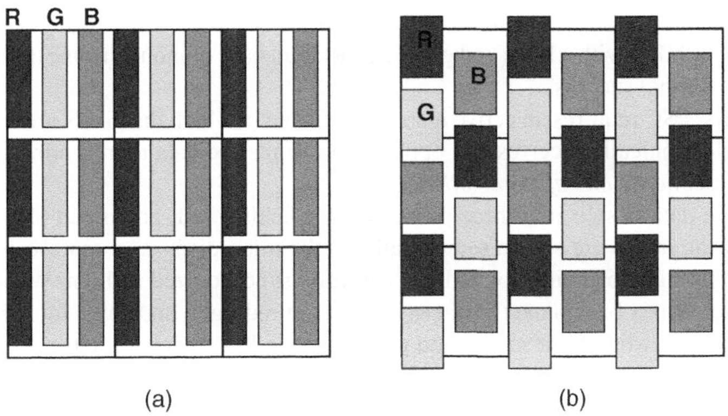

Figure 2.7 Generic pixel floor plans showing LCOS RGB sub-pixels: (a) RGB stripe; (b) RGB delta. Underlying square tiling is intended to illustrate pixel pitch

[ii]Several means of achieving sub-pixellation in an emissive microdisplay are described in Chapter 9.

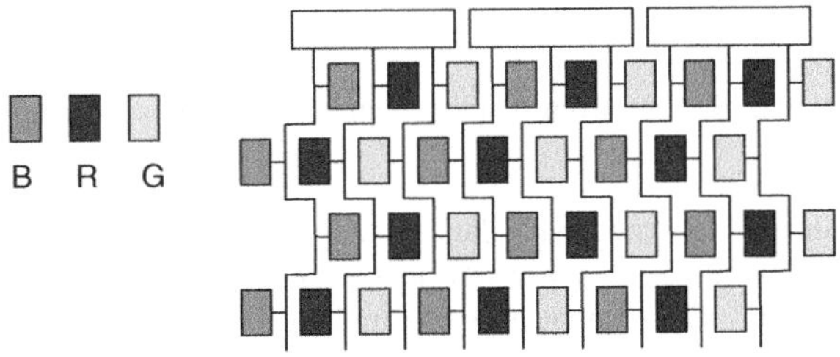

Figure 2.8 Example layout of RGB sub-pixels for miniature TFT LCD

in a direct-view LCD is by means of white backlight illumination with red, green and blue color filters covering each of the three sub-pixels. Historically the small dimensions of an LCOS microdisplay pixel has mitigated against this approach but one recent report by Kwok *et al.*[5] in 2005 shows promise.

Figure 2.7(a) implies that each of the three sub-pixels is activated from a single row address line. The layout of the pixel circuits may (but need not) be identical. In many LCOS designs, the pixel footprint can be minimized only by sharing elements between nearest- or close-neighbor pixels so that the repeat unit for pixel layout footprint can be 2×2 pixels or greater. In order for the three sub-pixels of Figure 2.7(b) to be activated from a single row address line the underlying sub-pixel circuits must either be laid out differently or the signals must be routed to the pixel electrodes using a local interconnect layer. Such possibilities exist with reflective LCOS backplanes manufactured in CMOS processes with multiple metal layers. This is not the case for transmissive LCOS nor for miniature TFT LCD. Figure 2.8 shows an example pixel floor plan for a transmissive miniature TFT LCD in which not all sub-pixels are accessed from the same row address line (and in which, incidentally, the notional RGB pixels are not square).

Field-sequential-color (FSC) technique

Field-sequential-color (FSC) involves the showing in quick succession of three sub-pictures or fields, usually red, green and blue; these contain respectively all of the red, green and blue information required to comprise the color picture. If cycled quickly enough, the eye sees only the full color picture – not the three component fields. The concept of field-sequential-color has been around since at least the beginning of the era of color television.[6] In 1950 (not for the first time) CBS proposed a FSC standard to the FCC. The standard was adopted but a combination of extended legal wrangling coupled with a lack of backward compatibility with a large and growing installed base of monochrome domestic televisions prevented its commercial success.

In microdisplays, FSC requires an active matrix backplane in which the pixels are not sub-pixelated. The pixel array electronically receives and optically transmits the data representing each of the three primary colors *in sequence one after the other.*

The color image data is split into three primary color fields – red, green and blue. The red, green and blue fields of the image are shown sequentially on the microdisplay, as represented in Figure 2.9, with appropriate illumination being applied in a carefully synchronized fashion. When the red field is stored in the active matrix backplane the display must emit red light, and so on for the green and blue. This is done by using a white light source and passing the light through a rotating color wheel, or the electronic equivalent such as a switchable liquid crystal filter. This technique is used to achieve low cost in single-microdisplay LCOS systems. This technique has been suggested for emissive microdisplays but the authors are not aware of any reported implementations. If used with a white emitter, this would remove any requirement to pattern the organic emissive layer(s) at the fine dimensions of a pixel.

Both DMD and FLC high-resolution microdisplays are currently available aimed at use in projection systems. Medium-resolution FLC microdisplays are in use for NTE systems. In 1996, Rankin *et al.*[7] reported an early demonstration of B-PWM grayscale, field-sequential-color using FLC and pulsed LED illumination.

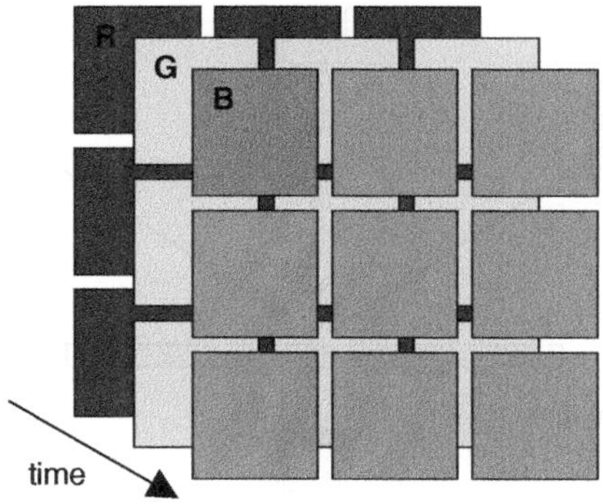

Figure 2.9 Representation of field-sequential-color

Scrolling-color mechanism

A variation on FSC is the scrolling-color system in which the illumination is split into bands of red, green and blue which scroll sequentially across the display from North to South (or vice versa) as represented in Figure 2.10 and described in Chapter 10 on projection. There is a gap between the bands to allow re-addressing and settling of the pixels prior to illumination by the next color band. The use of scrolling color has been reported[8,9] for LCOS systems. A single backplane design with the generic architecture of a monochrome microdisplay backplane is used.

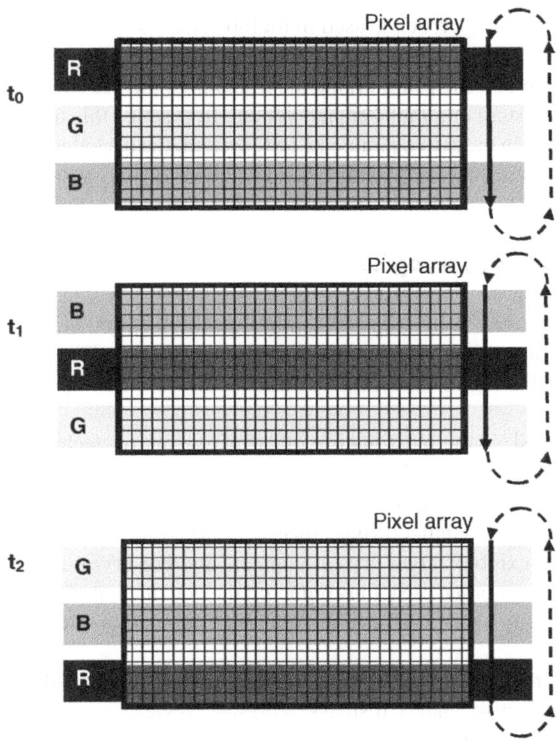

Figure 2.10 Time sequence representation of scrolling color

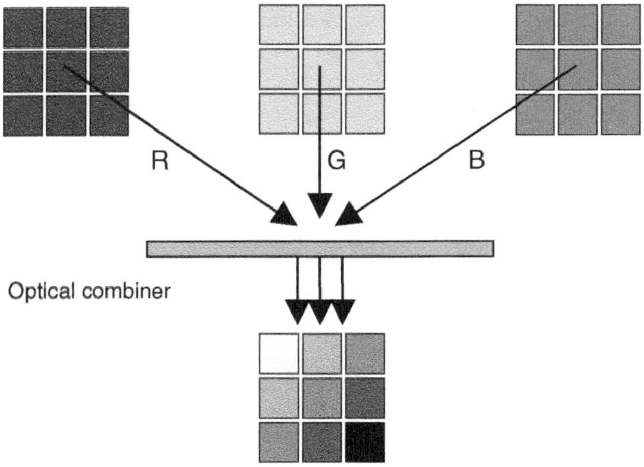

Figure 2.11 Representation of 3-backplane color system

Three-panel system

A three-panel system uses three identical displays in which pixels are not spatially sub-pixelated. Each display transmits one primary color field.

The color image data is split into three primary color fields – red, green and blue. The fields are each sent electronically to one of the displays. In a three-panel system the images from each of three carefully aligned monochrome microdisplays (of red, green and blue respectively) are optically overlaid as suggested in Figure 2.11. This approach is used in high-brightness microdisplay projection systems as described in Chapter 10.

In principle the same, or at least a similar, monochrome-architecture backplane design may be used for a single-engine FSC system or a multi-engine system. In practice this may involve optimization or compromise. For example, in order to operate in a FSC system, the backplane would have to be capable of 3× the bandwidth necessary for each backplane in a 3-microdisplay system.

2.1.5 Active Matrix Technologies

Medium- and large-area active matrix displays, or flat panel displays, utilize an inert glass plate as the starting substrate with active devices such as thin film transistors (TFTs) manufactured on the top surface. Various TFT technologies exist, largely differentiated in name by the crystal structure of the silicon.

Amorphous silicon (α-Si) is the most mature of the active matrix technologies but has the lowest carrier mobility and integration capability. *Polycrystalline silicon* (p-Si) is sub-differentiated into high-temperature polysilicon (HTPS) and low-temperature polysilicon (LTPS) according to the temperature range required in the manufacturing process. *Continuous Grain Silicon*[10] (CGS) is a proprietary variant of p-Si that exhibits larger crystal domains and preserves orientational continuity over greater distances than conventional p-Si, thus reducing the detrimental effect that domain boundaries have in conventional p-Si. In short, CGS has a higher mobility and greater integration potential than p-Si.

Den Boer[11] describes the device technology, fabrication technology, and active matrix addressing schemes of conventional active matrix displays. Voutsas[12] reviews the fundamentals of AMLCDs, whilst Soh[13] details the manufacturing technology behind TFT-LCDs.

Table 2.1 Some comparative properties of active matrix technologies[a]

	α-Si TFT	p-Si TFT	x-Si MOST
Electron mobility	Very low	Medium+	High+
Hole mobility	N/A	Medium −	High −
n-channel available	Yes	Yes	Yes
p-channel available	No	Yes	Yes
OFF leakage current(pA)	1	1	0.1
Example minimum feature size (μm)	5	0.6	0.18
Example minimum pixel(dot) size (μm)	<100	<25	<10
Minimum display size	Medium	Small	Tiny
Maximum display size	Huge	Large	Very small
Typical mask count	4 to 6	6 to 10	20+
Integration capability	Low	Medium	Very high
Stability (performance variation with time)	Poor	Good	Very good
Uniformity (unit to unit performance variation)	Good	Poor	Very good

[a]These are representative properties that are intended only to show the *relative* differences of the various technologies.

When considered as a potential active matrix substrate for a small display or microdisplay, the electronic properties of amorphous silicon are very limiting. Polycrystalline silicon is used for microdisplays, typically by companies that also manufacture larger area p-Si TFT LCDs.

For microdisplays (typically with a pixel-array area of, say, less than $1\,in^2$ or $6\,cm^2$) it is possible, even preferable, to use a crystalline silicon wafer as the starting substrate. The lower size limit suggested here is not a hard technological limit but a soft limit imposed by the economics of manufacture as described in Chapter 3. Complimentary metal oxide semiconductor (CMOS) is the ubiquitous microelectronic technology adapted from mainstream electronics using techniques described in Chapter 3 in order to optimize its use in microdisplays. In Table 2.1 some of the relevant characteristics of α-Si, p-Si and CMOS are compared.

Some notable features of most CMOS-based microdisplays are as follows: small circuit feature size leading to smaller pixel size or higher transistor count per pixel, and more transistors can mean more pixel functionality; small size and weight; high fill factor *in reflective mode* due to the mirror that forms the pixel aperture overlying the active matrix circuitry (rather than sitting side-by-side with it as in a transmissive TFT LCD microdisplay); reflective optics (more complex than transmissive optics for separation of illumination and viewing light paths); low power consumption due to the inherent efficiency and low voltage operation of CMOS electronics; low cost; high level of integration of driver circuits: RGB color filters have not been used until recently; instead, color has been produced using the time domain (field-sequential-color (FSC)) in low-cost single-panel systems or by three individual microdisplays – one for each color – in high-performance three-panel systems.

Mainstream CMOS technology evolves rapidly with time, moving to smaller dimensions, lower voltages and larger wafers for example. Table 2.2 indicates some of the parameters relevant to the electronic design of microdisplay backplanes that follow this scaling trend. LCOS does not reside at the leading edge of the trend. The need for sufficient voltage to switch the LC quickly and other constraints has meant that most LCOS backplanes of recent times have used CMOS in the range of 0.35–0.18 μm, sometimes with additional high-voltage capable components.

In a separate publication in the Wiley SID series in Display Technology, Lueder[14] analyzes the addressing of conventional LCDs by α-Si TFTs (Lueder, Chapter 14) and p-Si TFTs (Lueder, Chapter 15) and gives a very brief introduction to LCOS (Lueder, Chapter 16).

Table 2.2 Scaling factors in CMOS[84]

Parameter	Constant field	Constant voltage
Length	1/a	1/a
Width	1/a	1/a
Voltage	1/a	1
Gate oxide thickness	1/a	1/a
Current	1/a	a
Transconductance	1	a
Electric field	1	a
Load capacitance	1/a	1/a
Gate delay	1/a	$1/a^2$
Power	$1/a^2$	a
Power delay product	$1/a^3$	1/a
Power density	1	a^3
Current density	a	a^3

2.1.6 LCOS: The Early Days

The first report, by Ernstoff *et al.*,[15] of the use of an array of MOS transistors to drive a LC display dates from 1973. Thereafter, research and development activity flourished in the USA, Europe and Japan through the 1970s and early 1980s. The bibliography of this chapter lists some of the early publications on LCOS from that era. Figure 2.12 shows by way of example an early LCOS device.[16] It consists of 40 × 40 square pixels driven in binary mode to produce a binary monochrome text image.

Figure 2.13(a) shows a 100 mm wafer-scale device from around the same era; it consists of 240 × 240 pixels and is capable of showing 24 lines of text. At least one photograph of this device in operation

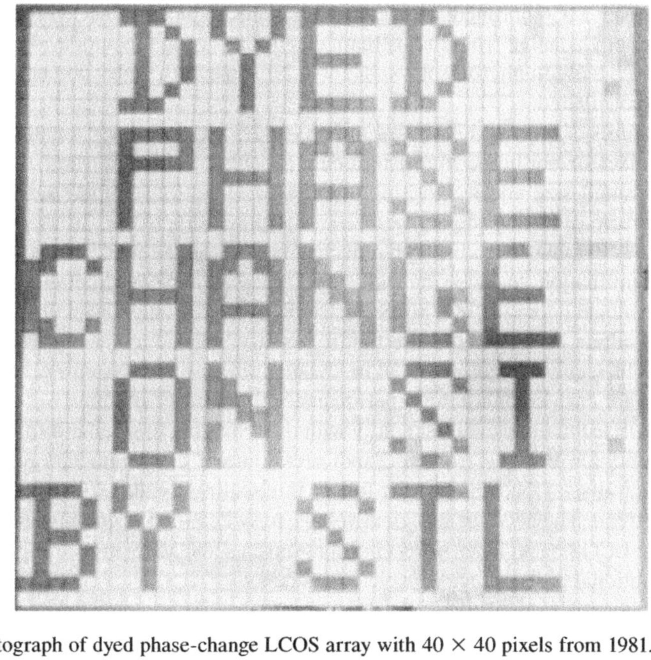

Figure 2.12 Photograph of dyed phase-change LCOS array with 40 × 40 pixels from 1981. Reprinted courtesy of Professor W.A. Crossland

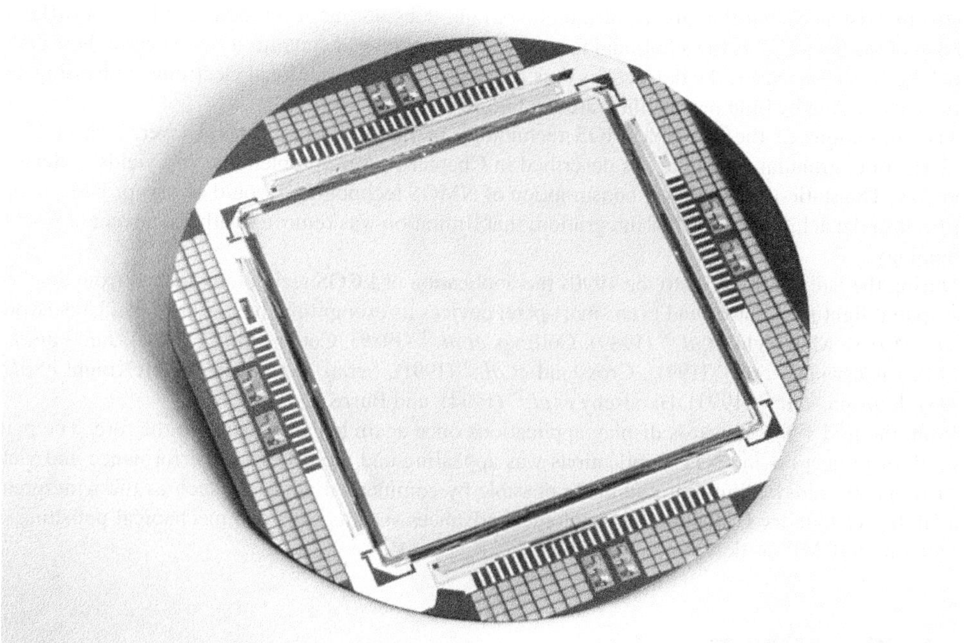

(a)

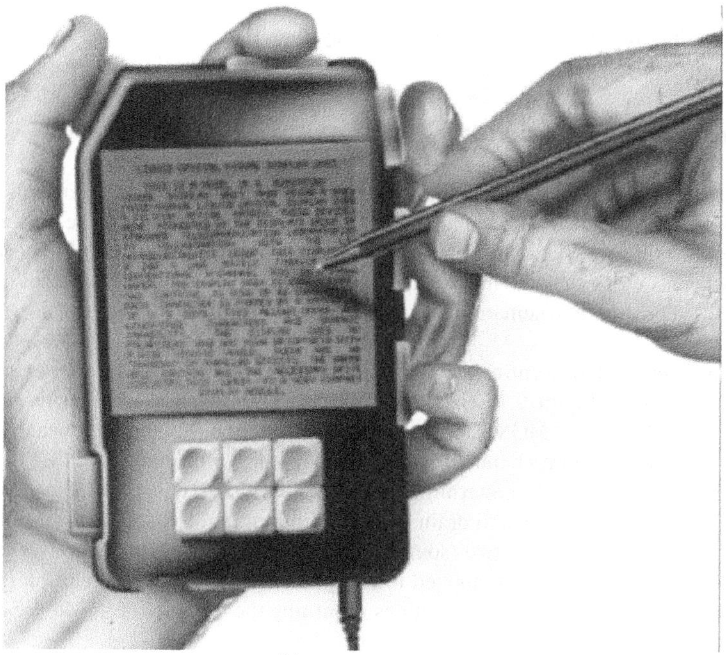

(b)

Figure 2.13 Dyed phase-change LCOS device from early 1980s: (a) photograph of 240 × 240 pixel backplane (access to sample courtesy of Professor W.A. Crossland; photograph by Peter Tuffy); (b) artist's impression of the backplane in use, circa 1981 (reprinted courtesy of Professor W.A. Crossland)

is still in existence[17] but the quality of the photograph does not merit reproduction here. A different version of the device[17] has the additional feature that each pixel pad contains a small central hole under which lies a photo-sensor; the device was thus capable of both conventional electronic addressing and optical addressing by light pen as illustrated in Figure 2.13(b).

The limitations of the available MOS technology of that era (single metal layer, lack of CMP planarization, granular metal etc., as described in Chapter 3) were problematic and yields of devices were low. The static-current power consumption of NMOS technology[18] would always provide an upper limit to the achievable level of integration; that limitation was removed with the advent of CMOS technology.

During the late 1980s and into the 1990s the application of LCOS technology moved from displays into spatial light modulators and even smart-pixel devices as exemplified by, for example, Underwood *et al.*[19] (1986), McKnight *et al.*[20] (1989), Collings *et al.*[21] (1989), Cotter *et al.*[22] (1990), Jared *et al.*[23] (1990), Underwood *et al.*[24] (1991), Crossland *et al.*[25] (1991), Serati *et al.*[26] (1993), McKnight *et al.*[27] (1993), Johnson *et al.*[28] (1993), Handschy *et al.*[29] (1994), and Burns *et al.*[30] (1995).

From the mid 1990s onwards display applications once again began to come to the fore. The market pull of projection and NTE applications was appealing and the necessary performance and yield improvements were beginning to be made possible by commercial advances such as the widespread availability of foundry CMOS and technological advances such as chemical mechanical polishing or planarization (CMP) as described in Chapter 3.

2.2 The MOS Transistor

The MOS transistor underpins all LCOS backplane circuits. A solid grounding in the characteristics of the MOS transistor is fundamental to a good understanding of the pixel drive circuits used to achieve the appropriate addressing schemes for LC microdisplays.

2.2.1 Characteristic Equations

The physical aspects of the metal oxide semiconductor (MOS) transistor and the overall CMOS process are described in Chapter 3. The name refers to the horizontal layers of the vertical capacitive structure at the heart of the device. More precisely, the name refers to the materials historically used in the structure, whereas today the gate material is almost always polycrystalline silicon, or polysilicon, rather than metal. The circuit schematics of n-channel and p-channel MOS transistors are shown in Figure 2.14.

The MOS transistor is a four-terminal device; the four terminals gate G, drain D, source S and bulk B or substrate are shown in Figure 2.15. Sze[31] and Tsividis[32] offer detailed accounts respectively of the physics and the modelling of the MOS transistor. The bulk or substrate potential is usually set at a fixed bias potential (either V_{SS} for an n-channel MOSFET or V_{DD} for a p-channel MOSFET) and thereafter the MOSFET is considered as a three-terminal device. The MOSFET is, in principle, a symmetrical structure so the definition of source and drain is by convention; in the case of an n-channel MOSFET the source is the less positive of the two (sometimes shorted to V_{SS}) while the converse is true for a p-MOSFET. In both cases the voltage applied to the gate controls the flow of current between drain and source. For an n-MOSFET the basic equations governing the behavior of the device are as follows (refer to Figure 2.15).

In the OFF region of operation (defined by $V_{GS} < V_T$) we can, to first order, approximate:

$$I_{DS} = 0\,. \tag{2.4}$$

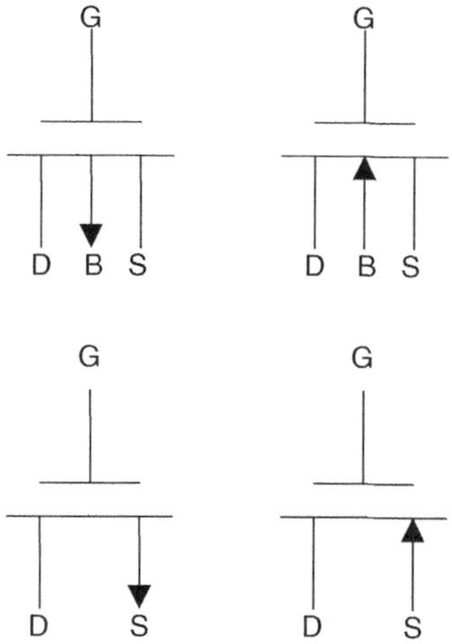

Figure 2.14 Schematic diagrams for n-channel (left) and p-channel (right) MOS transistors (top row) and simplified circuit schematics in common use (bottom row)

In the linear region of operation (defined by $V_{GS} \geq V_T$, $V_{DS} < V_{GS} - V_T$):

$$I_{DS} = \frac{W}{L}\mu C_0 (V_{GS} - V_T) V_{DS} . \tag{2.5}$$

In the saturation region of operation (defined by $V_{GS} \geq V_T$, $V_{DS} \geq V_{GS} - V_T$):

$$I_{DS,SAT} = \frac{1}{2}\frac{W}{L}\mu C_0 (V_{GS} - V_T)^2 (1 + \lambda) \tag{2.6}$$

In these equations:

I_{DS} is the current flowing from drain to source through the channel in the linear region of operation

$I_{DS,SAT}$ is the current flowing from drain to source through the channel in the saturation region of operation

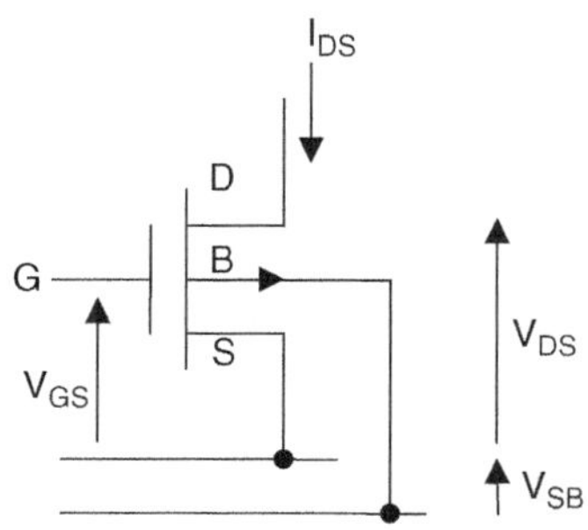

Figure 2.15 MOS transistor circuit configuration

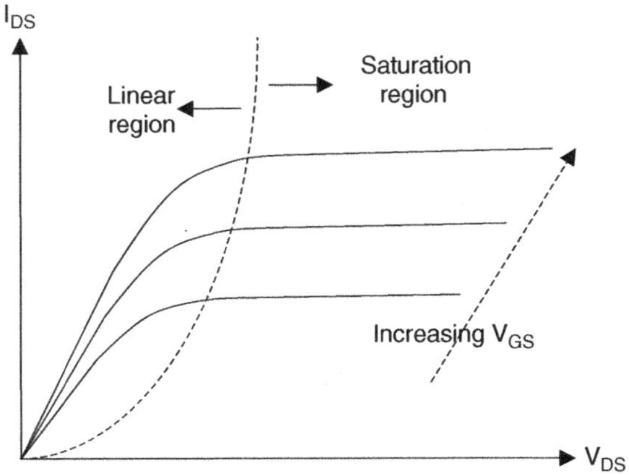

Figure 2.16 Characteristics of the NMOS transistor: I_{DS} as a function of V_{DS} for various V_{GS}

V_{DS} is the drain–source voltage, a signal voltage

V_{GS} is the gate–source voltage, a signal voltage

V_T is the threshold or turn-on voltage, a property of the device described below

W is the channel width

L is the channel length

μ is the carrier (electron or hole) mobility

C_0 is the normalized capacitance of the gate dielectric (capacitance per unit area)

λ is the empirical channel length modulation factor.

These characteristics are illustrated in Figures 2.16 and 2.17. Note that:

- μ, C_0, λ and V_T are process parameters – they are nominally fixed for a given CMOS process although random manufacturing variations in V_T affect LCOS pixel circuits and in V_T and μ affect OLED pixel circuits.

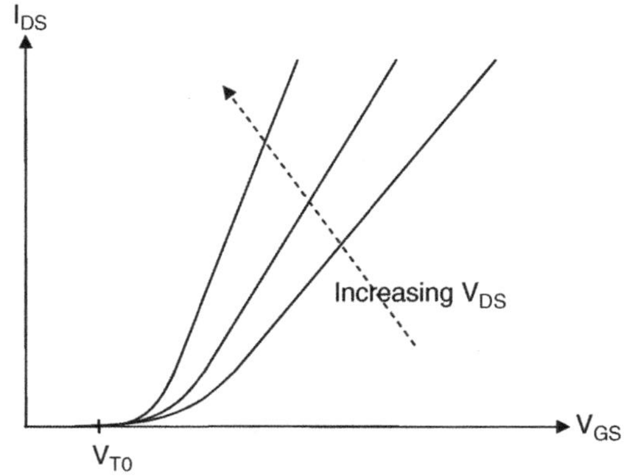

Figure 2.17 Characteristics of the NMOS transistor: I_{DS} as a function of V_{GS} for various V_{DS}

- W and L are circuit design parameters.
- In the linear region, current is proportional to gate voltage.
- In the saturation region, current is proportional to the square of gate voltage.

V_T is, in turn, given by

$$V_T = \Phi_{MS} - \frac{Q_{SS}}{C_0} + 2\Phi_B + \frac{(2\varepsilon_0 \varepsilon_{Si} q N_A)^{0.5}(2\Phi_B + |V_{SB}|)^{0.5}}{C_0} \tag{2.7}$$

where

Φ_{MS} is the difference in work function between the gate and the channel

Q_{SS} is the surface charge per unit area

C_0 is as above

Φ_B is the bulk or substrate potential

ε_0 is the permittivity of free space

ε_{Si} is the relative permittivity of crystalline silicon

N_A is the average dopant density in the channel

V_{SB} is the substrate bias (source to substrate) voltage.

When $V_{SB} = 0$, as is often the case in practice, (2.7) reduces to

$$V_{T0} = V_T(V_{SB} = 0) = \Phi_{MS} - \frac{Q_{SS}}{C_0} + 2\Phi_B + \frac{(2\varepsilon_0 \varepsilon_{Si} q N_A)^{0.5}(2\Phi_B)^{0.5}}{C_0}. \tag{2.8}$$

The first three terms on the right-hand side of (2.7) and (2.8) are process or device parameters whose values are pre-specified subject only to manufacturing variation. In a typical CMOS process, the threshold voltage is specified to be around 20% of the rated power supply voltage. So a 0.5 µm CMOS process with $V_{DD} = 5$ V might exhibit $V_{T0} = 1.0$ V, and a 0.35 µm CMOS process with $V_{DD} = 3.3$ V might exhibit $V_{T0} = 0.6$ V. This is a compromise determined by balancing performance factors such as switching speed, power consumption, and noise immunity.

The additional term in (2.7) is affected by a bias voltage applied between the bulk or substrate and the source of the device. Adding a bias to the substrate leads to an increase in the threshold voltage of the MOSFET. In the historical days of NMOS technology this bias voltage allowed circuits to be tuned for performance, e.g. high speed or low power. In CMOS technology it is highly unusual to apply a bias voltage to the substrate. However, the effect of the fourth term is noticed when a MOS transistor is used in the pass transistor configuration as explained below. This configuration is present in the pixel circuit of many microdisplays and plays a particularly key role in the so-called DRAM pixel architecture.

Equation (2.7) may be rewritten as

$$V_T = V_{T0} + \gamma\left[(2\Phi_B + |V_{SB}|)^{0.5} - (2\Phi_B)^{0.5}\right] \tag{2.9}$$

where $\gamma = (2\varepsilon_0 \varepsilon_{Si} q N_A)^{0.5}/C_0$ is called the body effect parameter. The variation of threshold voltage with substrate bias voltage is illustrated in Figure 2.18.

Taking a closer look at the region of operation in which $V_{GS} < V_T$, approximated above in Equation (2.4), the MOSFET is actually in its sub-threshold region of operation. The channel current in this re-

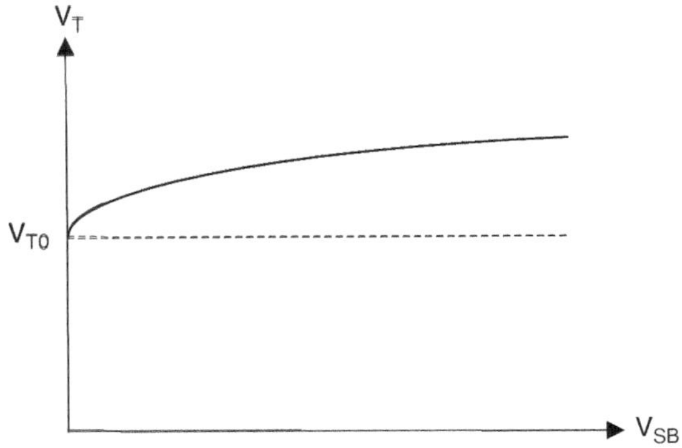

Figure 2.18 MOS threshold voltage variation with substrate bias voltage

gion is extremely small compared to that in the linear and saturation regions and varies exponentially with gate voltage. It may be described by

$$I_{DS} = \frac{W}{L} I_{DO} e^{-V_{BS}\left(\frac{1}{nV_t} - \frac{1}{V_t}\right)} \left(1 - e^{-\frac{V_{DS}}{V_t}}\right) e^{\frac{V_{GS} - V_T}{nV_t}} \tag{2.10}$$

where I_{DO} and n are process parameters whose values are typically around 20 nA and 2 respectively, and $V_t = kT/q$, where k is Boltzmann's constant.

When $V_{BS} = 0$ and $V_{DS} > 3V_t$, (2.10) simplifies to

$$I_{DS} = \frac{W}{L} I_{DO} e^{\frac{V_{GS} - V_T}{nV_t}}. \tag{2.11}$$

A typical sub-threshold characteristic is shown in Figure 2.19. Sub-threshold conduction is actively used in a relatively small number of applications in which power consumption must be kept to an absolute minimum, for example medical implants such as cardiac pacemakers in which battery life is a key concern. The very small currents involved limit the speed of operation of sub-threshold circuits. Other applications include those in which very small analog currents must be controlled, such as in determin-

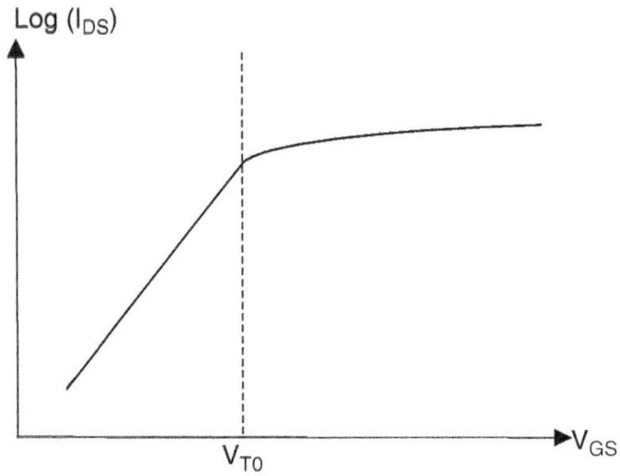

Figure 2.19 MOS sub-threshold characteristic

ing the light output from a single pixel of an OLED microdisplay. In this case a key issue is overcoming the pixel-to-pixel threshold voltage variation that is inherent in manufactured CMOS circuits.

In many other circumstances the sub-threshold channel current is considered as a leakage current that may, for example, limit the time for which signals may be stored capacitively in a circuit before deteriorating in quality or becoming corrupted. In many cases, such as the normal operation of digital circuits, the sub-threshold channel current can be legitimately considered to be zero as approximated in Equation (2.4).

2.2.2 MOS Capacitor

Some microdisplays use a MOSFET to provide some or all of the pixel storage capacitance. This is achieved by shorting together the source and drain and then using the capacitance between the gate and the source/drain as the storage capacitance. This gives a very high but bias-dependent capacitance per unit area. The bias dependence arises from the presence, at intermediate bias voltages, of a non-conducting layer (the depletion region) underlying the gate oxide and whose capacitance lies in series with that of the gate oxide. The bias-voltage dependent capacitance, C_{G0}, is given by

$$\frac{1}{C_{G0}} = \frac{t_{Ox}}{\varepsilon_{Ox}} + \frac{x_d}{\varepsilon_{Si}}. \tag{2.12}$$

x_d is, in turn, given by

$$x_d = \left(\frac{2\varepsilon_{Si}}{q}\frac{\Phi_S}{N_A}\right)^{0.5} \tag{2.13}$$

where Φ_S is the surface potential of the silicon. The effective capacitance of the MOS capacitor as a function of gate bias voltage is illustrated in Figure 2.20.

2.2.3 MOS Transistor Switches

The single n-channel MOSFET circuit of Figure 2.21 provides for switchable isolation or conduction between two circuit nodes. When $V_{GS} < V_T$ the MOSFET channel between source and drain is – to a good approximation - open circuit, otherwise it conducts. Typically V_{GS} is operated in digital fashion between two extreme voltages such as 0 V and V_{DD}. Note that there exists an asymmetry in the operation

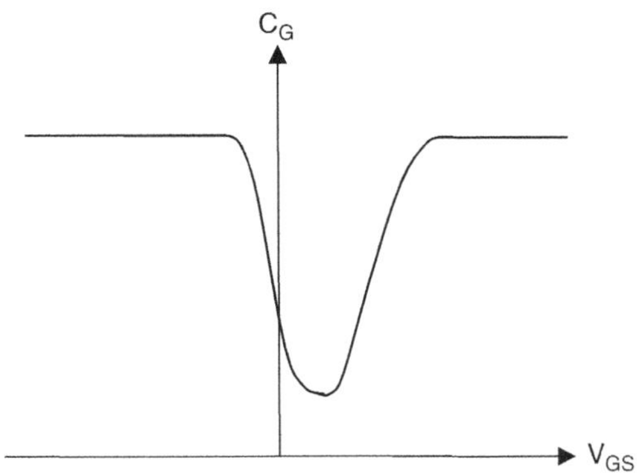

Figure 2.20 MOS gate capacitance variation with gate bias voltage

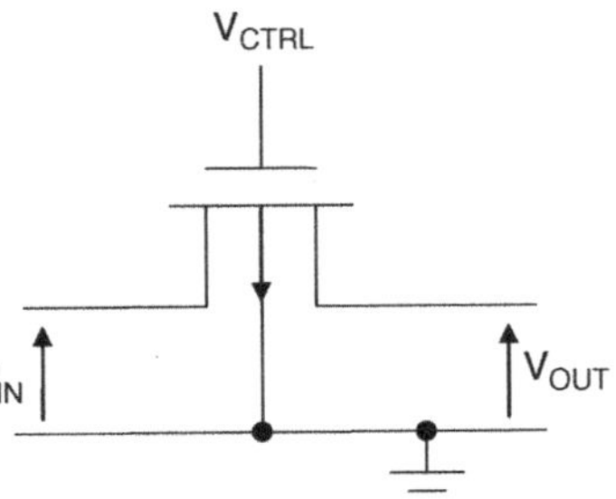

Figure 2.21 Circuit schematic of NMOS pass transistor

of this circuit. Low voltages are transmitted by the switch whilst attempts to transmit high voltages ($V_{IN} \geq V_{DD} - V_T$) cause the MOSFET to stop conducting according to the conditions applying for Equation (2.4). The body effect applies so that for $V_{CTRL} = V_{DD}$ the maximum transmitted voltage may typically be around (2/3) V_{DD} and may be as low as (1/2) V_{DD}. One possible solution to this limitation is to over-drive the gate of the switch so that $V_{CTRL} \gg V_{DD}$ but this requires a means of sourcing or generating the high gate voltage. Substituting a single p-channel MOSFET for the single n-channel device allows high voltages to be well transmitted whilst offering limited transmission of low voltage signals.

The transmission-gate circuit of Figure 2.22 contains n- and p-channel MOSFETs in a parallel configuration. This topology allows the full dynamic range of possible input signals (from 0 V to V_{DD}) to be transmitted. The price to be paid for the additional functionality includes higher area (two MOSFETs, not one), the need for a complimentary pair of control signals (instead of a single control signal) and the need for nearby substrate connections to both 0 V for n-well and V_{DD} for p-well. In short – this circuit exhibits larger area, more complex driving requirement and higher power consumption than a single-pass transistor.

2.2.4 CMOS Inverter

The CMOS inverter – a basic building block of many CMOS circuits, including the static memory pixels described later – is shown in Figure 2.23. This is a digital circuit that inverts the voltage input from a logic level low to high or vice versa. A high-voltage input, say 2.5 V or 3.3 V, turns on the n-channel

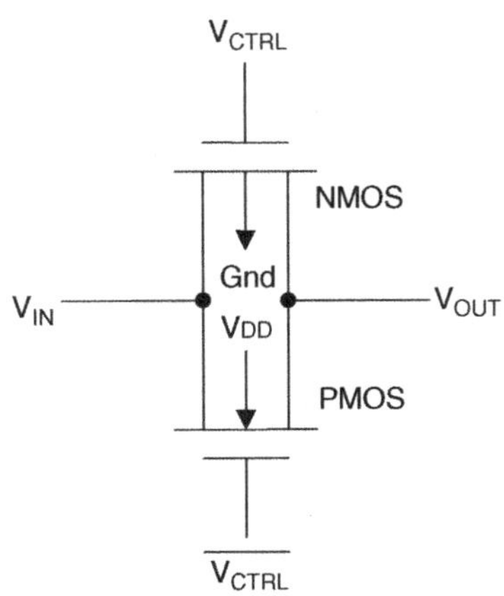

Figure 2.22 Circuit schematic of CMOS transmission gate

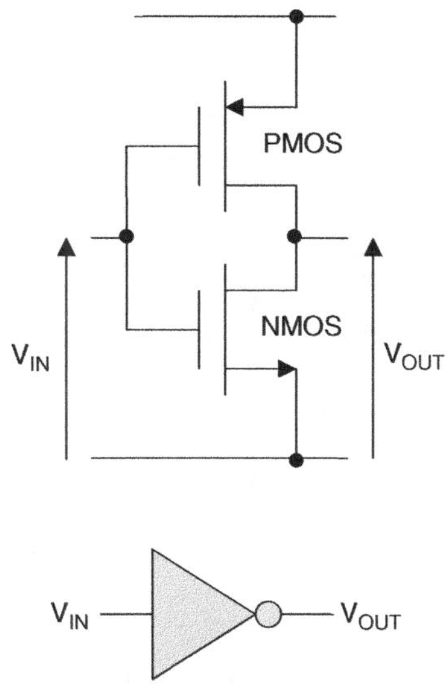

Figure 2.23 Circuit schematic (top) and logic schematic (bottom) of CMOS inverter

device alone, effectively shorting the output voltage to a 0 V. Conversely, a low-voltage input of 0 V turns on the p-channel device alone, effectively shorting the output voltage to the positive power rail (perhaps 2.5 V or 3.3 V as suggested above).

2.2.5 MOS Memory Circuits

In mainstream microelectronics, two memory types dominate the arena of volatile memory (memory that requires a power supply in order to retain data) – dynamic random access memory (DRAM) and static random access memory (SRAM).

A cell of DRAM stores 1-bit of data as charge on a capacitance as shown in Figure 2.24. Access to the stored signal is controlled by a single-pass transistor switch. Data is written into the cell by externally forcing a data value onto the column/data line whilst asserting the row/access line. Data is read from the cell by detecting small perturbations to the column/data line caused by charge flowing from the cell capacitor whilst asserting the row/access line. The read operation destroys the data value. Even in the absence of a read operation, leakage current would corrupt the data with time. So DRAM arrays must have a built-in refresh capability to ensure that the data is read and replaced at regular intervals.

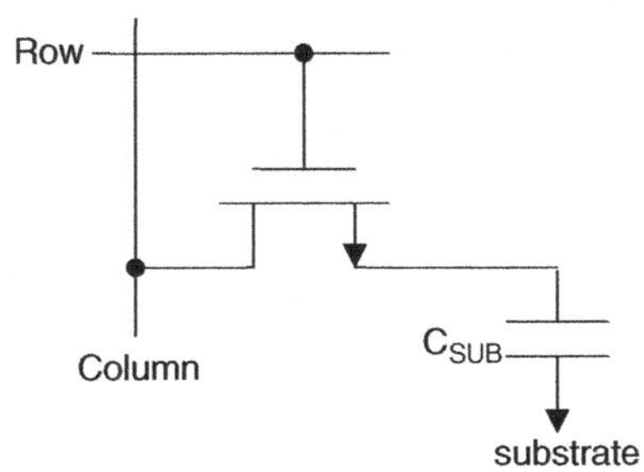

Figure 2.24 Circuit schematic of simple 1-T DRAM circuit

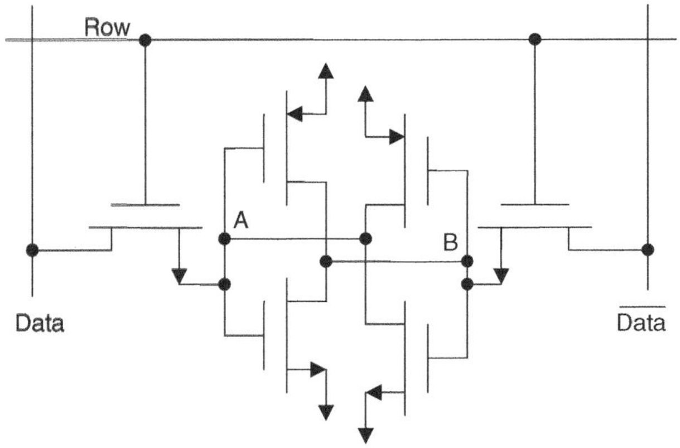

Figure 2.25 Circuit schematic of 6-T SRAM circuit

A single cell of SRAM stores 1-bit of data as two complimentary values on two linked storage nodes, A and B, of a bistable circuit – a pair of cross-coupled CMOS inverters as shown in Figure 2.25. The logic state of the call can arbitrarily be defined as the logic state of one of the nodes, say node A. So state [A, B] = [0, 1] defines a stored "0" whilst [A, B] = [1, 0] defines a stored "1." The cross-coupled inverters are accessed for both reading and writing by means of a pair of NMOS pass transistors. The bistability of the circuit means that SRAM has no refresh requirement – data is robust against both leakage and readout. A variant of the 6-T SRAM uses high-value polysilicon resistors in place of the two pull-up PMOS transistors. The primary advantage of the resulting 4-T SRAM is one of smaller footprint as the resistors can be fabricated on top of the remaining NMOS transistors.

Table 2.3 qualitatively summarizes the differences between DRAM and SRAM cells. As will be seen later, CMOS microdisplays have been designed using both DRAM and SRAM styles.

2.3 LCOS System Electronics Architecture

2.3.1 Overview and Classification

All LCOS microdisplays typically require a degree of support at the system or module level in order to function efficiently. The degree and nature of the support depends upon the application. Some ways in which LCOS systems can be classified are shown in Table 2.4.

Table 2.3 Brief comparison of SRAM and DRAM

Property	DRAM	SRAM
Cell area	Low	High
Active devices per cell	1	6 (4)
Switching speed	Slow	Fast
Switching power	Low	High
Refresh	Required	Not required
Data lines per cell	1	2
Access lines per cell	1	1
Power rails per cell	0	2

Table 2.4 Some factors that have a bearing on the electronic system architecture in LCOS

Application	Projection	Near-to-eye
Configuration (projection)	Rear-projection TV	Data projector
Configuration (near-to-eye)	Electronic viewfinder	Wearable headset
Number of microdisplays	Single microdisplay system 3 microdisplay system	2 microdisplay system
Monochrome/color	Monochrome	Color
Display aspect ratio	4:3	16:9
Display primary purpose	TV: SDTV, HDTV	PC: VGA, SXGA etc.
Picture data	Interlaced	Progressive
Data source	Digital	Analog
Display definition	Low definition	High definition

Examples of the effects of application on design include the preference to minimize power consumption in applications that are battery powered (e.g. EVF for digital cameras) versus the need to handle very high data rates for mains-powered high-definition projection systems or different data handling and sorting requirements for color depending on the format of the source data and the format of the display system (single-chip with RGB sub-pixelation versus single-chip field-sequential-color versus three-chip system). The following section describes some typical examples.

LCOS offers flexibility of system partitioning or integration that is not available with TFT LCD backplane technology. Although it will evolve with time, roughly speaking, α-Si TFT technology integrates the electronics of the pixel array only, p-Si TFT technology integrates in addition the line drive circuits into the backplane. In an LCOS system much of the electronic functionality normally associated with standard or custom interface ICs can be integrated onto the LCOS backplane. Just how much depends on a number of factors. These include technology node compatibility. If the LCOS backplane uses, for example, 0.35 µm CMOS, is that a suitable technology for the added functionality? It may well suffice for I^2C bus but not for MPEG decompression, which would be better done in a much smaller geometry in which the circuits probably already exist as Intellectual Property (IP) blocks and that, in addition, offers a smaller footprint and lower power consumption. Purpose is also important – added functionality may be beneficial for a single application but may limit, constrain or de-optimize the use of the backplane for other purposes. In order to justify the additional design cost of system integration, the final product may have to be sold at a premium or may have to have a large market potential. Finally, higher on-backplane integration is more important and more feasible for medium-resolution near-to-eye applications in which size, weight, power and cost budgets are more limited. To date OLED on CMOS appears to offer the greatest potential for overall integration, given the low CMOS compatible voltages and the integration of the light source and drivers onto the backplane. Another area of flexibility in LCOS design is the position of the digital to analog conversion process (if present) in the data path. This can be off-backplane (on an ASIC as is the case in the system described in section 2.3.2), on-chip as shown later in Figure 2.28, in-pixel as described in section 2.5, or even in-eye as described in section 2.6.

2.3.2 Interface and Support Architecture

Near-to-eye system example

A system-level schematic for a typical system based on a single LCOS microdisplay is illustrated in Figure 2.26(a). Specifically, the illustration is of a single-microdisplay-chip, field-sequential-color (FSC) LCOS system using LED illumination. This system comprises a display interface ASIC (application-specific integrated-circuit), an illumination control ASIC, an LED driver IC, red, green

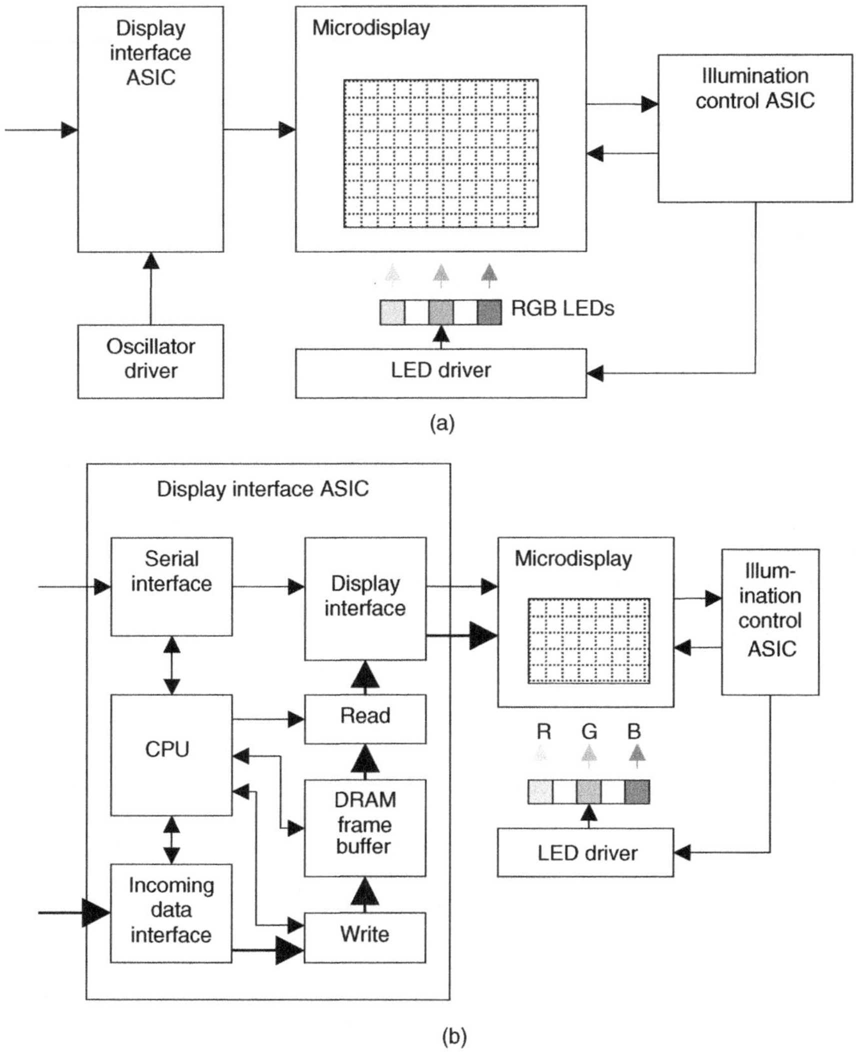

Figure 2.26 Schematic diagrams of example field-sequential-color near-to-eye LCOS system: (a) system level; (b) ASIC detail

and blue LEDs, and the microdisplay itself. The display interface ASIC is illustrated in more detail in Figure 2.26(b). The CPU controls its operation. The ASIC receives control and configuration signals, usually through a (slow) serial bus such as the I^2C bus described by Paret and Fenger,[33] and a representation of the data to be displayed on the microdisplay. Control signals may be used to set up the control parameters for the ASIC at power-up, to change mode during operation, or to monitor some aspects of the operation of the ASIC. Under CPU control data is passed into the (often dual) frame buffer. A dual frame buffer allows stored frame *n* to be written to the microdisplay whilst the next frame is incoming from the source. Frame data may be sorted from incoming format to microdisplay specific format on the way through the ASIC. The illumination control ASIC ensures that the timing of pulses of LED illumination is synchronized with the availability of appropriate data in the pixels of the microdisplay.

The control signals are used to set the values of parameters that do not change or change only rarely – interface and display modes and parameters such as clock rate, frame rate, brightness and image orientation (flip horizontal, vertical). This is done by setting values in dedicated or programmable storage registers in the display interface ASIC or in the microdisplay.

The data may be received in analog or digital form. If analog, it is often converted to digital as this allows more flexibility such as in the use of look-up tables for gamma correction.

The illumination control ASIC is responsible for synchronizing the timing of pulsed illumination with the availability of the appropriate image data on the pixel array and switching (if appropriate) of the polarity of the front electrode voltage. Depending upon the intensity of light required from the LED illumination, a separate LED driver chip may be necessary.

Projection system example

Figure 2.27(a) shows the detailed system layout for a projection display system employing three LCOS microdisplays. The microcontroller receives inputs that program the operation details of the system, such as gamma look-up tables and data management. Temperature environment of the microdisplays feeds into the microcontroller, setting the fan speed to maintain optimum temperature.

The system is under the supervision of the microcontroller which loads permanent data and code from the non-volatile EEPROMS at power-up. Supervision functions include control of the fan speed in response to a temperature monitor signal fed back from the imagers. In operation, incoming digital video data is fed via a DVI connection to the video digital signal processor that divides the incoming data into individual red, green, and blue data streams, serving output to the custom ASIC.

The custom ASIC is magnified in Figure 2.27(b). Color field data progresses into the ASIC, RGB in parallel through the data level adjusters (DLAs) or voltage-level shifters, and queues in the first-in first-out (FIFO) buffers prior to temporary storage in the double-data DRAM (DDRAM) frame buffer. On leaving the frame buffer, adjustment for device characteristics and display gamma are applied from a look-up table. Overall optical system nonuniformity is corrected, including microdisplays, by reference to measurements made and stored during factory final test. Digital to analog converters serve video voltages to output amplifiers feeding the microdisplay. The amplifiers are constantly monitored in order to adjust for drift and thereby maintain overall calibration.

2.3.3 Backplane Electronics

Backplane architecture and floor plan

An example of the architecture of an SXGA backplane reported by Gale *et al.*[34] is shown in Figure 2.28. Some features in particular are worthy of note, such as: the chip accepts analog video data; rows are driven from both ends. Another feature is that the display is designed for column inversion (see section 2.4.2) with high and low video driven alternately to odd and even columns. High data is always handled from the north of the array and low from the south. Control signals ensure that there are no drive conflicts. Jepsen *et al.*[35] offer a similar level of detail on an alternative SXGA backplane.

A schematic of a generic backplane floor plan is shown in Figure 2.29. Moving from the periphery to the centre of Figure 2.29 we encounter the following.

1. *Input/output pads.* These are the areas of the microdisplay through which individual electrical connections to the outside world are made, typically by gold wire or aluminium wedge bonding. The I/O pads carry
   - power rails
   - incoming and outgoing control signals via a serial bus

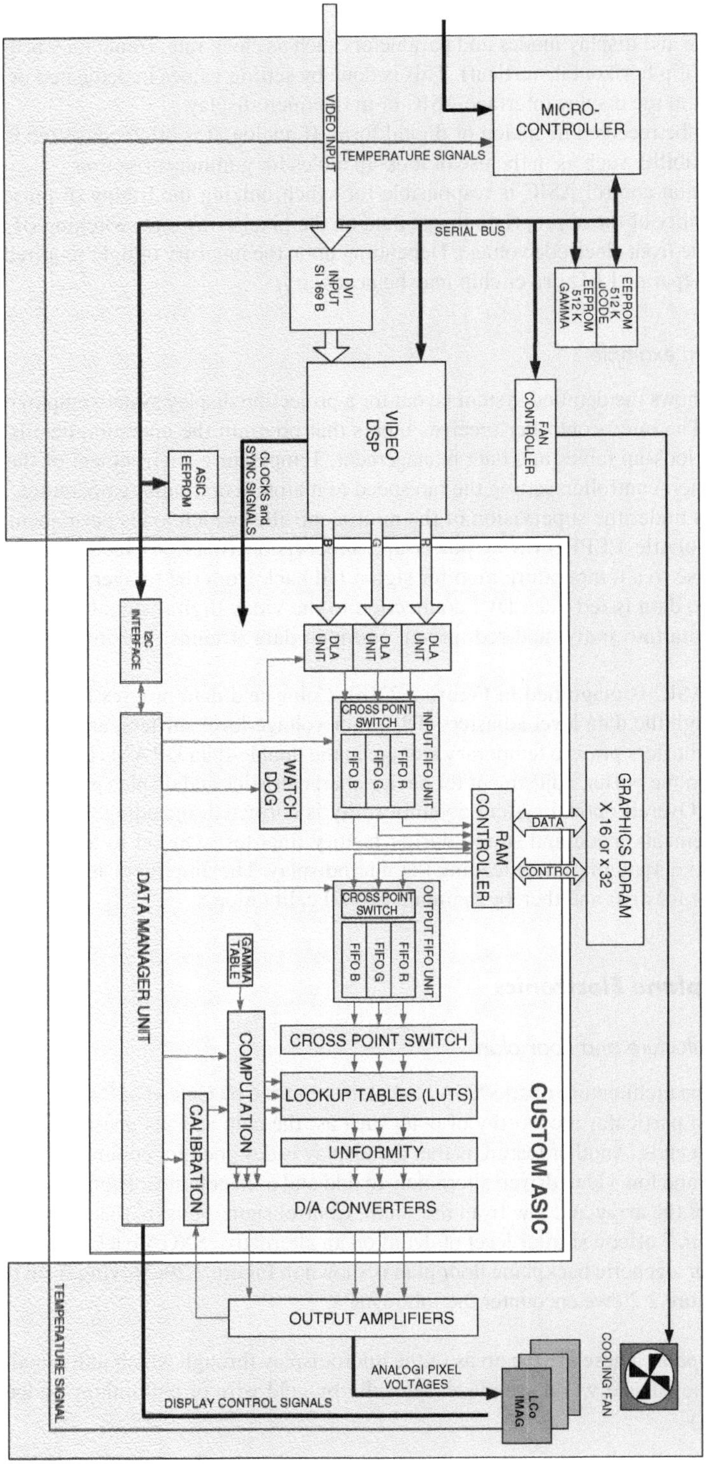

(a)

Figure 2.27 Schematic diagrams of example three-panel LCOS projection system: (a) overall system; (b) magnified detail of custom ASIC. Reprinted courtesy of Syntax-Brillian Corporation

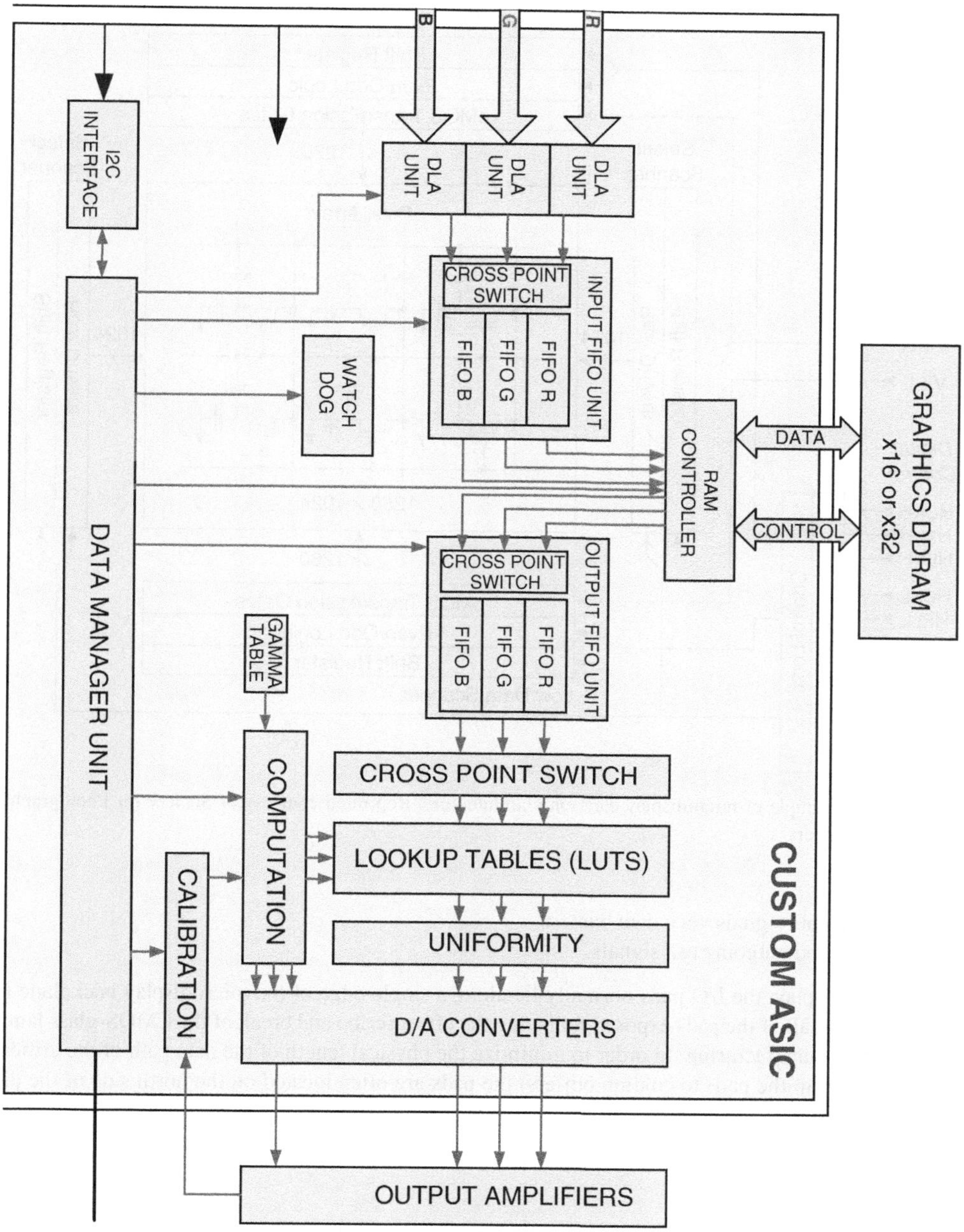

(b)

Figure 2.27 *(Continued)*

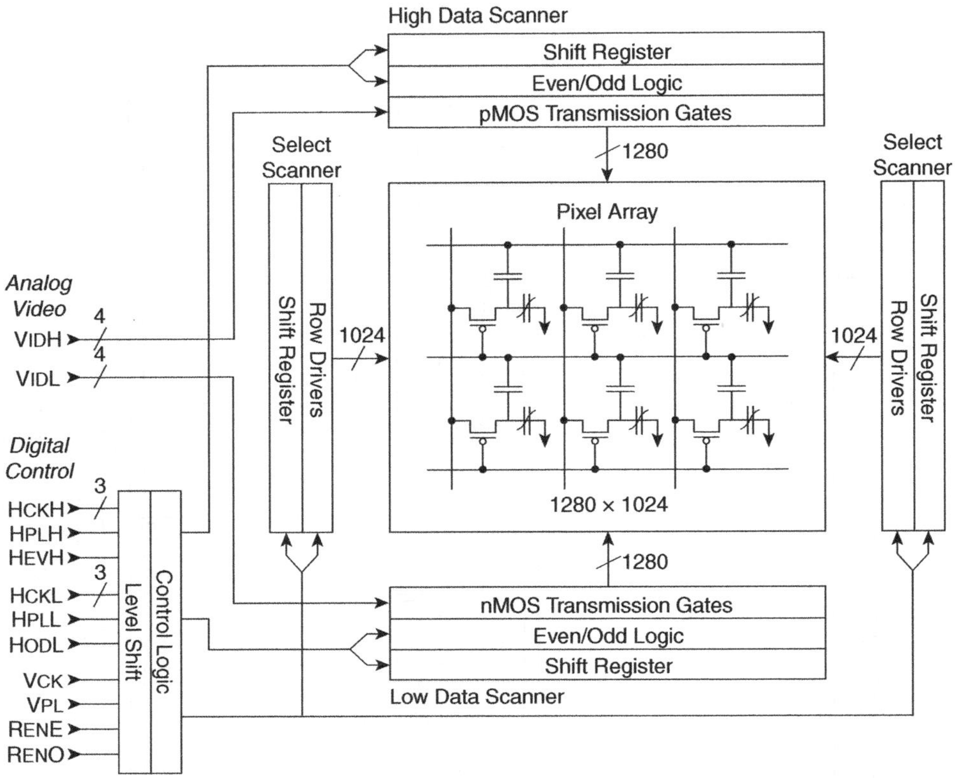

Figure 2.28 Example of microdisplay electronic architecture. Reprinted courtesy of Society for Photographic Instrument Engineers

- incoming data signals via a data bus
- incoming and outgoing test signals.

In a microdisplay, the I/O pads normally lie along a single edge of the microdisplay backplane in order to leave all of the pads exposed following the offset scribe and break of the CMOS-glass laminate during manufacturing. In order to minimize the physical length of the data path of the critical data flow (from the pads to column buffers) the pads are often located on the north side of the die as illustrated.

2. *Control block.* This circuit supervises the operation of the display, accepting data from the serial bus in order to do so.

3. *Data sorter block.* This circuit carries out the data manipulation determined by the control block.

4. *Column data driver block.* The column or data driver is the more complicated of the line addressing blocks, consisting typically of two stages as shown in Figure 2.30. The data to be delivered to a particular row is assembled here word by word. Data arriving serially must be distributed and stored in stage 1 of each column driver cell until the data for all columns is assembled and stored. At that time the data for all columns is transferred simultaneously from stage 1 to stage 2. This two-stage approach allows the data being written to row j to settle at the same time as the data for row $j + 1$ is being assembled. The precise nature of the circuitry depends on the data format and the chosen architecture. For example the incoming data can be distributed by means of a shift

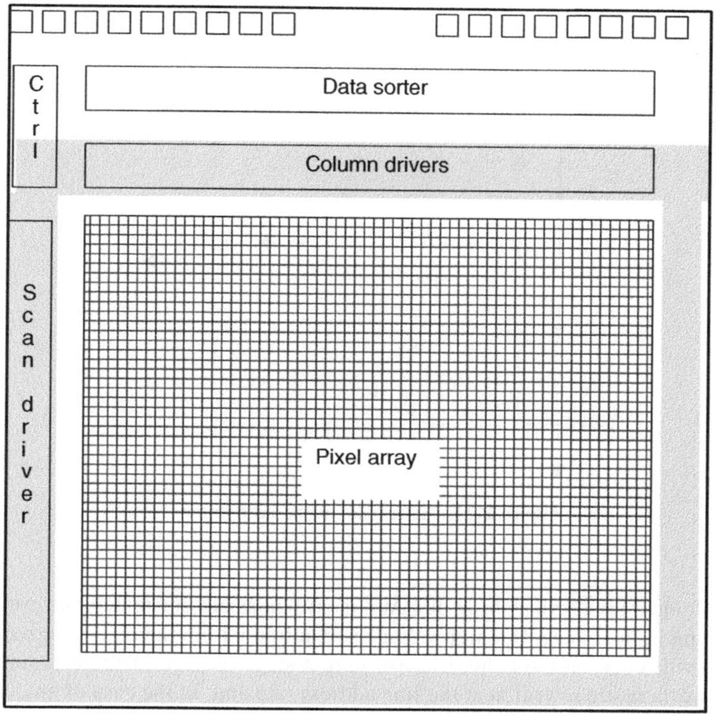

Figure 2.29 Microdisplay generic floor plan

register approach or a demultiplexor approach. The data received, stored and transmitted may be all-analog, all-digital (single-bit or multi-bit) or may undergo conversion (the most likely case being multi-bit digital to analog) within the column driver circuit. Actually, in the case of digital to analog conversion, a single digital-to-analog conveter (DAC) is likely to be shared among a

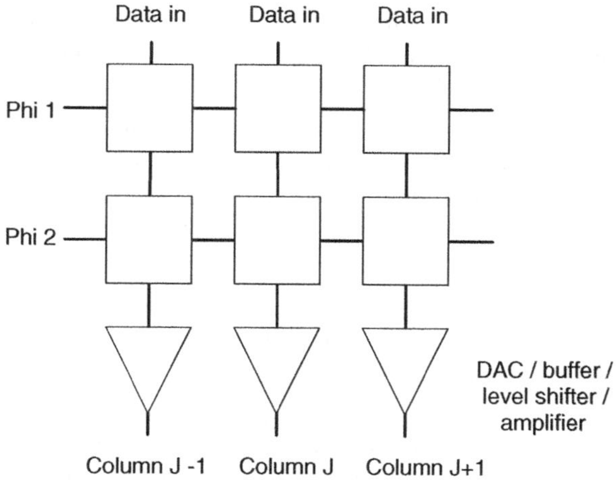

Figure 2.30 Schematic of three adjacent column driver cells

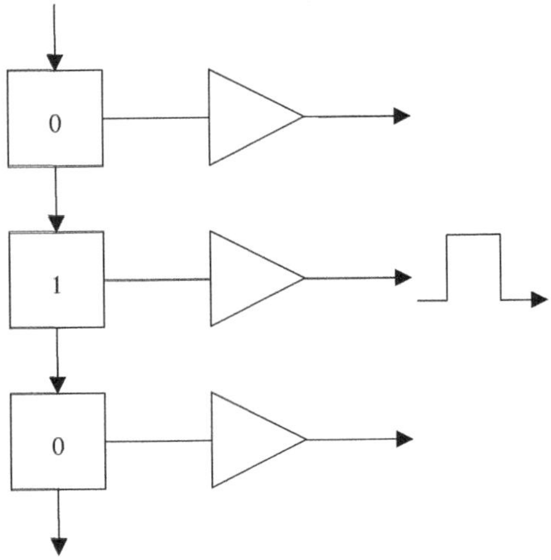

Figure 2.31 Schematic of row scan driver

number of columns. Precise matching of DACs is critical as any mismatch may cause noticeable vertical banding in the image. If a high voltage is required within the pixel, any required voltage amplification will occur at the column driver output stage. Design of the column driver stages is important as these are rewritten at the line address rate and, in the case of analog addressing, must settle to the required level (8- or 10-bit precision for example) very quickly. Analog drivers do not, in general, offer the full voltage range of the chip – some voltage is lost due to driver headroom or footroom requirements. In the case of digital addressing, precision of settling is less important but the rewrite rate is commensurately faster according to the number of bit-planes to be written.

5. *Row scan driver block.* The row or scan driver, illustrated in Figure 2.31, is the simpler of the line addressing blocks. Its function is usually to activate one row at a time, in sequence from top to bottom, in order to transfer assembled data from the column data lines into the pixels of the selected row. In the case of a SVGA pixel array with 600 rows, this can be achieved by a 10-bit decoder circuit (select 1 output from up to 2^{10} according to the value of the input 10-bit word) or by a token passing shift register arrangement. In a display in which row-sequential addressing is the norm, the shift register is often preferred. The decoder circuit offers more flexibility to cope with irregular addressing schemes. Low to high voltage level shifting may also occur here.

6. *Pixel array.* The pixel array constitutes the largest single area on the backplane. The pixels are arranged in a rectangular array on a square grid for some formats (e.g. VGA, SVGA), or on a rectangular grid for other formats (e.g. PAL, NTSC). In the case of color pixels consisting of RGB sub-pixels, the sub pixels may be arranged within the color pixel as thin stripes or in a delta pattern as illustrated in Figure 2.7.

7. *Periphery.* Shaded in light gray in Figure 2.29 is the area over which the glue seal ring is deposited. The glue seal ring area is shared with the circuit's peripherals to the pixel array such as line drivers and test circuits in order to minimize the required die area.

8. *Test circuits.* Figure 2.29 does not explicitly show any areas reserved for test circuits, but testability is important in establishing the electronic yield of the backplane. So the areas to the south and east of the pixel array of Figure 2.29 may be partly occupied by circuits used to test the functionality and yield of the addressing circuits, line drivers and pixel circuits as discussed later.

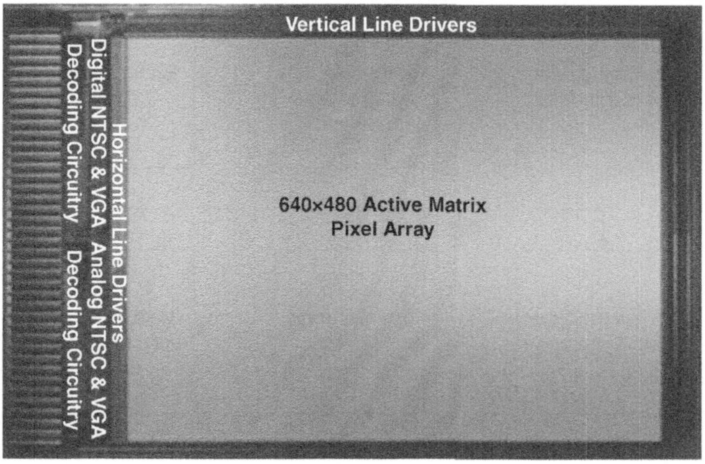

Figure 2.32 Photograph of vintage LCOS backplane. Reprinted courtesy of Microdisplay Corporation

Figure 2.32 is a photograph of a vintage monochrome LCOS backplane. The inherently monochrome nature of the display makes possible the inclusion of NTSC and VGA decoding circuitry for both analog and digital input data. Figure 2.33 is a photograph of a more recent microdisplay.[36] This device has input pads on the north side of the die; those on the south side are for test and characterization and need not be connected in normal operation.

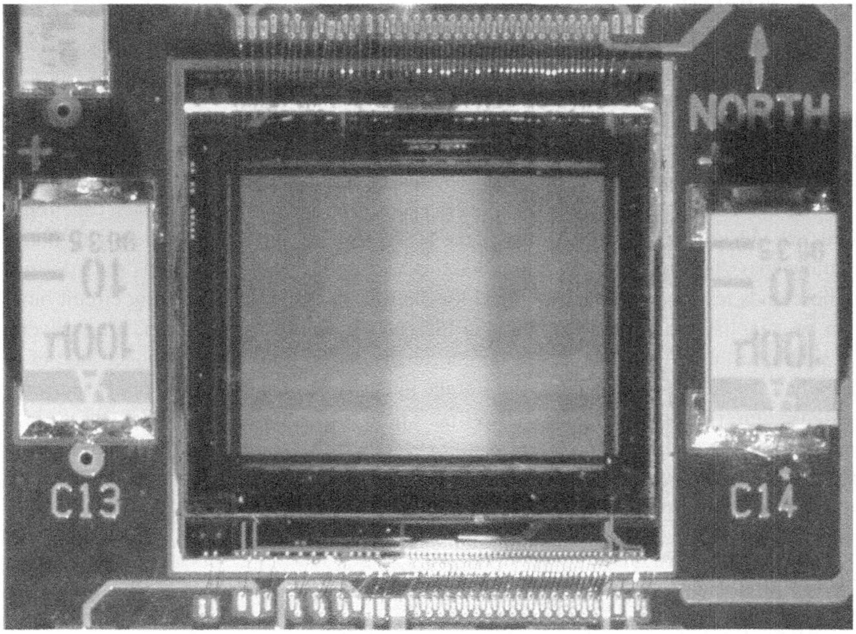

Figure 2.33 Photograph of DRAM-style XGA ferroelectric LCOS backplane from the late 1990s. Reprinted courtesy of the University of Edinburgh

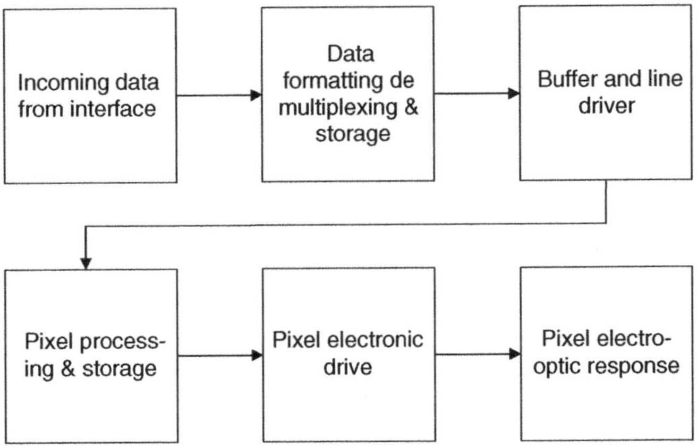

Figure 2.34 Data path through microdisplay

Data path

Figure 2.34 shows the stages of the data path through the backplane to the pixel. From arrival at the input bonding pads perhaps word-at-a-time, data is sorted, de-multiplexed and, if appropriate, converted from digital to analog format, in order to address the required number of columns. It is stored at the head of the column prior to transfer through and writing onto the column line. From there it is transferred into the pixel of the appropriate row. It is stored and may be further processed at the pixel where it elicits an electro-optical response from the LC. Figure 2.35 illustrates a small portion of a pixel array with the line drivers, and Figure 2.36 shows the data writing operation to the *j*th row of a portion of an array of DRAM pixels.

Test circuits

It is important in establishing the electronic functionality and yield of a backplane design that electronic testing be allowed to reveal as much data as possible. Functional testing is used in the early stages of backplane development to establish whether or not a given design is functioning or not as planned. It is used to debug the design. Yield testing is used to determine, for a given functional design in manufacture, whether an individual backplane is electronically working correctly or not. In other words it allows pass/fail selection of an individual device and thereby also yields statistics on the design. To those ends test strategies are required for the row and column addressing circuit blocks. Test points can be established intermediately and at the end of the addressing chain. Line drivers can be tested by simple test circuits placed at the far end of the driven line (at the opposite side of the pixel array). Ideally it would be possible to test the storage integrity of each individual pixel electronically by means of sense amplifiers on each column bus line. These could be used to test the storage of a digital signal or logic level at each pixel but there is no expectation that they could discriminate gray levels. The sense amplifiers and other electronic test circuits may be configured to be powered-up in test mode only and powered-down in normal operation.

Interfaces and data rates

The average data rate, D_{avge}, required to drive a color microdisplay system is

$$D_{avge} = 3MNF \tag{2.14}$$

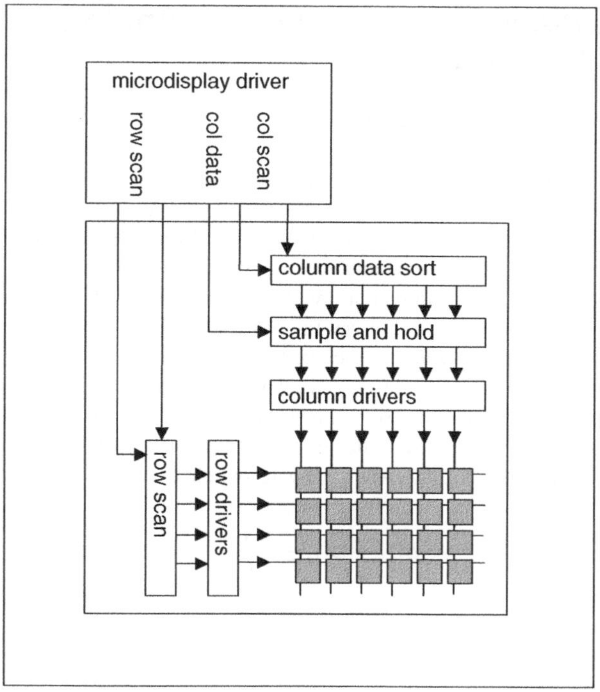

Figure 2.35 Simplified schematic showing row and column drivers

where F is the frame rate, and $3MN$ is the number of dots (assuming M columns, N rows and 3 sub-pixels, RGB, per color pixel). The actual or peak data rate may be significantly higher than this due to several factors. For example, the data may require to be sent twice in order to achieve DC-balancing in an LCOS system, or the data transmission duty cycle may be less than 100%. The latter is the case in systems in which, for example, the electronic addressing must pause during LC settling periods.

The high data rate generated by demanding applications such as high-definition TV (HDTV) implies considerable parallelism and storage as shown in Figure 2.27, and continued at the microdisplay level, particularly for color field sequential systems.

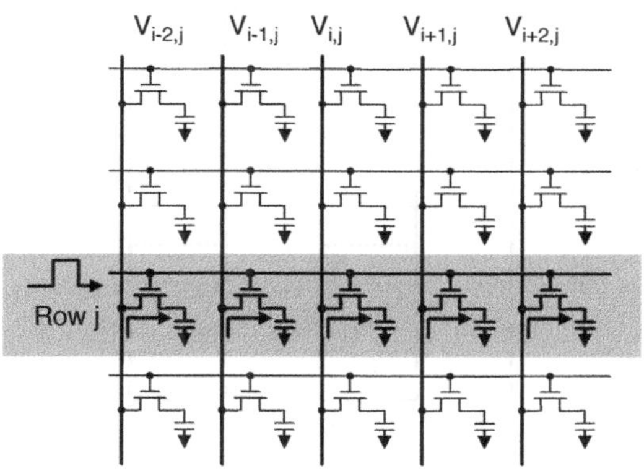

Figure 2.36 Addressing pixels of jth row

2.4 Analog Pixel Drive Schemes for Analog Electro-optic Response

2.4.1 Analog Voltage Addressing

Most analog drive schemes use a DRAM-style pixel circuit as described in section 2.4.3. Figure 2.43 shows an equivalent circuit of the pixel, including the gate-to-source capacitance and a resistor representing TFT and liquid crystal leakage current. However, leakage is dependent on temperature, illumination, and voltage; moreover, ion count and space charge limit ionic transport in the liquid crystal. The liquid crystal capacitance, C_{LC}, can increase with voltage by as much as a factor two or more, due to molecular reorientation of nematic liquid crystal. Setting the LC counter electrode voltage, V_C, to the mid point of the voltage range availabe at the pixel pad, $V_C = (V_{P1} + V_{P2})/2$, achieves an alternating amplitude $+V_{LC} = V_{P1} - V_C$ and $-V_{LC} = V_{P2} - V_C$, where numeric subscripts signify alternating odd and even frame cycles. The video signal pixel voltages V_{P1} or V_{P2} carry the grayscale information, while alternating the V_{LC} polarity. Figure 2.37 indicates the voltage waveforms over the addressing cycle, omitting a capacitive coupled term, ΔV_{LC}, when the gate switches through ΔV_G.

$$\Delta V_{LC} = \Delta V_G \frac{C_{GD}}{C_S + C_{LC} + C_{GD}}. \tag{2.15}$$

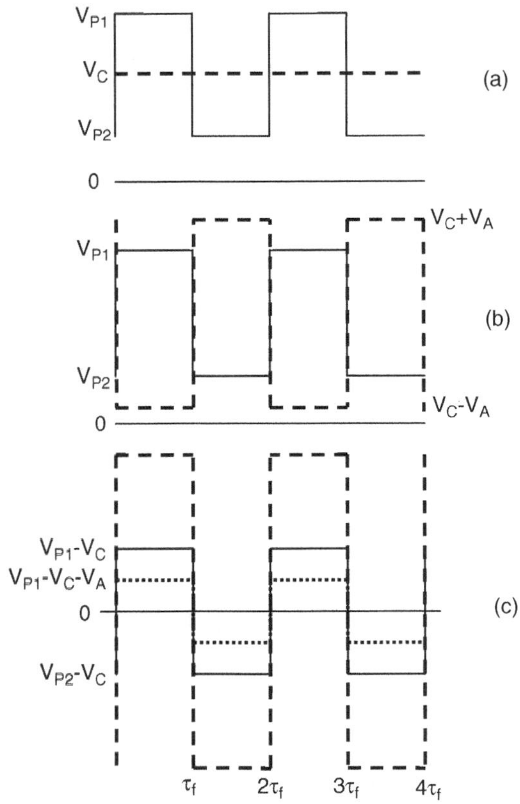

Figure 2.37 Addressing waveforms; τ_f = frame time. (a) Constant $V_C = (V_{P1} + V_{P2})/2$, pixel voltage alternates between V_{P1} and V_{P2}. (b) Modulated $V_C = (V_{P1} + V_{P2})/2 \pm V_A$, pixel voltage in anti-phase (or in phase). (c) Liquid crystal voltage for three cases: constant, anti-phase and in phase.

The gate voltage change ΔV_G is unidirectional when the gate switches off, while the liquid crystal voltage V_{LC} must be alternating; ΔV_{LC} makes it difficult to eliminate a DC component in V_{LC} for all values of grayscale voltage V_P, without compensation circuitry, or incorporating compensated values in a look-up table of grayscale voltages.

Correction is necessary in polysilicon microdisplays. A large storage capacitance minimizes cross-coupling effects from stray capacitance.

2.4.2 DC Balanced Driving of Liquid Crystal

Ionic current flow in liquid crystals mandates AC voltage drive in order to avoid space-charge effects and electrolytic degradation as described in Chapter 5. Nematic liquid crystal responds fundamentally to the square of the local electric field, making it independent of polarity. Avoidance of any long-term DC component in the waveform applied to the LC, particularly signal-dependent DC, is in general an essential requirement of any liquid crystal addressing scheme. The techniques of DC balancing are applicable to any drive scheme – analog drive (as described here) and digital drive (as described later) – and to all liquid crystal types (nematic, ferroelectric, etc.) and configurations (VAN, STN, SSFLC, etc.).

DC balancing with fixed front electrode voltage

Keeping the common front electrode voltage, V_C, fixed while alternating, on a frame-to-frame basis, the polarity of the liquid crystal voltage, V_{LC}, implies alternating the polarity of the voltage applied to each pixel pad, V_{PAD}. This is conventionally done either frame by frame (frame inversion) or line by line (line inversion) as illustrated in Figure 2.38.

DC balancing by frame inversion

The technique of *frame inversion* reverses the polarity of the AC drive voltage each time the pixel array is scanned or refreshed. Any modest DC component in the liquid crystal drive voltage introduces an optical flicker, as the direct voltage adds or subtracts on alternate frames with polarity reversal of the AC drive voltage, as shown in Figure 2.39. The flicker frequency is half the frame refresh rate, requiring frame rate to be raised to more than twice the critical fusion frequency in order to suppress visible flicker. In practice LCOS microdisplays are refreshed at 120 Hz, requiring each 60 Hz video frame to

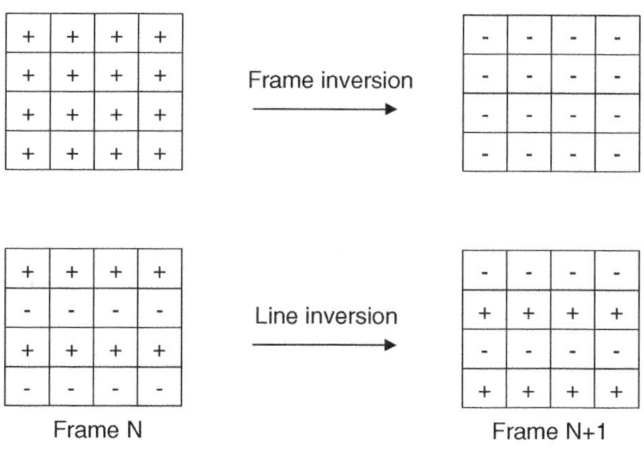

Figure 2.38 Illustration of frame inversion and row line inversion on small arrays of pixels. The "+" or "−" sign in each pixel represents the polarity of V_{LC} in that pixel

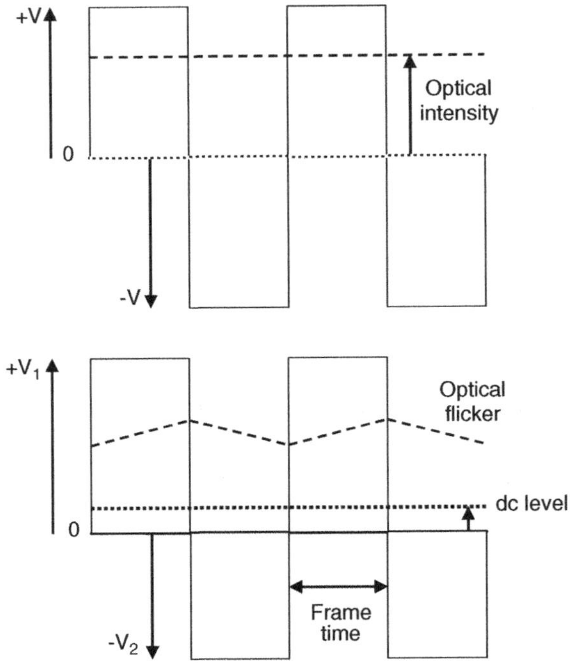

Figure 2.39 Effect of DC level $(V_1 - V_2)/2$ on optical flicker in nematic cells

be written twice with alternate positive and negative liquid crystal voltages. Higher frame refresh rate stresses the addressing rate problems, particularly for high pixel count displays.

Voltage droop between refresh cycles due to charge leakage introduces a further DC shift as some components of the charge leakage are unipolar, thus adding to the magnitude of any flicker.

The potential of the common transparent electrode (ITO) is adjusted to minimize the cell DC component and associated flicker averaged over the display active area. A residual DC level adds to the mean-square voltage, corrupting the gray-level values, and contributes to image sticking in conjunction with any video-signal-dependent DC term. Circuit design must minimize any signal-dependent DC components. Global DC drift is difficult to suppress in LCOS due to the inherent electrical asymmetry in the liquid crystal cell introduced by the use of two different electrode materials – aluminum alloy[iii] and ITO (see Chapter 5).

DC balancing by line inversion

TFT-addressed transmission-mode microdisplays generally employ *line inversion* to suppress flicker. Reversal in polarity of successive addressed lines averages out the flicker amplitude, pushing the flicker spatial frequency to the highest displayed value, attenuating flicker beyond the limits of perception. Polarity reversal in adjacent pixels at the same gray level imposes strong lateral fringe field effects. The low fill-factor inherent to transmission mode microdisplays hides the fringe field effects. In reflection mode, high fringing fields can reduce or defeat the high fill-factor advantage of LCOS as explained in Chapter 6. Recent developments that reduce pixel size, while maintaining substantial fill factor, favor

[iii]The electrode on the CMOS backplane may be aluminum or more likely some closely related alloy such as aluminum with small amounts of silicon and copper. The latter is more common in CMOS technology but may be customized for LCOS. In any case we will refer to it within this book simply as aluminum alloy.

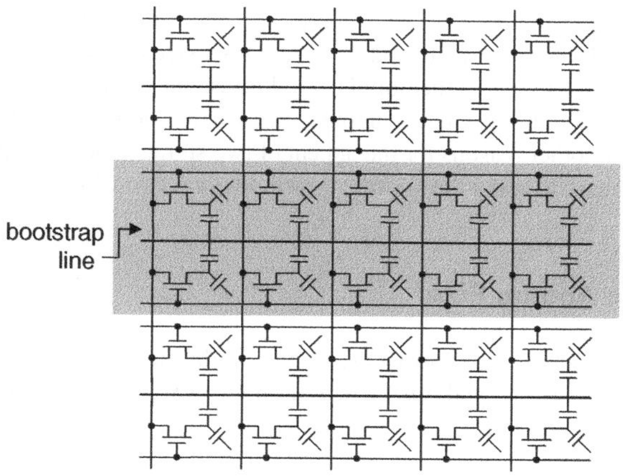

Figure 2.40 Pixel array schematic for bootstrapping. The "loose" ends of the in-pixel capacitors shown in the figure are all tied together to a common substrate voltage. The shaded dark band represents two rows of pixels sharing a single bootstrap line

scrolling-type line inversion to minimize fringe effects as described in Chapter 4. The advantages in flicker suppression outweigh the transmission losses in polysilicon microdisplays; however, in reflecting LCOS devices designed for high frame rate, the trade has less appeal and frame inversion is the norm.

Improved DC balancing by bootstrapping

Schlig and Sanford[37] report that the latest version of their LCOS projection light valves, of which an early version was described by Sanford *et al.*,[38] carry 12 V on chip and yet the voltage at the pixel pad is limited to the range 1.5 V to 8.5 V, i.e. a swing of 7 V peak-to-peak. Some of the reasons for this lie in the performance limitations of the DRAM-style pixel as described in section 2.4.3. They go on to describe a novel method of maximizing the voltage available using inversion (DC balancing) using bootstrapping. "Bootstrapping" is a technique for raising the output voltage of a driver circuit by changing the value of an internal circuit node via capacitive coupling of a changing voltage.

In the case of the DRAM-style pixel array shown in Figure 2.40, the signal that has been attenuated or restricted in reaching the pixel pad can be compensated or amplified using careful use of bootstrapping via an additional electrode shared between adjacent rows of pixels. This configuration supports frame or row-pair inversion. The results of bootstrapping showed a peak LC voltage of 6 V obtained on a chip using a 12 V supply with no adverse effect on luminance uniformity.

DC balancing by switching the front electrode voltage

Holding the counter electrode constant at V_C allows simple DRAM-like addressing, with the option of line inversion, but demands higher voltage switching transistors (MOSFETs). If V_C is modulated by V_A at the frame rate, the alternating pixel voltage (V_{P1}, V_{P2}) is in-phase or anti-phase with V_A, facilitating a lower MOSFET voltage:

$$V_C = \frac{V_{P1} + V_{P2}}{2} \pm V_A \tag{2.16}$$

$$V_P = V_{P1}, \ or \ V_{P2}, \ V_{P1} > V_{P2} \ \ V_A > V_{P1} \tag{2.17}$$

$$V_{LC} = V_C - V_P = \pm\frac{V_{P1} - V_{P2}}{2} \pm V_A \ \ or \ \ \pm\frac{V_{P2} - V_{P1}}{2} \pm V_A \,. \tag{2.18}$$

The maximum peak-to-peak amplitude is $2V_A + (V_{P1} - V_{P2})$, the minimum is $2V_A - (V_{P1} - V_{P2})$, with $2V_A$ the midrange value. The dynamic range $2(V_{P1} - V_{P2})$ is twice the constant V_C case; moreover, V_A adjusts the minimum voltage to match the liquid crystal threshold voltage.[39]

A V_C modulated system requires more complicated two-step addressing, employing two storage nodes per pixel. There are other reasons for the appeal of two storage nodes per pixel; these and examples of some pixels that have two storage nodes are included in section 2.4.4.

2.4.3 DRAM-style Analog Pixel

This is a widely prevalent pixel circuit for LCOS that has similarities to the purely electronic 1-T circuit used in DRAM and to the conventional pixel circuit used in active matrix TFT LCDs

Circuit description

The single-transistor DRAM (1-T DRAM or DRAM style) pixel circuit consists of a MOS transistor that operates as a switch to gate input signals and a capacitive or soft-node storage element. The MOS-T can be PMOS or NMOS. The gate of the MOS-T is connected to a bus line that is common to all pixels in a given *row* of the pixel array. One end of the channel (recalling the source–drain symmetry of the MOS transistor the nomenclature here is arbitrary so, for the sake of argument, the drain) is connected to a bus line that is common to all pixels in a *column* of the pixel array. The other end of the channel, the source, is one part of the floating internal storage node of the pixel. The various options for the storage element are described in Chapter 3. Figure 2.41 explicitly show a NMOS transistor and a NMOS capacitor.

This circuit is similar to that used in electronic DRAM. The main differences are that (a) electronic DRAM stores 1-bit of binary or digital data whereas, in the configuration described here, the 1-T

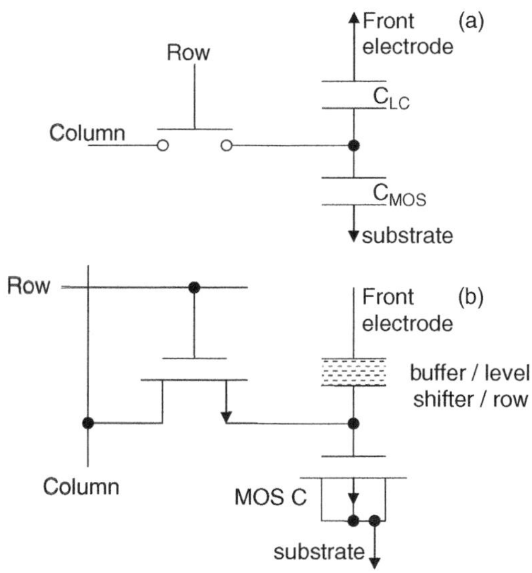

Figure 2.41 Simplified DRAM-style pixel circuit: (a) ideal representation; (b) schematic showing actual components

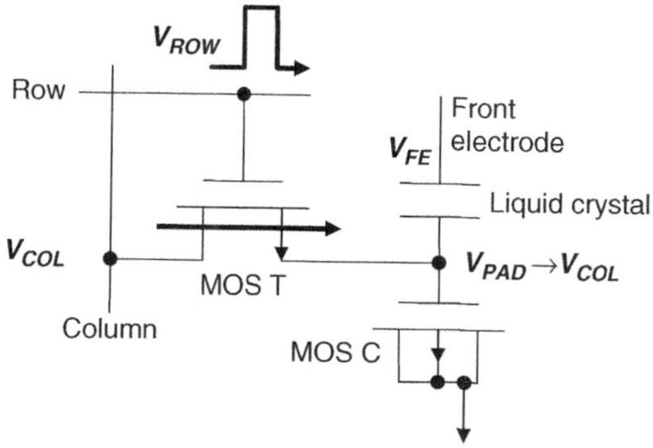

Figure 2.42 Addressing a pixel

DRAM pixel circuit must be designed to store a precise voltage value in order to control the gray level of the LC; and (b) the storage capacitance in electronic DRAM uses specialized elements that have a small footprint but a large capacitance value for maximum capacitance per unit area and maximum packing density of memory cells. This is achieved with elements such as trench capacitors that go deep into the substrate and contain large amounts of sidewall capacitance. Such elements require a manufacturing process that is highly customized and is not part of a normal LCOS CMOS process as described in Chapter 3.

This circuit is analogous to that used in the vast majority of active matrix LCDs. The pixels in the latter use a p-Si or α-Si switching TFT and various capacitive storage components as described by, for example, Voutsas[12] or Lueder.[14]

Circuit operation

The operation of the pixel circuit is illustrated in Figure 2.42. For most of the time the row voltage, V_{ROW}, is set to a value that causes the channel to be open circuit. During this phase, a pixel driving voltage, V_{COL}, is applied to the column bus line and allowed to settle to a stable value. V_{ROW} is then pulsed in order to transfer the V_{COL} into the internal storage node. In the case of a NMOS switching transistor, V_{ROW} undergoes a positive pulse as shown. This closes the circuit across the channel. The internal storage node voltage that was previously floating is now driven. Provided V_{COL} is externally controlled, V_{PAD} converges towards V_{COL}.

Circuit limitations

Some of the inherent limitations that affect the performance of this circuit include the following.

1. *Time constant to charge* V_{PAD}. The channel of the MOS-T has an effective resistance R_N and the internal storage node has an inherent capacitance C_{PAD}. So charging the node has an effective time constant $T_{PIX} = R_N C_{PAD}$. In practice this time constant tends to be very small (perhaps picoseconds) and is generally not a limitation; although it should be recalled that, in order for an analog signal to settle to a given level of precision, a settling time equal to a multiple of T_{PIX} is required. It is also worth noting that R_N varies with the applied bias signals.

2. *Nonlinearity of MOS capacitance.* As shown in Figure 2.20, the value of the MOS capacitance varies with gate to source voltage. In particular it drops significantly at intermediate values of V_{GS}. It is

particularly important to take this effect into consideration when C_G is a significant component of C_{PAD} or when the internal pixel capacitance is being used to drive a LC material with a spontaneous polarization that depleted the stored charge as it responds to being driven. C_{LC} is also non-linear.

3. *Fixed amount of stored charge.* Once the switching transistor is turned OFF the storage node is isolated with a fixed charge $Q = \Sigma C_i V_i$, where C_i are the various designed, incidental and parasitic capacitive elements associated with the storage node. As the LC begins to switch, an internal flow of charge associated with an induced polarization (in the case of nematic LC) or spontaneous polarization (in the case of FLC or AFLC) uses up the stored charge. Particularly in the latter case and with materials that have a high spontaneous polarization the design must ensure that there is enough charge to fully switch the state of the LC.

4. *Capacitive coupling between gate and source of switching MOSFET.* Switching OFF the MOSEFT involves applying a step down in voltage (assuming an n-channel MOSFET) to the gate which, through capacitive coupling inherent to the MOSFET, produces a reduction in the DC level of the voltage at the storage node in the pixel.

5. *Manufacturing variations in V_{T0}.* If the MOS-T is allowed to go into saturation whilst charging C_{PAD} then C_{PAD} will not charge to the full value of V_C. Rather, a threshold voltage drop will occur across the MOS-T. Pixel to pixel, die to die, wafer to wafer and batch to batch variations in V_{T0} will be reflected as variations in the achieved V_{PAD}.

6. *Body effect on V_T drop.* When charging C_{PAD} both the source and the drain of the MOS-T are above the substrate voltage and the effective threshold is no longer V_{T0} but the larger V_T of Equation 2.7. Whereas V_{T0} is typically less than 0.2 V_{DD} the body effect can raise the effective V_T closer to 0.5 V_{DD}.

One solution to the last two issues (5) and (6) above is to overdrive the row/gate line so that the MOS-T is never in saturation and never generates a V_T drop. However, this involves additional power consumption ($W = CV^2 f$) and will increase the effect of issue (4).

Parasitic components

Further issues arise due to the presence of parasitic and other elements in the circuit as illustrated in Figure 2.43. For example, charge leakage from the soft (capacitive) storage node between refresh cycles changes the stored charge and consequently the pixel voltage. Leakage is characterized by the pixel voltage holding ratio $(1 - \Delta V_{PAD})/V_{PAD}$, where ΔV_{PAD} is the change in pixel voltage V_{PAD} between refresh cycles. The polarity of the leakage through the LC and off MOSFET will depend upon the potential

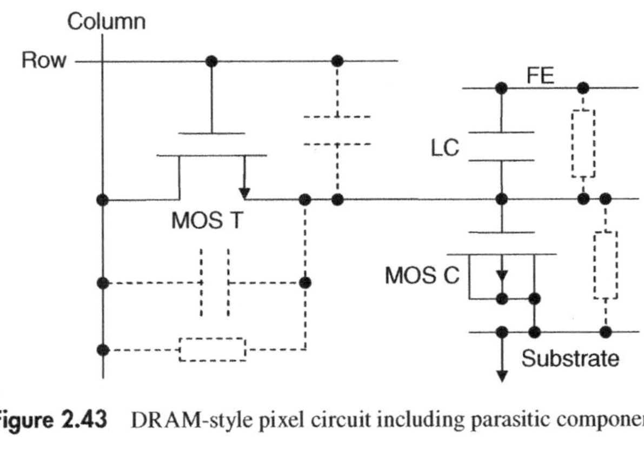

Figure 2.43 DRAM-style pixel circuit including parasitic components

of the front electrode and column bus with respect to the pixel pad and so may change during a frame or between frames whilst the polarity of any leakage to the substrate will not change. Voltage holding ratios >99% are normally achieved, and favored by maximizing the storage capacitance in relation to the leakage.

As with all low-level integrated circuit design, comprehensive simulation, using tools such as SPICE, is required in order to boost confidence in the eventual performance of a circuit. However, robust simulation of circuits such as the LCOS pixel and line drivers requires a knowledge of the electrical and electronic parameters of many things that are outside of the realm of standard circuit design, such as, for example, the load characteristics of the LC structure (with associated alignment layers and so on) and the effect of stray photons reaching the CMOS substrate.

2.4.4 Frame Buffer Pixels for Analog Drive

With the conventional 1-T DRAM, the pixel array is electronically updated line-at-a-time. So the updated image appears line-at-a-time on the display. This does not allow front electrode flipping to achieve DC balancing of the LC as described earlier. It also introduces issues for systems that use pulsed illumination for techniques such as field-sequential-color (FSC). Simplistically, assuming line sequential addressing from top (row 0) to bottom (row $N-1$), then once the LC in row 0 begins to switch into its new state the illumination should be off until the LC in row $N-1$ has settled into its new state. (Otherwise, for example, the blue LED may be illuminating pixels containing red data.) This – ($N \times$ row addressing time) + ($1 \times$ LC settling time) – is a long dead time for the illumination and was a reason for the adoption of alternative strategies such as the scrolling color technique previously described. At the backplane level, an option is to design two storage nodes into the pixel with a suitable arrangement of switches. Figure 2.44 shows two possible arrangements of the storage nodes – series and parallel; the clock signals may be, but need not necessarily be, complimentary.

One node maintains the current (frame F) voltage driving the LC whilst the second receives the next (frame $F+1$) signal ready to pass it on to drive the LC. Global switching of the new frame stored on the second set of pixel capacitors, to drive the liquid crystal, makes the first capacitors available for the succeeding frame addressing cycle, etc. In transmission-mode microdisplays, accommodating additional transistors and capacitors would require a significantly increased pixel pitch and/or reduced aperture ratio. In CMOS-based reflective-mode microdisplays the area required for the additional storage node and switching MOS transistor(s) can, in principle, be accommodated under the reflecting electrode of an LCOS microdisplay. In practice this is readily the case for digital addressing and the storage of digital data. The storage (and subsequent switching) of multiple analogue voltage values within a single pixel is less widely practised. Lee *et al.*[40] describe several circuits, one of which

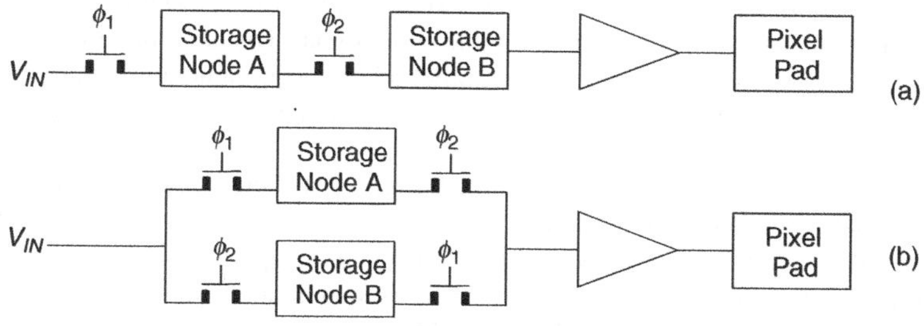

Figure 2.44 Block diagram of frame buffer pixel: (a) storage nodes in series; (b) storage nodes in parallel

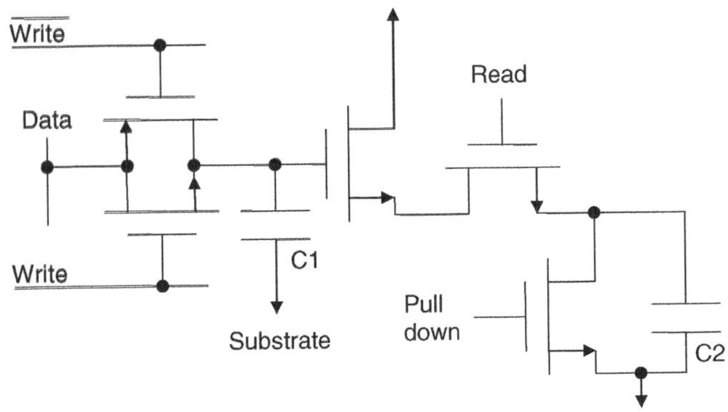

Figure 2.45 Analog-drive pixel circuit with frame buffer. Reprinted courtesy Society for Information Display

is redrawn in Figure 2.45, then go on to outline a mixed grayscale method that combines amplitude and time to achieve 8-bit grayscale. Tan and Sun[41] report a QVGA display capable of 4-bit grayscale in which the pixel has an intermediate storage node. Blalock *et al.*[42] report a pixel that uses a 2-node analog storage pixel in which the analog values are used to control the ON-time of a FLCOS pixel by PWM. This is described in more detail in section 2.6.

2.5 Digital Pixel Drive Schemes for Analog Electro-optic Response

2.5.1 Nematic Liquid Crystal

Display devices in general achieve grayscale by amplitude modulation or pulse width modulation (PWM) of the image light. Displays such as direct-view plasma displays, or the DMD used in projectors, are inherently on/off devices where PWM is the only grayscale option. The CRT has amplitude control, but the rapid decay of the phosphor introduces a pulse characteristic to light output that enhances clarity in moving images. The image light averaged over the integration period of the eye determines perceived gray level, while flicker arises from light pulses, or fluctuations, at inadequate repetition rate. Nematic liquid crystal (NLC) response is too slow for standard PWM as used in the DMD; however, the advantage of an all-digital electronic system favors the introduction of PWM addressing techniques in nematic LCOS devices.

The NLC in a conventional AMLCD receives an analog (grayscale) voltage refreshed at the frame-rate frequency, but remaining constant between refresh cycles. Digital/analog conversion provides the grayscale voltage from an input digital video signal. Distribution of the grayscale voltage over the array of pixels is susceptible to crosstalk and propagation nonuniformity. Spurious DC levels that arise from imperfect analog voltage distribution are a source of image sticking effects in LCDs.

Restricting the pixel voltage to a few well-defined levels, distributed globally and switched at the pixel location, avoids the analog voltage distribution difficulties. Switching the pixel voltage between the set voltage levels achieves an average pixel optical intensity appropriate to that pixel, exchanging amplitude modulation for time domain (pulse width) modulation as described by Hudson[43] and Shimitzu *et al.*[44,45] The nematic LC responds to the RMS value of the applied voltage waveform, when the nematic LC response time (τ) is much greater than the voltage periodicity (T). For the voltage levels

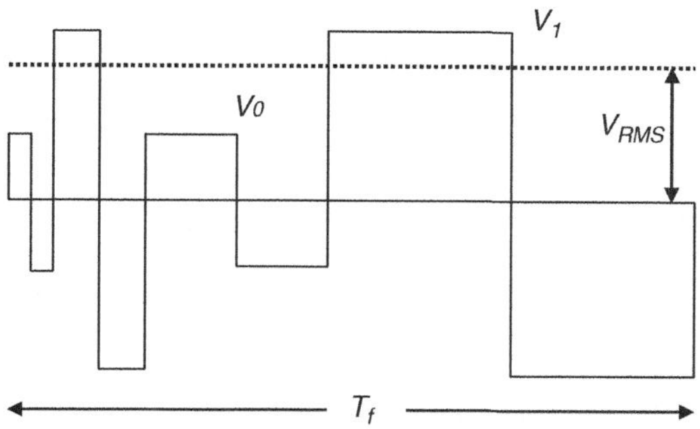

Figure 2.46 Addressing waveform for 4-bit data 0101

V_0 and V_1, applied for periods t_0 and t_1, the RMS value becomes

$$V_{RMS} = \sqrt{\frac{t_0 V_0^2 + t_1 V_1^2}{t_0 + t_1}} \tag{2.19}$$

where $T = t_0 + t_1$. By adjusting the time division (t_0/t_1) any value of V_{RMS} in the range V_0 to V_1 results.

Digital timing circuits supply the timing pulses propagated over xy pixel addressing lines to set the ratio t_0/t_1. The video frame time T divides into $(2^n - 1)$ equal time intervals by n-bit binary switching. A sub-frame addressing cycle (bit-plane) is required for each bit, implying an effective sub-frame address time $< T/2^n$. An n-bit binary number encodes the pixel gray level delivered by n-cycles of sub-frame addressing. Binary 0 signifies a switch to V_0, while binary 1 switches to V_1. Pixel V_{RMS} is given by (2.19), with the binary weighted bits $n(0)$ or $n(1)$ summed for the t_0 or t_1 values:

$$t_0 = \sum_{n(0)} 2^{n(0)} \text{ and } t_1 = \sum_{n(1)} 2^{n(1)} . \tag{2.20}$$

For example, a 4-bit binary word 0101 gives $t_0 = 1 + 4 = 5$, $t_1 = 2 + 8 = 10$, and putting $V_0 = 2$ and $V_1 = 5$ gives $V_{RMS} = 4.24$, as shown in Figure 2.46. In practice the modulation is applied at constant frequency, as shown in Figure 2.47, in order to reduce potential charging effects in the nematic LC that can shift the effective voltage level of a long pulse compared with that of a short pulse.

In standard 8-bit video, adequate grayscale requires 256 gray levels, provided the levels match the logarithmic response of the eye, giving a perceptually uniform grayscale. Binary time division generates uniform steps in $V_{RMS}{}^2$ given by (2.19), which together with nonlinear LC response results in a

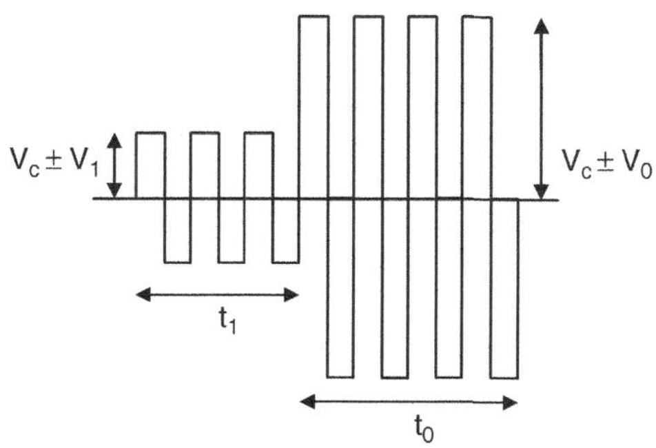

Figure 2.47 Nematic liquid crystal voltage in binary addressed system: V_C = common electrode voltage; V_0 or V_1 = pixel voltage

perceptually nonuniform grayscale. The grayscale depth must be increased to 10 bits or more, requiring a corresponding increase in sub-frame addressing, in order to provide the proper levels for a look-up table addressed by the 8-bit video signal.

The DC balance required by liquid crystals follows the methods established for analog voltage addressing and described in section 2.4.2. The common transparent electrode voltage V_C is constant while the pixel electrode is modulated $V_C \pm V_0$ or $V_C \pm V_1$. Alternatively, the voltage levels can be reduced by modulating V_C, making the pixel voltage modulation in-phase or anti-phase with V_C.

Figure 2.48 indicates the architecture of binary addressing applied to the LCOS consumer projection system of Shimitzu *et al.*[44,45] The off-chip voltages V_C, V_0 and V_1 allow the levels to be adjusted to optimize the performance of each device. Figure 2.47 illustrates the LC voltage waveform, alternating at a frequency of several kHz. The controller and look-up table process the input signal into bit-plane format appropriate to the target gray levels. The first bit-plane writes to the pixel SRAM through the frame buffer. The master pixel SRAM transfers the data to the slave, allowing simultaneous writing and display as per section 2.4.4. The drive voltages V_C, V_0 and V_1 are applied to activate the LC after

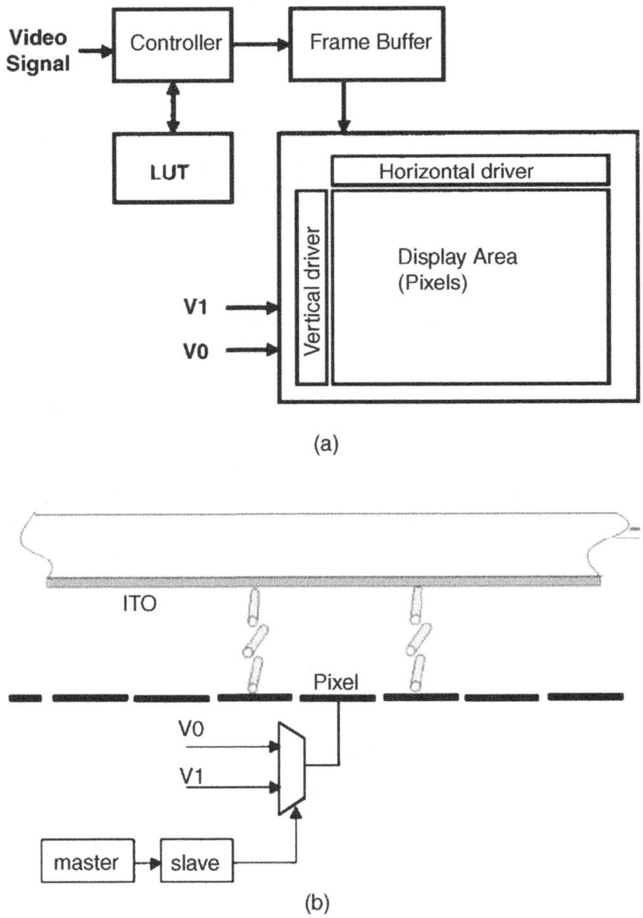

Figure 2.48 Schematics outlining a means of digitally addressing a nematic LC. The scheme employs off-chip voltages V_0 and V_1 routed to the pixel pad and a fixed V_C on the ITO electrode: (a) block level schematic; (b) cross-section of single pixel. Reprinted, courtesy of Society for Information Display

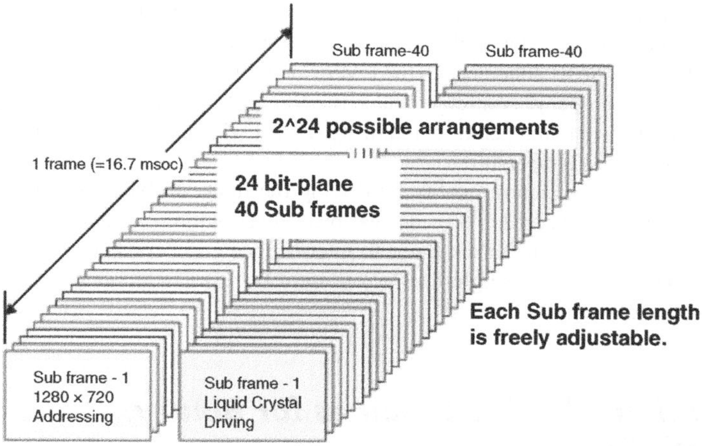

Figure 2.49 Sub-frame sequence for digitally addressing a nematic LC. Reprinted courtesy of Society for Information Display

data addressing. Since no charge is stored at the pixel, light shielding is unnecessary at consumer projector light levels. Addressing extends over 40 sub-frames, providing 24-bit depth, with each bit-plane of freely variable duration as shown in Figure 2.49. The LUT provides 1024 addressable states, sufficient for high-quality imaging.

2.5.2 Fringe Field Effects with Digital Drive

Fringe field effects in liquid crystals are inevitable when adjacent pixels are set to different voltages, disturbing the electric field near the pixel boundary. Reverse tilt walls induced by the field fringing enhance the optical distortion and compromise the resolution. As the gray-level difference between adjacent pixels declines, the analog voltage difference falls, reducing fringe field effects, which become negligible for small changes in gray level. In digital addressing large fringe fields are possible for small gray-level changes. The V_{RMS} controls gray level, while amplitude difference between the binary codes determines fringe field. For example, the 4-bit binary codes 0111 and 1000 differ by the LSB, i.e. one gray level, yet the voltage difference (1000) − (0111) $\rightarrow \pm (V_1 - V_0)$ is present at all times, and has maximum RMS value.

The fringe-field effect in digital addressing gives rise to objectionable image twinkling, requiring steps to limit the fringe field. A technique described by Worley *et al.*[46] as compound-binary addressing reduces field fringing, at the expense of increased sub-frame addressing. If the MSB is split into equal time intervals, expressed [00](111) = [01](011) and [11](000) for the above example, the voltage difference {[01](011)} − {[11](000)} exists for little over half the frame, compared with the previous full frame. Such bit-splitting can be extended as needed to attenuate fringe-field effects to the desired level. The maximum active time for the fringe field is the total binary time plus one interval of the split-bit. The DMD employs similar bit-splitting techniques to attenuate PWM artefacts by distributing the output light more uniformly over the frame period.

2.5.3 Response Time Considerations for Digital Drive

The response time of the NLC cell should be faster than the video frame rate to prevent smearing of fast-moving images. However, NLC response should be slow compared with the periodicity of the addressing waveform for the applicability of RMS voltage. The conflict can be resolved by making the addressing period small compared with the frame period, but this increases the demand on addressing speed and peak data rate. Strict adherence to the RMS condition is not essential since the LUT can be calibrated to provide the proper gray level, provided the NLC cell response is consistent. The response time of the NLC is sensitive to temperature, requiring careful selection of the driving pattern to make the gray levels insensitive to temperature.

Kurita[47] improved moving picture quality by driving the image to black within a frame period. PWM addressing can be engineered to provide near-black levels within and between frames as shown by Shimitzu *et al.*[44] Restricting the black insert to intermediate gray levels does not compromise the maximum luminance output. Image light fluctuations must be restricted in amplitude and periodicity to be below flicker perception.

2.6 Digital Pixel Drive Schemes for Binary Electro-optic Response

Digital systems have advantages that promote extension into all areas of electronics, HDTV being a relevant example. The display device may be capable of an analog response, e.g. CRT, but digital signal processing determines the analog voltage delivered to the electrodes. Pixelated devices, including all microdisplays, enhance the digital aspect of displays. Furthermore, some display systems, such as DMD and SSFLC, are capable of only binary or bistable response and are thus particularly suitable for digital (or PWM[iv]) pixel driving.

With PWM-based systems "digital-to-analog conversion" becomes "time-domain low-pass filtering" that takes place in-eye, which averages the intensity of the light pulse or pulses over its integration period as described by Hornbeck.[48] The pulse count must be high enough to accommodate the grayscale range and fast or frequent enough to prevent flicker, implying a pulse-frame rate that, from the perspective of active matrix addressing, mandates the use of a CMOS backplane. B-PWM has the advantage of allowing an all-digital system, but requires a higher bit depth than analog addressing or S-PWM, and is subject to B-PWM artefacts associated with eye motion.

Ferroelectric liquid crystal (FLC), in the surface stabilized configuration (SSFLC), provides pseudo-bistable[v] ON/OFF states, with enhanced switching speed over nematic LC configurations. Early FLC developments excited considerable interest in the display world. Direct-view passive matrix addressing of video displays appeared possible, with field-sequential-color within reach. Development efforts have failed to achieve switching accommodating FSC in large-area displays, and robust FLC alignment over a large area remains difficult. FLC research continues with material development, polymer dispersion, and V-shaped response in anti-ferroelectric LC (AFLC). The last has an influence on microdisplay development as discussed in Chapter 6.

Silicon backplane addressing of FLC must compete with DMD in B-PWM for projection applications although its competitiveness is compromised by the twin requirements of linearly polarized input illumination and DC balancing of the FLC. To that end, ongoing efforts continue to attempt to overcome these. Attempts to solve the latter include the use of a compensating element[49] that switches in synchronization with the DC balancing waveform, and drive schemes that do not fulfil DC balancing requirements.[50]

[iv]The reader is reminded of the various categories of PWM defined in section 2.1.3

[v]Due to the surface properties of the CMOS backplane, SSFLC alignment in microdisplays is typically in the chevron (not bookshelf) geometry and each pixel relies upon a persistent drive signal from the active matrix backplane to maintain a stable optical state

2.6.1 Single Pulse Width Modulation

FLC is competitive in near-eye applications where short durations of red, green and blue pulsed illumination are achieved by switching LEDs. Figure 2.50 shows a block diagram of a mixed-mode (analog/digital) pixel circuit used to produce S-PWM modulation in a FLC microdisplay. The modulation of the pulse width of the FLC drive waveform is generated by a comparator. The upper comparator input is a stored analog voltage corresponding to the gray level required at that pixel. The stored analog voltage is generated in this case from a digital video input signal through a digital to analog converter. The lower comparator input is a global rising-ramp waveform that is developed by digital means, incorporating gamma data, and delivered to all pixel comparators in the array. When the ramp voltage exceeds the pixel comparator voltage, the pixel pad switches to high voltage and remains high until the ramp voltage is returned to a low value. With a rising ramp signal, short pulse periods occur late in the cycle and duration is determined by an LED pulse within the pixel FLC on-period. Simple pulse width modulation is consistent with low power consumption which is important in battery-operated devices.

Figure 2.51 shows a related circuit by Blalock *et al.*[42] corresponding approximately to that part of Figure 2.50 enclosed by the dashed line. An inverse voltage cycle is imposed by residual ionic conduction in the FLC, and takes place with the illumination LEDs switched off. Handicapped with the optical duty cycle of 50%, the switching speed of the FLC still provides higher optical throughput efficiency than a nematic cell in field-sequential-color applications. The provision of two multiplexed storage nodes (A and B in Figure 2.50, color 1 and color 2 in Figure 2.51) allows simultaneous addressing of the array and displaying of the current image as described in section 2.4.4. Figure 2.52 illustrates the waveforms associated with the back-end comparator-to-pad part of the circuit. The ramp shown is linear but could in principle be adjusted to offer a more eye-friendly logarithmic response.

In the case of S-PWM the ON-period of the illumination source must precede (or coincide with) the earliest possible ON-time of the first-addressed row and must continue until or beyond the latest possible ON-time of the last-addressed row. The frame-update technique allowed by the frame buffer pixel is preferred over the line sequential update of the simple DRAM or SRAM pixel.

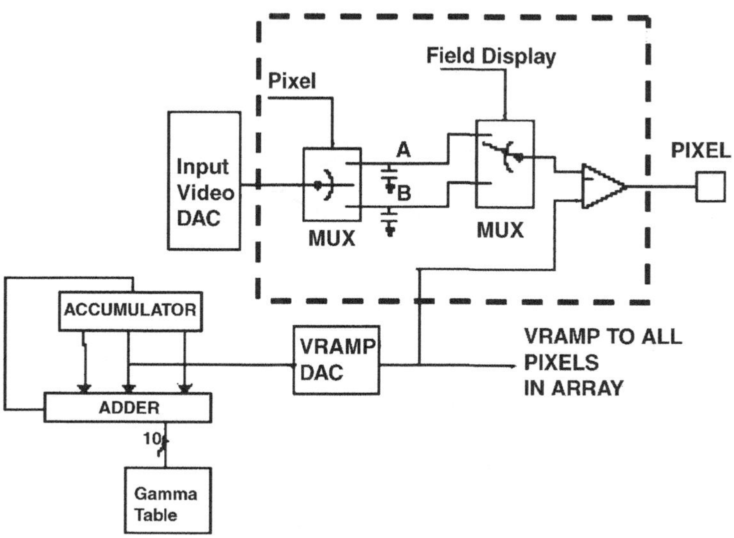

Figure 2.50 Block diagram of FLC electronics. Reprinted courtesy of Displaytech Inc.

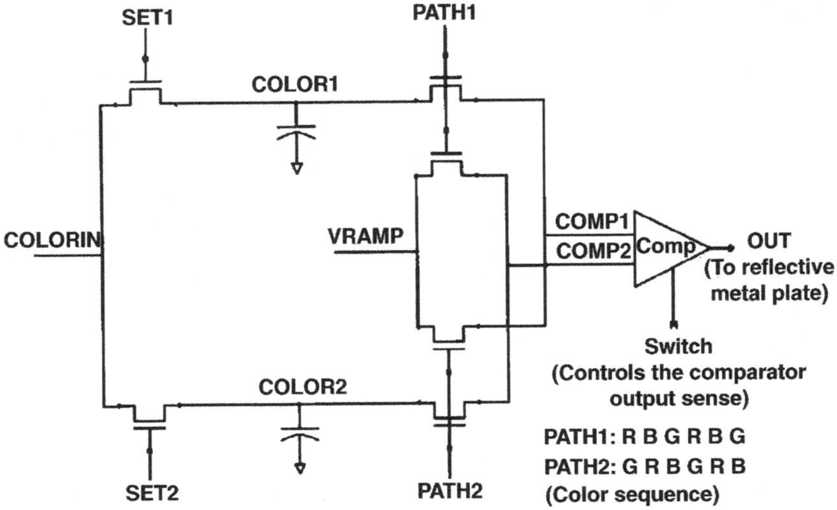

Figure 2.51 Circuit schematic of double-buffered pulse-width-modulation pixel for FLCOS microdisplay. ©2001 IEEE; reprinted with permission

2.6.2 Binary-Coded Pulse Width Modulation (B-PWM)

For binary-coded pulse width modulation (B-PWM), the maximum pulse length is limited by the image frame time (16.7 ms at 60 Hz); standard 8-bit binary encodes 255 levels of grayscale above zero, implying a minimum pulse width of 16.7/255 ms = 65 µs. Field-sequential-color requires three primary colors delivered in this time interval, reducing the minimum pulse to 22 µs. Returning to the case of monochrome or parallel color (three-microdisplay system), in binary code the LSB = 65 µs, and MSB = 8.37 ms, where PWM can be implemented by writing successive bits to each pixel, requiring eight "bit-planes" for the whole array. Eliminating the sample and hold circuitry in Figure 2.35 and accumulating binary video-bit data in the column shift register to activate the column drivers, delivers the

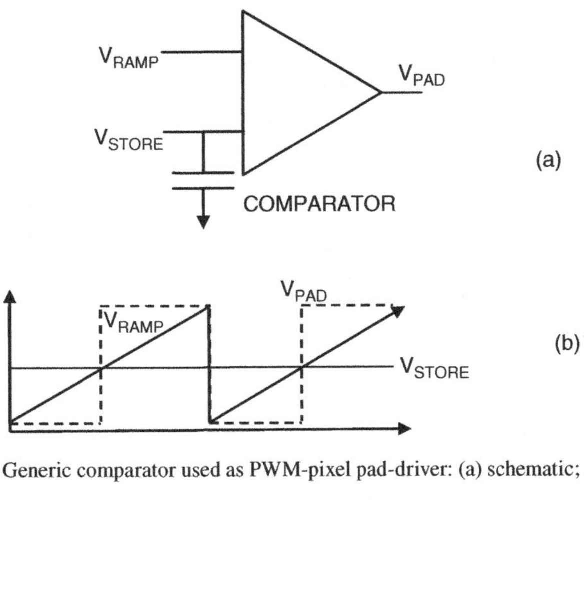

Figure 2.52 Generic comparator used as PWM-pixel pad-driver: (a) schematic; (b) waveforms

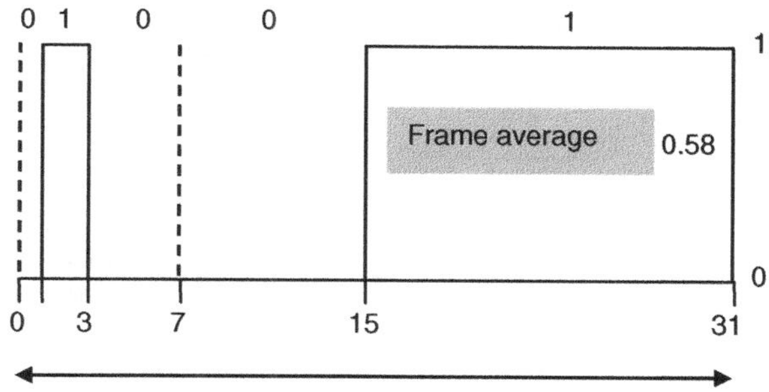

Figure 2.53 Binary-coded pulse width modulation 01001 = 18, grayscale averaged over frame time = 18/31 = 0.58

appropriate bit to the SRAM (static random access memory) in each pixel of the activated row. Repeating the bit-plane cycle for each bit writes the PWM version of the binary coded grayscale, as illustrated in Figure 2.53 for 5-bit pixel data 10010 = 18, averaged over the image frame time gives 10010/11111 = 18/31 = 0.58 grayscale. The grayscale luminance increases in uniform LSB steps from zero to 100%, making it a poor match to logarithmic human luminance sensitivity, implying a requirement for extended bit depth as discussed later in section 2.6.4. Drastically increased addressing speed reduces the advantages of an all-digital system, particularly for color sequential applications. Only CMOS backplane technology can provide the required addressing speed.

In the case of B-PWM the ON-period of a bit-plane may be defined either by the time that the LC is in the ON-state (with the illumination pulse commencing before, and finishing after, the LC is switched ON) or by the time that the light source is in the ON-state (with the LC maintained in the ON state for the full duration of the illumination pulse). In any case, the frame-update technique allowed by the frame buffer pixel is preferred over the line sequential update of the simple DRAM or SRAM pixel.

A further degree of freedom in pulsed and related systems is the potential to vary the intensity of pulses in a binary-weighted fashion between bit-planes rather than the time duration of pulses. This might be called binary-coded pulse intensity modulation (B-PIM). It follows, of course, that a combination of B-PWM and B-PIM may allow a greater dynamic range than either B-PWM or B-PIM alone.

2.6.3 B-PWM Pixel Circuits

DRAM-style pixel circuits

The 1-T DRAM-style pixel (usually so-called even when a further MOS structure is used as a component of the storage node) described in section 2.4.3 can be used in a B-PWM system. Driving and storing a digital signal into the pixel is a less onerous task than the equivalent task for a precise analog voltage. Historically such systems have been designed and used; see for example Vass *et al.*[51] and Wilkinson *et al.*[52] However, the demands made by modern materials with high spontaneous polarization (as outlined in section 2.4.3) eventually limited the appeal of the DRAM pixel whilst technological advances have allowed more capable pixel designs incorportaing many more transistors to be utilized as discussed below.

An evolution beyond the simple 1-T DRAM was reported by Jared *et al.*[23] in 1990. It utilized a gated CMOS inverter, as shown in Figure 2.54, to buffer the storage node from the pixel pad thus making available unlimited charge to deal with the spontaneous polarization of the FLC and ensuring a very precise and stable drive voltage to be supplied to the pixel pad. The MOS gate capacitance of the two MOSFETs of the CMOS inverter provided the soft-node storage. Like most dynamic logic circuits, provided the signal on the storage node remains within a threshold voltage of the power rails, this circuit behaves itself.

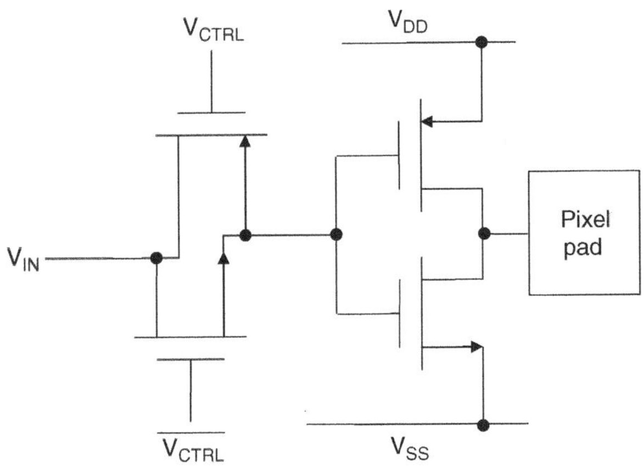

Figure 2.54 Circuit schematic of buffered dynamic-storage pixel circuit

SRAM-based circuits

Underwood *et al.*[19] reported the first SRAM-based pixel circuit implemented in NMOS technology. The device was configured as a spatial light modulator offering binary modulation of a guest–host LC configuration. Later evolutions of this pixel design scaled up in pixel number,[20] changed to CMOS technology,[30] and were used to drive SSFLC.[53] The pixel circuit is shown schematically in Figure 2.55. It incorporates a novel XNOR buffer between the SRAM and the pixel pad. The XNOR driver allows the logic state of the front electrode to be flipped simultaneously with those of all the pixels under the control of a single clock signal, thus simplifying greatly, and minimizing the electronic overhead associated with, the issue of DC balancing. In the case of nematic LC that responds to the RMS voltage, the logic state of the pixel storage node determines whether the pixel remains in-phase with the front electrode ($V_{LC} = 0$ V, LC is OFF) or in anti-phase with the front electrode ($V_{LC} = \pm V_{DD}$, LC is ON). In the case of SSFLC, the XNOR allows simultaneous switching of all pixels from the usable image to the inverted image; it may be used with a fixed front electrode or with a switching front electrode.

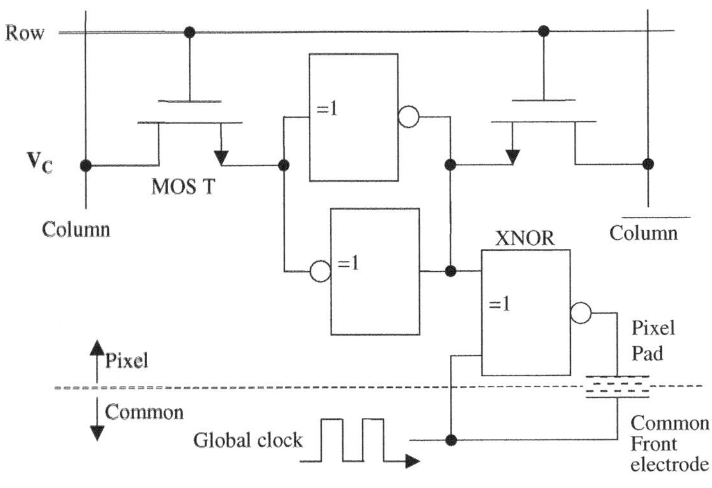

Figure 2.55 Mixed schematic of SRAM XNOR pixel

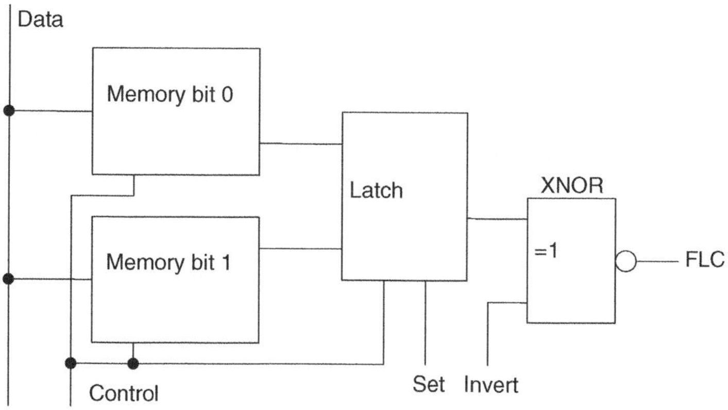

Figure 2.56 Schematic diagram of pixel with in-parallel memory bits

Cottes *et al.*[54] reported a pure SRAM-based SLM. There was concern that a pure SRAM pixel could be susceptible to corruption in the sense that effects related to the electronic switching of the FLC could feed back from the pixel pad to the SRAM and overwrite the programmed state of the SRAM. Drabik *et al.*[55] 1995 reported a SLM that added a CMOS inverter as a buffer between the SRAM cell and the pixel pad.

Frame buffer circuits for digital drive

Birch *et al.*[56] describe a FLC microdisplay of 1280 × 1024 pixels (SXGA) capable of 24-bit color depth at a frame rate of 60 frames per second. The least significant bit is displayed for only 20 μs. The two bits of memory provide a uniform rate of data delivery to the display. A block diagram of the pixel is shown in Figure 2.56. Note the presence of the latch which, added to the two memory blocks, actually implies three bits of storage per pixel, and the XNOR driver block to simplify DC balancing as mentioned in the previous section.

Akimoto *et al.*[57] and Yumoto *et al.*[58] describe a FLC microdisplay of 1920 × 1200 pixels (WUXGA) for NTE applications that achieves 10-bit-per-color B-PWM by modulating the pulse width of the illumination/readout LEDs in order to generate the four least significant bits per color. (These correspond to the shortest duration pulses for which the switching time of the FLC is too long to suffice alone.) The bit-planes of the bits above bit-6 are split and dispersed through the frame time (16.7 ms) in order to minimize bit-plane related visual artefacts. This increases the required number of bit-planes that must be electronically addressed per frame, giving a total of 36 bit-planes per color per frame time. The FLC switches in 50 μs, supporting a field-sequential-color bit-plane rate of 6.2 kHz, eliminating color break-up. Note that the FLC is driven by a fixed charge stored on capacitance C_S.

The system, shown in Figure 2.57, accommodates a variety of inputs with a digital video signal processor (DVSP). Gamma correction is applied by a programmable look-up table, converting from 8-bit input to 10-bit output. Data is transmitted to the LV module by gigabit video interface (GVIF) via a low-voltage differential signal (LVDS) link at 4 Gbit/sec. The bit-plane generator (BPGEN) transforms the stored frame image into bit-plane data driving the display, while accumulating the succeeding frame. The transfer rate from BPGEN to microdisplay is 76 MHz with 240 data input pins.

Figure 2.58 shows the FLC pixel drive circuit consisting of three stages – two 1-bit memories and a FLC driver. The memories use two capacitive soft-node storage nodes arranged in series as previously suggested in Figure 2.44(a). The row signal gates data into MEM-1; TRS gates the transfer of data into MEM-2. In preparation for FLC switching, the storage capacitor C_S of node 3 is pre-charged to V_{pchg} by the following sequence: (i) a pulse on the discharge line (DCHG) grounds node 2 via Q3, turning Q4

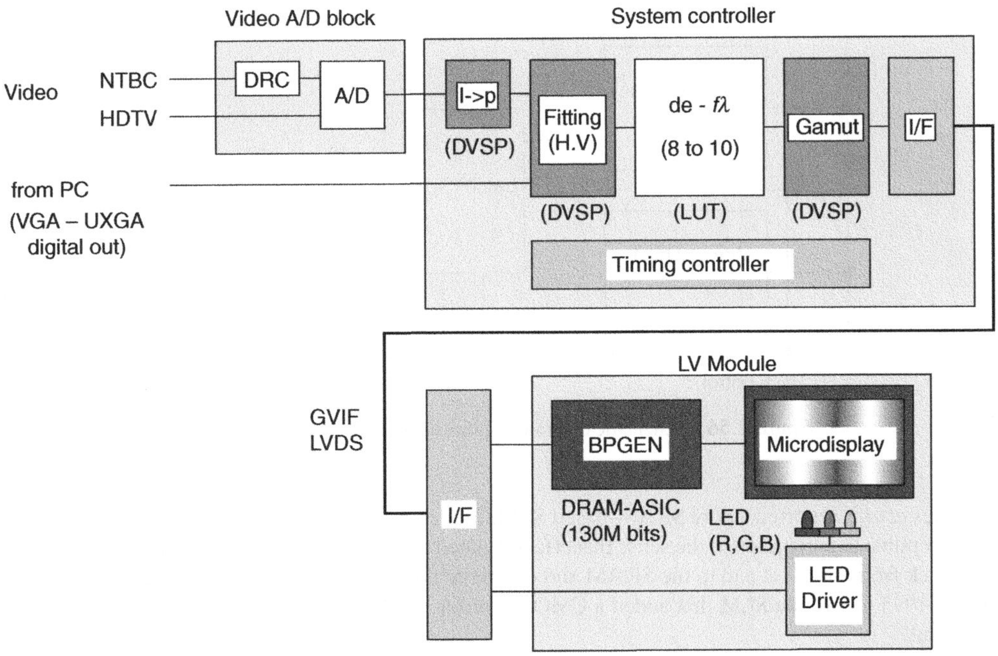

Figure 2.57 FLC system with 30-bit field-sequential-color and 1920 × 1200 resolution. Reprinted courtesy of Society for Information Display

off; then (ii) a pulse on the pre-charge line (PCHG) turns Q5 on. Q3 is turned off and TRS is pulsed to transfer the bit of node 1 into node 2, imposing its state on the drive circuit. A logic 1 on the gate of Q4 discharges node 3 through Q4 resulting in a low voltage at node 3, while a logic 0 on the gate of Q4 has no effect on the existing (pre-charge) voltage at node 3. PCHG is set to 14 V, driving node 3 to 12 V during pre-charge, while the FLC common electrode is set to 5 V. If the charge on C_S switches the

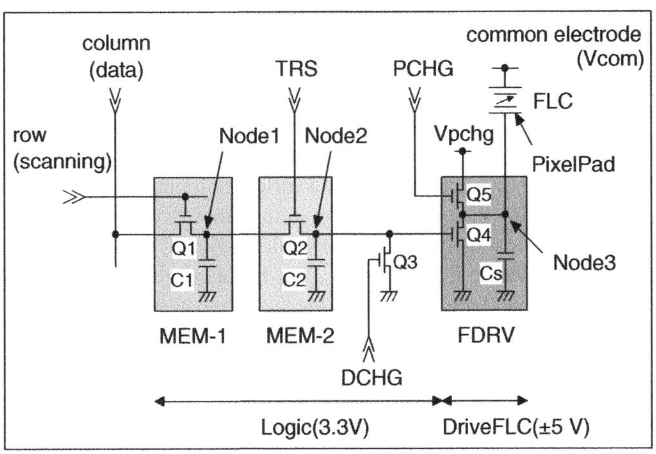

Figure 2.58 Dynamic logic double-buffered pixel. Reprinted courtesy of Society for Information Display

Figure 2.59 Perceptible grayscale steps compared with continuous grayscale

FLC, its voltage drops to 10 V due to the polarization reversal in the FLC. In the no-switching case, C_S discharges to 0 V through Q4. Thus, the FLC is driven effectively ±5 V, and the pre-charge pulse is too short to significantly affect the FLC.

2.6.4 Grayscale Contouring

When luminance is weakly dependent on spatial position, such as an image of the sky, digital grayscale steps can appear as contour lines as illustrated in Figure 2.59, unless the steps are below the limit of perception. The perception limit on grayscale step, estimated by the CIE luminance difference formula, gives the maximum tolerable grayscale step (ΔL) to avoid contour perception. That maximum tolerable grayscale step, imposed on the device electro-optic transfer characteristic, gives the maximum tolerable stimulus (voltage or current) step activating the device. The maximum tolerable voltage step sets the bit-depth of a digital system. Equations for luminance difference and electro-optic transfer are derived below from Brennesholtz,[59] where the eye is adapted to average perceptible luminance ($0.75L_{MAX}$) over the range zero to L_{MAX}, and luminance perception is set at one just noticeable difference (JND). The one-JND stringent luminance perception level protects against the greater color contouring (posterizing) sensitivity of the eye.

$$\textit{Luminance difference}\quad \Delta L = 0.00083L_{MAX} \quad \text{when } L \leq 6.64\times10^{-3}\ L_{MAX}. \tag{2.21}$$

$$\textit{Luminance difference}\quad \Delta L = 0.0235L^{2/3}L_{MAX}^{1/3} \quad \text{when } L > 6.64\times10^{-3}\ L_{MAX}. \tag{2.22}$$

$$\textit{Electro optic transfer}\quad L = f(V) \rightarrow \Delta V = \frac{\Delta L}{f'(V)}. \tag{2.23}$$

The maximum luminance step varies from $0.00083L_{MAX}$ to $0.0235L_{MAX}$; the darkest level steps are only around 3% of the brightest level steps.

The electro-optic transfer characteristic $f(V)$ has a strong influence on bit-depth. All B-PWM devices such as the DMD have a linear transfer characteristic, with pulse-time represented by an effective characteristic $L = V$. Consequently the minimum voltage step ΔV_{MIN} corresponds to the minimum luminance step $\Delta L_{MIN} = 0.00083L_{MAX}$, implying $\Delta V_{MIN} = 0.00083V_{MAX}$, requiring 11-bit depth to achieve the minimum voltage (time) step. In practice, 12-bit resolution is required as reported by Van

Kessel *et al.*[60] A characteristic $f(V)$ that closely matches the eye's response to luminance requires the minimum 7-bit depth. The CRT comes close with required 8-bit depth, while liquid crystal and OLED are of order 9-bits according to Brennesholtz.[59]

Standard 8-bit video includes an encoding $L = V^{\gamma}$ with gamma $\gamma \sim 1/2.2$ (matching CRT characteristic), and is interpolated by higher bit-depths to provide the proper grayscale. Higher bit-depth increases the cost of digital electronics; cheaper methods of contour reduction include dithering and addition of noise; see, for example, Daly and Feng.[61] In dark scenes the eye has time to adapt to a lower luminance level, equivalent to lower L_{MAX} and consequently smaller ΔL, making contouring more perceptible. A recent development reported by Toyooka *et al.*[62] modulates the physical light source (L_{MAX}) to match the required scene luminance, so the physical grayscale step falls with adaptation to lower luminance, compensating for reduced bit-depth elsewhere, while enhancing contrast ratio.

The eye has highest resolution over a narrow field of view, say around 1 degree, but eye motion provides the illusion of a high-resolution wide field of view. As the eye scans a PWM display, the short high-resolution dwell time on any small region may be insufficient to accumulate enough light to represent that region adequately. The resultant distortion, labelled a PWM artefact, is an annoying and puzzling distraction. Upon attracting attention, it disappears as the eye slows and focuses to clarify the defect. The more continuous the displayed light becomes during the frame time, the less chance of PWM artefacts. Bit-splitting techniques implement a progression towards optical continuity. In Figure 2.53 for example, most of the light emerges during the second half of the frame time; splitting that MSB into two equal fractions, evenly spaced during the frame, improves the continuity, but at the expense of increased effective bit depth from the perspective of electronic addressing. Van Kessel *et al.*[60] and Critchley *et al.*[63] describe this in more detail.

Systematic or periodic fluctuations in the output of the light source approximately synchronized with the frame rate of the microdisplay can introduce flicker artefacts as the optical pulses occur at different times in the light source cycle. It may be necessary to synchronize an AC arc-lamp to the display frame rate.

2.7 DMD Microdisplay Electronics

Figure 2.60 is a functional block diagram of a 1024 × 768 DMD. Binary data is loaded one row at a time via the 64-bit data bus D(0–63). For each row, the data for mirrors (pixels) 15, 31, 47···1023 is loaded first, and mirrors (pixels) 0, 16, 32···1008 last. The data clock (DCLK), data mirror control signals DMC(0/1), and toggle rate control (TRC) control data transfer. The row-address shift-register selects one of the 768 rows. The rows are grouped into 16 individually controlled blocks of 48, each block with is own mirror bias reset voltage input MBRST(0:15), which holds or releases the pixel mirror. The data mirror control signal DMC(2/3) activates block control to route the bias reset voltage and mirror electrode stepped voltage VCC2. Division into blocks facilitates phased array addressing, exploiting the shorter block addressing time in generating the LSB, improving thoughput efficiency. Chapter 8 gives details of the pixel addressing cycle.

In color sequential systems, three colors have to fit into the image frame time; moreover, color break-up imposes enhanced frame rate, resulting in 540 color frames per second. Therefore, each color frame occupies 1.85 ms, implying shortest pulse length 7.3 μs, comparable with the DMD switching time. Activating low-level bits below frame rate compensates for switching time limitations. The addressing time for the entire array is much longer than the shortest pulse; early DMDs achieved the LSB by storing the data on the pixel SRAMs and globally switching ON/OFF, maintaining the OFF state during a subsequent addressing cycle. The original method sacrificed ~2% optical thoughput due to the addressing dark state, prompting replacement by "phased array addressing" that breaks the addressing into small blocks with short addressing time and phased sequence, essentially eliminating the dark period.[64]

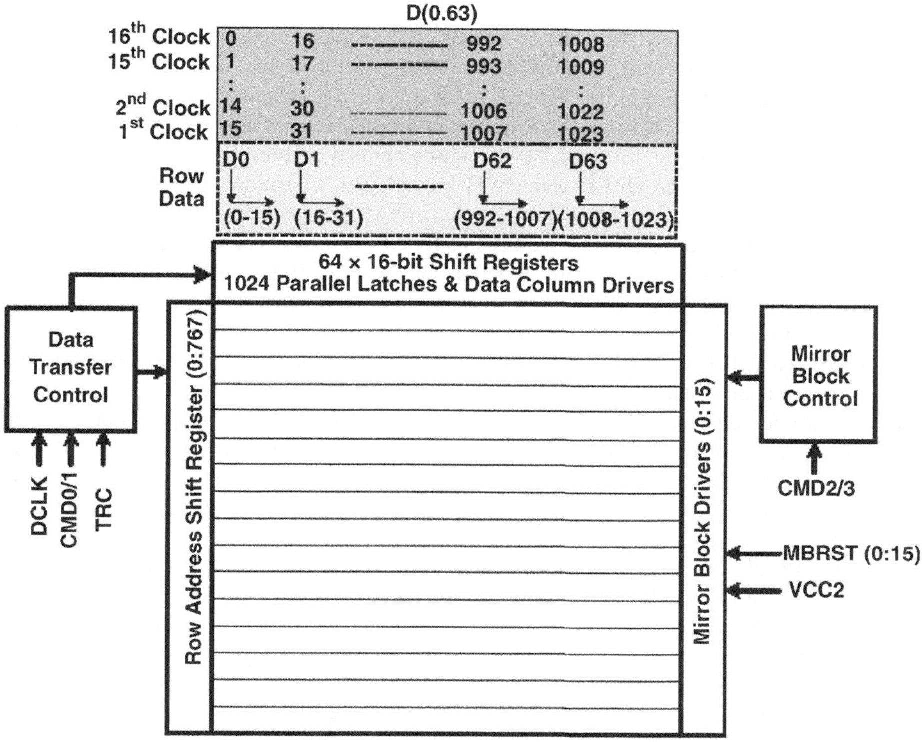

Figure 2.60 Functional diagram of DMD electronics. Reprinted courtesy of Texas Instruments Inc.

The classical DMD pixel circuit is that of the SRAM previously described in section 2.2.5 and illustrated schematically in Figure 2.25. In the DMD pixel each of the complimentary storage nodes is electrically connected into the overlying DMD structure so as to provide electrically driven pads as described in section 8.2.6 and illustrated in Figure 8.17.

Recent developments, including a third-generation DMD design, are discussed in Chapter 8. The new generation employs hybrid circuitry; with higher resolution CMOS operating at low voltage for the addressing circuit, while retaining higher voltage circuitry for the pixel drive voltage.

2.8 OLED Microdisplay Electronics

2.8.1 OLED Microdisplay System Overview

OLED elements respond in an analog fashion to an analog stimulus, so analog addressing is possible. OLED elements in general switch ON and OFF much more quickly (some nanoseconds) than nematic LC elements (some milliseconds) or even ferroelectric LC elements (some microseconds), so time domain grayscale (PWM) techniques may also be used. Furthermore, OLED elements do not require an addressing waveform to be DC-balanced, although in some cases occasional reverse biasing of the element has been used to improve lifetime.[65] At the time of writing, the use of OLED microdisplays is restricted largely to near-to-eye applications. Again, at the time of writing, the implementation of color in commercially available products is restricted to RGB spatial sub-pixelation (by means of white emission and RGB color filters).

In an LCOS microdisplay, the state of the LC, and hence the brightness of each pixel, is controlled by an electric field supplied by the driving circuit as a voltage between the individual pixel pad and the common counter-electrode. The LC cell can be modelled to first order as a capacitance. In an OLED microdisplay the brightness of each pixel is typically proportional to the magnitude of current passing through the OLED layer from the individual reflective pixel pad to the transparent common counter-electrode. Most OLED displays employ a current source in the pixel in order to control the brightness. The OLED element is modelled to first order as a diode or nonlinear resistor.

The main challenges of active matrix circuit design for OLED displays in general are (a) minimizing the perceptible effect of any manufacturing variations that occur in (i) the OLED structure and (ii) the active matrix circuits; and (b) maximizing the usable life of the display by minimizing (i) the effect of increasing turn-on voltage with use, (ii) the effect of OLED efficiency degradation with use, and (iii) the appearance of differential aging of materials in which the efficiencies of the differently colored sub-pixels degrade at different rates.

In the case of microdisplays (a)(i) may be assisted by the small size of the display, (a)(ii) is exacerbated by the extremely small currents involved (sometimes sub-nanoampere per pixel), (b)(i) is aided by the use of a constant current (rather than constant voltage) drive, (b)(ii) may, in future, be reduced with the use of in-pixel optical feedback, and (b)(iii) is not yet an issue for microdisplays as available microdisplays use white-emitting OLED material with RGB color filters.

Taking all of the above into account, the electronic system architecture of an OLED microdisplay will most likely be similar to that of an equivalent LCOS (or other) microdisplay with the following modifications: provision for the adoption of constant current drive; provision for the integration of electronic elements that support the inclusion of the OLED layer (for example a bias circuit to control the voltage at which the transparent top electrode is held which may, in some circumstances, assist as a brightness control); and the addition of any circuits that address the specific challenges of OLED as outlined above. An example from 2002 by Tan and Sun[66] is shown in Figure 2.61.

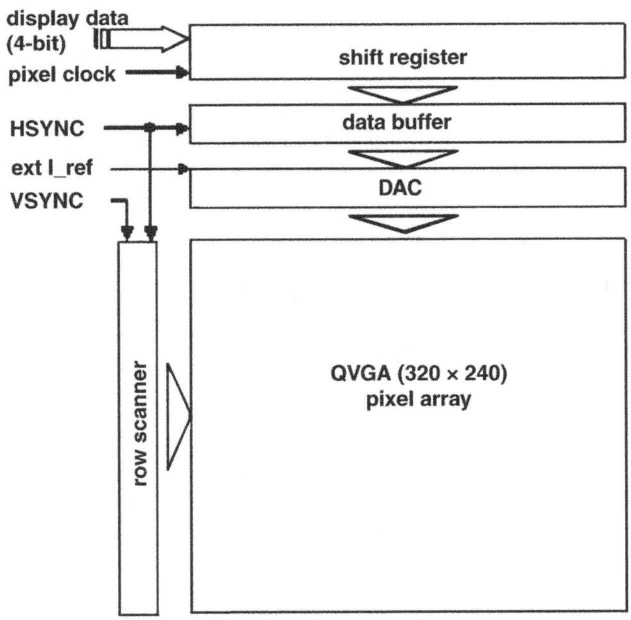

Figure 2.61 Example of OLED microdisplay backplane. Reprinted courtesy of Society for Information Display

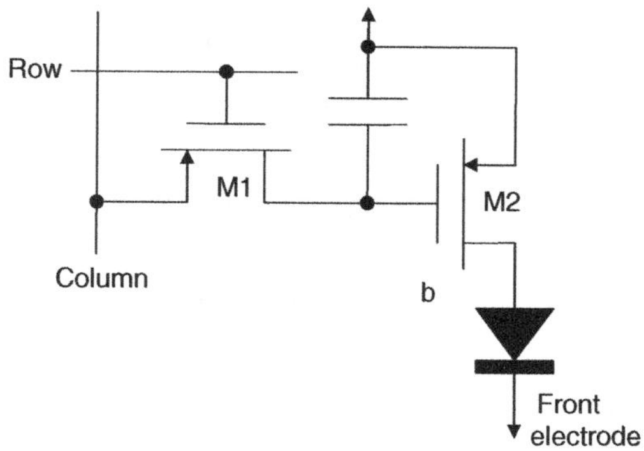

Figure 2.62 Generic two-transistor active matrix OLED pixel circuit

Compared to that of a p-Si TFT-based OLED display, the pixel circuit of an OLED microdisplay must control a much smaller (possibly sub-threshold) magnitude of current but has the potential to incorporate a significant degree of CMOS functionality into a pixel or sub-pixel.

2.8.2 OLED Pixel Circuits using TFTs

Before considering OLED microdisplay circuits, we shall take a qualitative look at some exemplar TFT-based circuits for driving direct-view OLED displays as these have benefited from a longer duration and higher level of development effort. Reviews of this topic include those of Tam *et al.*,[67] Tam *et al.*,[68] Fish *et al.*,[69] and Baynes and Smith.[70]

Figure 2.62 shows a basic voltage-programmed current-drive two-transistor circuit for driving an OLED load that derives from the work of Brody *et al.*[71] in 1975. Transistor M1 acts as a sample-and-hold switch that allows the drive signal to be stored as a voltage on soft-node C. M2 then converts the stored voltage to a current through the OLED structure. The primary shortcoming of this circuit is that process related variations in MOS transistor parameters, primarily threshold voltage and carrier mobility, result in pixel-to-pixel variations in the conversion factor of stored-voltage to drive-current as described in section 2.2. Equation (2.5), Equation (2.6) or Equation (2.10) applies depending on the bias region of the drive transistor. Variations from display to display within a wafer, from wafer to wafer within a batch of wafers, or from batch to batch are small enough to be of little concern. Within a single display pixel-to-pixel variation in brightness (fixed salt and pepper noise) is a concern.

A number of circuits have been proposed that offer some compensation for threshold voltage variations of the drive transistor in a voltage-programmed current-drive configuration. Figure 2.63 shows one example of such a circuit by Dawson *et al.*[72] and Dawson *et al.*[73] The column line is set to zero and the row line is activated. Control 1 is activated, discharging the gate of M2 to the threshold voltage. With M3 and M4 open circuit, any signal now applied via the column data line and M1 to C1 produces a signal at the gate of M2 that is offset by the threshold voltage. Dawson *et al.*[74] report on a QVGA display that uses this pixel circuit.

Perhaps a more robust approach to addressing a pixel circuit, in which the aim (after all) is to control the current through the OLED load, is to use a current to program the pixel. Dawson *et al.*[73] examine a possible circuit. Figure 2.64 by Hunter *et al.*[75] shows an example of a current-programmed current-drive pixel circuit. The row line is activated to address the pixel, resulting in M4 going open circuit. Any mismatch between the current sourced by M2 and that sunk by the driver of the column line charges the capacitor until soon the currents are balanced. The row is then deactivated and the appropriate current flows through the OLED. Small values of programming current can lead to excessive settling times that

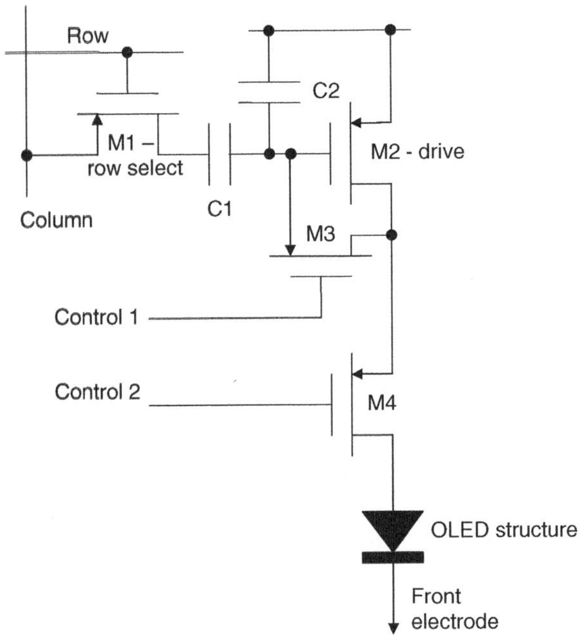

Figure 2.63 Voltage–programmed current-drive OLED pixel circuit with compensation for V_T variation. Reprinted courtesy of Society for Information Display

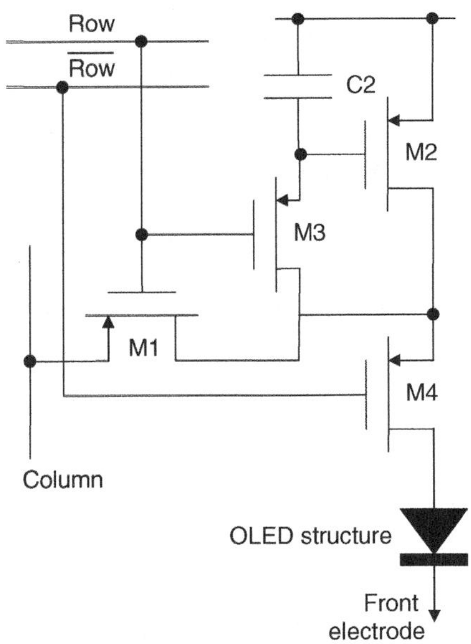

Figure 2.64 Current–programmed current-drive OLED pixel circuit with current drive. Reprinted courtesy of Society for Information Display

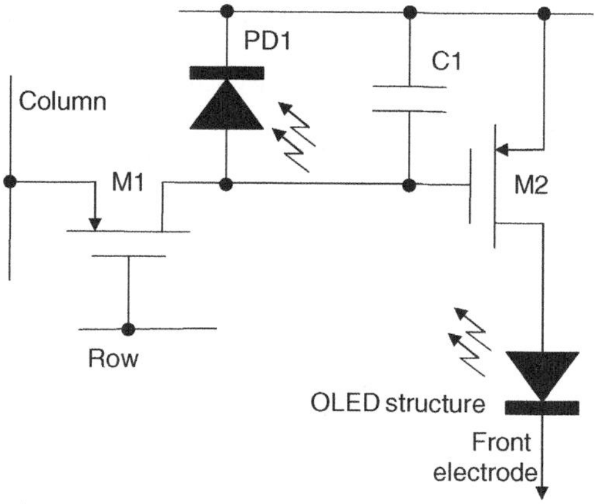

Figure 2.65 Active matrix OLED pixel circuit with optical feedback. Reprinted courtesy of Society for Information Display

become more problematic as array size (number of lines) increases. Whereas Hunter *et al.*[75] use M4 to gate a row of pixels, Fish *et al.*[69] suggest globally gating the cathode, citing a more efficient OLED operating point, higher programming currents and improved motion portrayal as justification.

Fish *et al.*[69] propose a pixel circuit that incorporates optical feedback, shown in Figure 2.65. The row line is activated and the programming data voltage is stored on C1. A proportion of the light emitted from the OLED impinges on the photodiode, causing C1 to discharge. Provided the capacitor discharges sufficiently to switch off M2 during the frame period then, to first order, the light output by the pixel depends only on the stored signal and is independent of any spatial variation, time variation or deterioration in the OLED. This approach continues to be refined by, for example, Fish *et al.*[76]

The relative benefits of various driving schemes are listed in Table 2.5. The relative effects of voltage drive, current drive and optical feedback drive on the luminance of an OLED pixel over time are illustrated notionally in Figure 2.66.

2.8.3 OLED Microdisplay with Digital Addressing: Example

Sanford and Schlig[77] first briefly reported, then subsequently Sanford *et al.*[65] reported in more detail, on an early miniature OLED display using a CMOS backplane. Strictly speaking this was not a

Table 2.5 Pros and cons of various addressing schemes for OLED pixels

Cause and effect	CMOS process variation	OLED process variation	Reduced lifetime due to resistive effects	Reduced lifetime due to reduced conversion efficiency
Voltage program Voltage drive	Very small	Medium	Medium	Medium
Voltage program Current drive	Medium	Small	Very small	Medium
Voltage program VT compensated current drive	Small	Small	Very small	Medium
Current programmed current drive	Small	Small	Very small	Medium
Optical feedback	Medium	Small	Very small	Very small

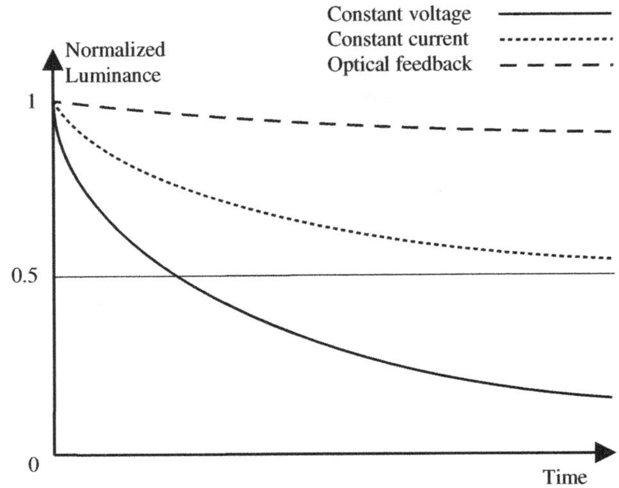

Figure 2.66 Generic lifetime characteristic of OLED structure for different drive scheme

"microdisplay" as it measured 27.4 mm = 1.08 inches (more than the conventionally accepted micro-display maximum size of 25.4 mm = 1 inch in diagonal) and it was intended for direct (unmagnified) viewing in a wristwatch computer. However, it is clearly relevant to the topic of this chapter. Some of the display parameters are summarized in Table 2.6. The device used a fully digital approach with 1-bit SRAM storage per-pixel and achieved grayscale by binary-coded pulse width modulation (B-PWM). The 1-bit SRAM could be used to maintain the appearance of a binary-level monochrome still image without recourse to any external refresh mechanism – a neat power saving feature for the kind of very slow-moving images typical of a watch face. The same authors also describe the inclusion of a control block to allow reverse-biasing of the OLED in order to remove forward-bias trapped charge.

Figure 2.67 shows the electronic architecture of the backplane for which no further explanation is offered here, and Figure 2.68 shows the pixel circuit. This SRAM configuration uses a single shared

Table 2.6 Summary of VGA OLED microdisplay parameters

Parameter	Value
Format	VGA (640 × 480) monochrome
Pixel pitch	34.3 μm
Picture diagonal	27.4 mm (1.08 inches)
Pixel architecture	1-bit SRAM plus memory bus interface and OLED driver
Emissive aperture ratio (EAR)	89%
Luminance range	O to 500 cd/m^2 in steps of 4 cd/m^2
Backplane technology node	0.35 μm 3.6 V CMOS
Grayscale	8-bit (256 levels)
Grayscale encoding	Binary-coded pulse width modulation
OLED color	Yellow
OLED efficiency	Approx 2 lm/W

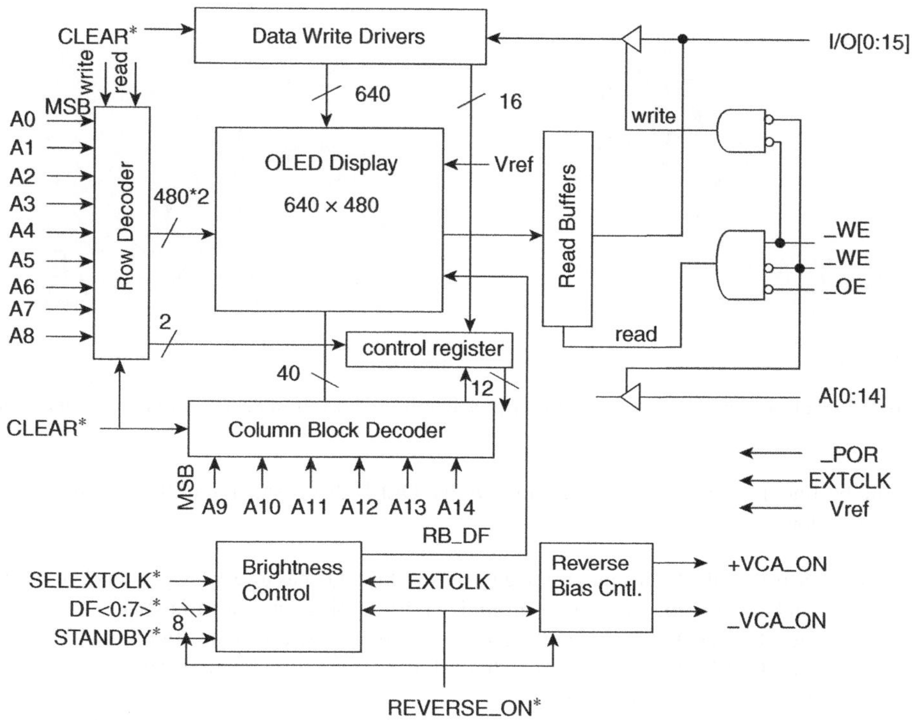

Figure 2.67 VGA OLED display architecture. Reprinted courtesy of Society for Information Display

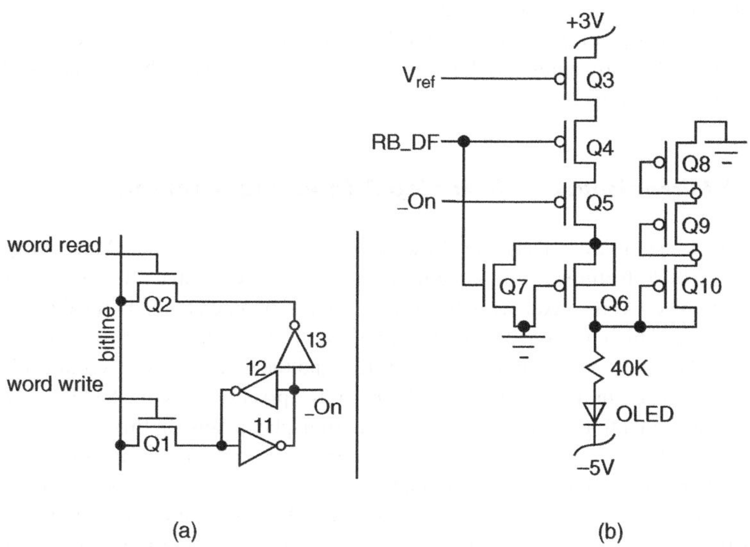

Figure 2.68 VGA OLED display pixel circuit: (a) pixel buffered SRAM; (b) OLED driver stage. Non-standard schematic symbols have been used. Reprinted courtesy of Society for Information Display

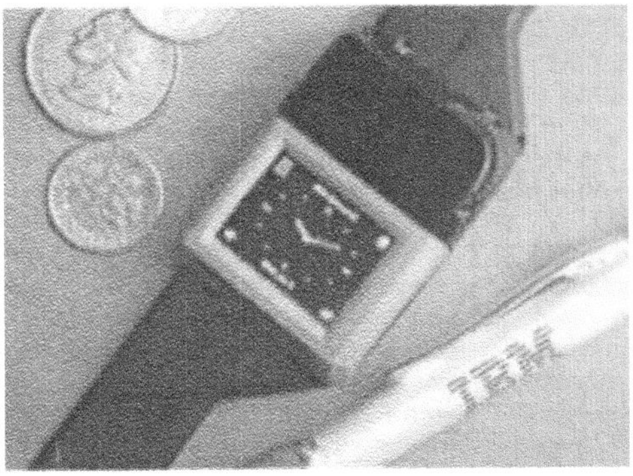

Figure 2.69 Photographs of the VGA OLED display. Reprinted courtesy of Society for Information Display

read/write bit-line; readout is via a buffer (I3). On the OLED driver side the PMOS transistors functions are as follows:

Q3	Sets the OLED current via the gate bias voltage. Q3 has a very long channel ($W/L \ll 1$) and operates in saturation to minimize pixel-to-pixel current variations.
Q4	Sets OLED ON/OFF according to the global pulse width modulation signal.
Q5	Sets OLED ON/OFF according to the binary value stored in the pixel.
Q6	Cascode/switching transistor.
Q7	Stray capacitance discharge during normal operation; also has a role in reverse bias discharge.
Q8, Q9, Q10	Clamp the source voltage of Q6.

It is likely that the approach used in the design of the VGA OLED backplane benefited from experience gained during the previous work of Sanford, Schlig and colleagues in the design of LCOS backplanes for light valves[37] and spatial light modulators[78].

The working display is pictured in Figure 2.69. There are no subsequent reports on the commercialization of this device. The comparatively large physical size, for a CMOS backplane, is likely to have led to low yield and high unit cost.

2.8.4 OLED Microdisplay with Analog Addressing: Example

Figure 2.70 shows a system-level electronic block diagram of an SVGA + OLED microdisplay reported by Levy *et al.*[79] Notable features of the system include analog RGB inputs that are converted to currents for distribution across the pixel array (in contrast to the approach taken in the device described in section 2.8.3 above in which the conversion to current is carried out in the back-end of the pixel), array column drivers at the north and south of the pixel array, and the clock recovery circuit (PLL etc.).

Figure 2.71 shows a circuit implemented in CMOS for the pixel of the same SVGA + OLED microdisplay. As previously mentioned, the small size of microdisplay pixels (in this case $225\,\mu m^2$ and soon perhaps less than $100\,\mu m^2$) demands the use of very small currents and transistors biased for sub-threshold conduction. One problem with very small currents is the excessive time taken to, for example, charge and discharge large capacitive nodes such as row and column buses. The circuit of Figure 2.71 uses the logarithmic response of the MOS sub-threshold characteristic to produce a current scaling solution that allows a relatively high current source (×100) to produce a small in-pixel current. The key enhancement over the circuit of Figure 2.64 is the inclusion of the programmable row reference

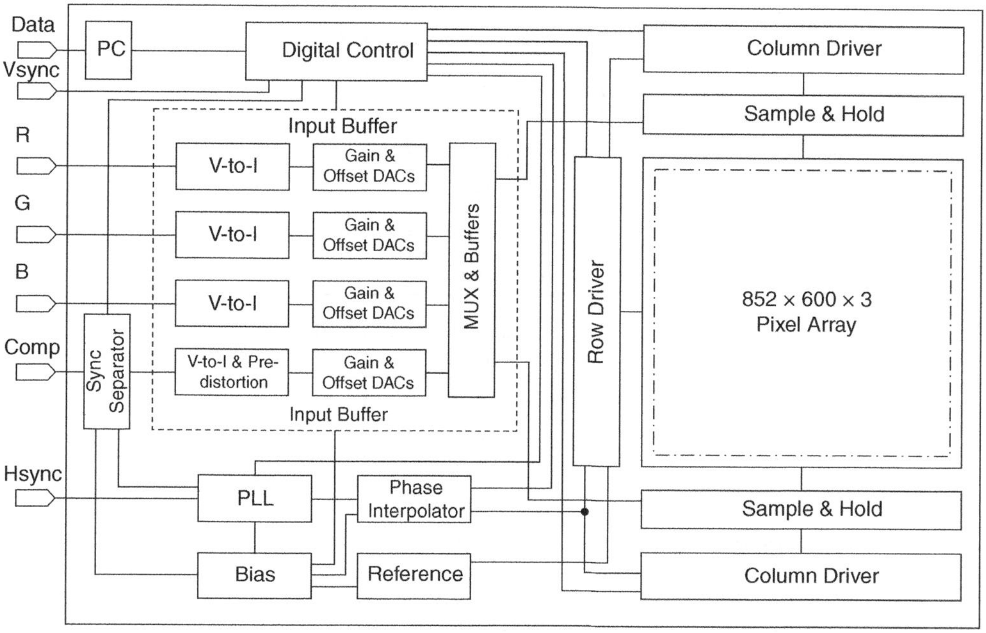

Figure 2.70 System-level electronic block diagram for SVGA+ OLED microdisplay. ©2002 IEEE; reprinted with permission

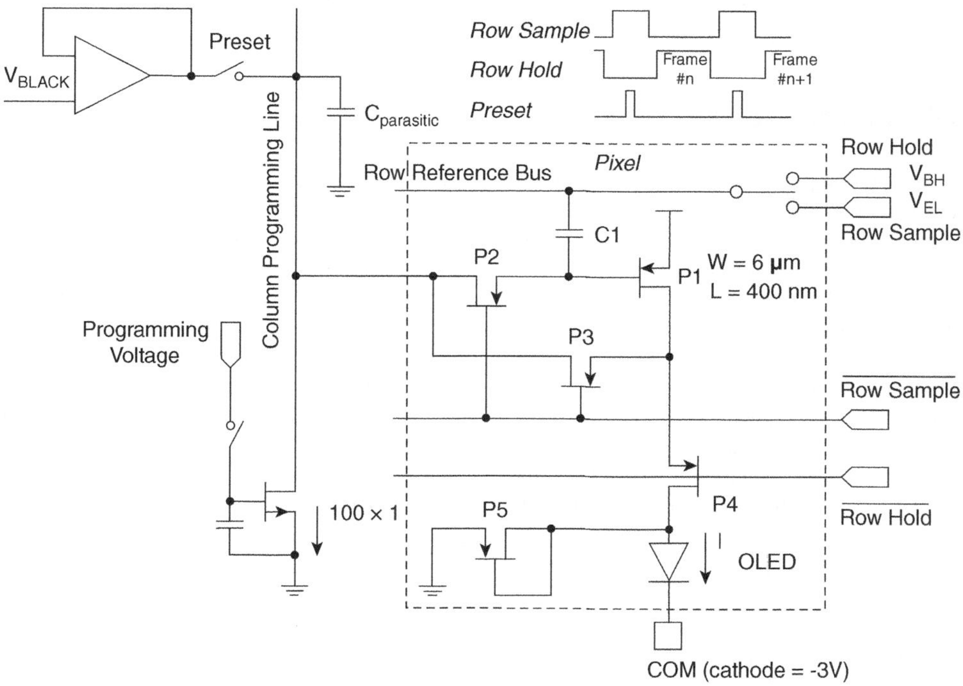

Figure 2.71 Sub-threshold voltage-scaling current source microdisplay pixel circuit. ©2002 IEEE; reprinted with permission

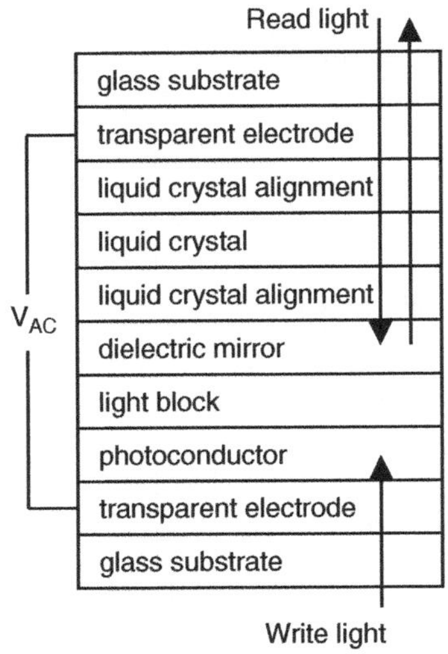

Figure 2.72 Thin film structure of photo-addressed microdisplay

bus. During row sampling the circuit is addressed in much the same way as the circuit of Figure 2.64. Upon moving to row hold, the signal on the voltage reference bus is switched, leading to an offset on the gate of P1 that reduces the current through P1 by a factor of 100. MOSFET P1 is moved from normal conduction (during addressing) to sub-threshold (during hold). Additional enhancements shown here include the data bus preset to reduce settling time and the P5 clamp to prevent the anode of the OLED from going too negative.

2.9 Photo-addressing

Photo-excitation provides a simple means of addressing an electro-optic element, but involves optical addressing complexity. Figure 2.72 illustrates the structure of a photo-addressed liquid crystal device. A broadband dielectric mirror, reinforced by light blocking layer, protects the photoconductor from read-out light. An applied alternating voltage $V_{AC} \sim 10\,V$ at frequency $\sim 3\,kHz$, applied to the outer transparent electrodes (ITO), distributes over the various films according to their impedance, with a significant component across the photoconductor. A write light beam probing the photoconductor raises the local conductivity, enhancing the local liquid crystal voltage at the expense of photoconductor voltage.

The equivalent circuit approximation in Figure 2.73 indicates local liquid-crystal voltage V_{LC} increasing, as local photoconductor resistance R_{PC} falls with rising illumination. Spatial gradient in the write light gives rise to transverse electric fields and currents represented by a transverse resistance R_T, limiting the spatial resolution. Electric field fringing also limits the optical resolution. The equivalent capacitance of the dielectric mirror and light block, C_M, attenuates the voltage transfer to the liquid crystal, while their thickness enhances electric field fringing. For simplicity, we ignore rectifying effects due to non-ohmic semiconductor contacts or silicon photodiode structures.

The resistance levels are normally sufficiently high that capacitive voltage division applies in the dark state, while high illumination can do no more than short out the photoconductor, implying the liquid-crystal voltage range

$$\frac{1/C_{LC}}{1/C_{LC} + 1/C_M + 1/C_{PC}} < \frac{V_{LC}}{V_{AC}} < \frac{1/C_{LC}}{1/C_{LC} + 1/C_M}. \qquad (2.24)$$

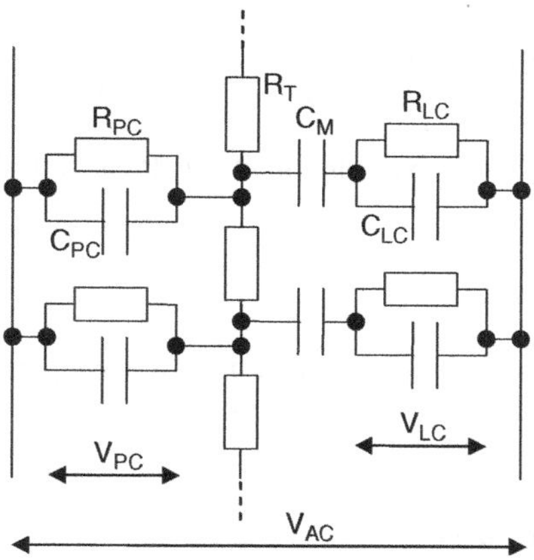

Figure 2.73 Equivalent circuit approximation of photo-addressed microdisplay

The reduction of any or all layer thicknesses attenuates field fringing, improving resolution. Lowering the thickness increases capacitance, with consequent increase in required charge, higher photocurrent, and write light intensity. The design of a photo-addressed microdisplay adjusts the various tradeoffs according to application and cost.

A high switching ratio indicated in (2.24) requires a high dark-resistance photoconductor; moreover, the dark-resistance recovery time limits the device response time. Quantum efficiency and wavelength sensitivity are other factors in photoconductor choice. Photoconductive materials of interest include cadmium sulfide, α-silicon, gallium arsenide, and crystalline silicon photodiode. The latter requires additional structure to inhibit lateral charge transfer.[80–82]

The first cinema-quality projectors used photo-addressed liquid-crystal light valves activated by light from closely coupled CRTs.[83] The light valves, at 2-inch diagonal, were hardly microdisplays. The complexity of addressing, and small production volumes, led to declining interest and program termination in favor of electronically addressed microdisplays. Photo-addressing may compete at very high resolution, employing laser-scanned addressing, but the area of such displays is beyond the microdisplay category.

A photo-addressed microdisplay has advantage in optical systems requiring the manipulation of optical arrays or images. Incoherent-to-coherent image conversion follows from writing with incoherent light, while reading with an appropriate laser. The joint transform correlator achieves optical pattern recognition by superimposed photo-addressing of Fourier transformed images.

2.10 Bibliography

IBM

IBM J. Res. Dev. **42**, 3/4, May/July 1998: contains a number of papers relating to the design and fabrication of LCOS devices and systems at IBM. This offers an almost tutorial-like introduction to the field of LCOS.

VLSI Circuit Design

Allen, P. E. and Holberg, D. R. (1987) *CMOS Analog Circuit Design*. HRW.

Bakoglu, H. B. (1990) *Circuit Inteconnections and Packaging for VLSI*. Addison–Wesley.

Glasser, L. A. and Dobberpuhl, D. W. (1985) *The Design and Analysis of VLSI Circuits*. Addison–Wesley.

Mukherjee, A. (1986) *Introduction to NMOS & CMOS VLSI Systems Design*. Prentice-Hall.

Smith, M. J. S (1997) *Application-Specific Integrated Circuits*. Addison–Wesley.
Troutman (1986) *Latchup in CMOS Technology*. Kluwer Academic Press.
Weste, N. H. and Eshraghian, K. (1994) *Principles of CMOS VLSI Design: A Systems Perspective*, 2nd edn. Addison–Wesley.
Wolf, W. (1998) Modern VLSI Design: Systems on Silicon. Prentice-Hall.

Semiconductor Memories

Prince, B. (1991) *Semiconductor Memories*, 2nd edn. John Wiley. ISBN 0-471-94-295-2
Prince, B. (1996) *High Performance Memories*. John Wiley. ISBN 0-471-95-646-5

Some Early LCOS Publications

Ernstoff, M. N., Leupp, A. M., Little, M. J. and Peterson, H. T. (1973) Liquid crystal pictorial display. *IEEE Intl Electron Devices Meeting Technical Digest*, 548.
Lipton, L. T., Meyer, M. A. and Massetti, D. O. (1975) *SID Digest* **16**.
Stephens, C. P. and Lipton, L. T. (1976) *SID Digest* **17**.
Lipton, L. T., Stephens, C. P. and Lloyd, R. B. (1977) A fully integrated MOS liquid crystal video-rate matrix display. *SID Digest* **18**, 64–5.
Kasahara, K., Yanagisawa, T., Sakai, K. and Adachi, T. (1980) *Digest of Biennial International Display Research Conference* (IDRC), 96.
Yanagisawa, T. *et al.* (1981) *SID Digest* **22**, 110–11.
Crossland, W. A. and Ayliffe, P. J. (1981) A dyed phase change liquid crystal display over a MOSFET switching array. *SID Digest* **22**, 112–13.
Hosokawa, M., Oguchi, K., Ikeda, M. *et al.* (1981) Dichroic guest-host active matrix video display. *SID Digest* **22**, 114–15.
Kasahara, K., Yanagisawa, T., Sakai, K. and Adachi, T. (1981) A liquid crystal display panel using a MOS array with gate-bus drivers. *SID Digest* **22**, 318–19.
Yamasake, T., Kawahara, Y., Motle, S. *et al.* (1982) A liquid crystal TV display panel with drivers. *SID Digest* **23**, 48–9.
Kasahara, K., Sakai, K., Komatsubara, Y. *et al.* (1983) A 480 × 480 element dichroic dye MOS-LCD. *SID Digest* **24**, 150–1.
Shields, S. E., Fletcher, B. G. and Bleha, W. P. (1983) A 240 × 320 element MOS-addressed high resolution LCD. *SID Digest* **24**, 178–9.
Crossland, W. A., Needham, B. and Ross, P. W. (1985) Single-crystal silicon active-matrix displays. *Proc. SID* **26**, 175–81.

Microdisplay And Spatial Light Modulator Design

Collings, N., Crossland, W. A., Ayliffe, P. J. *et al.* (1989) Evolutionary development of advanced liquid crystal spatial light modulators. *Applied Optics* **28**, 4740–7.
De Smet, H., Cuypers, D., Van Calster, A. *et al.* (2002) Design, fabrication and evaluation of a high-performance XGA VAN-LCOS microdisplay. *Displays* **23**, 89–98.
Howard, W. E. and Prache, O. F. (2001) Microdisplays based on organic light emitting diodes. *IBM J. Res. Dev.* **45**, 1, 115–28.
Mao, C. C., McKnight, D. J. and Johnson, K. M. (1995) High-speed liquid-crystal on silicon spatial light modulators. *Institute of Physics Conference Series* **139**, 539–42.
Underwood, I. *et al.* (1990) A high performance spatial light modulator. *Proc. SPIE*, **1562**, 107–15.
Underwood, I. (1997) Liquid crystal over silicon spatial light modulators: principles, practice and prospects. In *OSA Trends in Optics and Photonics*, Vol. 14: *Spatial Light Modulators*. Optical Society of America, Washington, DC, 76–88.

Wilkinson, T. D., Crossland, W. A., Coker, T. *et al.* (1998) Ferroelectric liquid crystal on silicon spatial light modulator designed for high yield and low cost fabrication: the fast bitplane SLM. *Ferroelectrics* **213**, 613–17.

Wilkinson, T. D., Crossland, W. A., Coker, T. *et al.* (1997) The fast bitplane SLM: a new ferroelectric liquid crystal on silicon spatial light modulator. In *OSA Trends in Optics and Photonics*, Vol. 14: *Spatial Light Modulators*. Optical Society of America, Washington, DC, 154–6.

2.11 References

1 Matsumoto, S. (ed.) (1990) Liquid crystal displays. In *Electronic Display Devices*. John Wiley.

2 Alt and Pleschko (1974) Scanning limitations of liquid-crystal displays. *IEEE Trans. Electron. Dev.* **21**, 146–55.

3 Clark, N. A., Crandall, C., Handschy, M. A. *et al.* (2000) FLC microdisplays. *Ferroelectrics* **246**, 97–110.

4 Jepsen, M. L. (2005) Why analog may be best for LCOS digital tv. *SID Digest* **26**, 1358–61.

5 Huang, H. C., Zhang, B. L. and Kwok, H. S. (2005) Color-filter liquid-crystal-on-silicon microdisplays. *SID Int Symp. Dig.* **26**, 880–3.

6 www.novia.net/~ereitan/

7 Rankin, I. D., Underwood, I., Vass, D. G. *et al.* (1996) Full colour images on a binary spatial light modulator. *Ferroelectrics* **181**, 139–46.

8 Shimizu, J. A. (2001) Philips' scrolling color LCOS engine for HDTV rear projection. *Information Display* **11**, 14–19.

9 Janssen, P., Shimizu, J. A., Dean, J. and Albu, R. (2002) Design aspects of a scrolling color LCOS display. *Displays* (UK) **23**, 99–108.

10 Voutsas, T. (2003) Poly-Si TFT technology, status and applications. *Digest of the December 2003 Meeting of the Pacific North West Chapter of the SID.*

11 den Boer, W. (2005) *Active Matrix Liquid Crystal Displays: Fundamentals and Applications*. Elsevier.

12 Voutsas, T. (2002) Fundamentals of AMLCDs. *SID Seminar Lecture Notes*, seminar M-3.

13 Soh, H-S. (2002) TFT-LCD manufacturing technology. *SID Seminar Lecture Notes*, seminar M-4.

14 Lueder, E. (2001) *Liquid Crystal Displays*. John Wiley.

15 Ernstoff, M. N., Leupp, A. M., Little, M. J. and Peterson, H. T. (1973) Liquid crystal pictorial display. *IEEE Intl Electron Devices Meeting Technical Digest*, 548.

16 Crossland, W. A. and Ayliffe, P. J. (1981) A dyed phase change liquid crystal display over a MOSFET switching array. *SID Digest* **22**, 112–13.

17 Private communication with Prof. W. A. Crossland, University of Cambridge, UK.

18 Mavor, J., Jack, M. A. and Denyer, P. B. (1982) *Introduction to MOS LSI Design*. Addison Wesley.

19 Underwood, I., Vass, D. G. and Sillitto, R. M. (1986) Evaluation of an nMOS VLSI array for an adaptive liquid crystal spatial light modulator. *IEE Proc. J.* **133**, 77–82.

20 McKnight, D. J., Vass, D.G. and Sillitto, R. M. (1989) Development of a spatial light modulator: a randomly addressed liquid-crystal-over nMOS array. *Applied Optics* **28**, 4757–62.

21 Collings, N., Crossland, W. A., Ayliffe, P. J. *et al.* (1989) Evolutionary development of advanced liquid crystal spatial light modulators. *Applied Optics* **28**, 4740–7.

22 Cotter, L. K. *et al.* (1990) Ferroelectric liquid crystal silicon integrated circuit spatial light modulator. *Optics Letters* **15**, 291–3.

23 Jared, D. A., Turner, R. and Johnson, K. M. (1991) Electrically addressed spatial light modulator that uses a dynamic memory. *Optics Letters* **16**, 1785–7.

24 Underwood, I. *et al.* A high performance spatial light modulator. *Proc. SPIE*, **1562**, 107–115.

25 Crossland, W. A., Birch, M. J., Davey, A. B. and Vass, D. G. (1991) Ferroelectric liquid crystal/silicon VLSI backplane technology for smart spatial light modulators, *Digest of IEE Colloquium on Two-Dimensional Optoelectronic Device Arrays*, London, 7/1–7/4.

26 Serati, S. A., Ewing, T. K., Serati, R. A. *et al.* (1993) Programmable 128 × 128 ferroelectric liquid crystal spatial light modulator compact correlator. *Proc. SPIE*, **1959**, 55–68.

27 McKnight, D. J., Serati, R. A. and Johnson, K. M. (1993) Electrically addressed 256 by 256 liquid-crystal-on-silicon spatial light modulator. *Optics Letters* **18**, 2159.

28 Johnson, K. M., McKnight, D. J. and Underwood, I. (1993) Smart spatial light modulators using liquid crystals on silicon. *IEEE J. Quantum Electronics* **29**, 699–714.
29 Handschy, M. A. *et al.* (1994) One-transistor DRAM ferroelectric-liquid-crystal spatial light modulator. *Proc. SPIE*, **2237**, 432–8.
30 Burns, D. C., Underwood, I., Gourlay, J. *et al.* (1995) A 256 × 256 SRAM-XOR pixel ferroelectric liquid crystal over silicon spatial light modulator. *Optics Communications* **119**, 623–32.
31 Sze, S. M. (1981) *Physics of Semiconductor Devices*, 2nd edn. John Wiley.
32 Tsividis, Y. P. (2003) *Operation and Modeling of the MOS Transistor.* McGraw-Hill.
33 Paret, D. and Fenger, C. (2000) *The I2C Bus from Theory to Practice.* John Wiley.
34 Gale, R., Herrmann, F., Lo, J., Mteras *et al.* (1999) Miniature 1280 by 1024 active matrix liquid crystal displays. *Proc SPIE*, **3689**, 231–7.
35 Jepsen, M. L., Ammer, M. J., Bolotski, M. *et al.* (2002) High resolution LCOS microdisplay for single-, double- or triple-panel projection systems. *Displays* **23**, 109–14.
36 Worboys, M. R., Radcliffe, S. and Griffith, M. S. (1998) A high-resolution miniature colour liquid crystal display. *GEC J. Techn.* **15**, 1, 56–9.
37 Schlig, E. S. and Sanford, J. L. (2000) An SXGA reflective liquid crystal reflective light valve incorporating inversion by pixel bootstrapping. *IBM J. Res. Dev.* **44**, 909–18.
38 Sanford, J. L., Schlig, E. S., Tomooka, T. *et al.* (1998) Silicon light-valve array chip for high resolution reflective liquid crystal projection displays. *IBM J. Res. Dev.* **42**, 347–58.
39 Shields, S. (1989) AC activated liquid crystal display cell employing dual switching devices. US Patent 4,870,396.
40 Lee, S., Morizio, J. C. and Johnson, K. M., (2002) A LCOS microdisplay driver with frame buffering pixels. *SID Digest* **33**, 292–5.
41 Tan, S. C. and Sun, X. W. (2002) Generic design of silicon backplane for LCOS microdisplays. *SID Digest* **33**, 200–3.
42 Blalock, T. N., Gaddis, N. B., Nishimura, K. A. and Knotts, T. A. (2001) True color 1024 × 768 microdisplay with analog in-pixel pulse width modulation and retinal averaging offset correction. *IEEE J. Solid State Circuits* **36**, 838–45.
43 Hudson, E. L. *et al.* (1999) Display with multiplexed pixels for achieving modulation between saturation and threshold voltages. US Patent 6,005,558.
44 Shimitzu, S. *et al.* (2004) Fully digital D-ILATM device for consumer applications. *SID Digest* **35**, 72–5.
45 Shimitzu, S. *et al.* (2005) Fully digital LCOS device for consumer applications. *J. SID* **13**, 1, 3–8.
46 Worley, W. S. *et al.* (2000) System and method for using compound data words to reduce the data phase difference between adjacent pixel electrodes. US Patent 6,151,011.
47 Kurita, T. (2001) Moving picture quality improvement for hold-type AM-LCDs. *SID Digest* **32**, 986–9.
48 Hornbeck, L. J. (1996) Digital light processing and MEMS: reflecting the digital display needs of the networked society. *Proc. SPIE*, **2783**, 2–13.
49 McKnight, D. J. (1994) Continuous view of dc-balanced images on a ferroelectric liquid-crystal spatial light modulator. *Optics Letters* **19**, 1471–3.
50 Takanashi, H. *et al.* (2000) A novel driving technique for a ferroelectric liquid crystal microdisplay using non DC-balanced waveforms. *SID Digest* **31**, 992–5.
51 Vass, D. G. *et al.* (1998) A high resolution, full colour, head mounted ferroelectric liquid crystal-over-silicon display. *Ferroelectrics* **213**, 209–18.
52 Wilkinson, T. D., Crossland, W. A., Coker, T. *et al.* (1998) Ferroelectric liquid crystal on silicon spatial light modulator designed for high yield and low cost fabrication: the fast bitplane SLM. *Ferroelectrics* **213**, 613–17.
53 Burns, D. C., Begbie, M. L., Underwood, I. and Vass, D. G. (1996) Ferroelectric liquid crystal over silicon spatial light modulator viewable under continuous illumination. *Ferroelectrics* **181**, 93–6.
54 Cotter, L. K., Drabik, T. J., Dillon, R. J. and Handschy, M. A. (1990) Ferroelectric-liquid-crystal/silicon-integrated-circuit spatial light modulator. *Optics Letters* **15**, 291–3.
55 Drabik, T. J., Titus, A. H., Handschy, M. A. *et al.* (1995) 2D silicon/ferroelectric liquid crystal spatial light modulators. *IEEE Micro*, August, 67–76.
56 Birch, M., Kruerke, D., Yates, C. *et al.* (2002) SXGA resolution FLC microdisplays. *SID Digest* **33**, 954–7.

57 Akimoto, O. and Hashimoto, S-i. (2000) A 0.9-inch UXGA/HDTV FLC microdisplay. *SID Digest* **31**, 194–7.
58 Yumoto, A., Mano, M. and Yamagishi, M. (2000) A silicon backplane for UXGA/HDTV microdisplay. *Int. Display Workshop*, Japan, 183–6.
59 Brennersholtz, M. S. (1998) Luminance contouring at the display in digital television. *Proc. SPIE*, **3296**, 62–70.
60 Van Kessel, P. F. *et al.* (1998) A MEMS-based projection display. *Proc. IEEE* **86**, 1687–704.
61 Daly, S. and Feng, X-F. (2004) Bit-depth-extension: overcoming LCD-driver limitations by using models of the equivalent input noise of the visual system. *SID Digest* **35**, 1470–3.
62 Toyooka, T. *et al.* (2004) Illumination control system for adaptive dynamic range control. *SID Digest* **35**, 174–7.
63 Critchley, B. *et al.* (1995) Picture quality in large-screen projectors using the digital mirror device. *SID Digest* **26**, 524–7.
64 Doherty, D. and Hewlett, Phased reset timing for improved Digital Micromirror Device™ (DMD™) brightness. *SID Digest* **29**, 125–8.
65 Sanford, J. L., Schlig, E. S., Prache, O. *et al.* (2002) Technology and design of an active matrix OLED on crystalline silicon direct view display for a wrist watch computer. *Proc SPIE*, **4464**, 11–22.
66 Tan, S. C. and Sun, X. W. (2002) Designing of circuit building blocks for OLED-on-silicon microdisplays. *SID Digest* **33**, 980–3.
67 Tam, SW-B., Matsueda, Y., Maeda, H. *et al.* (1999) Polysilicon TFT drivers for light emitting polymer displays. *Technical Digest of IDW* **6**, 171–4.
68 Tam, SW-B., Matsueda, Y., Kimura, M. *et al.* (2001) Poly-Si driving circuits for organic EL displays. *Proc SPIE*, **4925**, 125–33.
69 Fish, D., Young, N., Childs, M. *et al.* (2000) A comparison of pixel circuits for active matrix polymer/organic LED displays. *SID Digest* **21**, paper 32.1.
70 Baynes, N. and Smith E. (2006) Organic electroluminescent displays. In *Handbook of Optoelectronics*. CRC Press.
71 Brody, T. P., Asars, J. A. and Dixon, G. D. (1975) A 6-inch × 6-inch 20 lpi electroluminescent display panel. *IEEE Trans. Electron Devices*, 739–48.
72 Dawson, R. M. A., Shen, Z., Furst, D. A. *et al.* (1998) Design of an improved pixel for a polysilicon active-matrix organic LED display. *SID Digest* **19**, 11–14.
73 Dawson, R. M. A., Shen, Z., Furst, D. A. *et al.* (1998) The impact of the transient response of organic light emitting diodes on the design of active matrix OLED displays. *IEEE Intl Electron Devices Meeting Technical Digest*, 875–8.
74 Dawson, R. M. A., Shen, Z., Furst, D. A. *et al.* (1999) A polysilicon active matrix light emitting diode display with integrated drivers. *SID Digest* **30**, 438–41.
75 Hunter, I. M., Young, N. D., Johnson, M. T. and Young, E. W. A. (1999) Design of an active matrix polymer LED display with reduced horizontal cross-talk. *Proc. IDW*, 1095–6.
76 Fish, D., Childs, M. J., Deane, S. C. *et al.* (2005) Improved optical feedback for OLED differential ageing correction. *J. SID* **13**, 131–8.
77 Sanford, J. L. and Schlig, E. S. (2001) Direct view active matrix VGA OLED-on-crystalline silicon display. *SID Digest* **32**, 376–9.
78 Sanford, J. L., Greier, P. F., Yang, K. H. and Lu, M. (1998) A one-megapixel reflective spatial light modulator system for holographic storage. *IBM J. Res Dev.*, 411–26.
79 Levy, G. B., Evans, W., Ebner, J. *et al.* (2002) An 852 × 600 pixel OLED-on-silicon color microdisplay using CMOS subthreshold-voltage-scaling current drivers. *IEEE J. Solid State Circuits* **37**, 1879–89.
80 Barbier, P. R. *et al.* Thin-film photosensor design for liquid crystal spatial light modulators. *Optics Eng.* **33**, 1322–9.
81 Sayyah, K. *et al.* (1989) High performance single crystal silicon liquid crystal light valve with good image uniformity. *Applied Optics* **28**, 4748–55.
82 Armitage, D. *et al.* (1989) Photoaddressed liquid crystal spatial light modulators. *Applied Optics* **28**, 4763–71.
83 Sterling, R. D. and Bleha, W. P. (1999) Electronic cinema using ILA projector technology. *SID Digest* **38**, 216–19.
84 Weste, N. H. and Eshraghian, K. (1994) *Principles of CMOS VLSI Design: A Systems Perspective*, 2nd edn. Addison Wesley.

3

CMOS Backplane Technology

3.1 Introduction

Silicon wafer technology encompasses the electronic materials, devices, device structures and manufacturing processes involved in the realization of a wide range of integrated circuits such as microprocessors, application-specific integrated circuits (ASICs), field programmable gate arrays (FP-GAs), random access memories (RAMs), electrically erasable programmable read only memories (EEPROMs), charge coupled devices (CCDs), CMOS image sensors and others. Generic technologies include complimentary metal oxide semiconductor (CMOS) and bipolar. Often specialized structures are added to a generic technology in order to optimize it for a more specific application; e.g. the addition of trench capacitors for dynamic RAM, the addition of thick gate oxide for programmable ROM, and the addition of micromechanical structures for technologies such as DLP. This chapter deals primarily with the structures and processes added to generic CMOS in order to optimize it for LCOS microdisplays.

Reflection-mode liquid crystal microdisplays utilize CMOS circuitry on a semiconducting crystalline silicon (x-Si) substrate for active matrix addressing. Within a pixel, the circuitry is "hidden" under a top-metal mirror-electrode that occupies almost all of the area of the pixel. Transmission-mode liquid crystal microdisplays, on the other hand, utilize either polysilicon (p-Si) thin film transistor (TFT) circuitry on glass or crystalline silicon (x-Si) on insulator (SOI) for active matrix addressing. In the latter case a transparent glass substrate supports the circuitry and transparent pixel electrodes, forming the lower boundary to the liquid crystal cell. The counter electrode, ITO on glass, forming the upper or opposite cell boundary, is identical to the LCOS case. The structure and fabrication of transmission microdisplays is described in Chapter 4. The pixel area is shared between necessary opaque circuit elements and a portion covered by a transparent electrode – the optically active part of the pixel. In the ongoing evolution of the technology to microdisplays with larger numbers of smaller-sized pixels the need to maximize (or at least maintain) the fraction of optically active area (or fill-factor) forces a lower limit to the pixel pitch.

Introduction to Microdisplays D. Armitage, I. Underwood and S-T. Wu

In a reflection-mode microdisplay, the CMOS silicon substrate has several functions. As is the case for any mainstream integrated circuit (IC), it contains the electronic circuitry that determines the state of the system. Furthermore, it provides a suitable substrate or backplane upon which to construct the optical structures of the microdisplay. (In this context, the term "backplane" refers to the silicon chip that is the active matrix substrate for the microdisplay.) Thus, any microdisplay carries in its specification electronic, optical and mechanical constraints in addition to the customary constraints applicable to a mainstream electronic IC.

"Standard" CMOS technology is the basic starting point for many current microdisplays. Following a brief description of generic CMOS technology, the focus of the remainder of this chapter is on a description, specification and realization of these microdisplay-specific optical, mechanical and other material constraints.

3.2 CMOS Technology

3.2.1 Background

The basic building block of CMOS technology is the metal oxide semiconductor field effect transistor (MOSFET). A simple description of the electronic functionality of this device, and some simple circuits utilizing it, are contained in Chapter 2. Sze[1] provides a sound introduction to CMOS technology. It is worth pointing out that the term "MOS" is historical in that early devices utilized a sandwich structure comprising a thin layer of silicon dioxide (O) between an overlying aluminum metal gate (M) and underlying crystalline silicon channel (S). Today the vast majority of "MOS" devices actually use a gate formed of polycrystalline silicon (called polysilicon or just poly), but the term "MOS" remains.

The "complimentary" nature of CMOS circuits follows from the symmetry of the two primary types of active devices (or transistors) involved, n-type in which the majority carriers are negative electrons, and p-type in which the majority carriers are positive holes. The sign of the carriers determines the polarity of the gate on/off voltage. Historically, circuits (and in particular digital circuits) combining both types of MOSFET simplified circuit design. The prevalence of CMOS technology in general derives from a number of distinct but related technical attributes:

- *Planar technology.* The active devices or transistors lie on, or close to, the surface of the substrate so that many hundreds (or indeed thousands or millions) of transistors can be produced simultaneously on a single substrate (integrated circuit or chip) and linked together to form complex circuits.
- *Low-power technology.* It is typically the case that, for a given functionality, a CMOS circuit dissipates less power than an equivalent circuit in an alternative electronic technology such as silicon bipolar technology. It is also the case that many CMOS circuits, particularly CMOS digital circuits, dissipate power only when switching state and not in between times.
- *Small technology.* CMOS transistors and other active devices are manufactured with very small feature sizes. The state of the art today is somewhere in the region of 0.13 μm (130 nm) to 0.09 μm (90 nm), although microdisplay CMOS (MD-CMOS) tends to be in the region 0.35 μm to 0.18 μm for reasons that will be explained later in this chapter.

Furthermore, there are some practical attributes associated with the ubiquity and maturity of CMOS as a technology and an industry that make it self-perpetuating:

- *IC design infrastructure.* There is a large and growing infrastructure associated with the system, circuit and layout design of CMOS ICs. This includes, for example, computer aided design software and companies specializing in CMOS design. This makes possible the fast and efficient design of ICs of a complexity not previously possible.

- *Wafer fabrication infrastructure*. There are a large number of wafer foundries that specialize in manufacturing CMOS wafers in large volume at low price and with a minimum number of defects on a wafer.
- *Back-end infrastructure*. All of the processes between the completion of wafer fabrication and the finished product, such as wafer dicing, packaging, electrical testing and burn-in (if necessary), are widely used and available.
- *Research and development*. The whole of CMOS technology is subject to a huge investment in improving and evolving the capability of the technology.

Many of the technical attributes have exhibited an inexorable evolution with time for many years past, as first noted by Moore,[2] and are predicted to continue to do so for many years hence. That is to say, the real estate or footprint, energy and time required to perform a given function all reduce with time while packing density (MOS transistors per unit area), maximum chip size and maximum clock frequency all increase. While the evolution is relatively steady, in some cases its manifestation becomes quantized. As an example, CMOS technology tends to cluster around specific "process nodes" or minimum feature sizes. As time has passed, state-of-the-art CMOS has stepped in recent years through the following minimum feature sizes: 0.35 µm; 0.25 µm; 0.18 µm; 0.13 µm; 0.09 µm (90 nm) and so on into the future.

Some aspects of this evolution (known as Moore's Law although it is more of an empirical observation and prediction than a law) are shown in Table 3.1. The future evolution is these days predicted, with a high degree of accuracy, by the International Technology Roadmap for Semiconductors[3] or ITRS (formerly known as the Silicon Industry Association, or SIA, Roadmap). Up-to-date information on the ITRS can be readily found on the internet.

3.2.2 MOS Transistor Structure

A CMOS IC consists of MOS transistors that are partly (source and drain) sunk into the surface of the active substrate and partly (gate) above it. This is in contrast to other thin film technologies such as amorphous silicon (α-Si) or polysilicon (p-Si) that generally start with an inert substrate, usually glass, and build all of the active layers on top of, and upwards from, the substrate.

In almost all CMOS processes, the substrate is an active substrate consisting of doped silicon. In some advanced processes the substrate is an insulating material (such as sapphire) with a thin uniform surface layer of doped silicon into which the CMOS devices are manufactured. Wafers manufactured in such processes offer properties, such as lower parasitic capacitance (leading to faster switching and/or lower power) and radiation hardness (leading to robust operation in harsher conditions), that are advantageous for some applications. They are, however, more expensive to produce.

Table 3.1 The evolution of CMOS technology

Year	2001	2004	2007	2010
Node (µm)	0.18	0.13	0.09	0.065
V_{DD} (V)	1.8	1.5	1.2	0.9
Max chip size (mm^2)	750	900	1100	1400
Wafer diameter (mm)	200	300	400	400
Typical #metal layers	5	6	7	8
Defect density (m^{-2})	140	120	100	80
Minimum mask levels	20	22	22	24
Maximum number of transistors per chip	7M	12M	25M	40M

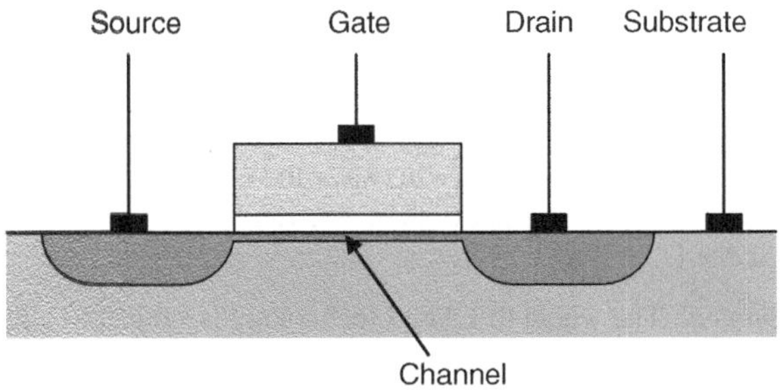

Figure 3.1 Schematic cross-section through MOS transistor. Interconnect layers are not shown

The basic n-channel MOS field effect transistor (FET) is shown in schematic cross-section in Figure 3.1. The source and drain are doped together in opposition to the substrate so that a p–n junction is formed between substrate and source/drain. This p–n junction is always zero-biased or reverse-biased in order to maintain electrical isolation between substrate and source/drain. In CMOS technology, both n-channel and complimentary p-channel MOSFETs are available. These are illustrated in cross-section in Figure 3.2, which also explicitly highlights the p–n junction isolation.

The footprint (or the pitch) of a minimum geometry MOS transistor in any given process is determined by a combination of the minimum gate width, the minimum active area (drain or source) width and the minimum active area to active area separation. Typically, the gate width is the smallest feature in a CMOS process and so defines the process node. For example, a 0.35 μm CMOS process means, in effect, a process with a polysilicon or gate width of 0.35 μm. It is worth noting that, at small geometries, the physical dimensions of features on the wafer may differ from the drawn or designed size. Maly[5]

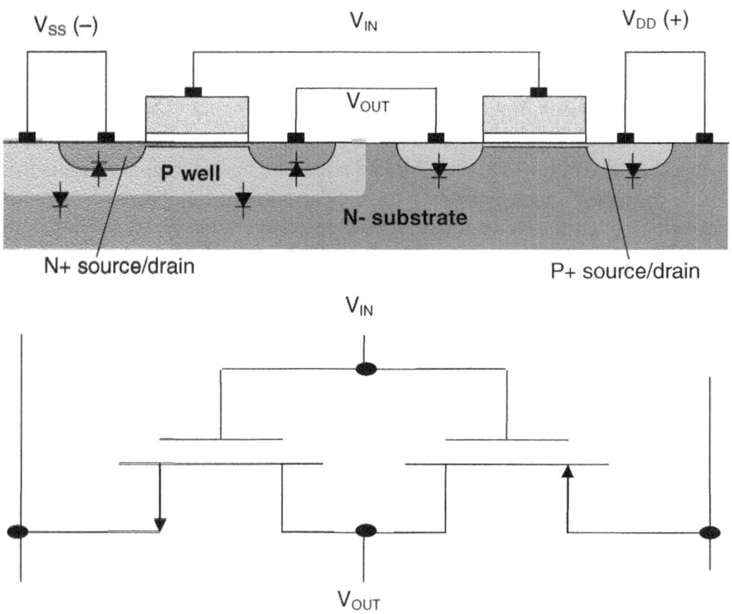

Figure 3.2 Circuit schematic (bottom) and cross-section (top) through CMOS inverter structure explicitly showing reverse-biased p–n junctions used for isolation

illustrates this. This offset is generally well characterized for a given CMOS process. The minimum lateral dimensions of the source/drain regions and the minimum separation between adjacent source/drain regions presents one of the major limitations to shrinking the size of a CMOS circuit. This, coupled with increased leakage current at low dimensions, has led to the development of trench-oxide structures as an alternative to the p–n junction for superior device isolation in the substrate in some advanced processes.

The MOS transistor can be configured to perform many functions. It can operate as, for example, a capacitor, a switch, an amplifier or a current source. Some of these functions that are most relevant to microdisplays are described in Chapter 2. Many more functions can be achieved by various combinations of MOS transistors.

3.2.3 MOS Integrated Circuit Structure

Interconnection of the combination of MOS transistors in an integrated circuit is a key part of the whole structure. This is achieved between neighboring active devices using heavily doped active areas of low resistance and polycrystalline silicon (known as polysilicon). The relatively high sheet resistance of these materials is illustrated for a typical 0.25 μm CMOS process in Table 3.2. Long-distance interconnection of devices is carried out by patterned metal interconnect layers built above the MOS transistors. Insulation between the conducting metal layers, in contrast to isolation within the substrate, is by means of deposited and patterned dielectric layers. As we will see later, among the most important features for a microdisplay are the minimum metal width and metal-to-metal spacing as these are key factors in determining the pitch and aperture ratio for the pixel. Table 3.3 shows the typical minimum metal-1 feature size on a variety of CMOS processes. Wolf[6,7] describes the MOS integrated structure in detail.

3.2.4 CMOS Fabrication Process

A mainstream CMOS fabrication process has three main sequential stages: substrate preparation; active device formation; interconnection and passivation. Within each stage, a sequence of layers (conductor, semiconductor or insulator) are deposited or otherwise formed and patterned.

Table 3.2 CMOS conducting materials and typical parameters for a 0.25 μm CMOS process

Material or feature	Purpose	Thickness (μm)	Sheet resistance (Ω/square)	Deposition method	Minimum feature size (μm)
Source, drain	(depth)	–	50 to 200	Ion implantation	0.35
Silicided active area	Source, drain	–	3 to 10	Ion implantation	0.35
Polysilicon	Gate, local interconnect	0.15	50 to 400	CVD	0.25
Silicided polysilicon	Gate, Local interconnect	0.15	3 to 10	CVD	0.25
Metal 1 aluminum	Local and other interconnect	0.70	0.02	Sputter	0.35
Top metal[a] aluminum	Global interconnect, power rails	2.0	0.01	Sputter	1.0

[a]The early or lower metal layers are typically used for local interconnect, whilst the later or upper metal layers used for global interconnect and power lines tend to be both thicker and wider in order to facilitate higher current density and lower resistance.

Table 3.3 Minimum metal-1 width, gap and pitch (typical values)

Process node (μm)	0.35	0.25	0.18	0.13	0.09
V_{DD} (V)	3.3	2.5	1.8	1.5	1.2
Min metal width (μm)	0.5	0.35	0.25	0.18	0.12
Min metal gap (μm)	0.5	0.35	0.25	0.18	0.12
Min metal pitch (μm)	1	0.7	0.5	0.36	0.24

- *Substrate preparation* includes doping the substrate to the appropriate level at all locations and, in particular, in the vicinity of where active devices will be formed. It also includes the formation of a thick (~μm) oxide layer to cover areas where no active devices will be present. The main techniques used are ion implantation, thermal diffusion and thermal oxidation of silicon.
- *Active device formation* includes thin gate dielectric formation, deposition and patterning of transistor gate and sinking of source and drain regions. The main techniques used in layer formation are as before plus chemical vapor deposition (CVD) for gate deposition. Patterning is carried out by dry etching.
- *Interconnection* involves the deposition and patterning of alternate layers of dielectric and metal using mainly physical and chemical vapor deposition techniques plus sputtering. Historical wet etching has been replaced by dry etching. In the case of advanced processes with multiple metal layers, layer flattening is necessary by chemical mechanical planarization or polishing (CMP).

Maly[5] offers a pictorial description of a basic CMOS fabrication process, whilst Tung *et al.*[8] give a more detailed and quantitative treatment of a more advanced CMOS process.

3.3 CMOS for Microdisplays

3.3.1 Background

In contrast to most conventional LC displays that operate in transmission mode in conjunction with a backlight, all LCOS microdisplays, with the exception of those based on transferred silicon, operate in reflection mode. One ongoing issue with transmission-mode active matrix LCDs, and miniature AMLCDs in particular, is that the pixel real estate must be shared between opaque circuitry and a transparent optical window. The resultant limitation on the size of the optical window as a proportion of the area limits the optical efficiency of the pixel. In a reflective LCOS microdisplay made on an advanced CMOS process with multiple metal layers, the top metal layer has a dual function as both the activating electrode to switch the LC and the reflective optical window. All of the active matrix circuitry is hidden in underlying layers of the CMOS and the only limitation to the size of the optical window is that nearest-neighbor electrode/mirrors must be separated by a small gap to ensure electrical isolation. Microdisplays make particular and unusual (possibly unique) demands on CMOS technology in the following areas.

- *Optical characteristics.* In a microdisplay, the optical as well as electrical characteristics of the upper layers must be carefully considered. The top surface of LCOS CMOS, for example, must exhibit optical quality to specifications of flatness and reflectivity that are beyond those of standard CMOS.
- *Optical defects.* The occurrence of optical defects must be tightly controlled. Due to the high magnification under which microdisplays are viewed, very small defects may become visible in use.

Thus, the production of potentially visible defects must be carefully controlled during manufacture. For example, a defect of 20 μm diameter would be invisible in a conventional LCD but might cover several pixels in a microdisplay and, therefore, be highly and irritatingly visible in a magnified image. Furthermore, an optically significant blemish may not be picked up by electronic testing. For example, a hillock or protruding grain in the top metal layer would have no direct electrical consequence for an IC but may cause an optical defect in an LCOS microdisplay.

- *Exposure to light.* The VLSI design/process combination must minimize the exposure of the substrate to incident light. Whilst most ICs are packaged in such a way as to avoid light ever reaching them, microdisplays must, in order to carry out their function, be exposed to light. Precautions must be taken to avoid any detrimental effects on performance resulting from the exposure to light.
- *Redundancy.* The nature of a microdisplay prevents the use of some standard redundancy techniques to maximize the yield of a backplane. Many silicon chips such as memory chips (that bear some architectural resemblance to microdisplay backplanes) use redundancy techniques to increase yield. For example, surplus blocks of memory may be added to the chip such that if any of the core blocks fails a functional test, the failed block can be substituted by a functional surplus block. In the case of a microdisplay, such techniques are not possible. A surplus row or column at the edge of the pixel array cannot be used to replace a faulty row or column of pixels within the main central area of the pixel array.

The optical aims of a CMOS substrate for a reflective LCOS (or organic electroluminescent layer) microdisplay are as follows.

- *Flatness.* The top surface of the substrate should be optically flat in order to minimize distortion or aberration in reflected light; maximize reflectivity by minimizing scattering of reflected light; facilitate liquid crystal alignment and uniform layer thickness[i]; and allow smooth flow of the LC during the cell filling process.
- *Reflectivity.* The aim is to maximize the amount of light reflected, by maximizing the fill-factor or aperture ratio of the pixel mirror, and reflectivity of the mirror material.
- *Light leakage.* The aim is to minimize light reaching the CMOS substrate where photo-excitation of carriers has a detrimental effect on circuit performance.
- *Electro-optical material.* The substrate must be amenable to the application of an appropriate electro-optical material or device structure such as, for example, a LC alignment layer and well-aligned LC layer or organic conducting and light-emitting layers. This involves controlling the mechanical, chemical and/or electrical properties of the surface.

The ideal CMOS substrate for a microdisplay would exhibit perfect surface flatness over a wide range of lateral dimensions. There should be no wafer bow during or after wafer processing (~20 cm lateral dimension); there should be no backplane warpage after the wafer is diced and no CMP-induced variations due to variations in circuit density across a die (>5 mm lateral dimension); there should be no residual IC topography or inter-mirror trench apparent from the standard CMOS process (~10 μm lateral dimension); there should be no mirror convexity or concavity (~10 μm lateral dimension) and no vertical step between the edge of a mirror and the adjacent dielectric; the mirror surface should be smooth with no hillocks or obtrusive grain structure (<μm lateral dimension); finally the mirror surface should be highly reflective so as to exhibit insignificant absorption or scattering in the visible region of the spectrum (>nm lateral dimension).

---

[i]To an extent, uniformity of LC layer thickness can be achieved independent of absolute flatness provided the inner surfaces of the CMOS backplane and cover glass remain parallel.

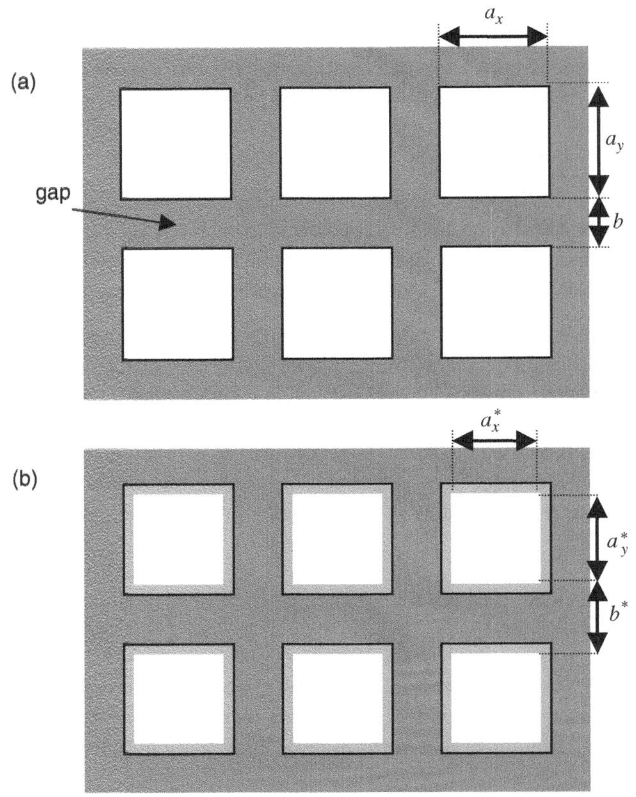

Figure 3.3 Schematic diagram illustrating fill-factor: (a) backplane fill-factor (white area); (b) lower pixel fill-factor including loss due to edge fringe effects

3.3.2 Pixel Aperture Ratio

Pixel fill-factor (PFF), or pixel aperture ratio, F_P, is defined as the fraction of the overall area of the pixel that correctly passes incident light, as illustrated schematically in Figure 3.3. The product of PFF and reflectivity of the pixel mirror surface gives an upper limit to the optical throughput efficiency of a reflective device.

The backplane fill-factor (BFF or F_B), also illustrated in Figure 3.3, is the proportion of the pixel area in the backplane that is occupied by the pixel mirror and is given by

$$F_B = \frac{a_x a_y}{(a_x + b)(a_y + b)} \tag{3.1}$$

where a_x and a_y are the mirror edge dimensions for a rectangular pixel, and b is the distance between neighboring pixel mirrors.

From an optical viewpoint, b should be minimized. However, in any CMOS manufacturing process there is specified a minimum metal-to-metal gap. This "design rule" is defined as the smallest dimension that allows reliable electrical performance of the circuit without *undue* risk of accidental nearest-neighbor short-circuiting. In a 0.35 μm CMOS process (which is typical for a microdisplay in the early 2000s), these would be approximately as shown in Table 3.3. Thus the value of b is set by the CMOS process and is the same in the x and y directions. In the simplified case of a square pixel

$$F_B = \frac{a^2}{(a+b)^2} \tag{3.2}$$

where a is the common edge dimension of the square mirror.

Returning to the general case of a rectangular pixel, the overall reflection efficiency, H_R, of the pixel or pixel array on the backplane is

$$H_R = \eta(\lambda) F_B = \frac{\eta(\lambda) a_x a_y}{(a_x + b)(a_y + b)} \tag{3.3}$$

where $\eta(\lambda)$ is the reflectivity of the pixel mirror surface, which is a function of the wavelength of the incident light as well as various properties of the mirror including the mirror material composition, thickness, surface roughness etc., and the refractive index of the material in contact with the reflecting mirror surface.

In an LCOS microdisplay, the overall pixel fill-factor (F_P) is reduced in comparison to F_B, as illustrated in Figure 3.3. This is due, for example, to fringing fields in the LC layer at the edge of the pixel such that $F_P \leq F_B$. Verweire and Defever,[9] Armitage[10] and De Smet *et al.*[11] independently explore the issues of diffraction loss with increasing pixel count, reducing pixel pitch and electric field fringing at pixel edges. For adjacent pixels at the same voltage, separated by a gap $<0.5\,\mu m$, the fringe field effect is negligible. However, substantial differences in voltage between adjacent pixels (such as are produced by some of the digital addressing schemes described in Chapter 2) produce fringe-field effects, discussed in detail in Chapter 6.

A further factor reducing the overall efficiency of an LCOS microdisplay comes from the attenuation factor, $\upsilon(\lambda)$, of the various layers (glass, ITO, liquid crystal etc.) of the structure as the light passes through in both directions. Then the overall reflection efficiency, H_R^*, of the pixel or pixel array on an LCOS microdisplay is

$$H_R^* = \upsilon(\lambda)\eta(\lambda) F_P = \frac{\upsilon(\lambda)\eta(\lambda) a_x^* a_y^*}{(a_x + b)(a_y + b)}. \tag{3.4}$$

3.3.3 Metal Layer Count

In the early days of LCOS, available NMOS and CMOS processes had only a single metal layer. The implications of this included the following.

- *Low fill-factor.* The pixel mirror had to sit alongside metal interconnect features (both local interconnections and global bus lines).
- *Bus lines across the pixel array.* These could run in uninterrupted metal in only one direction. If metal ran continuously along rows of pixels, then column bus lines had to be run at least partly in other (higher resistance) materials such as polysilicon tracks or diffused silicon (source/drain regions) in the substrate. This meant resistive drops in power rails and *RC* time-constant limitations in signal lines.
- *Optical noise.* This was created when signals on the non-mirror metal structures activated the directly overlying liquid crystal.
- *Light blocking failure.* This was the inability to effectively block light from reaching the silicon substrate.
- *Low reflectivity.* Poor optical quality was typical as the metal layer was optimized primarily for its electrical properties at the expense of its optical properties.

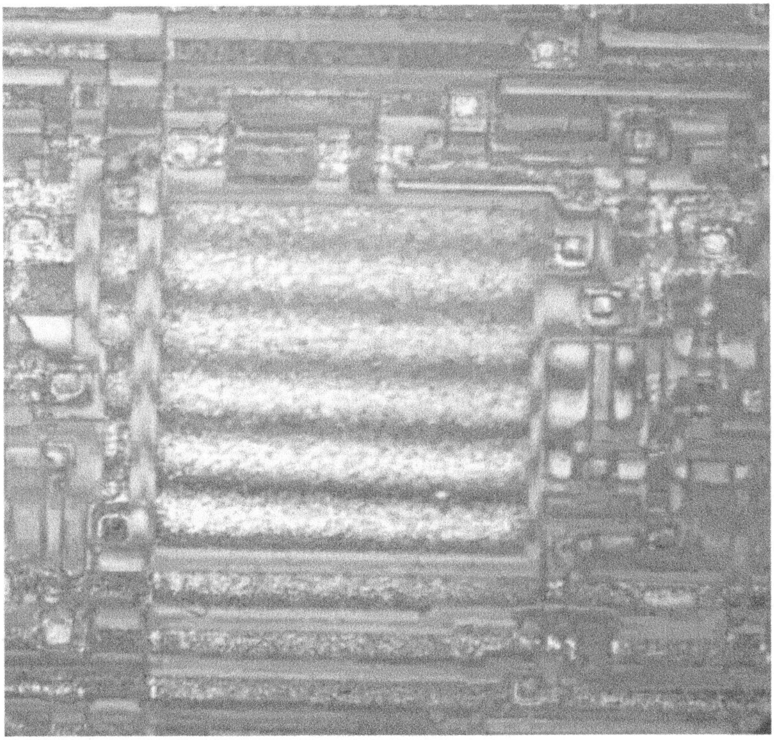

Figure 3.4 White-light Linnik interferogram of one pixel showing the poor optical surface quality of MOS wafers for use in an early LCOS spatial light modulator. Device designed and fabricated at the University of Edinburgh, with backplane technology 6 μm NMOS, pixel pitch 200 μm and the fill factor 30%. Reprinted courtesy of the University of Edinburgh

Whilst it was possible to make liquid crystal spatial light modulators and liquid crystal microdisplays using single-metal NMOS and CMOS, the above factors placed stringent limitations on optical quality, efficiency and electrical performance. The advent of multi-level metallization in CMOS offered a step improvement in both.

This situation is illustrated very clearly in Figure 3.4, which shows a white-light Linnik interferogram of the surface of a single pixel of a vintage single-metal SRAM-based NMOS backplane[12,13] in which a pixel with a 200 μm pitch has a mirror size of $110 \times 110\,\mu m^2$. Any gaps between metal structures represent possible light paths into the substrate. The metal mirrors are noticeably grainy with hillock-like structures on most mirrors. The hillocks cause light scattering, disruption to LC alignment and, in extreme cases, can cause a vertical short-circuit across the LC layer. O'Hara *et al.*[14] describe an early effort to improve the situation, utilizing techniques for fill-factor improvement, planarization and hillock reduction. Colgan and Udal[15] describe subsequent efforts.

Today, a CMOS IC can utilize four or more (up to ten) metal layers making solutions to the above problems relatively straightforward. The move from single metal to two and later more metal layers took place in the mainstream silicon industry as a solution to the problems of (a) reducing *RC* time delays in lengthy interconnects, and (b) interconnecting more and more active devices in a substrate. In parallel, a small number of research groups and companies worked on developing multiple metal solutions for microdisplays. The term "post-processing" was coined to refer to the continued processing of CMOS wafers after completion of the standard CMOS process (sometimes, although not ideally, in a separate factory or foundry). The key difference between mainstream CMOS and LCOS CMOS is the required degree of planarization of the top surface of the processed wafer.

The availability of multiple metal layers allows the use of overlapping metal layers to block light from reaching the active devices in the CMOS substrate. Furthermore, it offers the possibility for the "top" or final metal layer to be dedicated for use solely as a mirror/electrode.

- *High backplane fill-factor.* Mirror/electrode are the only required functions of the top metal layer and so can occupy the maximum possible fraction of the pixel.
- *High pixel reflectivity.* Optimization of the pixel metal properties for optical quality is possible. In a standard CMOS IC, the upper metal layers are reserved for global interconnect and distribution of power rails where the primary requirements are low sheet resistance and high electromigration resistance; these properties are achieved by thick layers that exhibit substantial granularity. In a CMOS backplane for a microdisplay, the top-most metal layer is reserved as the pixel mirror layer that carries very little current so it can be thin with minimal grain size.

As multi-level metallization was developed, the topography on which the upper or later metals were deposited became problematic. As each metal layer is added to the wafer, the features on that layer (tracks and gaps) must be superimposed upon the existing surface whose topography is determined by the patterns of all of the underlying layers. Vertical edges and step heights can build up to the point where the yield and reliability of the higher metal layers are compromised. The primary cause of this is the shallow depth of focus of the lithography tools used in the patterning process.

In order to allow reliable patterning of metal layers under the restriction of very limited depth of focus, it became necessary to flatten or "planarize" the inter-layer dielectric (ILD), deposited on the wafer surface to provide isolation between successive metal layers, prior to the deposition and patterning of each of the later or higher metal layers. Several techniques were developed including resist-etchback, spin-on-glass and electron cyclotron resonance (ECR) oxide deposition. These enjoyed varying degrees of success. However, the most successful technique has been chemical mechanical polishing (CMP). Today CMP is widely used. Sivaram *et al.*[16] offer an early review of chemical mechanical polishing (also called chemical mechanical planarization) whilst the technique is described comprehensively by Steigerwald *et al.*[17]

As previously stated, the flatness requirements of the reflective top surface of the backplane of a microdisplay are more stringent than those of mainstream multi-layer metal CMOS. However, the widespread deployment of CMP in advanced mainstream CMOS processes has made possible the development and spread of LCOS-friendly or LCOS-compatible CMOS processes.

3.3.4 High-Voltage Structures

It can be seen from an examination of Table 3.1 that the typical rail voltage (V_{DD}) of a CMOS process scales in proportion to the geometry of the process. Smaller geometry processes support lower supply rail and signal voltages. This corresponds to maintaining approximately a constant electric field strength within the isolation layers in the electronic structures. The LC materials and configurations typically used in LCOS, coupled with the short switching times required for viewing full-motion video, sometimes require a higher drive voltage than is available from a standard CMOS process. This is particularly the case for field-sequential-color systems (see Chapter 10). In order to support this and other similarly demanding applications, some CMOS processes incorporate optional high-voltage circuit elements. In order to avoid various breakdown mechanisms that could be triggered by high electric field strength, such high-voltage structures tend to use thicker dielectric layers and, more importantly, occupy more real estate than conventional low-voltage structures. In addition, any increase in switching voltage is accompanied by an increase in on-chip electronic power dissipation according to

$$W = CV^2 f \tag{3.5}$$

where W is the electronic power, C is the capacitance being switched, V is the switching voltage, and f is the switching frequency. Thus, where high-voltage circuits are required, it is often the case that power and real estate can be saved by using standard-voltage front-end circuitry with a conversion to high voltage either at the line drivers for the pixel array or even within the pixels.

Some examples of backplanes that utilize high-voltage drive include those of Sanford *et al.*[18] who describe modifications to a 1.2 μm single-poly double-metal 5 V CMOS process to allow 10 V operation, whilst Van Calster[19] describes the use of DMOS[20] structures embedded in the CMOS backplane to give 15 V switching capability.

A constant potential maintained at the ITO counter electrode, which is common to all pixels, requires the pixel mirror/electrode to alternate about that common potential. A liquid crystal cell requiring an upper limit of, for example, 4 V drive (full scale) may require an 8 V swing in pixel potential to satisfy polarity reversal. The pixel transistor then may be required to accommodate 12 V or more, taking into account the potential lost due to the body effect (as described in Chapter 2). The large footprint required by such a high voltage defines a larger-than-otherwise lower limit to the pixel pitch.[19,21] Chapter 2 includes a description of methods of switching the common electrode voltage to reduce the pixel electrode voltage swing, facilitating smaller pixels at high liquid crystal voltage.

3.3.5 LCOS Microdisplays

Even in the days of single-metal CMOS processes, it was possible to make LCOS devices. In the 1980s and early 1990s, LCOS technology was used for spatial light modulators (SLMs) as well as displays. Whilst the illumination for SLMs was frequently coherent, it was also often structured – that is to say small spots of light could be applied to the pixel mirrors with no overspill into the gaps. So, single or double metal was sufficient for the purpose, such as in the examples of Underwood *et al.*[22] and Wilkinson *et al.*[23] The advent of multi-level metallization enabled a major increase in LCOS capability by allowing the top layer of metal to be reserved as a mirror/electrode layer having a very high fill-factor, and optimized for LC alignment and optical properties. Flood illumination of the microdisplay was possible. Additional metal light shields under the pixel electrode gaps allowed optical readout even at projection arc-lamp intensities.

A schematic of an idealized modern LCOS backplane employing DRAM-type pixel voltage store is shown in Figure 3.5. A minimum of one transistor per pixel is required in a DRAM system, along with a pixel storage capacitor (often comprised at least partly of a MOS capacitor) in parallel with the LC capacitance. Pixel pitch in the range 8–12 μm is typical, while 7 μm and below is already projected.

3.4 Wafer and Die Bow

The manufacturing process for CMOS wafers involves the deposition and patterning of a sequence of layers of different materials (with different coefficients of thermal expansion) by different techniques at vastly different process temperatures. Although these layers are thin (typically ~1 μm) in comparison with the wafer substrate, nevertheless thermal strains build up resulting in stress causing wafer bow or warpage. As with many aspects of CMOS wafers, this phenomenon is well known in the mainstream of CMOS but is of more critical importance for LCOS.

Following wafer dicing, the separated dice typically exhibit noticeably greater warpage than pre-dicing. The additional degrees of freedom in a separated die allow further deformation of the die in response to surface stresses accumulated in fabrication. This is one reason in favor of carrying out wafer scale LC cell fabrication prior to dicing as discussed in Chapter 7.

In a transmissive LC device, uniformity of LC cell thickness is prioritized over substrate flatness. Small curvatures of the substrate have little effect as suggested in Figure 3.6. In a reflective LC device, both flatness and uniformity of LC cell thickness are important. Substrate bow can cause significant aberrations in the reflected light.

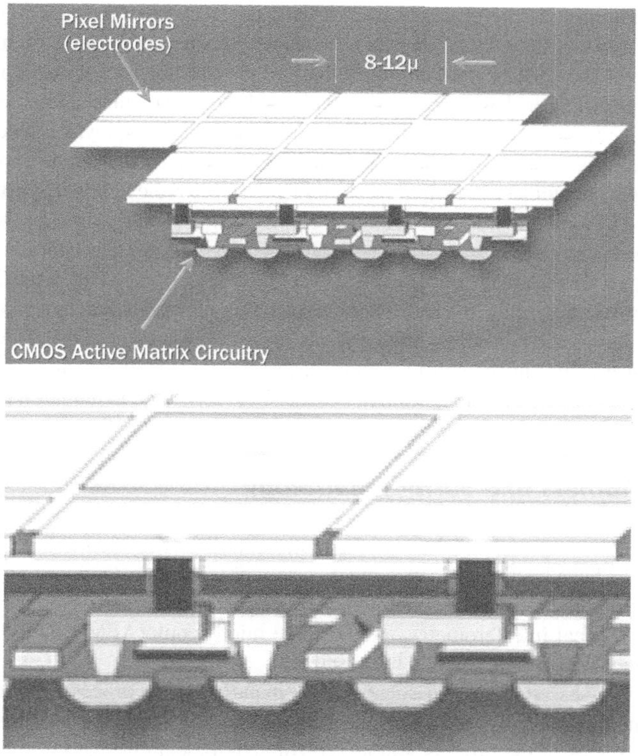

Figure 3.5 Pseudo 3-D schematic of LCOS backplane showing pixelated surface and cross-section of underlying CMOS circuitry. Reprinted courtesy of Brillian Corp.

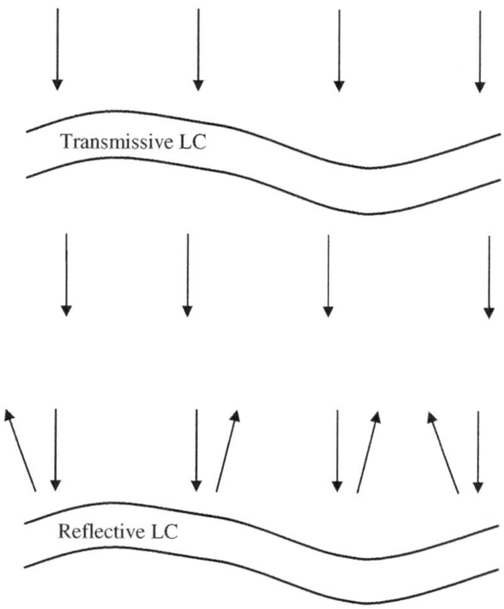

Figure 3.6 Qualitative comparison of the effect of wafer bow in transmissive and reflective LC configurations

Several strategies minimize the problems caused by wafer warpage. Designed-in dummy structures within the various layers of the CMOS substrate minimize the stress. Dummy structures are structures within one or more of the CMOS material layers that are present for mechanical, optical or other purposes, but fulfill no electrical or electronic function. This method often requires a fairly empirical approach and several mask set revisions. Post-processing techniques have been employed to counter the warpage. Miremont *et al.*[24] describe a technique that involves depositing a patterned oxide layer on the back surface of the wafer in order to counteract a known stress. Bonding the CMOS die tightly to a flat substrate is an approach that was used in some early devices prior to the introduction of wafer scale assembly. However, the flattening process precedes cell assembly making it too expensive. Finally, to accept some level of warpage then match the warpage of a relatively thin, flexible cover glass, conformal to that of the CMOS substrate so that, even if the reflective face of the microdisplay structure is not perfectly flat, nevertheless the cell thickness is uniform. However, the conformal approach has the disadvantage of internal spacers and requires negative differential pressure in the cell to assure spacer contact. Manufacturing cost and yield favors adequate flatness in CMOS wafers and optimizing design and fabrication to maintain flatness.

3.4.1 Wafer Flatness and Surface Metrology

The figures used in this chapter, and in the source material to illustrate the surface quality and flatness of LCOS backplanes, have been produced using a variety of techniques including mechanical surface profilometry, scanning electron microscopy (SEM), optical interferometry, optical interference microscopy, scanning optical interferometry, atomic force microscopy (AFM), focused ion beam (FIB) cross-sectioning, and conventional optical microscopy.

Surface quality is assessed at several stages of the manufacturing process. Optical interferometry is a non-contact method used to assess surface flatness, starting with the virgin silicon wafer, and repeated at critical stages of the process. Manufacturing equipment designed for on-line testing operates with minimum interruption to the process flow. However, any handling procedure is apt to increase surface defects. Upon attaining adequate manufacturing yield it is cost effective to limit testing to quality assurance tests that identify the stage responsible for any unexpected dip in yield.

Wafer bow or even die bow is a large-scale property, that must be kept to a minimum during the manufacturing cycle, but small-scale defects are also critical and require other test methods. Dark-field microscopy often reveals small-scale surface defects. Damage to the wafer surface contributes to electronic defects in the active matrix addressing circuits, and in the LCOS optical readout.

Walker and Handschy[25] utilized optical interferometry on existing microdisplay devices to conclude that "from wafer to wafer the fringe pattern is the same" for a given backplane design fabricated on a given CMOS process. But the prediction of the level of bow likely on a new backplane design or process is not an exact science.

The output from the CMOS foundry is a processed silicon wafer containing many dice, manufactured according to the specification of the chosen CMOS process (e.g. 0.35 μm 3-metal mixed-signal twin-well CMOS) and containing circuits according to the design specification of the particular LCOS backplane. A typical CMOS wafer of diameter 200 mm may contain from less than 100 to upwards of 1000 backplanes.

Standard foundry practice facilitates the electronic circuit assessment of the finished wafer by probe station tests. LCOS also requires optical surface inspection to assess flatness, surface defects and staining before shipping the product. More details on assembly and testing are given in Chapter 7.

3.5 Wafer Surface Planarization

3.5.1 Introduction to Wafer Planarization

As previously stated in section 3.4, the manufacturing process for CMOS wafers involves the deposition and patterning of a sequence of layers of different materials. Although these layers are thin (typically

~1 μm), nevertheless in the absence of techniques to minimize the effect, significant height differences can build up between sites where most of the layers are present and adjacent or other sites where most of the layers have been etched away. As with many aspects of CMOS wafers, this phenomenon is well known in the mainstream of CMOS but is of critical importance for LCOS.

In mainstream CMOS the primary concern is to ensure that the maximum height difference does not exceed the depth of focus of the wafer stepper that images the pattern for the next level of material. Otherwise the image quality of the pattern will be reduced leading to a reduction in the faithfulness of reproduction of the pattern in the mask with a consequent reduction in yield and/or performance of the final circuit. This issue arose in particular with the development of multi-level metallization processes in the 1990s, necessitating the planarization of the inter-metal dielectric layers. Note that in mainstream CMOS it is the planarity or otherwise of the intermediate layers that is of primary concern. In LCOS CMOS, the planarity of the final metal layer is most important, as that layer constitutes the pixel mirror/electrodes for the microdisplay. The degree of planarity achieved in the intermediate layers is only of concern for the same reasons as in mainstream CMOS and for any effect a lack of underlying planarity may have on the planarity of the top surface.

The primary concerns arising from non-flat surface topography in an LCOS substrate are long and short-range variation of LC cell gap within the active area, variation of LC alignment and scattering in the reflected wavefront from small prominent structures.

The purposes of planarization techniques[26] include, as shown in Figure 3.7, *gap filling* that involves filling the "gaps" or trenches that exists between adjacent features in a given layer, for example, between adjacent metal lines separated by the minimum design rule dimension. *Topography smoothing* is the process of turning sharp vertical steps in the surface into smooth sloping transitions. *Step height reduction* intends to reduce the vertical distance between the highest and lowest surface features.

The degree of planarization achieved is quantified by two parameters as shown in Figure 3.8: the relaxation distance R and the angle θ. Perfect planarization would give $\theta = 0$ and $R = \infty$.

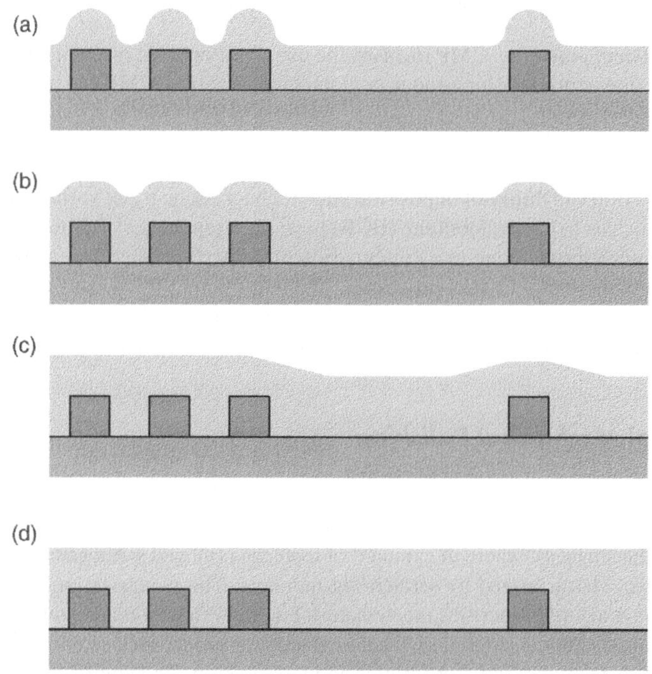

Figure 3.7 Scales of dielectric planarization shown in schematic cross-section: (a) no planarization; (b) gap-fill; (c) local planarization; (d) global planarization

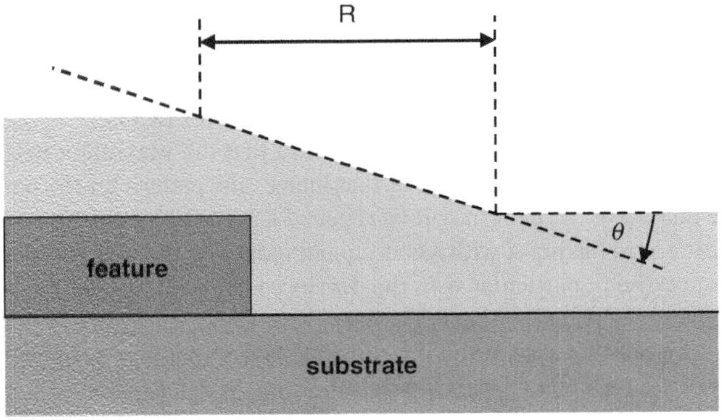

Figure 3.8 Schematic cross-section illustrating parameters allowing a quantitative definition of planarization. R is the relaxation distance; θ is the planarization angle

A number of potential planarization techniques have been investigated, including the following.

- *Spin-on glass.* This involves the use of material that is spin-coated onto the wafer in a liquid state and then subsequently solidified. Surface tension in the liquid state favors a planar surface. It fills or partly fills gaps and smoothes sharp steps. It works best for short-range planarization and for a small number of metal layers.
- *Deposition and etchback.* As the name suggests "deposition & etchback" techniques such as electron cyclotron resonance plasma enhanced chemical vapor deposition (ECR-PECVD) combine the processes of depositing new material and etching existing prominent features either sequentially or simultaneously. The primary use of this technique is for gap filling.
- *Chemical mechanical polishing.* CMP involves the use of wafer-scale mechanical polishing assisted by chemical reaction to produce long-range or global planarization. CMP can be applied to dielectric or metal.

The effective range and angle of each of the above is summarized in Figure 3.9. In the context of LCOS, the planarization capability of deposition and etchback alone is not sufficient. Attempts to use spin-on materials such as benzocyclobutane (BCB) have been reported.[27,28] However, dielectric CMP is the technique of choice today due to a combination of its performance and its high level of adoption in mainstream multi-metal CMOS processes. The latter point allows the LCOS planarization to be an integral part of the LCOS CMOS process in the foundry rather than, as was sometimes the case in the early days, a completely separate post-processing step carried out in a different foundry.[29]

3.5.2 Chemical Mechanical Polishing

CMP of dielectrics is a complex process whose full explanation is beyond the scope of this book. Tomazawa[30] and Cook[31] offer descriptions of the process. Abrasive polishing has long been used to fabricate optically flat surfaces, where the amount of material removed is not critical, and some surface damage is acceptable, characterized by scratch/dig numbers. The planarization polish of a dielectric insulating layer cannot afford to sacrifice much material without risking inadequate insulation, and surface damage is an important consideration. Chemical etching avoids surface damage, and gives better control over loss of surface material, but little or no planarization. The combination of weak abrasion and weak chemical erosion in CMP gives planarity with a minimum of surface damage, and adequate

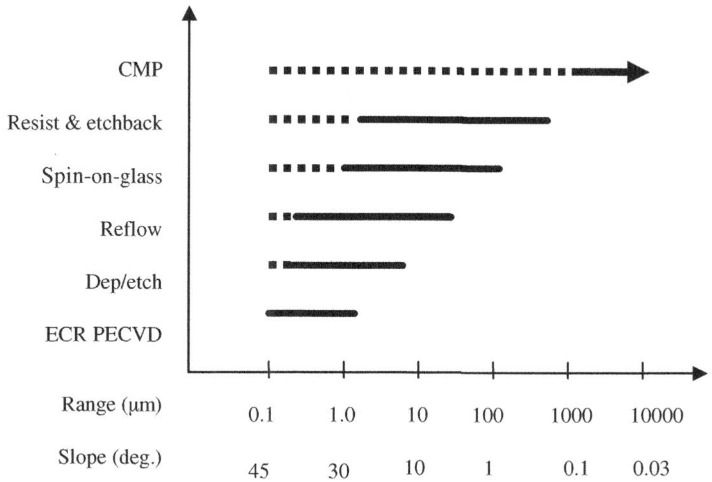

Figure 3.9 The effective range and planarization angle of various planarization techniques

control over material removal rate. Figure 3.10 shows a schematic diagram of a CMP tool, whilst Figure 3.11 is a photograph of a Presi Mecapol E460 CMP tool.[32] A process flow (based on schematic cross-sections) for planarization by CMP followed by deposition and patterning of an additional metal layer is shown in Figure 3.12.

The benefits of CMP are illustrated by example. Figure 3.13 shows cross-sections (of lateral dimension one or two pixels) of an LCOS CMOS backplane at the end of a conventional two-metal non-CMP process, and following the post-processing with CMP of an additional metal layer. Figure 3.14 shows

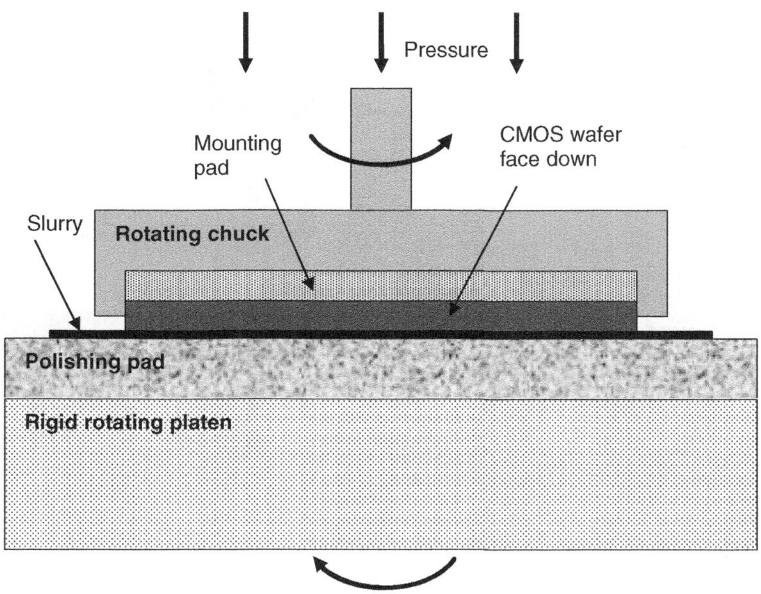

Figure 3.10 Schematic diagram of CMP tool

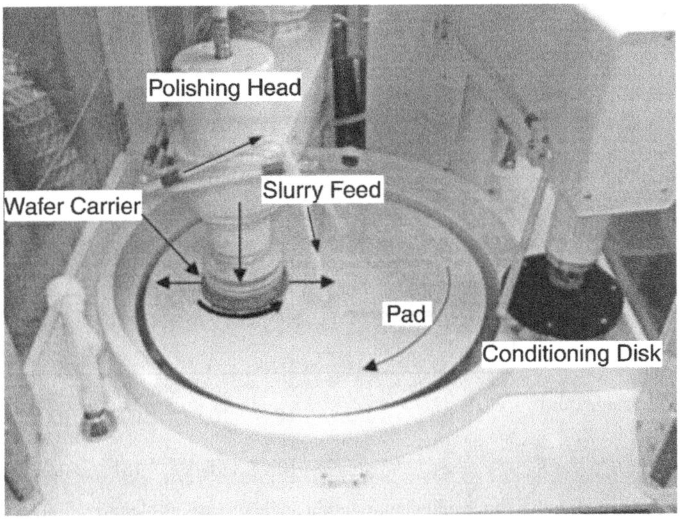

Figure 3.11 Photograph of CMP tool. Reprinted courtesy of the University of Edinburgh

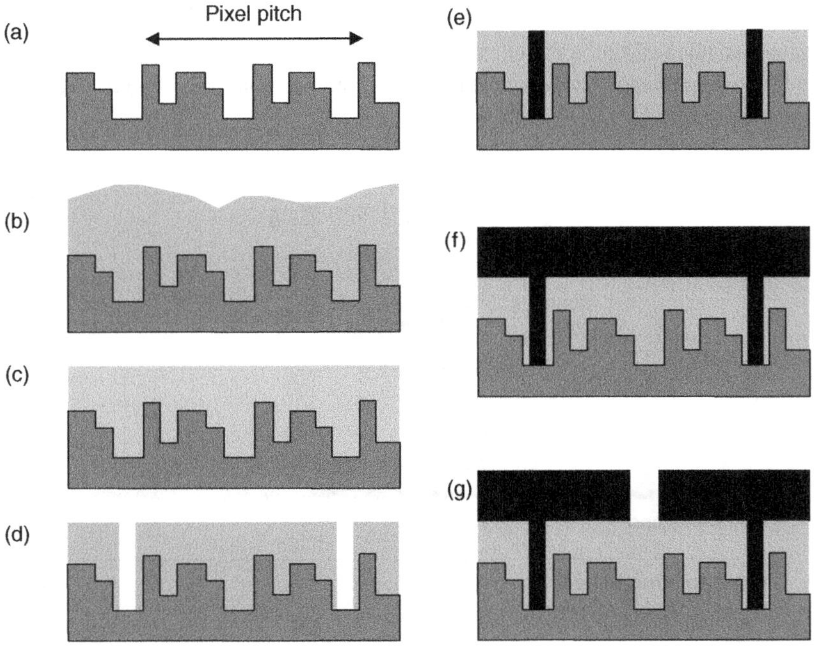

Figure 3.12 Flow of a chemical mechanical polishing process for microdisplay backplane planarization showing schematic cross-section at each stage. For clarity, the vertical dimension has been exaggerated. Pixel pitch is of the order of 10 μm whilst vertical dimensions are of the order of 1 μm. (a) Underlying substrate with existing circuit features. (b) Conformal dielectric deposition. (c) Dielectric polished flat by CMP. (d) Via hole etched. (e) Via plug deposited and polished flat. (f) Metal deposited. (g) Metal patterned to define pixel mirrors

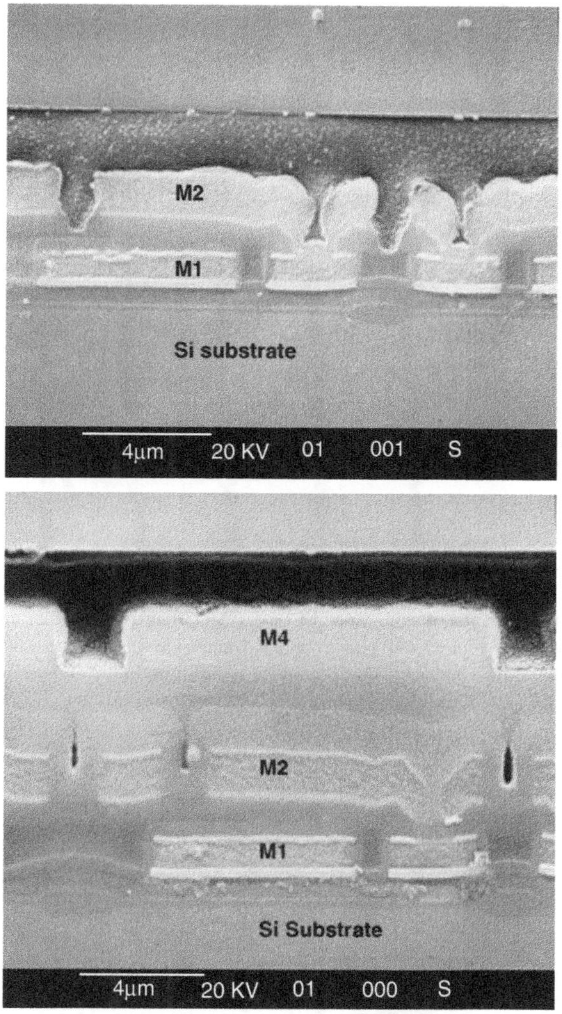

Figure 3.13 Scanning electron micrographs showing cross-sections through LCOS CMOS backplanes. (a) Backplane with two metal layers fabricated at foundry and no CMP planarization. (b) Backplane with three metal layers. This is the metallization present on Devices 76 and 83 in Figure 3.25. Inter-metal dielectric between M2 and M4 was planarized by CMP post-processing at separate foundry from standard CMOS foundry. Light shield metal M3 (see Section 3.7 and Figure 3.24) is not present. Reprinted courtesy of the University of Edinburgh

optical micrographs of a single-metal LCOS CMOS backplane that illustrate the small fill-factor with a single metal layer and the higher fill-factor after CMP, deposition and patterning of metal 2. Figure 3.15 contains scanning electron micrographs of a single-metal LCOS CMOS backplane[33] subsequently post-processed with CMP-oxide and an additional metal layer. A white light Linnik interferogram of the backplane of Figure 3.15 is shown in Figure 3.16. Upper and lower pictures in Figure 3.17 correspond respectively to Figure 3.15(a) and Figure 3.15(b) devices and show a binary optical image on the LCOS device after it is filled with liquid crystal. The optical and image quality benefits of the higher fill factor and improved mirror flatness are obvious.

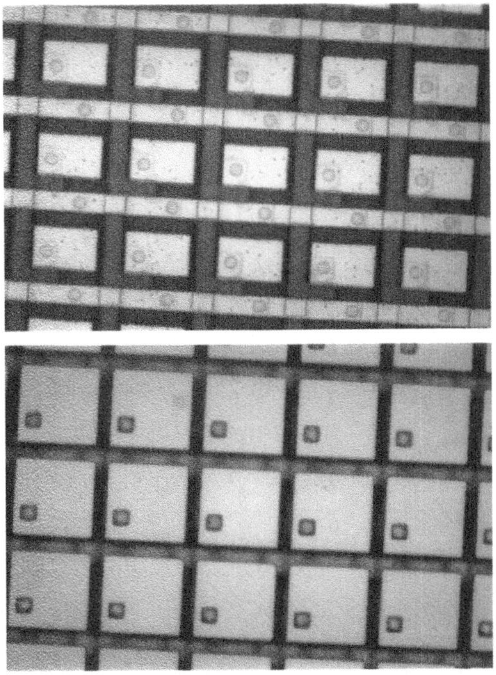

Figure 3.14 Optical micrographs of single-metal LCOS CMOS backplane before and after deposition and patterning of additional metal layer with CMP planarization. The device was designed, fabricated and post-processed at the University of Edinburgh. The backplane technology is 3 μm single-metal p-well CMOS with a DRAM style pixel and a pixel pitch of 30 μm. Reprinted courtesy of the University of Edinburgh

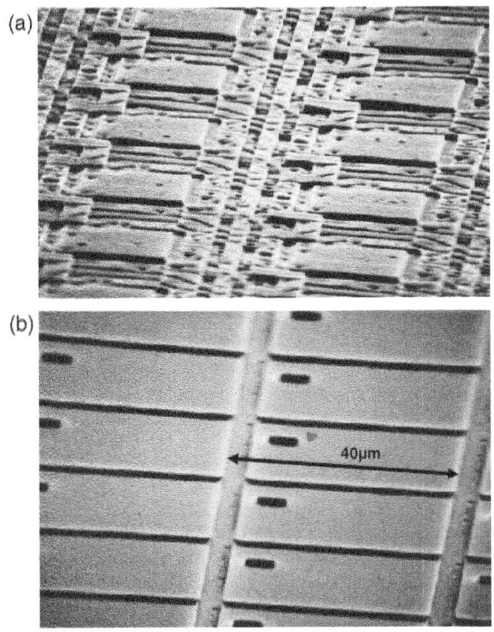

Figure 3.15 Scanning electron micrographs of LCOS CMOS backplanes illustrating the benefits of CMP. This device was designed and fabricated at the University of Edinburgh. The technology is 1.2 μm single metal CMOS with an SRAM-XOR pixel, the pixel pitch is 40 μm. (a) Original single-metal solution shows low fill-factor and poor metal surface quality with visible hillocks. (b) Post-processed wafer exhibits much higher fill factor and improved surface smoothness and flatness. Reprinted courtesy of the University of Edinburgh

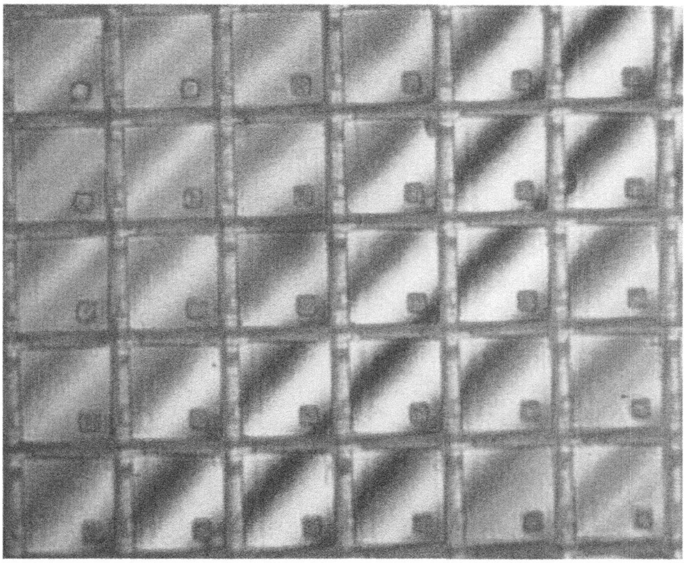

Figure 3.16 White-light Linnik interferogram of early post-processed CMOS backplane showing relatively high fill-factor and good optical quality of mirror/electrode. Note the via dimple in the corner of the mirror (as this process pre-dates via filling techniques) and the reflectivity of the patterned metal lying under the gaps which should be minimized in practice. Reprinted courtesy of the University of Edinburgh

(a)

(b)

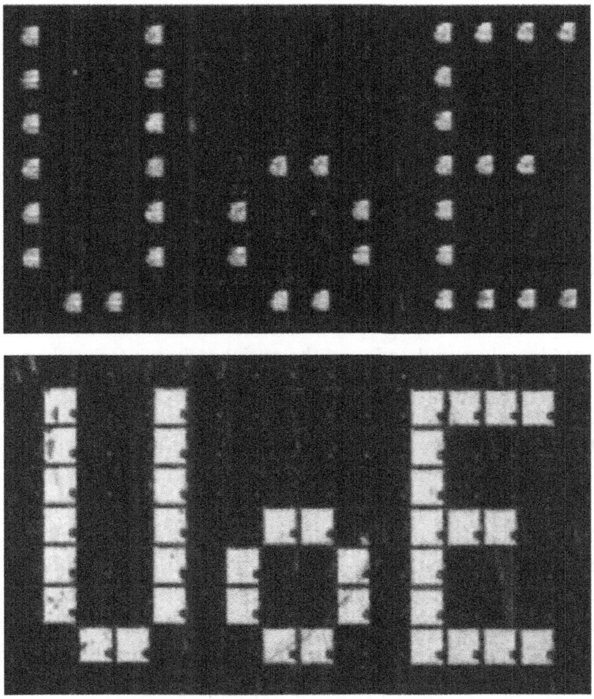

Figure 3.17 Photomicrographs of working LCOS devices that illustrate the benefits of CMP. Devices were designed and fabricated at the University of Edinburgh. The technology is 1.2 μm CMOS, the pixel pitch is 40μm. The LC is BDH E7 in the hybrid field effect configuration. (a) Original single-metal solution shows low fill-factor and poor uniformity. (b) Post-processed wafer exhibits much higher fill factor and improved surface uniformity. Note the black spot in each pixel caused by the via-hole seen correspondingly in Figure 3.14. Reprinted courtesy of the University of Edinburgh

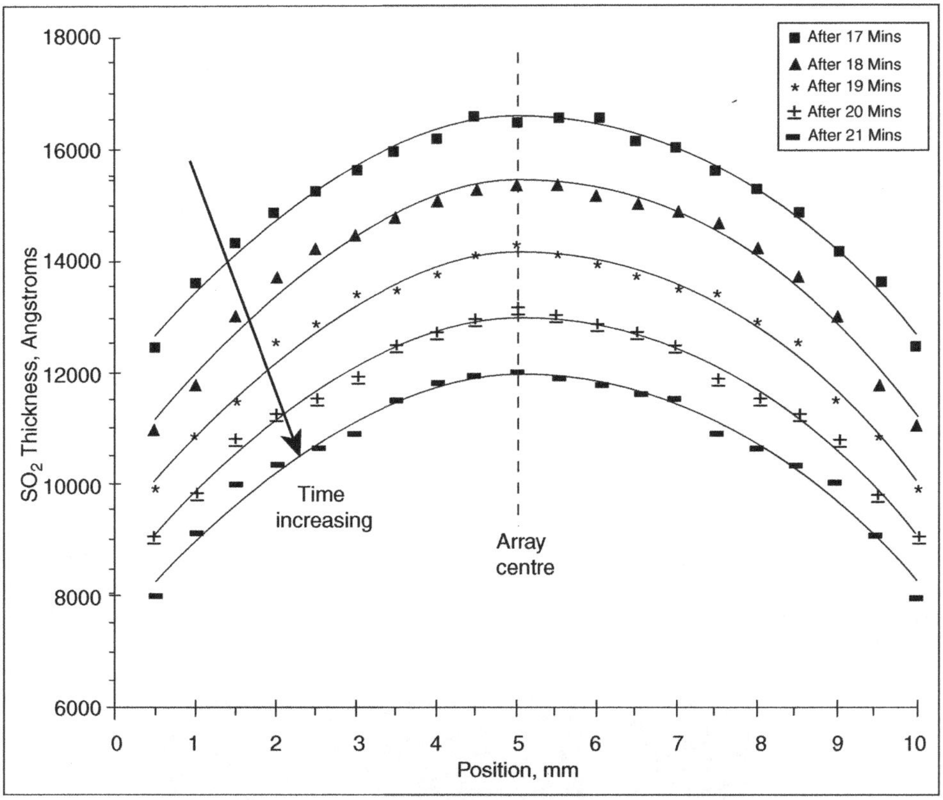

Figure 3.18 Measured SiO_2 thickness over the pixel array in a CMOS backplane after various polishing times. Reprinted courtesy of the University of Edinburgh

The ability of a CMP process to perfectly planarize a surface is hampered by sensitivity to the underlying circuit layout, in particular the circuit density and the topography as illustrated in Figure 3.7(c). Differences in circuit density and topography can be severe in an LCOS backplane, for example, between the pixel array (densely populated) and the peripheral area (sparsely populated). The effect of such differences is shown in Figure 3.18 in which the oxide above the pixel array takes on a dome-like shape after CMP. Chen *et al.*[34] demonstrate the same issue.

A process-oriented compensation technique was developed by Seunarine *et al.*[35] It involves selective etching of the oxide plateau above the pixel array prior to the CMP process step. A process-flow is shown in cross-section view in Figure 3.19. Figure 3.20 shows the resultant surface-profile of a planarized backplane for various amounts of etchback indicating that the etchback technique can result in an adequate level of planarization.

A design-oriented compensation technique[36] that has been adopted in modern microdisplay backplanes, in order to avoid custom process steps, is to design electrical dummy structures that fill gaps in the electronic circuitry in such a way that all parts of the backplane contain approximately equal topography and density of underlying circuitry. This optimizes the degree of global planarization achieved by CMP. Today, this is done routinely and automatically in CMOS foundries in a manner that is typically invisible to the designer. In the case of an LCOS design, it is important for the designer to be aware of, monitor and perhaps actively control the use and effect of such structures in order to avoid detrimental side-effects for LCOS optical performance.

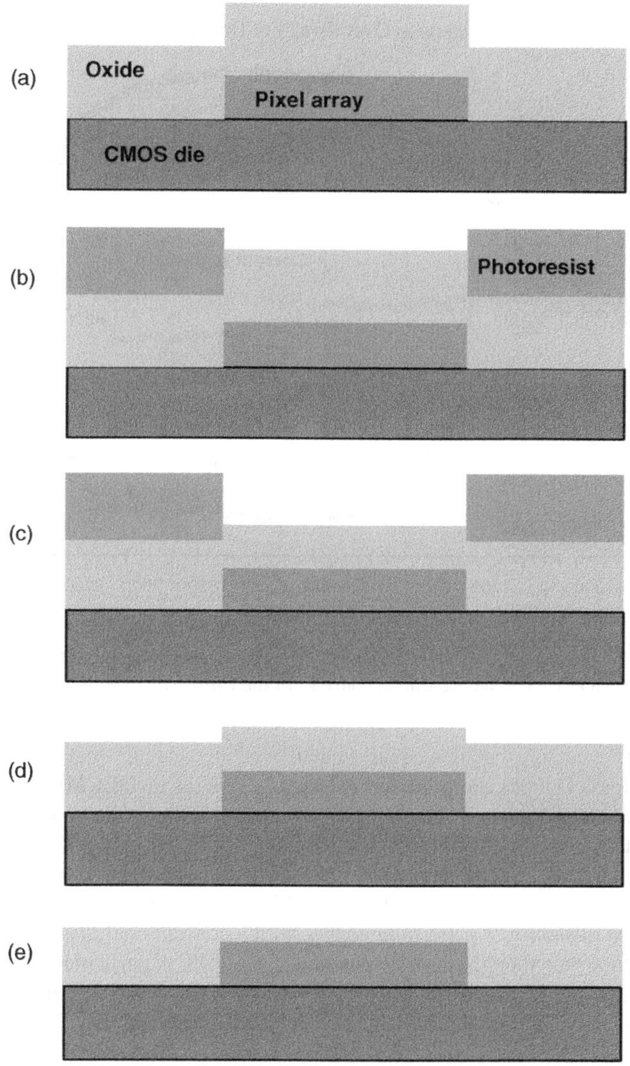

Figure 3.19 Process flow for CMP etch back process. (a) Deposited oxide makes platform above pixel array. (b) Pattern photoresist to isolate area above pixel array. (c) Partially etch exposed oxide. (d) Remove photoresist. (e) CMP

In the case of either compensation technique it is likely that case-specific customization of the compensation technique will be necessary in order to optimize the eventual flatness of a given backplane design.

3.5.3 Damascene Polishing

"Damascene" is a term inherited from the art of working precious metals into grooves cut in base metal, followed by polishing to create an interesting pattern, which originated in 16th century Damascus.

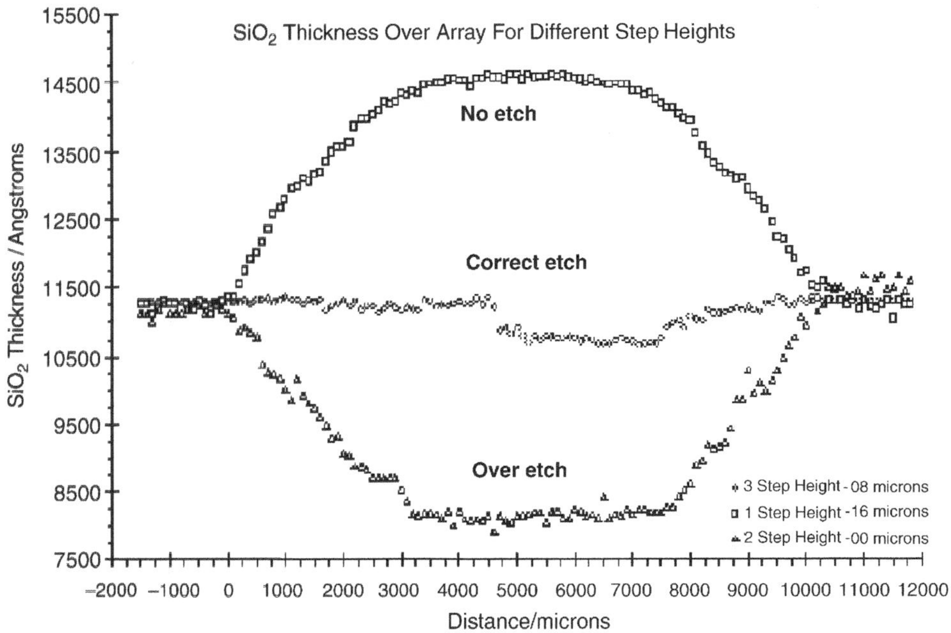

Figure 3.20 Results illustrating the use of etchback technique to minimize the dome effect of CMP oxide over the pixel array in a CMOS backplane. Reprinted courtesy of the University of Edinburgh

In silicon technology, Damascene polishing refers to a type of metal CMP, employing a recess patterned into a dielectric layer. In plan view, the recess pattern is that of the required interconnect. Metal is blanket-deposited so that it sits proud of the recess and subsequently polished back until flush with the surrounding dielectric. The process flow is shown using schematic cross-sections in Figure 3.21. Damascene polishing is widely used in advanced CMOS processes for the patterning of copper interconnects due to the difficulty of patterning copper with a conventional dry etch process. At one time, Damascene polishing was thought to show potential for LCOS. A particular advantage is the inherent co-planarity of the mirror surface and inter-mirror gap surface at the end of the polishing step. This

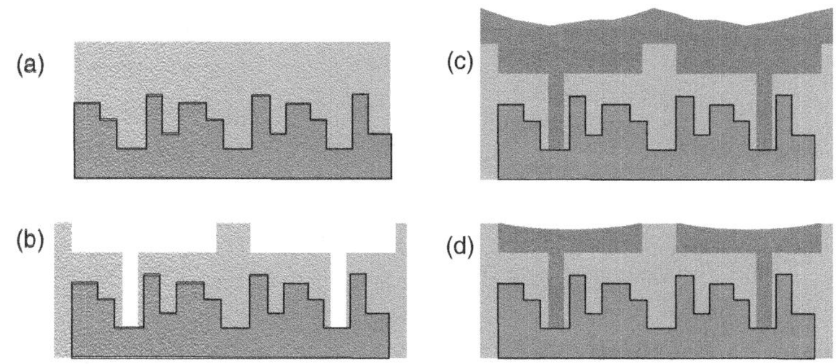

Figure 3.21 Process flow for Damascene polishing. Starting point is equivalent to Figure 3.11(c). (a) Dielectric deposited and polished flat. (b) Mirror recess etched then via hole etched. (c) Via filled then metal blanket deposition. (d) Metal polished flush with dielectric surface

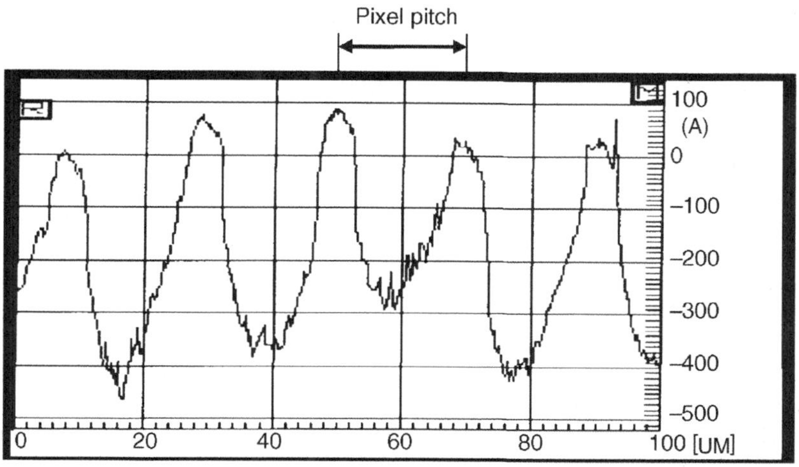

Figure 3.22 Surface trace illustrating mirror dishing on Damascene polished backplane. Reprinted courtesy of the University of Edinburgh

contrasts with conventional CMP followed by pixel metal, which leaves a recessed gap that requires an additional gap-fill process. However, practical difficulties including dishing of the pixel mirrors as illustrated in Figure 3.22 (arising from their relative softness in comparison with the inter-pixel dielectric) and polishing striations on the mirror surface have prevented the widespread use of Damascene polishing on LCOS wafers.

3.6 Pixel Storage

The active matrix addressing methods presented in Chapter 2 show that a charge is typically stored on a capacitance at the pixel between refresh cycles, creating a voltage. In the case of LCOS, the voltage directly generates an electric field to drive the LC. In DMD microdisplays, electrostatic attraction causes movement of a micro-mirror. In the case of OLED microdisplays, the voltage is converted to a current that drives the OLED. Static storage uses a bistable circuit to set the pixel to a well-defined state according to the addressing signal, usually by tying or shorting the pixel electrode to a fixed potential, but is inherently restricted to binary values or logic levels – in display terms to two states, "ON" and "OFF." Single pulse width modulation (S-PWM) or binary-coded pulse width modulation (B-PWM) is then required in order to achieve grayscale. Static storage is not very sensitive to light-induced charge leakage, as the pixel electrode switches to a potential derived directly from the power supplies of the LCOS backplane, implying robustness of stored signal integrity.

Dynamic storage schemes utilize capacitive storage of an analog or digital voltage between refresh cycles. The liquid crystal cell presents a capacitive load to the pixel and acts as the DRAM storage capacitance to some extent. However, the normalized capacitance (or capacitance per unit area, C_0) of the LC is small and voltage dependent, while suffering charge leakage by residual ionic conductivity in the liquid crystal. Furthermore, some LC materials such as ferroelectric (FLC) make for additional complications. Additional capacitance is necessary to maintain the voltage level and attenuate cross-talk through parasitic coupling. The total storage capacitance at the pixel must provide a discharge time-constant that is much greater than the frame refresh period. It is essential to minimize isolation transistor leakage and LC ionic leakage to achieve a high voltage-holding ratio at the pixel. Any voltage dependence of the storage capacitance is not significant, provided the capacitance is always high enough to maintain a voltage-holding ratio approaching 100%.

The capacitance of the liquid crystal in a single pixel, $C_{\mathrm{LC,\,pix}}$, is given by

$$C_{\mathrm{LC,pix}} = \frac{\varepsilon_0 \varepsilon_{\mathrm{LC}} A}{d} \tag{3.6}$$

where A is the area of the pixel electrode, ε_0 is the permittivity of free space, $\varepsilon_{\mathrm{LC}}$ is the (orientation dependent) relative permittivity of the LC, and d is the spacer height or cell gap.

The area is limited by small pixel dimensions, and higher capacitance is usually achieved by reducing d, although silicon nitride provides an increase in dielectric constant over silicon oxide.

There exist a number of capacitive elements or options in a standard CMOS process that may be used for charge storage in a DRAM-style LCOS pixel.

- *MOS (gate oxide) capacitance* (C_G). The gate oxide is thin and reliable, since it determines the transistor performance, and provides the most effective route to high storage capacitance with low leakage. This offers a very high capacitance per unit area whose value changes in a nonlinear fashion with gate voltage.
- *Source/drain to substrate reverse bias p–n junction capacitance* (C_{PN}). This offers a high capacitance per unit area that increases nonlinearly with voltage.
- *Poly/poly-capacitance* (C_{PP}). Analog and mixed-mode (analog/digital) processes often have two polysilicon layers between which precise fixed parallel-plate capacitances may be defined.
- *Row/column line overlap capacitance* (C_B). The row and column address lines run through the pixel array and, depending on the nature of the propagating signals, may represent a source of parasitic capacitive coupling or offer some potential as designed-in capacitance.
- *LC cell capacitance* (C_{LC}). This is the parallel plate capacitance between the pixel electrode/mirror and the transparent counter-electrode of the LC cell through the LC medium.

Figure 3.23(a) shows a schematic cross-section through the active CMOS part of a DRAM-style pixel, which has been designed to use primarily C_G. Figure 3.23(b) labels the electronic components of the equivalent circuit diagram. Figure 3.23(c) shows most of the above capacitive elements. A charge storage node in a DRAM-style pixel is likely to use a combination of more than one of the above.

Whilst the specialist three-dimensional high-capacitance small-footprint structures used in mainstream DRAM cells, such as trench capacitors, would be very appropriate for DRAM-style LCOS pixels, they are not available in standard CMOS or LCOS CMOS processes, and DRAM-specific processes have typically say two metal layers and are thus unsuitable for modern microdisplays.

3.7 Light Blocking

As stated previously in section 3.2, the CMOS structure makes use of reverse-biased p–n junctions to isolate independent nodes in electronic circuits. Furthermore many LCOS microdisplays use soft-node storage at the pixel level, the so-called DRAM pixel architecture, such as that detailed in Chapter 2, in which reliable performance depends upon charge being stored robustly and effectively on a circuit node that contains one or more of these junctions.

Should the isolation properties of any junction be compromised at a storage node, the stored charge leaks away too quickly, corrupting the electronic drive signal and potentially degrading the optical image data at that pixel. The overall result is likely to fall between loss of contrast in the image and partial or complete corruption of the image. In extreme cases, latch-up[37] may be induced in a CMOS circuit leading to irreversible damage to the backplane.

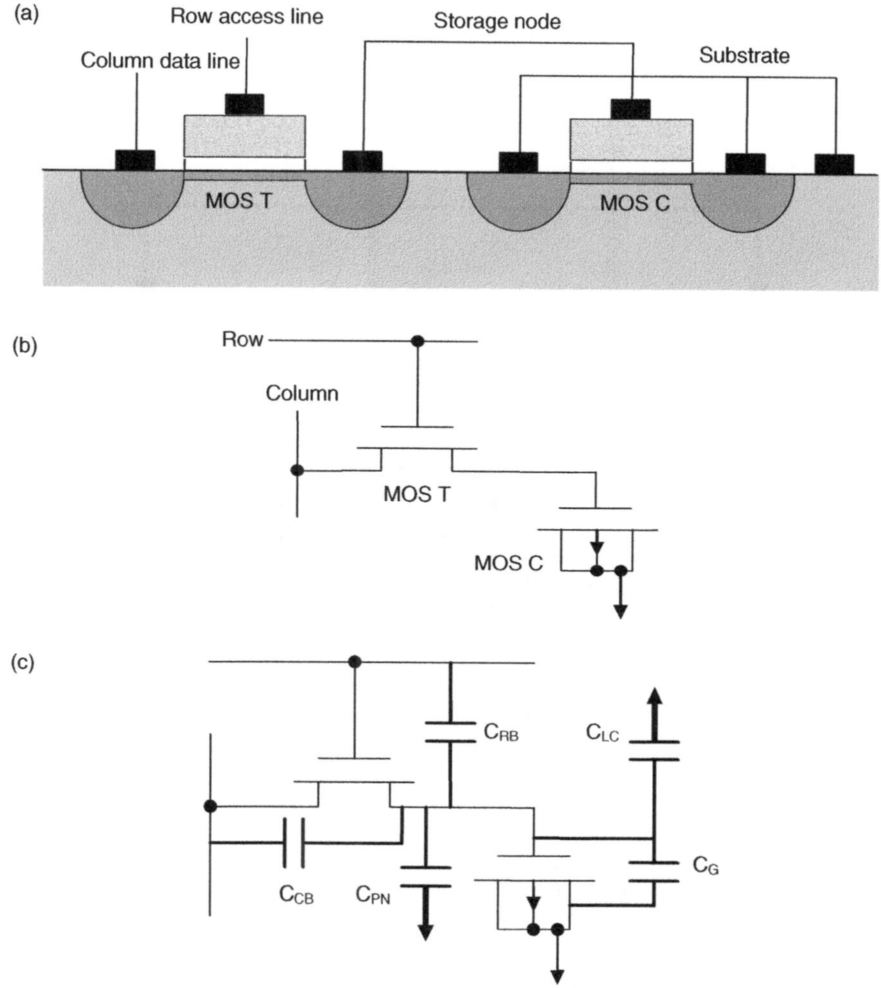

Figure 3.23 Schematic cross-section and circuit diagram of active devices in LCOS CMOS pixel. See text for explanation

In the context of microdisplays, one important mechanism for charge leakage across the p–n junction is the absorption of light. Other optoelectronic devices,[38] such as CMOS image sensors, make use of the current generated by the absorption of photons in a p–n junction in order to convert an optical image into a signal in the electronic domain. A detailed treatment of the physics[39] of the absorption mechanism is not attempted in the context of this book. We restrict the discussion to describing why the absorption of light should be minimized and how to achieve that.

The main key to the prevention is that modern CMOS processes use multiple levels of metal interconnect. This allows the metal features in consecutive layers to overlap such that there are no gaps through which light, incident on the top of the CMOS, can be transmitted directly into the substrate. With careful design, the metal layers can be made to overlap such that there are no direct light paths and that any light path to the substrate is significantly convoluted. This is illustrated schematically in Figure 3.24 and described in detail by Sanford *et al.*[18] Furthermore, the small dielectric thickness

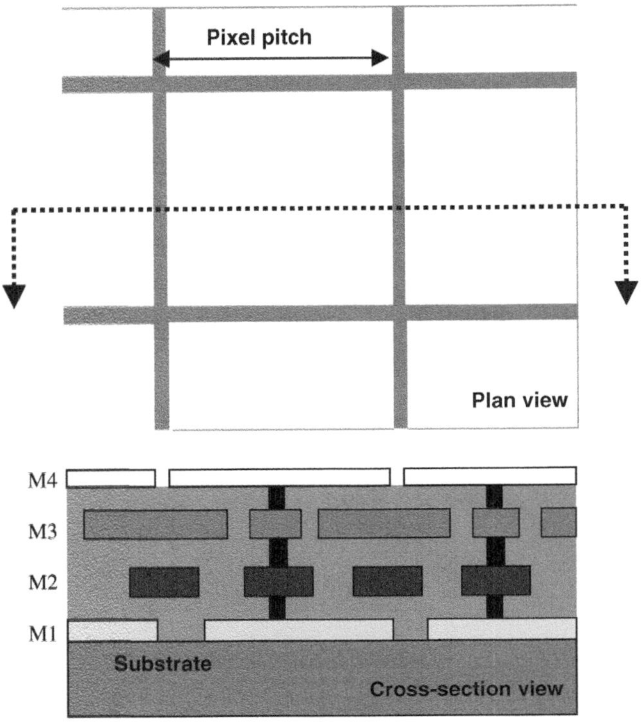

Figure 3.24 Schematic plan view and cross-section of part of pixel array showing multiple metal levels used for light blocking (see also Figure 3.13)

between consecutive metal layers is typically below the cutoff wavelength for optical propagation in a planar waveguide, leading to rapid attenuation of optical radiation.

Figure 3.25 shows an early result of the performance benefits of adding a light-blocking metal layer. This result was taken from the SLIMDIS[40,41] microdisplay. The SLIMDIS backplane used a DRAM-style pixel circuit, digital voltage and pulse code modulation to drive a chevron-structure ferroelectric liquid crystal; the backplane utilized four metal layers as follows: metal 1 – row bus line; metal 2 – column bus line; metal 3 – light blocking; metal 4 – pixel mirror. The devices fabricated without the light-blocking layer in Figure 3.24 show severe deterioration of contrast with increasing intensity of incident light, while those that have the light-blocking layer show no such deterioration in contrast.

Further measures that have been used to minimize any adverse effects of the light that is incident on the backplane include those below.

- *Absorbing coating* of intermediate metal surfaces in order to attenuate light undergoing multiple reflections. This is part of the standard CMOS process in order to avoid spurious reflections during the photolithographic patterning of multiple metal layers.
- *Absorbing dielectric layers* as described by Huang *et al.*[42] and Colgan and Udal[15] in order to attenuate light traveling within the dielectric layers.
- *Guard rings and substrate taps* that collect or channel stray charge carriers away from storage nodes and other problem areas.
- *Single-well CMOS processes*, ensuring that key storage nodes (e.g. pixel capacitors) lie within the well rather than the substrate. The mean free path of photons within the CMOS substrate is typically

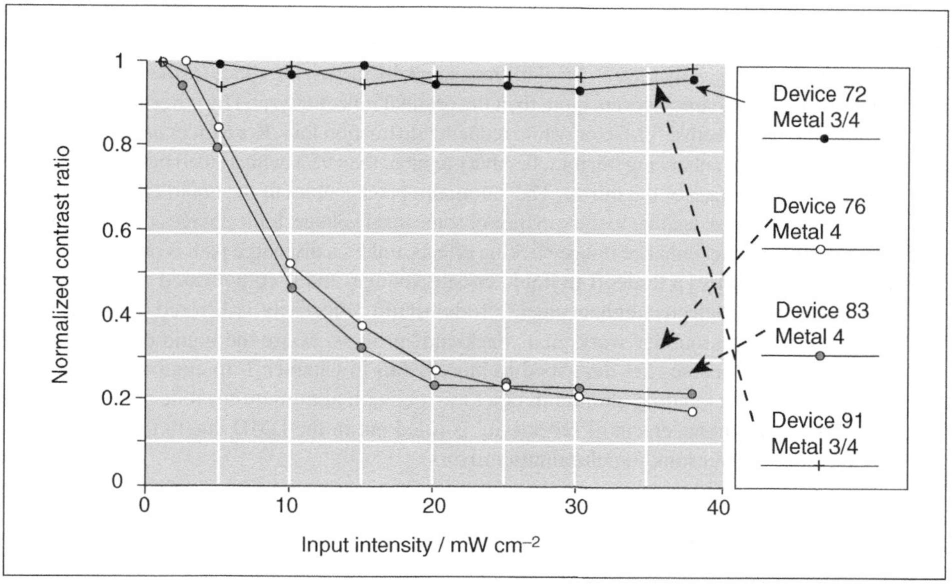

Figure 3.25 Normalized contrast ratio versus input light intensity for several DRAM-based microdisplay samples. Samples 76 and 83 are unprotected. Samples 72 and 91 have overlapping metal layers that provide the substrate some protection from the incident light. Reprinted courtesy of M. Warboys, GEC Marconi

greater than the well depth, meaning that much of the electron–hole pair generation happens not in the well where it may cause problems but in the underlying substrate where it may not.

In the case of a twin-well CMOS process, again the pixel storage nodes should lie within the well that is isolated from the substrate rather than the well that is ohmically connected to the substrate. Of course, SOI CMOS technology is the most robust with respect to light-induced charge leakage as explained in section 3.12.1.

3.8 Mirror Quality

The pixel electrode requires modest lateral electrical conductivity in comparison with most other CMOS metal structures, but maximum reflectivity. Silver has the highest reflectivity at visible wavelengths and obviously fulfills the conductivity requirement. However, apart from the expense, silver quickly tarnishes and is not suited to microelectronics. The optics industry prefers aluminum for high reflectivity and durability promoted by the strength of its native oxide. The choice of aluminum pixel electrodes concurs with its dominant role as interconnect material in microelectronics. However, microelectronic applications alloy aluminum with small components of copper and silicon to improve a number of properties,[43] including electromigration resistance, but degrade the reflectivity slightly.

Assuming the pixel mirror is aluminum deposited on an essentially flat substrate, the primary determinant of reflectivity is surface roughness, which introduces a scattering loss. Reducing the grain size increases the reflectivity. Furthermore, thinner aluminum layers tend to have smaller grain size as demonstrated by Calton *et al.*[44] Chen *et al.*[34] demonstrated a mirror reflectance $R = 92\%$ by depositing a thin layer of aluminum using cold deposition coupled with a dielectric enhancement coating.

The reflectivity of a plain aluminum mirror falls on contact with the liquid crystal bulk having a refractive index ~1.5, rather than unity refraction of air. In an LCOS microdisplay the reflectivity of the aluminum/liquid crystal interface is typically <80%. Pixel fill-factor and diffraction introduce further loss. A dielectric mirror coating on the aluminum recovers more than the original reflectivity, and by covering the aluminum pixel gaps enhances the effective fill-factor, while reducing diffraction loss. Kozakai *et al.*[45] and Katayama *et al.*[46] describe methods of enhancing mirror reflectivity to more than 95% with a small number of dielectric layers deposited on the surface of the mirror. The aluminum pixel couples to the liquid crystal through the dielectric mirror capacitance, leading to the sacrifice of some pixel voltage. Ionic charge accumulation at the dielectric mirror surface may enhance image-sticking effects, unless a discharge path is provided.

The mirror design requires a tradeoff in liquid crystal voltage, and is complicated by ionic and contact potential effects. If dielectric enhancement of aluminum reflectivity is favored, the coatings are generally applied after the foundry work, in a "back-end" process, as are the liquid crystal alignment layers. Such back-end processes are described in more detail in Chapter 7. In comparison, the DMD microdisplay employs freestanding aluminum pixel mirrors with aluminum/air reflectivity of approximately 88%. Dielectric enhancement of reflectivity is ruled out in the DMD due to the distortion that would be induced in the freestanding aluminum mirrors.

3.9 Pixel Gap Fill

The liquid crystal alignment is improved by filling in the inter-pixel gaps or trenches thus further planarizing the surface of the backplane. Kozakai *et al.*[45] have demonstrated an increase in overall efficiency by the use of gap filling. Lee *et al.*[47] have demonstrated the superiority of ECR deposited oxide over pyrolitic CVD oxide for trench filling in microdisplay backplanes.

An example of a process capable of inter-pixel gap filling is the self-aligned insulator-filled trench (SIFT) process of Calton *et al.*[48] The SIFT process flow is shown in Figure 3.26. The same photoresist pattern that is used to etch the metal and pattern the pixel mirrors is left in place for deposition of the

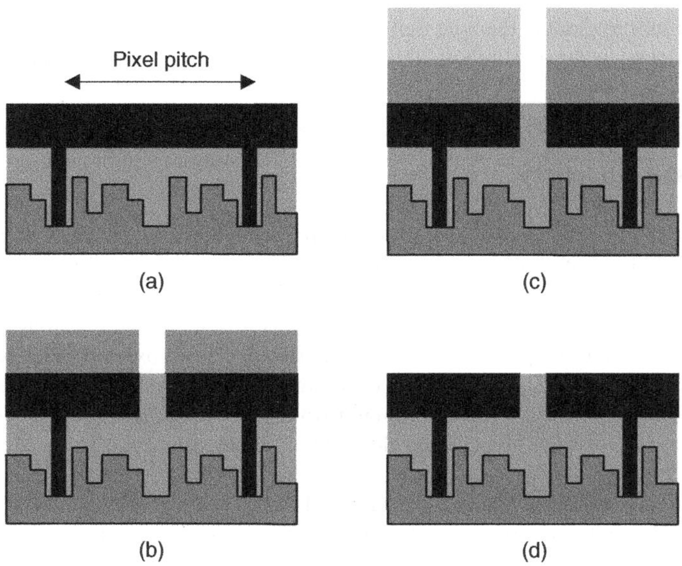

Figure 3.26 Process flow for SIFT process. Starting point is equivalent to that of Figure 3.11(f). (a) Blanket metal deposition. (b) Deposit and pattern photoresist; etch metal. (c) Deposit low-temperature oxide; oxide fills the inter-mirror gap. (d) Lift-off photoresist and overlying oxide leaving behind gap-filling oxide

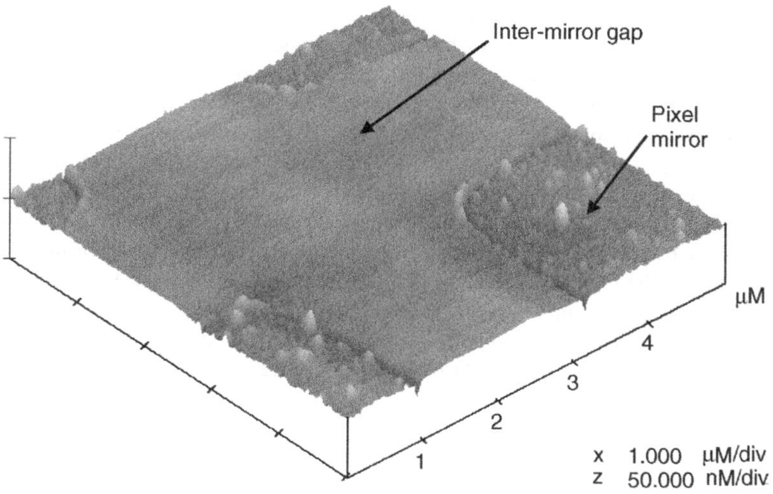

Figure 3.27 Atomic-force microscope image showing the effectiveness of the SIFT process in gap filling between pixel mirrors. Reprinted courtesy of the University of Edinburgh

gap-filling dielectric, thus ensuring perfect self-alignment of the filling material. Figures 3.27 and 3.28 illustrate the beneficial effect of a trench or gap-filling process.

3.10 LC Cell Thickness

In almost all LC configurations for LCDs it is necessary to set and maintain a precise and uniform LC cell thickness in order to achieve uniformity of color, luminance and contrast across the display. LC cell thickness uniformity is critical in achieving display color uniformity, requiring uniformity within 3%

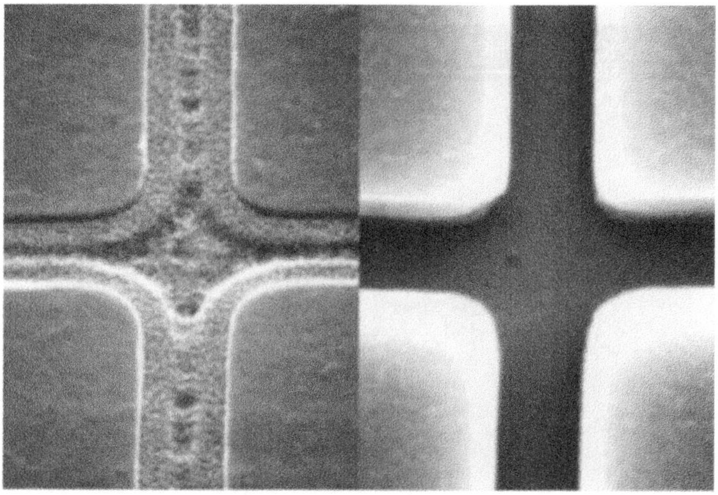

Figure 3.28 SEM pictures: gap-filled (left) and before gap-filling (right). Reprinted courtesy of Society for Information Display

over the microdisplay area,[49] even when implementing voltage compensation of cell gap variation. Thus the manufacturing yield of microdisplays is substantially dependent on thickness uniformity, ensuring detailed attention to wafer flatness and the cell assembly process. Conformal assembly of the substrates can, in principle, achieve LC thickness uniformity, with the cell gap set by internal spacers. However, wafer bow in an LCOS substrate involves compromises as described earlier in section 3.4. Thus, in a reflective LCOS microdisplay, liquid crystal cell thickness uniformity relies on the assembly of two flat substrates.

In many conventional LCDs the required LC cell gap is achieved by the use of "spacer" particles in the shape of spheres or rods that are distributed throughout the display substrate prior to lamination. Although these were used in some early microdisplays, they are not so suitable for microdisplays for several reasons. Under the high magnification that microdisplays are viewed, spacers may be visible and appear as optical defects of up to pixel size or greater. Even when spacers are not directly visible, they can degrade image quality by introducing haze. Under the stress of cell construction, spacers may sink into the "soft" aluminum mirrors, as illustrated in Figure 3.29, and thereby fail to fulfill their function of setting the required cell gap.

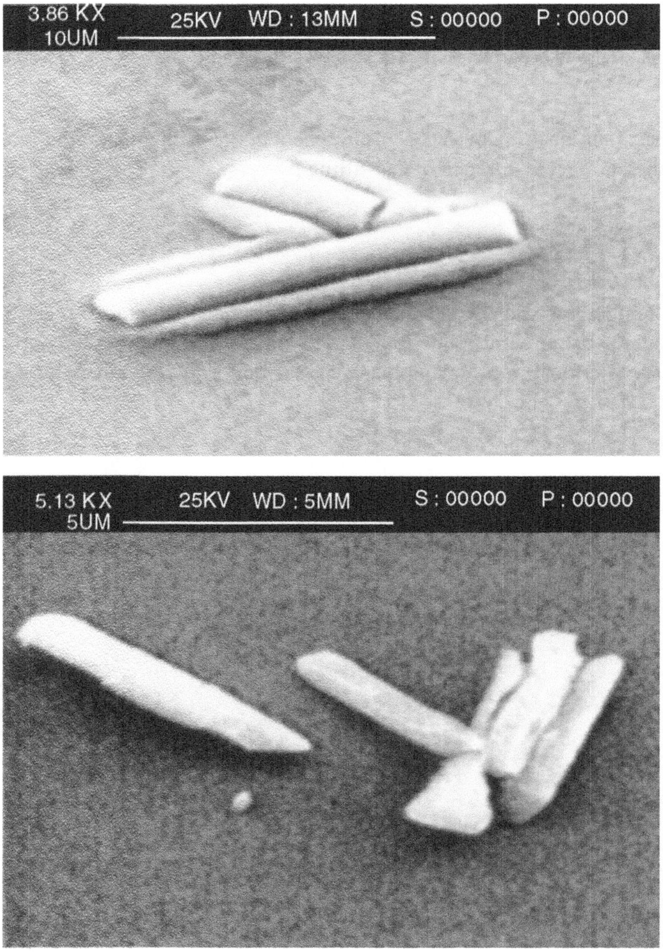

Figure 3.29 Scanning electron micrographs showing conventional spacer rods sinking into aluminum pads on a microdisplay. Reprinted courtesy of Mark Begbie and the University of Edinburgh

An accepted solution for some microdisplays has been the development of micro-fabricated spacer pillars.[50] These are structures that are added to the surface of the microdisplay towards the end of wafer fabrication. Micro-fabricated spacer pillars have the following advantages over conventional spacer particles: the footprint, shape, absolute height, height uniformity and profile of the spacers can be precisely controlled by the spacer fabrication process; the position of the spacer within a pixel can be set with precision. Typically, a spacer will be placed at the gap between the corners of four adjacent pixel mirrors in order to minimize its optical effect. The density of spacers can be precisely set. Typically spacers will appear either on every pixel or on every *n*th pixel. However, it has to be at every pixel to avoid a fixed pattern noise at lower spatial frequency than the pixel period, although spacers at twice the pixel period are claimed not to degrade image quality significantly.

A potential disadvantage of micro-fabricated spacer pillars is the issue of compatibility with techniques for LC alignment. In the case of rubbed-film alignment, rubbing would take place after pillar formation; rubbing may cause damage to the pillars and the pillars may have "shadows" which would remain un-rubbed. Similarly in the case of oblique evaporation techniques for producing alignment layers, pillars may create shadows in which the alignment layer is absent. These restrictions limit the use of built-in spacers to thin cell gaps of the order of 1 μm.[51]

However the optical path length, l, for light traveling in and out through the LC in a *reflective* microdisplay is

$$l = 2nd \tag{3.7}$$

where n is the appropriate (ordinary or extraordinary) refractive index of the LC, and d is the cell gap as set by the spacer. Thus the LC layer is thinner in reflective devices than in transmissive ones, improving compatibility with micro-fabricated spacers.

One possible process for the production of spacer pillars involves the deposition of a layer of ECR oxide. The layer is then patterned using a lift-off technique. The flow for one such process, which uses a lift-off technique, is illustrated in schematic cross-section in Figure 3.30. An example of a micro-fabricated spacer pillar is shown in Figure 3.31.

Recent developments favor the use of particle spacers restricted to the boundary seal, where they control the boundary cell gap without influencing the active area optics. Spacer particles may be mixed with the seal polymer before dispensing the seal. The necessary flatness of the substrates maintains the gap set at the boundary throughout the pixel array.

3.11 LCOS CMOS Summary

Chen *et al.*[34] summarize some of the developments in LCOS CMOS including combined BPSG reflow and CMP for maximum planarity, optimization of mirror reflectance, gap fill, passivation and spacer pillars.

CMOS development has followed a predictable path for many years, shrinking circuitry and increasing speed, while reducing defects to support vast arrays of transistors. Liquid crystal technology also advanced rapidly as the display industry appreciated its potential. Crystalline silicon and liquid crystals flirted for many years, before planarization provided a firm basis for the marriage. The past few years have seen LCOS microdisplays become competitive in the projector marketplace, and now produce image quality beyond any competing display device. LCOS will continue to benefit from advances in silicon technology. However, CMOS voltage has dropped in step with the shrinkage in geometry, falling short of the liquid crystal requirements, requiring hybrid structure to provide adequate drive voltage. SRAM pixel storage may gain in popularity as small footprint and high transistor count become cheaper. DRAM addressing can look forward to lower manufacturing costs to improve its competitive position.

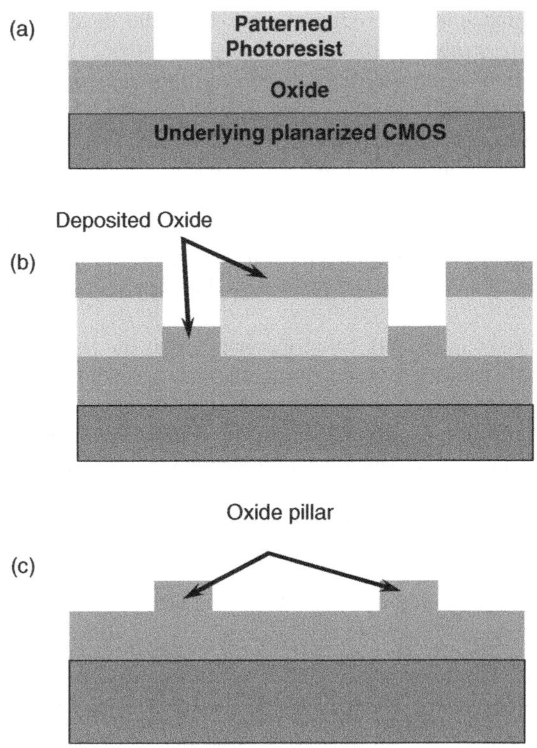

Figure 3.30 Generic process flow for spacer pillar manufacture using ECR oxide and lift-off. (a) Substrate with patterned photoresist over oxide (but could be mirror metal). (b) Following low-temperature deposition of oxide, filling gaps between photoresist features, or lying atop the photoresist. (c) When the photoresist is removed, overlying oxide is also removed (lift-off), leaving oxide features or pillars

Figure 3.31 Scanning electron micrograph showing deposited spacer pillar, at the corner of four pixel mirrors. Bar = 1 μm. Reprinted courtesy of Elsevier Publishing

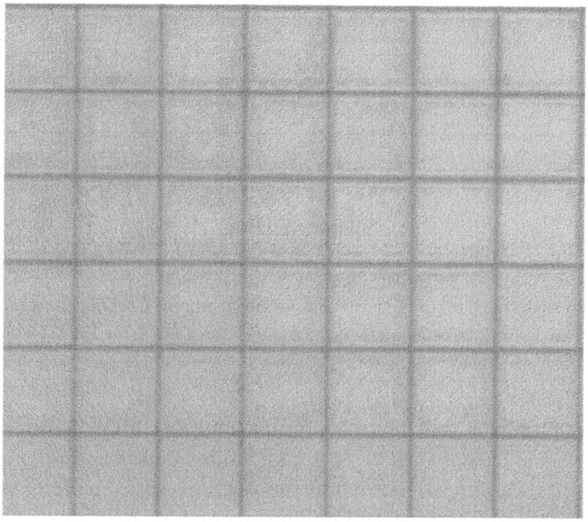

Figure 3.32 Scanning electron micrograph of state-of-the-art LCOS backplane illustrating high fill-factor and good optical quality at 12 μm pixel pitch. Reprinted courtesy of Syntax-Brillian Corp.

LCOS fabrication follows a fairly standard CMOS process optimized for LCOS, and available from several foundries. The electronic requirements of LCOS do not stretch CMOS capability, but optical flatness continues to test manufacturing skill. A study of foundry web pages shows the progress made in providing LCOS CMOS as a foundry service. UMC,[52] for example, now offers passivation of the pixel mirror that also provides dielectric enhancement of the pixel reflectivity. Inter-pixel gap filling is now standard, proving a completely smooth surface to the liquid crystal alignment layer, and built-in spacer process is an option.

Figure 3.32 shows a state-of-the-art microdisplay backplane including very high fill-factor, with a mirror that shows a high degree of flatness and an absence of the dimple usually associated with the underlying via.

3.12 Backplane Technology for Other Microdisplays

3.12.1 Transmissive LCOS

A transmissive LCOS (T-LCOS) microdisplay is similar in principle to a miniature active matrix p-Si TFT LCD. In practice it allows smaller pixels, and the higher mobility of single crystal silicon provides superior transistor characteristics. Moreover, manufacture of the silicon backplane of a liquid crystal microdisplay by a silicon foundry is a further advantage. Miniature p-Si TFT-based microdisplays lack foundry support, implying high manufacturing investment. The transferred silicon process removes the silicon addressing circuits from the wafer together with a thin layer of silicon and transfers it to a transparent substrate. The assembly of a microdisplay then follows the same route as polysilicon addressed devices. This method of creating a transmission-mode microdisplay from a foundry-produced silicon backplane has proved successful in NTE applications.[53] Projection applications, on the other hand, require high light output and it is a matter of conjecture that low optical efficiency is a reason for current T-LCOS devices not being used in projection applications.

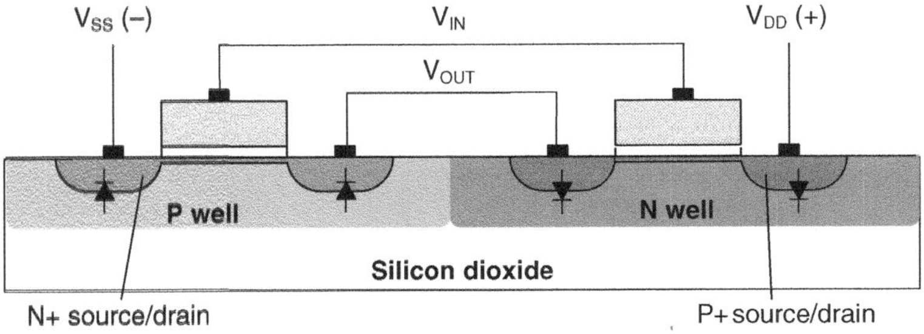

Figure 3.33 Schematic cross-section of generic silicon-on-insulator CMOS inverter showing buried oxide isolation

A variant of standard CMOS called silicon-on-insulator (SOI) CMOS is a prerequisite for T-LCOS. Wolfe[53] offers a background to SOI CMOS. SOI CMOS uses dielectric insulation to replace reverse p–n junction isolation in the substrate normal to the surface. An example is shown in Figure 3.33. SOI CMOS was originally developed for military and other highly demanding purposes due its inherent radiation hardness; ionizing radiation can produce soft-errors by the production of charge carrier pairs in p–n junction isolation. So it is inherently much more robust to any light in the substrate than CMOS. Other SOI advantages include elimination of both latchup and parasitic field oxide FETs, lower parasitic capacitance and capacitive coupling, lower power consumption and higher maximum clock frequency. A revival of interest in silicon-on-sapphire for high-speed transmission microdisplays may be underway.[55] Transmission microdisplays are discussed in detail in Chapter 4.

3.12.2 Micro-optical-electromechanical Systems

Micro-electromechanical systems (MEMS) have moved from research and development to commercial applications such as pressure sensors, accelerometers, electronic filters, etc. MEMS are three-dimensional structures produced by sophisticated photolithography techniques involving sacrificial layers and etch-stop processes. In developing products, various mechanical fatigue and stiction issues have been resolved.

Micro-optical-electromechanical systems (MOEMS) have the added complication of optical readout. The optical issues discussed for LCOS – reflectivity, scattering, and flatness – apply to MOEMS. The simplest MOEMS is a single scanning mirror that can write a scanned image from a beam source, or function as a switching device in an optical network. The digital mirror device (DMD) is the most successful MOEMS in large volume production. The importance of MOEMS in projection systems, particularly the DMD, is our reason for Chapter 8 devoted to MOEMS function and fabrication.

3.12.3 OLED CMOS

To date, little of detail or substance has been published on CMOS technology for OLED (either SMOLED or P-OLED) microdisplays. It is reasonable to speculate that the optical and mechanical requirements on the CMOS substrate for use in an OLED microdisplay are similar to those of LCOS CMOS. In short, the aim is a very flat top surface upon which the OLED material may be deposited. One constraint over and above that of LCOS is that the work function of the surface metal is important in determining the efficiency of the light generation in the microdisplay as discussed in Chapter 9. Thus, if conventional CMOS metal (an aluminum alloy AlSiCu or similar) is not suitable it would be necessary to add an additional custom metal layer or to surface-treat the existing metal layer.

In LCOS a consequence of mirror granularity or other surface topography is disruption to the LC alignment; in OLED microdisplays prominent metal grains and other sharp topographic features may produce pin-holes in the organic layers leading to short-circuiting between cathode and anode and consequent hot-spots.

In all microdisplays, a high pixel aperture ratio helps the appearance of the display. The higher the pixel aperture ratio, the less likely that the pixelation of the display will be apparent to the viewer. In a transmissive or reflective modulating microdisplay technology such as LCOS, pixel aperture ratio is also important in determining optical efficiency as described in section 3.3. In an organic emissive microdisplay, a high pixel aperture ratio is also important for reasons of optical efficiency as described in detail in Chapter 9, but the small inter-pixel gap can lead to unwanted lateral conduction between adjacent pixels.

3.13 Silicon Technology Roadmap

This section mentions some aspects of the ongoing evolution of CMOS technology that have been, and will continue to be, of relevance to microdisplays.

Smaller feature size

Smaller feature sizes allow a smaller footprint for a given circuit or function or more functionality in a given area. This means that a pixel can become smaller while maintaining its functionality or remain the same size with increased functionality. The factors that limit pixel shrinkage in LCOS microdisplays today are diffraction effects and fringing fields. As the pixel becomes smaller and, in particular, as the lateral dimension of the pixel begins to approach the wavelength of the illumination, the proportion of light lost to pixel diffraction rises significantly. Verweire and Defever[9] explore the limits imposed upon pixel dimensions by diffraction effects. Fringing fields reduce the effective pixel fill-factor more as the pixel pitch reduces; a detailed analysis is presented in Chapter 6.

Reduced thickness of material layers

Reduced layer thickness leads to higher interlayer normalized capacitance. This means that "designed-in" storage capacitance values can be achieved in a smaller footprint, but also that undesirable parasitic capacitance becomes more of an issue.

Increased leakage current density

Leakage current increases with decreasing CMOS feature size, thus putting greater demands on the refresh of soft-node storage elements such as those used in DRAM-style LCOS pixels. Examples of increased leakage paths include channel leakage in the MOSFET off state and reverse-bias p–n junction diode leakage.

Increasing number of interconnect layers

This generally reduces or at least alleviates bottlenecks in data flow. However, uniquely in the case of microdisplays, every single pixel includes a "vertical" connection that extends from the bottom (the source/drain storage node in the substrate) to the top (the mirror/electrode structure in the top metal layer) of the pixel cross-section. Thus, the structure of a microdisplay pixel is necessarily made more complicated and convoluted by the addition of extra metal layers.

Smaller supply rail voltage

As physical dimensions shrink both vertically and laterally, supply rail voltage must similarly shrink in order to avoid excessive electric fields leading to dielectric and/or p–n junction breakdown within the CMOS structures. This means that the voltage available to drive the optical material (LC, OLED etc.) becomes more limited or that the pixel must contain non-standard high-voltage structures. Reduced voltage also means reduced power consumption as explained in Chapter 2.

Larger maximum die size

Ordinarily the maximum die size on a CMOS wafer is limited to that of the field size of the wafer stepper. The stepper is the machine that produces the required illumination pattern on the photoresist in order to produce each layer of the circuit, while the field size is the maximum rectangle that can be illuminated in a single exposure by the stepper. With each generation of technology, the field size typically increases. This allows for the manufacture of microdisplays with larger numbers of pixels, albeit at the cost of lower yield and higher cost per die. De Smet *et al.*[56] describe the production of microdisplays that exceed even the available field size by means of a technique called "field stitching." This technique involves the careful mutual alignment and combining of adjacent fields to allow a single backplane design to spread across multiple fields.

Larger wafer size

Generally speaking (in high volumes and all other things being equal), the larger the wafer diameter, the lower the cost per die. That is, it is probably more economical to fabricate microdisplays on a 200 mm wafer process than a 150 mm wafer process.

Increased tooling cost

At smaller geometries, the capital cost of a "mask set" to allow manufacture of a particular CMOS design is typically greater.

3.14 Cost of Silicon

3.14.1 Wafer Cost

For any microdisplay technology, the cost of the silicon backplane represents a significant fraction of the overall cost of manufacture. Here we present a simple means of modeling the cost of a CMOS backplane. The basic unit of purchase is the wafer and the cost of a wafer depends on many factors, including:

- *Volume* – procurement in wafers per month.
- *Wafer diameter* – options are 150 mm, 200 mm or 300 mm.
- *Process node* – including the maturity and complexity of the wafer manufacturing process: 0.35 μm, 0.25 μm, 0.18 μm.
- *Masks* – the number of standard and non-standard mask levels. The number of mask levels rises with, say, the number of metal layers used in a standard process.
- *Any requirement for non-standard or additional processing.* Spacer posts, for example, represent an additional non-standard mask level. During the development of LCOS as a technology, additional "custom" process steps were developed and added to the basic CMOS process. These are now mature enough and widely enough used that "LCOS CMOS" has become a "standard" variant of mainstream CMOS.

Microdisplays are typically manufactured on 150 mm (6-inch) and 200 mm (8-inch) wafers in 0.35 μm to 0.25 μm technologies although some microdisplays are already manufactured in smaller geometry processes. More modern processes typically involve a combination of larger wafer, smaller feature size, more mask levels and lower defect density. Larger wafers cost more per wafer but the combination of the reduction in feature size (leading to a smaller chip) and the larger wafer area leads to an increase in the number of dice available on a wafer and a reduced cost per die.

3.14.2 Yield

The manufacture of CMOS wafers, like that of many mass manufactured objects, is subject to the occurrence of random and other defects that cause individual dice to partially or fully fail to perform as required. It is generally safe to assume that, if a particular design conforms to the design rules of the appropriate process, random defects will dominate. In mainstream CMOS, the statistics are well understood and generally well characterized for a given CMOS process and foundry.

Ferris-Prabhu[57] offers a good introduction to the topic of yield in semiconductors. Yield is defined as the percentage of dice on the wafer that perform to a level pre-defined as acceptable. Maximization of yield is important in minimizing the cost of production of any integrated circuit or integrated circuit-based device such as a microdisplay. Yield is determined by factors such as: typical density of defects in the fabrication process; die area, or more accurately die *critical* area, i.e. the proportion of the die area in which a process defect is likely to cause a chip to fail; and the density and nature of the circuitry on the die.

A number of yield models exist. The most appropriate in any situation depends on factors such as known yield statistics of the process, and the nature and defect sensitivity of the circuitry on the die. We will follow through a simplified yield and cost analysis based upon a Murphy formulation.

A simple and widely accepted model is the Murphy yield equation:[57]

$$Y = \left(\frac{1 - e^{-AD_0}}{AD_0} \right)^2 \tag{3.8}$$

where Y is the fraction of candidate dice with zero defects, A is the die area (usually expressed in mm^2), and D_0 is the defect density (defects per unit area usually expressed as defects per square cm). In the case of a microdisplay backplane, in which a proportion of the area, A_C, comprising say the pixel array and addressing circuitry, may consist of dense circuitry, A_C constitutes the critical area with regard to yield. The remaining die area $A - A_C$ (such as areas under the seal ring) may be sparsely populated with very little susceptibility to conventional microfabricated defects. The Murphy equation may be approximated to

$$Y = \left(\frac{1 - e^{-A_C D_0}}{A_C D_0} \right)^2 . \tag{3.9}$$

For a given D_0, when $A_C \ll 1/D_0$, $Y \to 1$, and when $A_C \gg 1/D_0$, $Y \to 1/(A_C D_0)^2 \ll 1$.

3.14.3 Qualitative Yield Comparisons

Defect density: LCOS CMOS vs conventional CMOS

LCOS CMOS defect density is always higher than the defect density for standard CMOS, for two reasons. LCOS CMOS has more process steps and therefore more opportunities for creation of defects. Top-surface optical abnormalities that have no effect on the electrical properties or yield of a standard CMOS device may cause optical or other failure of an LCOS backplane. Thus, the yield of LCOS CMOS is lower than that of equivalent conventional CMOS.

Defect density: LCOS CMOS vs conventional active matrix LCD

The high magnification under which LCOS microdisplays are viewed compared to conventional active matrix LCDs means that many small optical defects or abnormalities that would be small enough to be insignificant in a typical active matrix LCD backplane are significant enough to count as defects in an LCOS CMOS backplane.

Yield: LCOS CMOS vs SRAM or DRAM

There is an architectural similarity between an LCOS CMOS pixel array and a SRAM or DRAM cell array. In some LCOS backplanes, the similarity extends to the pixel level. Microdisplays using SRAM- and DRAM-based pixel circuits have been reported. Both microdisplay and memory backplanes contain arrays of close packed cells addressed by row and column lines. Yet the physical location of a memory cell within an array is unimportant as it is read electronically whilst the physical location of a microdisplay pixel within the array is very important as the readout process involves looking at the microdisplay. Thus the redundancy techniques typically used in memory chips[58] in order to maximize yield for a given defect density are not available to microdisplays.

Yield: LCOS CMOS vs conventional CMOS

The conventional yield analysis above assumes that a single defect breaks a chip. Sometimes one or more defects are tolerated to an extent. For example, a processor with a faulty cache may be sold at a lower price as a cache-less processor provided the cache can be disabled. In the case of a microdisplay, it may be permissible to set the specification such that one or more broken pixels (say for example two) are allowed within the pixel array. In this case, the backplane yield will be higher as it includes all backplanes with 0, 1 or 2 pixel defects rather than simply the backplanes with zero defects as previously assumed above. A quantitative estimate of the increased yield is usually possible.

3.14.4 Good Dice per Wafer

We wish to maximize the number of good dice, G, per wafer:

$$G = NY \tag{3.10}$$

where N is the number of candidate dice per wafer, and Y is the yield (%), explained above. The usable wafer area is slightly smaller than the wafer radius due to a small area around the perimeter of the wafer from which no working devices are taken. Because we are fitting rectangular dice into a circular area, there is also some unused area around the wafer perimeter as illustrated in Figure 3.34.

For large N, an approximation to N is given by

$$N = \pi\left(\frac{r^2}{A} - \frac{2r}{\sqrt{2A}}\right) \tag{3.11}$$

where r is the radius of the usable wafer area. Factors influencing A are, for example, pixel pitch, number of pixels (display definition), other necessary circuits, bonding pads, seal ring (for LCOS) etc. The imperative is that the die, and the critical area in particular, should be no larger than necessary. A reduction in the non-critical area improves N whilst a reduction in the critical area improves both N and Y.

3.14.5 Cost per Good Die

The "cost per good die," C (\$), is given by

$$C = \frac{W}{G} = \frac{W}{NY} = \frac{W}{\pi}\left[\frac{A_C D_0}{\left(1 - e^{-A_C D_0}\right)}\right]^2 \frac{A}{\left(r - \sqrt{2A}\right)r} \tag{3.12}$$

where W is the wafer cost. Considering, for now, the simplified case of a microdisplay with a small non-critical area ($A_C \approx A$) due to minimization of the seal ring etc., produced on a mature process with a well-characterized D_0. Then when $A \ll 1/D_0$, $Y \to 1$ and cost is approximately proportional to area. When $A \gg 1/D_0$, $Y \to 1/(AD_0)^2$ and cost rises very rapidly with A.

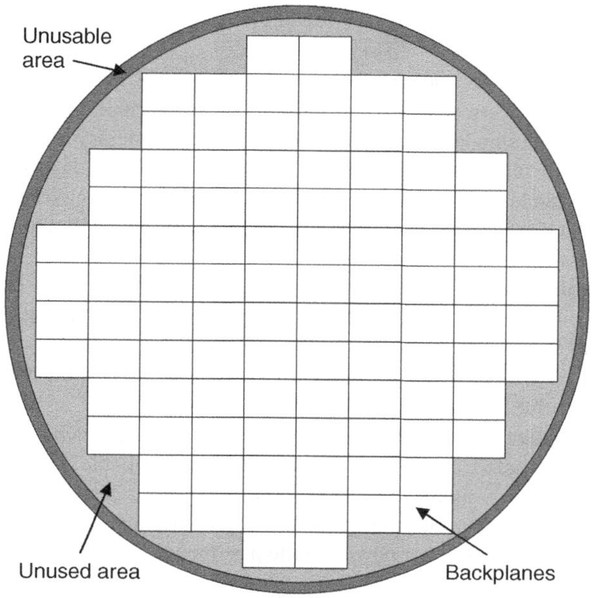

Figure 3.34 Schematic of silicon wafer layout. Shows candidate backplanes as white rectangles, unused wafer area in light gray and unusable wafer area in mid gray

As an example, we calculate the cost per good die for three backplanes all manufactured on the same CMOS process. Process details common to each are: 0.35 μm CMOS process; 200 mm wafer; periphery to pixel array 2 mm on each side; wafer cost $1000. Table 3.4 illustrates the yield and cost implications of a larger die size purely for the yield and cost of the bare silicon.

This yield and cost model can be extended from the silicon backplane alone to cover the complete microdisplay device including display fabrication on top of CMOS, packaging and final test provided adequate details are known of the cost and yield of each of the process steps.

Table 3.4 Computed die cost estimates for various devices

Definition	QVGA	SVGA	WUXGA
Columns	320	800	1920
Rows	240	600	1200
Pixels	76800	480000	2304000
Pixel pitch (μm)	15	12	9
Array width (mm)	4.8	9.6	17.28
Array length (mm)	3.6	7.2	10.8
Pixel array area (mm^2)	17.28	69.12	186.624
Critical area (mm^2)	19.008	76.032	205.2864
Defect density (mm^{-2})	0.005	0.005	0.005
Yield (%)	91.0%	69.2%	39.1%
Die area (mm^2)	38.08	106.72	246.784
Wafer radius (mm)	95	95	95
Dice per wafer	676	225	88
Good dice	615	156	34
Cost per good dice ($)	$1.63	$6.43	$29.07

3.15 Summary

The proprietary DMD CMOS technology, described in Chapter 8, developed over a period of approximately twenty years, involves the extension of CMOS to include highly sophisticated surface MEMS structures above the standard interconnect layers. The proprietary transferred silicon process, described in Chapter 5, involves the radical step of removing the CMOS layers from the surface of a SOI wafer and bonding the laminate to a new glass substrate, thus allowing transmissive LCOS. Most other microdisplay CMOS processes rely on relatively minor tweaks and additions to an otherwise standard CMOS process.

Most LCOS CMOS processes differ from mainstream CMOS primarily in the optimization of the top layers for surface flatness, smoothness, fill-factor, LC cell gap and LC alignment. OLED microdisplays, as described in Chapter 9, may additionally require optimization of the electronic properties of the top surface for work function in order to minimize turn-on voltage and maximize efficiency of the organic electroluminescent structure.

Like many other derivative technologies, LCOS CMOS has matured such that the required process tweaks and additions are bundled as a process option pack that is available from a number of foundries. LCOS CMOS is, or has been, available from foundries such as, for example, UMC, TSMC, Chartered Semiconductor, Fujitsu Microelectronics and Amkor. Other vertically integrated companies with internal CMOS foundry capabilities either now or in the past include included IBM, Sony, Philips, Intel, Samsung and JVC.

As requirements on microdisplay definition and performance continue to grow, further demands will be made of the CMOS. Some examples are introduced below. As the market for microdisplay CMOS increases, the hope is that foundries will be willing to make the necessary additional investment in process development.

At smaller process geometries, charge leakage from soft nodes becomes more significant; proper DRAM-style storage structures, for example trench capacitors, such as those reviewed by Wolf,[59] would provide larger capacitance, smaller footprint and better leakage resistance. Typical DRAM processes are today restricted to around two metal layers, not enough for a robust LCOS device. Therefore, this approach would require DRAM structures and many metal layers to be integrated into the same process. Silicon on insulator (SOI) substrates would also be beneficial with regard to robust soft-node charge storage but typically add a small premium to the wafer cost.

LC drive voltages are unlikely to shrink as fast as CMOS supply rails, leading to an increased reliance upon high-voltage drive capability within LCOS CMOS.

Advanced CMOS processes (typically at and below the 0.13 μm node) utilize copper interconnects. Copper has a sheet resistance of 1.7 Ωcm compared to aluminum with a sheet resistance of 2.8 Ωcm, thus allowing faster signal transmission along, and lower ohmic loss in, tracks. However, the optical properties of copper are not suitable for the top metal layer in microdisplay backplanes. Future microdisplays built on CMOS processes that utilize copper interconnects will require the development of an optically specified top metal layer.

3.16 Bibliography

Bodammer, G., Calton, D. W. and Underwood, I. (2001) Investigation of the bow in silicon backplanes for microdisplay applications. *SID Digest* **32**, 439–41.

Calton, D. W., Seunarine, K., Bodammer, G. and Underwood, I. (2000) Liquid crystal flow control using microfabrication techniques. *IEE Proc. Optoelectronics* **147**, 163–70.

Gourlay, J., O'Hara, A., Stevens, A. J. and Vass, D. G. (1996) A comparative investigation into planarized liquid crystal over silicon spatial light modulators. *J. Mod. Opt.* **43**, 181–98.

Lee, Y. M. (2005) Improving the phase modulating properties of electrically addressed liquid crystal on silicon devices. PhD thesis, University of Edinburgh.

Miremont, C. (2006) Developments in the manufacturability of ferroelectric liquid crystal on silcon microdisplays. PhD thesis, University of Edinburgh.

Miremont, C., Calton, D. W., Bodammer, G. and Underwood, I. (2000) Alignment of ferroelectric liquid crystals over CMOS-based microdisplay. *Proc. SPIE*, **4207**, 2000.

O'Hara, A., Underwood, I. and Vass, D. G. (1994) Planarization of spatial light modulator silicon backplanes using chemical-mechanical polishing. *IEE Coll. Digest on Micro-engineering and Optics*, 1994/043, 5/1–6.

O'Hara, A., Vass, D. G. and Underwood, I. (1994) Interlevel dielectric planarization to optical tolerances using chemical-mechanical polishing in the fabrication of spatial light modulator backplanes. *Proc. Conf. on Advanced Metallization for ULSI Applications*, 194–9.

O'Hara, A., Begbie, M., Rankin, I. D. *et al.* (1995) Protection of DRAM spatial light modulators from incident light using a multi-level metallization procedure. *Proc. 25th European Solid State Devices Research Conf.*, 270–7.

O'Hara, A. *et al.* (1995) Post-processing using microfabrication techniques to improve the optical performance of liquid crystal over silicon backplane spatial light modulators. *Proc. SPIE*, **2641**, 129–39.

O'Hara, A. *et al.* (1997) Investigation of novel structures on silicon backplane SLMs. In *OSA Trends in Optics and Photonics*, Vol. 14: *Spatial Light Modulators*. Optical Society of America, Washington, DC, 154–6.

O'Hara, A., Bodammer, G., Vass, D. G. *et al.* (1997) Investigation of novel structures on silicon backplane SLMs to improve the device performance. In *OSA Trends in Optics and Photonics*, Vol. 14: *Spatial Light Modulators*. Optical Society of America, Washington, DC, 111–17.

O'Hara, A., Bodammer, G., Vass, D. G. *et al.* (1998) Fabrication of spatial light modulator backplanes using damascene processing. *Proc. SPIE*, **3292**, 107–15.

Schuck, M. H., McKnight, D. J. and Johnson, K. M. (1997) A planarized LCOS display for projection applications. In *OSA Trends in Optics and Photonics*, Vol. 14: *Spatial Light Modulators*. Optical Society of America, Washington, DC, 74–6.

Semiconductor Industry Association (1999) *International Technology Roadmap for Semiconductors*. Available at www.itrs.net/1999_SIA_Roadmap/Home.htm

SLIMDIS technical data sheet, available from Dr M. R. Worboys, BAe Systems, Chelmsford, England.

Seunarine, K. (1999) Techniques to improve the optical quality of liquid crystal over silicon spatial light modulators. PhD thesis, University of Edinburgh.

Seunarine, K., Calton, D. W. and Underwood, I. (1999) A novel thin-mirror trench-fill technique for the manufacture of high optical quality LCoS spatial light modulators. *OSA Technical Digest on Spatial Light Modulators and Integrated Optoelectronic Arrays*, PD3.1-3.3.

Stefanov, M. E., Ghandi, J., Worthington, S. *et al.* (2000) Volume manufacturing of LCoS microdisplays. *Proc. of IDMC.*

Tsaur, B. Y., Johnston, J., Baczko, L. and Plutnicki, R. (2000) Lessons learned in establishing mass production of transmissive LCoS microdisplays. *Proc. SPIE*, **4207**, 39–44.

Underwood, I., Vass, D. G., O'Hara, A. *et al.* (1994) Improving the optical quality of liquid crystal over silicon spatial light modulators: issues and achievements. *Applied Optics* **33**, 2768–74.

Underwood, I. *et al.* (1995) Fabrication issues for silicon backplane active matrix miniature liquid crystal displays. *Proc. SPIE*, **2641**, 105–16.

Underwood, I. *et al.* (1999) Recent advances in silicon micromachining for liquid crystal microdisplays. *IEE Coll. Digest on Micro-engineering in Optics and Optoelectronics*, 99/187, 10/1–6.

Underwood, I. (2000) VLSI design and fabrication for liquid crystal on silicon. *IEE Coll. Digest on Microdisplay and Smart Pixel Technologies*, 00/006, 7/1–6.

Underwood, I. *et al.* (2000) The successful application of micromachining technique to microdisplays. *IEE Coll. Digest on Demonstrated Micromachining Techniques for Industry*, 00/032, 7/1–8.

Van den Steen, J. (2006) Design of LCoS microdisplay backplanes for projection applications. PhD thesis, Universiteit Gent, Belgium.

Wilkinson, T. D., Crossland, W. A., Coker, T. *et al.* (1998) Ferroelectric liquid crystal on silicon spatial light modulator designed for high yield and low cost fabrication: the fast bitplane SLM. *Ferroelectrics* **213**, 613–17.

Wolf, S. and Tauber, R. N. (1986) *Silicon Processing for the VLSI Era*. Vol. 1: *Process Technology*. Lattice Press.

3.17 References

1 Sze, S. M. (2002) *Semiconductor Devices: Physics and Technology*, 2nd edn. John Wiley.
2 Moore, G. E. (1965) Cramming more components onto integrated circuits. *Electronics* **38**, 8.
3 http://public.itrs.net
4 Processes of the future. *Solid State Tech.* 42–48, February 1995.
5 Maly, W. (1987) *Atlas of IC Technologies*. Benjamin Cummings.
6 Wolf, S. (1986) CMOS process integration. In *Silicon Processing for the VLSI Era*, Vol. 2. Lattice Press.
7 Wolf, S. (1994) The submicron MOSFET. In *Silicon Processing for the VLSI Era*, Vol. 3. Lattice Press.
8 Chih-Hang Tung, Sheng, G. T. T and Chih-Yuan Lu (2003) *ULSI Semiconductor Technology Atlas*. John Wiley.
9 Verweire, B. and Defever, R. (1999) Limitation of resolution of LCoS-based projection displays by diffraction effects. *Proc. 19th IDRC* (Eurodisplay 99), 489–92.
10 Armitage, D. (1999) Resolution issues in reflective microdisplays. *Proc SPIE*, **3634**, 10–19.
11 De Smet, H., den Steen, J. V. and Van Calster, A. (2001) Microdisplays with high pixel counts. *SID Digest* **32**, 968–71.
12 Underwood, I., Vass, D. G. and Sillitto, R. M. (1986) Evaluation of an nMOS VLSI array for an adaptive liquid-crystal spatial light modulator. *IEE Proc. J Optoelectronics*, 77–82.
13 Underwood, I. (1987) An NMOS addressed liquid crystal spatial light modulator. PhD thesis, University of Edinburgh.
14 O'Hara, A., Hannah, J. R., Underwood, I. *et al.* (1993) Mirror quality and efficiency improvements of reflective spatial light modulators by the use of dielectric coatings and chemical-mechanical polishing. *Applied Optics* **32**, 5549–56.
15 Colgan, E. G. and Udal, M. (1998) On-chip metallization layers for reflective light valves. *IBM J Res. Dev.* **42**, 339–46.
16 Sivaram, S., Bath, H., Leggett, R., Maury, A., Monnig, K. and Tolles, R. (1992) Planarizing interlevel dielectrics by chemical mechanical polishing. *Solid State Tech.*, May, 87–91.
17 Joseph, M. *et al.* (1997) *Chemical Mechanical Planarization of Microelectronic Materials*. John Wiley.
18 Sanford, J. L., Schlig, E. S., Tomooka, T. *et al.* (1998) Silicon light-valve array chip for high-resolution reflective liquid crystal projection displays. *IBM J. Res. Dev.* **42**, 347–58.
19 Van Calster, A. (1999) Design and fabrication of high resolution reflective light valves with silicon backplanes. *Proc SPIE*, **3634**, 64–70.
20 Sze, S. M. (1981) *Physics of Semiconductor Devices*. John Wiley.
21 De Smet, H. *et al.* (2002) Design, fabrication and evaluation of a high-performance XGA VAN-LCOS microdisplay. *Displays* **23**, 89–98.
22 Underwood, I., Vass, D. G. and Sillitto, R. M. (1986) Evaluation of an nMOS VLSI array for an adaptive liquid crystal spatial light modulator. *IEE Proc. J.* **133**, 77–82.
23 Wilkinson, T. D., Crossland, W. A., Coker, T. *et al.* (1987) The fast bitplane SLM: a new ferroelectric liquid crystal on silicon spatial light modulator. *Spatial Light Modulators*, Technical Digest (OSA, Washington, DC), 154–6.
24 Miremont, C., Bodammer, G., Calton, D. W. *et al.* (2002) Improving the flatness of silicon backplanes for high quality FLCoS microdisplays. *Displays* **23**, 115–20.
25 Walker, C. M. and Handschy, M. A. (2000) Microdisplay silicon wafer flatness metrology by optical interferometry. *Digest of Microdisplay 2000*, UEF, August.
26 Olsen, J. and Moghadam, F. (1993) Planarization techniques. In S. R. Wilson, C. J. Tracy and J. L. Freeman (eds) *Handbook of Multilevel Metallization for Integrated Circuits*. Noyes.
27 Schuck, M. H., McKnight, D. J. and Johnson, K. M. (1997) Spin-cast planarization of liquid-crystal-on-silicon microdisplays. *Optics Letters* **22**, 1512–14.
28 Hermanns, A., Shirey, L. M., Geer, R. E. *et al.* (1999) A post-processing sequence for liquid-crystal-on-silicon microdisplays. *Proc. SPIE*, **3635**, 103–11.
29 Walton, A. J. *et al.* (1999) A review of the history and technology of micromachined miniature displays using foundry produced silicon backplanes. *Proc. SPIE*, **3891**, 25–38.

30 Tomazawa, M. (1997) Oxide CMP mechanisms. *Solid State Tech.*, 169, July.
31 Cook, L. M. (1990) Chemical processes in glass polishing. *J. Non-cryst. Solids* **120**, 152.
32 Calton, D. W. (2001) Improving the flatness of microdisplay backplanes using chemical mechanical polishing, PhD thesis, University of Edinburgh.
33 Burns, D., Gourlay, C. J., O'Hara, A., Underwood, I. and Vass, D. G. (1995) A 256x256 SRAM-XOR pixel ferroelectric liquid crystal over silicon spatial light modulator. *Optics Communications* **119**, 623–32.
34 Chen, W. S., Wang, P., Liu, M. T. *et al.* (2000) A novel modification mirror layer to fabricate a silicon-based light valve display technology. *Digest of Microdisplay 2000*, Session IV. United Engineering Federation.
35 Seunarine, K. *et al.* (1999) Techniques to improve the flatness of reflective micro-optical arrays. *Sensors and Actuators* **78**, 18.
36 Stine, B. E. *et al.* (1998) The physical and electrical effects of metal-fill patterning practices for oxide chemical-mechanical polishing processes. *IEEE Trans. Elec. Dev.* **45**, 665–79.
37 Troutman (1986) *Latchup in CMOS Technology*. Kluwer Academic Press.
38 Zimmermann, H. (2004) *Silicon Optoelectronic Integrated Circuits*. Springer.
39 Chuang, S. L. (1995) *Physics of Optoelectronic Devices*. John Wiley.
40 Worboys, M. R., Radcliffe, S. and Griffith, M. S. (1998) A high-resolution miniature colour liquid crystal display. *GEC J. Technol.* **15**, 56–9.
41 Underwood, I. *et al.* (1998) A high resolution, high frame rate spatial light modulator. *Digest of LEOS Summer Topical Meetings on Smart Pixels*, Monterey, USA, July, 57–8.
42 Huang, H. C., Wong, Y. T., Nguyen, C. T and Kwok, H. S. (1996) Planarization of liquid-crystal-on-silicon projection display with multi-level metallization and checmical-mechanical polishing. *Proc. SID*, **27**, 685–8.
43 Wilson, S. R., Tracy, C. J. and Freeman, J. L. (eds) (1993) Section 4.3 – Aluminum alloys, in *Handbook of Multilevel Metallization for Integrated Circuits*. Noyes.
44 Calton, D. W., Seunarine, K., Stevenson, J. T. M. and Underwood, I. (2000) Liquid crystal over silicon SLM micro-mirror fabrication using chemical mechanical planarisation. *IEEE Trans. Elec. Dev.*,
45 Kozakai, T., Hori, Y., Sugihara, M. *et al.* (2001) Reflection improvement for D-ILA. *SID Digest* **32**, 910–13.
46 Katayama, T., Natsuhori, H., Moroboshi, T. *et al.* (2001) D-ILA device for top-end projection display (QXGA). *SID Digest* **32**, 976–9.
47 Lee, Y. M., Parker, W., Bodammer, G. and Underwood, I. (2004) Characterisation of inter-metal dielectric deposition processes on CMOS backplanes for liquid crystal on silicon microdisplays. *IEE Proc. J. Optoelectronics* **151**, 53–9.
48 Calton, D. W., Seunarine, K., Bodammer, G. and Underwood, I. (2000) Liquid crystal flow control using microfabrication techniques. *IEE Proc. Optoelectronics* **147**, 163–70.
49 Personal communication in relation to Hashimoto *et al.* (ref. 51).
50 Jannen, P., Shimizu, J. A., Dean, J. and Albu, R. (2002) Design aspects of a scrolling color LCoS display. *Displays* **23**, 99–108.
51 Hashimoto, S. *et al.* (2005) SXRD (silicon X-tal reflective display): a new display device for projection displays. *SID Digest* **37**, 1362–65.
52 www.umc.com
53 Gale, R. *et al.* (1999) Miniature 1280 by 1024 active matrix liquid crystal displays. *Proc. SPIE*, **3689**, 231–37.
54 Wolf, S. (1986) Isolation technologies. In *Silicon Processing for the VLSI Era*, Vol. 2. Lattice Press.
55 Article on Radiant Images. *Projection Monthly* **1**, 8, August 2003, 10–11.
56 De Smet, H., Van Steen, J. V. and Colson, P. (2002) Use of stitching in microdisplay fabrication. *Proc. SPIE*, **4657**, 23–30.
57 Ferris-Prabhu, A. V. (1992) *Introduction to Semiconductor Device Yield Modeling*. Artech House.
58 Prince, E. (1992) *Semiconductor Memories*, 2nd edn. John Wiley & Sons, Chichester.
59 Wolf, S. (1986) Section 8.3 – Dynamic random access memories, in *Silicon Processing for the VLSI Era*, Vol. 2. Lattice Press.

4

Transmission Microdisplay Structure

4.1 Background

Transmission microdisplays imply optical modulating devices, essentially LCD, since transmission MEMS are at a primitive stage of development. In a transmission device, the input and output rays (or beams) have independent polarisation states, while the polarisation states are common in reflection devices, as illustrated in Figure 4.1. Separating the input and output spaces is generally an advantage, allowing independent polarization settings (P_1, P_2), and shorter back-focal length in optical magnification. The 90-degree twisted nematic configuration provides performance advantages restricted to transmission, as analyzed in a later chapter. However, development of the vertical aligned nematic (VAN) cell and commitment to inorganic alignment has eroded the advantage of the 90-degree TN transmission cell.

Reflection provides the compactness of a folded optical system, with heat sinking on the non-optical side, but requires a beam-splitter or off-axis optical system. Throughput efficiency mandates a polarizing beam-splitter in projection systems. The optical path length in the liquid crystal doubles on reflection, making the transmission cell gap twice the reflective cell gap for the same optical modulation. Nematic liquid crystal response time is inversely proportional to the square of the cell gap, conferring an important response speed advantage to the reflective cell. Electrode symmetry (ITO/ITO) of the transmission cell is an advantage over the asymmetry (ITO/aluminum) of the reflective cell, where contact potential and ionic effects may introduce spurious shifts in DC potential.

The first commercial active matrix microdisplay projector used polysilicon-addressed nematic liquid crystal transmission devices. The pace of LCOS development followed advances in silicon technology in general; initially, poor light shielding restricted activity to miniature direct-view and near-to-eye systems. LCOS is now fully competitive in all applications, challenging the long-term supremacy of transmission technologies.

Introduction to Microdisplays D. Armitage, I. Underwood and S-T. Wu

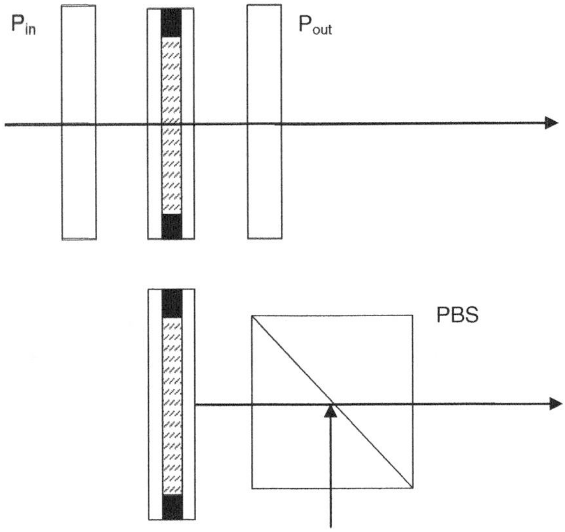

Figure 4.1 Input and output polarization settings P_{in} and P_{out} are independent in transmission, compared to the polarizing beam-splitter used in reflection

High-temperature polysilicon (HTPS) addressing is still the dominant transmission technology, but greatly improved over the pioneering devices. The demand for higher pixel count, reduced pixel size, and high throughput efficiency has prompted clever circuit layout maximizing the pixel aperture ratio. Moreover, a microlens array focuses light into the pixel aperture, increasing throughput efficiency. Overall system cost dictates the success of competing microdisplays, not just the microdisplay cost.

The success of polysilicon transmission microdisplays has prompted alternative transmission technologies. Transferred silicon technology employs CMOS addressing circuitry fabricated in a standard foundry, subsequently transferred to glass by a proprietary process. Recent developments include a revival of transmission microdisplays employing single-crystal silicon-on-sapphire addressing.

4.2 Thin Film Transistors

Thin film transistors (TFTs) are transistors fabricated by thin film deposition methods, the only option for large-area flat panel displays.[1] TFT operation is similar to c-silicon MOSFETs employed in reflective microdisplays, discussed in Chapter 2. Similar equations apply, but the TFT carrier mobility is always inferior to the mono-crystalline equivalent device, with consequent loss of performance; moreover, higher trap density and interface states enhance leakage current. Figure 4.2 indicates the basic TFT structure, with the gate electrode uppermost. Inverting the gate structure, putting the gate electrode in contact with the substrate, has advantages in some processes. In n-channel TFTs, the source and drain are heavily doped n^+, and for p-channel TFTs, the source and drain are heavily doped p^+, while the channel remains close to intrinsic. Light p-doping of the n-channel region close to the n^+ source and drain (lightly doped drain, LDD) enhances voltage levels and reduces leakage current in n-channel devices. The insulating substrate provides electrical isolation of each TFT, avoiding the complication of p–n junction isolation inherent to c-silicon wafer technology.

Applied voltages distinguish drain and source, drain identifying with the highest positive potential. MOSFETS isolated by p–n junctions need to maintain reverse bias, restricting the polarity of the applied voltage. The greater bulk of TFTs compared to MOSFETs increases stray capacitance, and the

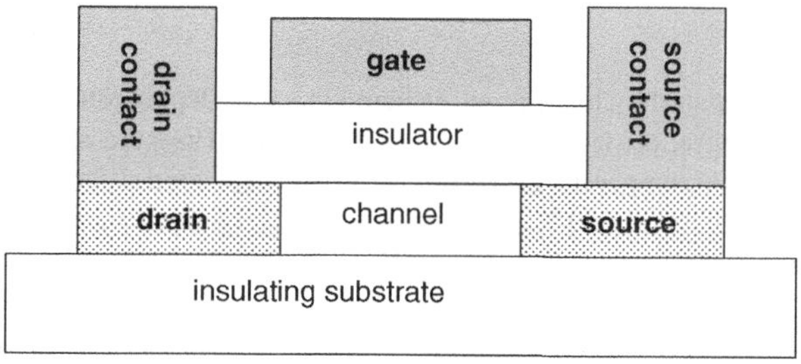

Figure 4.2 Basic structure of thin film transistor

lower channel mobility enhances gate pulse feed-through.[1] TFT matrix addressing circuits are more susceptible to crosstalk and gate pulse feed-through, and generally require compensation circuits.

Large-area flat panels require a TFT technology with adequate performance at low cost. Early competition between cadmium selenide and hydrogenated amorphous silicon (α-silicon) has long settled in favor of α-silicon TFTs. The dominance of silicon technology in general influenced the choice, although low electron mobility $<1\,\text{cm}^2/\text{Vs}$ and even lower hole mobility are a handicap. Inadequate hole mobility in α-silicon prohibits the fabrication of CMOS circuitry, while electron mobility limits the response speed, preventing application to peripheral addressing circuitry. Threshold voltage stability is another issue, tolerated by the simple on/off switching demands of the pixel transistor. External addressing circuitry connected to the α-silicon active-matrix rows and columns requires thousands of interconnects, prompting the development of technologies such as anisotropic conducting film (ACF), chip-on-glass (COG), gold bumps, and thermal-compression eutectic bonding.

Much effort has gone into the development of a TFT technology that could provide all the peripheral circuit addressing and matrix line drivers, in addition to pixel isolation, drastically reducing the interconnections between the display and video source. Cadmium selenide attracted some attention, but the dominance of silicon technology prompted much greater interest in polycrystalline silicon (polysilicon, poly-Si). Polysilicon grain size $\sim 0.1\,\mu\text{m}$ is small compared with the TFT channel gap, but supports electron mobility $>100\,\text{cm}^2/\text{Vs}$, and hole mobility $>50\,\text{cm}^2/\text{Vs}$. The development of polysilicon TFT technology to address direct-view displays has proved difficult. High-temperature polysilicon processing is effective, but requires a quartz substrate, making it too expensive for substantial areas. However, the cost of quartz is minor for microdisplays, making HTPS the initial and continuing choice for transmission microdisplays.

4.3 Polysilicon

4.3.1 Background

Polysilicon matrix addressing, complete with peripheral support circuitry, has been the holy grail of LCD flat panels for many years, while systems-on-glass in general are a further incentive for polysilicon development. However, the success of α-silicon at surprisingly large display areas, and the packaging technique for drivers and interconnects, blunts the incentive for polysilicon. The arrival of OLEDs requiring current drive rather than voltage favors the higher carrier mobility of polysilicon, bolstering interest in polysilicon development; however, superior TFT uniformity in α-silicon is at present an advantage in achieving current uniformity in OLEDs. In microdisplays, small pixel size limits the choice to c-silicon or polysilicon. The aperture ratio advantage of polysilicon in transmission microdisplays does not apply to OLED microdisplays, since the emissive layer covers the pixel circuitry. The absence of polysilicon-addressed OLED microdisplays reinforces the view that c-silicon technology has technical and cost advantages.

4.3.2 Preparation

Polysilicon is generally prepared by low-pressure chemical vapor deposition (LPCVD) from silane gas (SiH_4) at elevated temperature, providing uniformity and purity. Pressure and temperature determine growth rate and grain size. At temperatures below 580 °C, the deposited silicon film is essentially amorphous, while temperatures beyond 650 °C give rough and loosely adhering films. Grain size increases with deposition temperature, along with surface roughness; amorphous depositions have the smoothest surface. Crystallization from the amorphous film state at ~1000 °C gives the grain size ~100 nm, uniformity and surface smoothness.

Fabrication of an HTPS TFT begins with the deposition of an amorphous silicon film ~15 nm thick on a quartz substrate, followed by photolithography defining the TFT semiconductor region. Annealing at ~1000 °C converts the silicon film to polysilicon; addition of an oxidizing atmosphere grows a silicon oxide layer ~10 nm forming the gate insulation. A polysilicon deposition followed by photolithography forms the TFT gate and circuit interconnections. Further high-temperature annealing lowers the gate interface states. The quality of polysilicon, thermal oxide, and gate interface justifies the expense of HTPS. Ion implantation forms the lightly doped regions of source and drain, with the gate electrode acting as a mask protecting the channel region. Patterned photoresist protects the lightly doped regions adjacent to the gate electrode, while further ion implantation forms the heavily doped source and drain regions, with the channel protected by the gate electrode. A protective CVD oxide layer covers the TFT structure, with the source and drain connections formed by standard lithographic methods, before a final protective oxide layer.

The first microdisplay projector employed TFTs formed from HTPS, and continuous development maintains a strong market presence.[2] Low-temperature polysilicon (LTPS) processes, consistent with inexpensive glass, are developing to meet the cost requirements of large area displays. Laser pulse heating briefly raises the temperature of the silicon layer into the recrystallization region, before the substrate mass can respond. Laser recrystallization and annealing methods have some success, but with difficulty in maintaining uniformity over large areas in a manufacturing environment. An alternative method of solid-phase crystallization using metal catalysis produces continuous-grain polysilicon (CGP) having uniformly oriented polycrystallinity with enhanced mobility.[3,4] Modest area LTPS direct-view displays, now marketed, are expected to grow in area and sales volume for both liquid crystal and OLED applications. The cost saving of LTPS is significant in microdisplays, prompting development efforts.[5]

4.4 Polysilicon LC Microdisplay

4.4.1 Matrix Addressing

Figure 4.3 indicates the circuitry incorporated in a commercial microdisplay.[6] The control ASIC performs serial-to-parallel and digital-to-analog conversion, followed by polarity inversion, feeding a six-fold parallel analog data signal to the microdisplay. Ghost reduction and crosstalk reduction circuits are included, in addition to the standard matrix-addressing structure. Leakage current in polysilicon TFTs influences the matrix-addressing design. In standard DRAM-type addressing the pixel is isolated from the column addressing line by a TFT (see Chapter 2). Leakage through that TFT influences the pixel voltage between refresh cycles. Frame inversion is sensitive to leakage current, imposing line inversion on TFT matrix addressing.

Frame inversion achieves AC drive by reversing the liquid crystal voltage on alternate frames. Therefore, the column address voltage changes polarity relative to the liquid crystal common ITO electrode, on alternate frames. Consequently, the leakage current through the TFT charges the pixel towards the average potential of the column. However, the average potential depends on the row location of the pixel. Rows early in the row scanning cycle will experience the same column polarity for

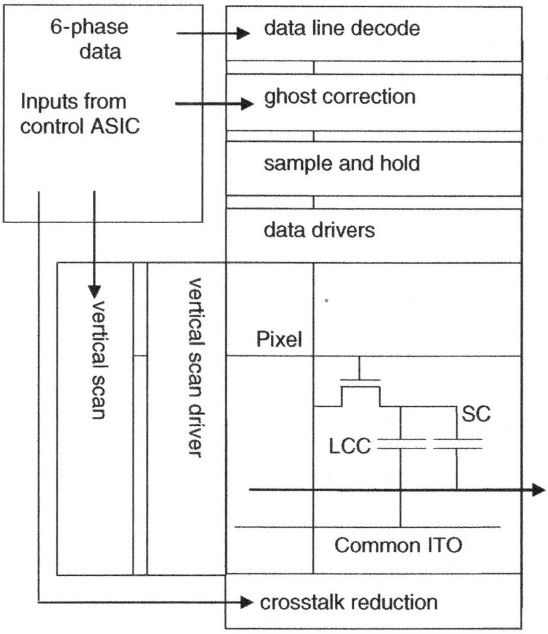

Figure 4.3 On-chip circuitry includes standard address decoders, latches, and drivers; in addition, ghost correction and crosstalk reduction circuitry. Storage capacitance (SC) returns to a controlled potential, and is effectively in parallel with the liquid crystal capacitance (LCC). Reprinted courtesy of Society for Information Display

most of the frame, but rows late in the cycle will experience the reverse column polarity for most of the cycle. Consequently leakage produces a noticeable luminance gradient from top to bottom of a uniform image. Moreover, vertical crosstalk among pixels is introduced by TFT leakage. Frame inversion of voltage is never perfect, and one polarity RMS voltage is significantly higher than the alternate RMS voltage, resulting in flicker at half the frame frequency. The standard video frame rate of 60 Hz must be doubled to 120 Hz to suppress flicker in frame inversion addressing.

Line inversion reverses the liquid crystal voltage polarity on alternate rows, so the column voltage averages to the liquid crystal common ITO electrode potential. Therefore TFT leakage charges the pixel towards common potential for all pixels. Consequently, luminance uniformity is maintained and pixel crosstalk minimized. Imperfect line inversion gives rise to flicker, but it is at the highest spatial frequency that the display can present. A low flicker at the highest spatial frequency is well below visual perception. Line inversion has the disadvantage that adjacent pixels at the same gray level have opposite voltage polarity, generating large fringe field effects (see Chapter 5). The fringe field effects are hidden by the small aperture ratio inherent to polysilicon transmission microdisplays, making line inversion the favored addressing mode. Recent developments, shrinking the pixel size to 10 μm, require fringe field suppression achieved by "alternate frame scan," a modification of line inversion (see section 4.4.6).

A sustained image may persist as a "ghost image," well beyond the addressing period, due to ionic charging in the liquid crystal.[7] Eliminating a signal-dependent DC component, and DC voltages in general applied to the liquid crystal, avoids ionic charging and consequent ghost images. Gate voltage feed-through by way of gate–drain capacitance introduces DC, as do TFT leakage and crosstalk.[1] Adjustment of the common electrode potential cancels the DC level to first approximation. Ghost circuitry injects appropriate voltage into the data signal to cancel residual DC, preventing ghost development. Crosstalk reduction shifts the data voltage to cancel crosstalk errors. Combining three images, by say a color cube, requires the formation of a mirror image on one of the microdisplays to match the reflected images of the other two microdisplays (see projection chapter). Circuitry built into the microdisplay performs right/left and up/down inversion of the image, according to preset controls.

Adjustments in pixel voltages compensate cell gap variations over the device area. The voltage adjustments are written into look-up tables in the companion ASIC during test of color uniformity. For this purpose, the addressing matrix is subdivided into 32 horizontal and 24 vertical regions.[6] The look-up table with the help of interpolation determines the data voltage correction in each region. More details appear in the assembly and packaging chapter.

4.4.2 Physical Layout

The development of polysilicon TFTs to address liquid crystals resulted in the early demonstration of transmission microdisplays.[2,8] Continuing development of polysilicon sustains its prominent role in microdisplay projectors.[6,9,10] Commercial pressure to reduce costs and enhance performance demands a reduction in pixel size. Opaque addressing circuitry limits the clear aperture of transmission microdisplays, as illustrated in Figure 4.4. Stacked capacitor geometry minimizes the storage capacitance area, which is substantially larger than the TFT area; moreover, locating the capacitor on the addressing lines further conserves pixel aperture. A black matrix defines the active pixel area, eliminating pixel boundary effects such as fringe field distortion of the liquid crystal. Finally, a microlens array focuses light into the pixel aperture to enhance throughput. Building the microlens into the common electrode substrate maximizes numerical aperture to collect more light.

Figure 4.5 indicates the pixel cross-section, showing the TFT fabricated in a substrate recess. A video data line connects to the source, and the drain connects to the pixel electrode. The gate connects to line-enable, controlling the on/off state of the TFT. It is essential to screen the TFT from projector intense readout light, requiring opaque metal shields overlapping above and below the TFT area. A substrate embedded shield below the TFT, together with a shield in the ITO counter electrode above

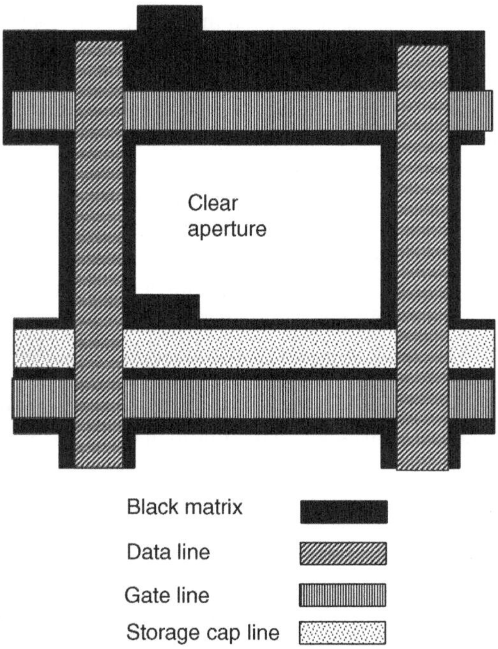

Figure 4.4 Transmission microdisplay pixel layout. Reprinted courtesy of Society for Information Display

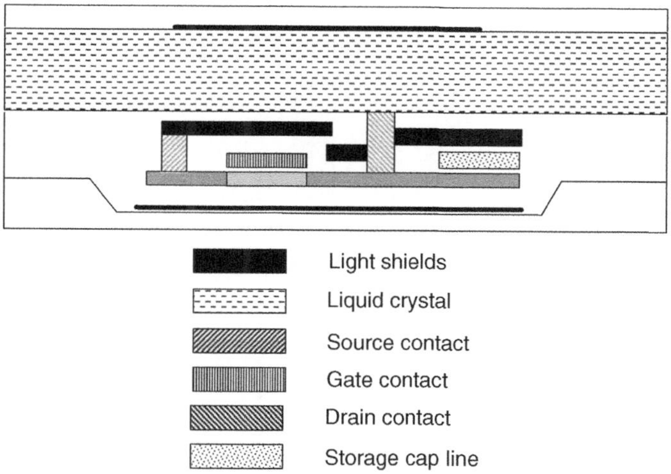

Figure 4.5 Transmission microdisplay pixel TFT cross-section. Reprinted courtesy of Society for Information Display

the TFT, form the primary light shield. The TFT source and drain aluminum connections extend to provide secondary screening of the polysilicon channel. TFT photocurrent leakage compromises the voltage-holding ratio between refresh cycles, and introduces undesirable signal-voltage dependent DC potentials into the liquid crystal.

Pixel storage capacitance connects to the capacitance return line as indicated in Figures 4.3 and 4.4; storage capacitance enhances holding ratio and reduces crosstalk. The pixel area devoted to storage capacitance is much higher than that occupied by the TFT. Capacitance is proportional to area, dielectric permittivity and inverse thickness. Extending the TFT gate thermal oxidation process to the storage capacitance dielectric minimizes the dielectric thickness, and benefits from the reliability of thermal oxidation. Successive capacitance layers stacked with interleaving electrodes reduce the area sacrificed by the pixel, but require successive processing steps. The higher dielectric constant of silicon nitride makes it an attractive dielectric alternative. Burying the storage capacitor under the data and enable lines frees more pixel area.

Successive design cycles of the polysilicon transmission microdisplay incorporate advances in silicon technology. Recessed substrate accommodation of circuitry and addressing lines, together with chemical mechanical polishing, produce the flat surface illustrated in Figure 4.6, which compares successive designs. A flat surface promotes uniform liquid crystal alignment, generating device contrast ratio >500. Design rules of 0.6 μm provide 12 μm pixel pitch with 59% aperture ratio. Figure 4.7 illustrates success in maintaining or improving aperture ratio with shrinking pixel pitch over numerous design cycles dating from the first commercial polysilicon microdisplay projector.

4.4.3 Aperture Ratio

The reflective nature of LCOS hides the pixel circuitry behind the pixel mirror, allowing the aperture ratio to approach 100%. Transmission microdisplays cannot compete on aperture ratio; however, Figure 4.7 shows that with clever design AR > 50% at 10 μm pixel pitch can be achieved. The transmission loss includes successive passes through an air/glass interface and ITO interface. Such losses are duplicated in a reflective device, which suffers an additional reflective loss at the mirror/dielectric interface. LCOS throughput efficiency wins at sufficiently small pixel pitch, but detailed comparison should include the microlens array incorporated in transmission, and dielectric enhancement of aluminum reflectivity in LCOS.

Figure 4.6 Flat surface technology improvement comparing D3 and D4 microdisplay designs; bar = 5 μm. Reprinted courtesy of Seiko Epson Corp.

4.4.4 Microlens Array

A microlens array expands the effective pixel aperture, as shown in Figure 4.8. The *f*-number of the microlens should be as small as possible to gather maximum light. Low f-number follows from fabricating the ITO counter electrode and pixel black matrix on 30 μm glass or quartz, before cementing to form the microlens array.[6] Pixel pitch 12 μm accommodates $\sim 30/12\sqrt{2} = f$-1.78 for the microlens. Choosing a high-index glass for the microlens plate, and a low-index optical adhesive, minimizes the focal length $F = nr/\Delta n$, where n is refractive index, Δn index difference, and r radius of curvature. The required low *f*-number implies an aspherical microlens.

Optical throughput efficiency depends on the microdisplay area, and optical divergence is limited by the projection lens *f*-number; details are available in the projection display chapter. Incorporating

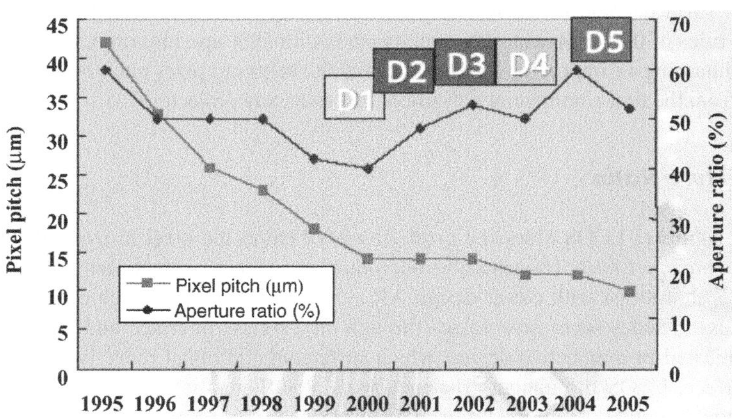

Figure 4.7 Maintenance of aperture ratio with reduction in pixel pitch following improved design over the commercial history of polysilicon microdisplays. Reprinted courtesy of Seiko Epson Corp.

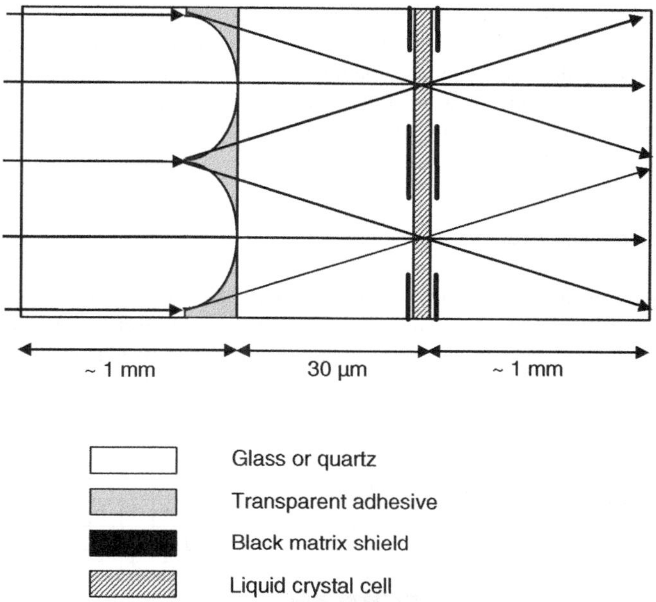

Figure 4.8 Effective increase in pixel aperture ratio by microlens array. Reprinted courtesy of Society for Information Display

a microlens increases the optical divergence, sacrificing some light, or demanding a more expensive low *f*-number projection lens. The gain in focusing the light into the clear pixel aperture is balanced against optical divergence loss. In a practical system, throughput efficiency improves 50% or more by the addition of a microlens array.[6]

4.4.5 Performance

Figure 4.9 is a picture of a polysilicon transmission microdisplay with 1280 × 720 resolution and 0.7-inch diagonal. A twisted nematic with 2.5 μm cell gap and 5 V drive ensure rapid response to video,

Figure 4.9 Polysilicon transmission microdisplay with 720 × 1280 resolution and 0.7-inch diagonal. Reprinted courtesy of Seiko Epson Corp.

particularly at the elevated operating temperature of the microdisplay. Early microdisplays employed plastic packaging, but increasing power density at high resolution requires metallic packaging to help dissipate heat. A standard flex cable connects the microdisplay to control circuits. Transmission liquid crystal microdisplays have the advantage of simpler optics, since the input and output lights follow different paths, but are restricted to analog voltage addressing due to the circuit complexity and response speed required in digital methods.

4.4.6 Recent Developments

Recent developments include increased resolution 1080×1920, at 0.87-inch diagonal, with $10\,\mu m$ pixel pitch and 51% aperture ratio.[11,12] The size of the fringe field disclination contracts in order to maximize the aperture ratio and contrast ratio, as shown in Figure 4.10. Changing from line inversion

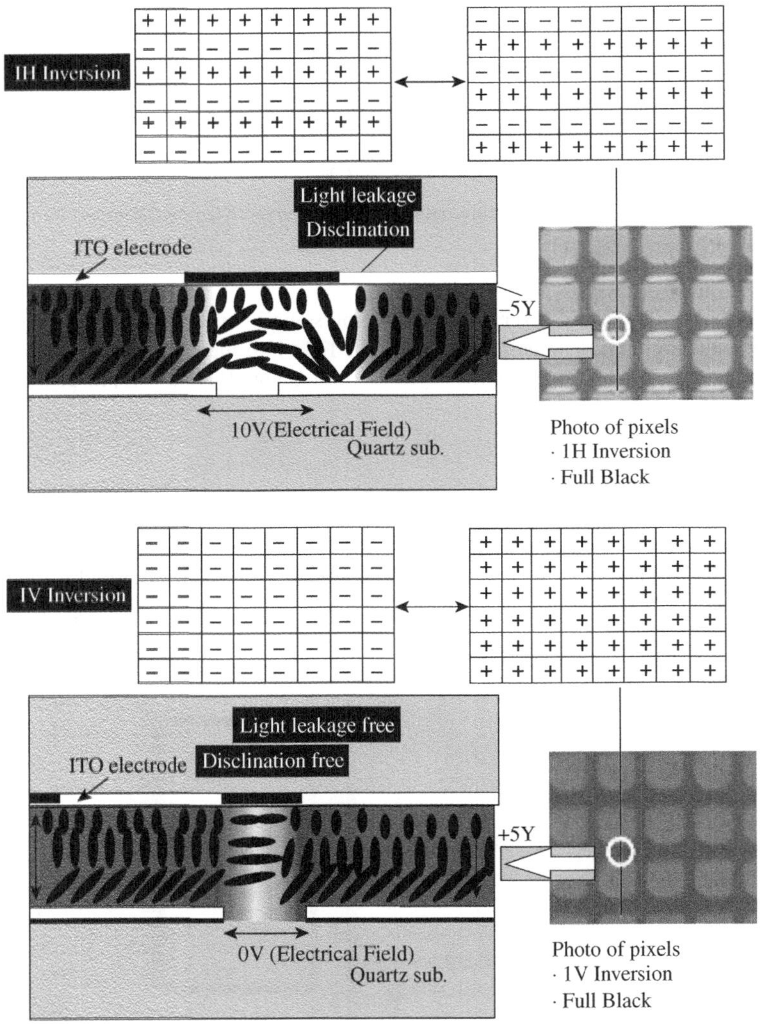

Figure 4.10 Comparing fringe field effect in line inversion (upper) and frame inversion (lower). Reprinted courtesy of Society for Information Display

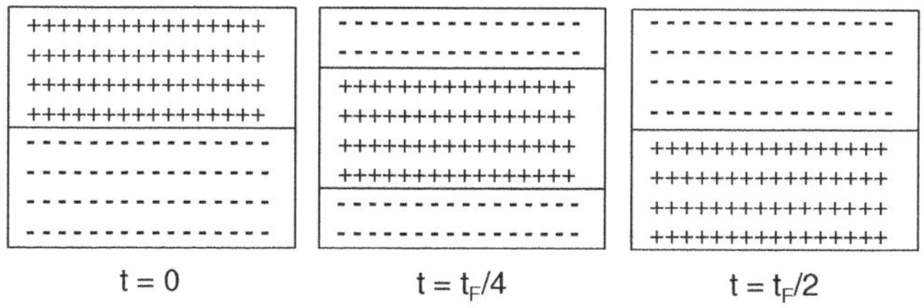

Figure 4.11 Alternate frame scan addressing. Plus signs indicate areas where positive polarity is applied to liquid crystal, while negative signs indicate negative polarity. Addressing field rate 120 Hz, video frame rate 60 Hz has period $t_F = 16.7$ ms

to frame inversion eliminates the potential difference between dark pixels, minimizing the fringe effect in the liquid crystal. The black matrix hiding the fringing becomes narrower, increasing the aperture ratio.

Frame inversion eliminates polarity cancellation in vertical crosstalk, amplifying crosstalk to an intolerable level in polysilicon addressing. The alternate frame scan (AFS) addressing scheme illustrated in Figure 4.11, combines the advantages of line inversion and frame inversion.[12] Horizontal lines do not follow simple geometric progression, but jump forward from N_y to $N_y + N/2$ then backward $N/2 - 1$ to line $N_y + 1$, etc., where horizontal line number N_y is modulo N, the total number of horizontal lines. Addressing polarity alternates across the liquid crystal at each jump, providing the alternating potential advantage of line inversion, while adjacent dark pixels are essentially at the same potential. Two high-fringe-field pixel lines scroll with the polarity reversal boundary. For any given horizontal line, the high transverse field is only present for two line scan periods ~15 μs at refresh, which is below the nematic response time. The pixels are addressed at 120 Hz, twice the video frame rate 60 Hz, having period $t_F = 16.7$ ms. Any asymmetry in polarity inversion produces a flicker at half the addressing rate, requiring the higher 120 Hz addressing to attenuate flicker.

4.5 Transferred Silicon

4.5.1 Concept

The desire to combine the performance of c-silicon CMOS with transmission liquid crystal microdisplay optics led to the development of transferred silicon methods. The wafer input to a foundry provides the semiconducting material for CMOS fabrication and the supporting substrate. After the fabrication is completed, the wafer substrate is not essential, yet it is surprising that the circuitry can be separated and transferred to a glass substrate in a reproducible manner. The initial Kopin process took a standard silicon wafer, oxidized the surface, but revealed regions of the silicon crystal to provide seed surface, deposited a polysilicon layer, and then recrystallized that layer into near single crystal form.[13] A standard foundry processed the wafer, producing the desired CMOS active matrix addressing circuits. A selective etching process lifted the silicon circuitry, including a thin layer of silicon, from the wafer, allowing transfer to a transparent glass substrate.[14] The layer of silicon is thin enough to form transparent pixel electrodes. The remaining assembly steps follow the pattern of polysilicon transmission microdisplays.

There is considerable experimental background to transferred silicon. It is often achieved through a release layer (e.g. oxide) facilitating chemical or mechanical separation of the desired silicon layer, while recycling the original wafer. Alternatively, the unwanted silicon can be etched away using the implanted oxide as an etch-stop layer, sacrificing the starting wafer. The new glass substrate attaches to the silicon, either before or after transfer. Prior attachment gives easier mechanics, but inhibits conformation to the new substrate, and may restrict access to the release layer.

The high mobility of single-crystal silicon provides superior transistor characteristics and low leakage; moreover, manufacture of the silicon backplane of a liquid crystal microdisplay by a silicon foundry is a further advantage. Polysilicon-based microdisplays lack foundry support, implying high manufacturing investment. The Kopin method of creating a transmission microdisplay from a foundry-produced silicon backplane has led to commercial success in NTE applications.

4.5.2 Process

The starting point of the Kopin process has evolved to a silicon wafer with an ion implanted oxide layer buried under 0.3 μm c-silicon. Ion implanting the starting wafer with oxygen atoms achieves the same result as recrystallization, but with improved crystalline silicon, following anneal. A standard foundry processes the wafer to form CMOS circuitry in the 0.3 μm c-silicon surface layer. The silicon layer is thin enough to provide adequate optical transmission for pixel electrodes; improved transmission requires ITO deposition and removal of silicon. A high-yielding proprietary manufacturing process transfers the CMOS circuit to a glass substrate.[15] The transfer adhesive limits further high-temperature processing. Figure 4.12 lists the process flow from incoming wafer to final device.

Color display favors RGB color pixels. Figure 4.13 shows pixel color filters incorporated under the circuitry prior to transfer.[16]

4.5.3 Performance

Low transmission efficiency ~27%, exacerbated by silicon-electrode transmission losses, rules out projection applications.[16] The Kopin process has achieved a pixel pitch of 6 × 11 μm to date, supporting a 15 × 15 μm tri-color pixel for NTE applications. Aperture ratio achieves >45%, and response time <5 ms at 35 °C. Response speed is marginal for field-sequential-color methods where the aperture ratio increases to ~50%.

4.6 Silicon-on-Sapphire

Crystal silicon-on-insulator technology provides dielectric isolation of circuit elements, where lower parasitic capacitance enhances the frequency response over p–n junction isolation in c-silicon. Growth

Silicon wafer
Implant oxide
Fabricate CMOS circuits
Liftoff CMOS & silicon layer
Transfer to glass
Assemble with ITO glass
Singulate
Fill with LC and plug cell
Package and test

Figure 4.12 Transferred silicon process flow

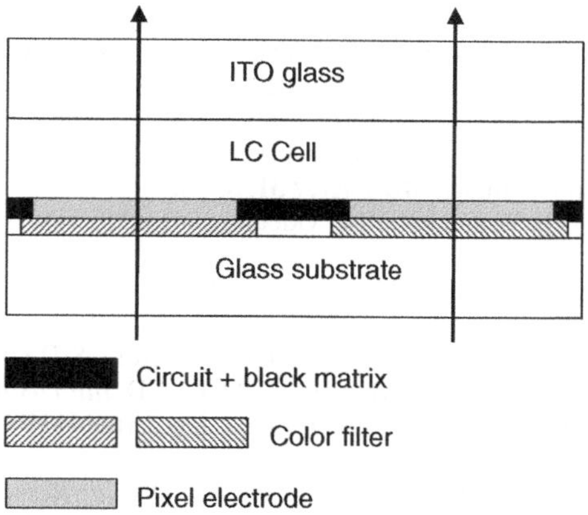

Figure 4.13 Transferred silicon transmission microdisplay incorporating RGB color filter pixels

of a c-silicon film on an insulating substrate requires a close match in lattice constant. Sapphire matches the silicon lattice sufficiently to promote single-crystal formation. Silicon-on-sapphire (SOS) has been in service for many years, offering radiation hardness and speed advantage at a premium price over standard CMOS. SOS crystal defects were higher than standard silicon wafers, resulting in lower yield and consequent higher price, compounded by the cost of sapphire. The application of SOS to liquid crystal displays was explored in early work on matrix addressing.[17]

The march of technology has reduced the cost of SOS sufficiently to compete with polysilicon and transferred silicon microdisplays.[18] The high thermal conductivity of sapphire provides a bonus in thermal dissipation, effectively heat sinking the device. SOS supports minimum circuit features comparable with standard CMOS, while the speed can be faster. Addressing easily supports 120 Hz frame inversion, lowering fringe field and consequent nematic disclinations, allowing reduced black masking of disclinations to increase aperture. Birefingence $\Delta n = 0.008$ is a downside to sapphire, requiring optical polarization to be oriented along the optic axis, and compensation at low f-number. An ITO counter electrode on optical glass that matches sapphire thermal expansion completes the liquid crystal cell.

4.7 Closing Comment

The advantages of transmission microdisplays have been identified in the first section of this chapter. SOS addressing speed raises the prospect of a competitive transmission microdisplay for field-sequential-color projection. However, adequate response speed is difficult to achieve in nematic liquid crystals, and reflective cells have an intrinsic speed advantage due to smaller cell gap for given retardation modulation.

4.8 References

1 Lueder, E. (2001) *Liquid Crystal Displays*. John Wiley.
2 Morozumi, S. *et al.* (1986) LCD full-color video projector. *SID Digest* **12**, 375–8.
3 Ohtani, H. *et al.* (1998) A 60-in HDTV rear-projector with continuous-grain-silicon technology. *SID Digest* **29**, 467–70.
4 Matsuo, T. and Muramatsu, T. (2004) CG silicon technology and development of system on panel. *SID Digest* **35**, 856–9.

5 Tomihari, Y. *et al.* (2004) A low temperature poly-Si TFT liquid crystal light valve (LCLV) with a novel light-shielding structure for high performance projection displays. *SID Digest* **35**, 972–5.
6 Aruga, S. (2002) High temperature poly-Si TFT LCD for projection. *Intl Display Manufacturing Conference*, 551–6.
7 Vershueren, A. R. M. *et al.* (2000) DC current measurements on liquid crystal cells as indicative tool for image retention phenomena, *SID Digest* **31**, 55–8.
8 Morozumi, S. *et al.* (1983) B/W and color LC video display addressed by poly-Si TFTs. *SID Digest* **9**, 156–7.
9 Matsuo, M. *et al.* (1994) Full-color VGA poly-Si TFT-LCDs with completely integrated drivers. *SID Digest* **25**, 87–90.
10 Okamoto, N. (2001) Developments in p-Si TFT LCD projectors. *SID Digest* **32**, 1176–9.
11 Koike, S. (2005) New technological trend in HTPS TFT LCD. *Intl Display Works*, Japan, 1883–5.
12 Hosaka, H. *et al.* (2005) New driving method for next-generation HTPS TFT light valves. *Intl Display Works*, Japan, 1935–8.
13 Salerno, J. P. (1994) Single-crystal silicon AMLCDs. *Intl Display Research Conference*, 39–44.
14 Salerno, J. P. *et al.* (1992) Single-Crystal Silicon Transmissive AMLCD. *SID Digest* **22**, 63–6.
15 Gale, R. *et al.* (1999) Miniature 1280 by 1024 active matrix liquid crystal displays. *Proc. SPIE*, **3689**, 231–7.
16 Woodard, O., Gale, R. and Reese, C. (2003) The colorization of solid state AMLCDs for military head mounted displays (HMDs) and other viewer applications. *Proc. SPIE*, **5079**, 19–30.
17 Lipton, L. T. *et al.* (1975) A liquid crystal television display using a silicon-on-sapphire switching array. *SID Digest* **6**, 78–9.
18 Shimabukuro, R. L., Russell, S. D. and Offord, B. W. (2003) Method for fabricating an electrically addressable silicon-on-sapphire light valve. US Patent 6,617,187.

5

Transmissive Liquid Crystal Microdisplays

5.1 Introduction

Both direct-view and projection methods have been developed for large screen displays. Direct-view high-definition televisions (HDTVs) using thin film transistor liquid crystal display (TFT-LCD) with diagonal sizes larger than 104 inches have been demonstrated by Sharp, LG Philips, and Samsung.[1] By today's technology, to fabricate a direct-view TFT LCD panel larger than 104 inches is feasible, but is still quite expensive. As the manufacturing technology advances, the panel size will further increase and the cost will gradually decrease. To obtain screen sizes beyond 104 inches for high-definition TVs, boardroom presentation, and electronic cinema, projection is a more economic approach. Two types of projection displays have been configured: front and rear projections. Front projection utilizes a distant screen to view the magnified image. The viewers are on the same side as the projector. On the other hand, rear projection configurations enclose the magnification optics behind an imaging screen to produce a self-contained system. For a 60- to 100-inch screen size, the display should have 500 to 2000 lumens so that the images are still viewable under room light conditions.

Both transmissive and reflective liquid crystal microdisplays have been employed in projectors.[2] In this chapter, we focus on transmission-type liquid crystal microdisplays. Reflective liquid crystal microdisplays, especially liquid-crystal-on-silicon (LCOS), will be described in Chapter 6.

5.2 TFT-LCD

Figure 5.1 shows the device structure of a transmissive TFT-LCD using amorphous silicon (α-Si) transistors for large screen display. An LCD does not emit light. Instead, it functions as a spatial light modulator. Thus, a backlight is needed. A diffuser is used to homogenize the backlight. Since most LCDs

Introduction to Microdisplays D. Armitage, I. Underwood and S-T. Wu

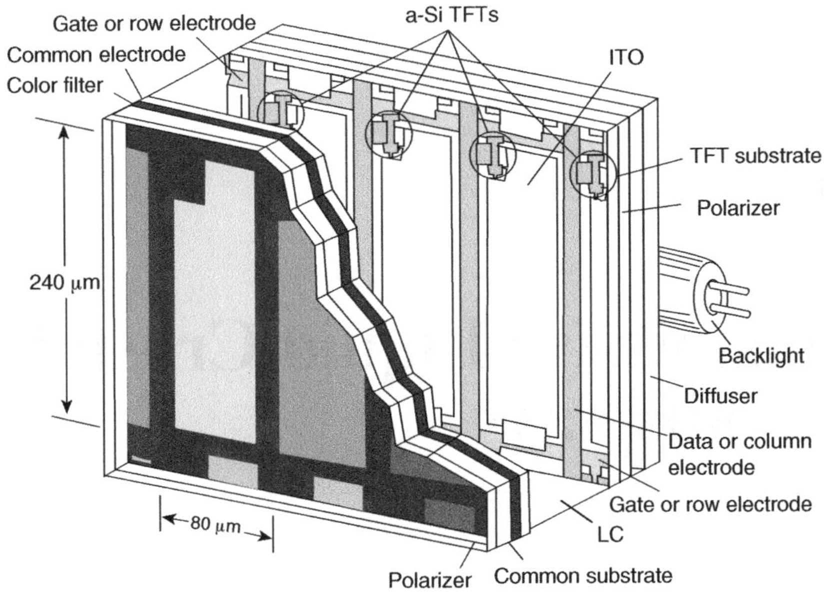

Figure 5.1 Device structure of a transmissive TFT-LCD

require a linearly polarized light for achieving high contrast ratio, two sheets of stretched dichroic polarizers are commonly used for large-screen direct-view displays. The first glass substrate contains TFT arrays, which serve as light switches. Each display pixel is independently controlled by a TFT. Since TFTs need to be protected from light exposure, the actual aperture ratio (the transparent indium-tin-oxide electrode area) varies from 80 to 50%, depending on the device resolution and panel size. The LC layer is sandwiched between two substrates. The cell gap is usually controlled at around 4 μm for transmissive LCDs and 2 μm for reflective LCDs. The performance of the display such as light throughput, response time, and viewing angle are all determined by the LC mode employed.

For direct-view displays, such as notebook computers and desktop monitors, compact size and light weight are critically important. Under such circumstances, color filters are imbedded on the inner side of the second substrate. Three sub-pixels (red, green, and blue) form a color pixel. Each sub-pixel transmits only one color; the rest is absorbed. Thus, the transmittance of the color filters alone is less than 33%, and is usually ~25%. After having considered polarizers, color filters, and TFT aperture ratio, the overall system optical efficiency is about 6–7% for a direct-view LCD panel. If wide-view technology is implemented,[3] the total light efficiency is decreased to ~5%. Low optical efficiency implies high power consumption. For portable displays, low power consumption is desirable because it lengthens the battery operating hours.

For projection displays using an arc lamp, there is more space available for color filters and polarization conversion. The red, green, and blue colors are separated by the dichroic mirrors. Three monochrome LCD panels are used. As a result, the resolution is tripled and optical efficiency enhanced. However, the corresponding pixels in each LCD panel need to be aligned.

Both amorphous silicon (a-Si),[4] poly-silicon (p-Si)[5] and crystalline-silicon (c-Si)[6] thin film transistor (TFT) technologies have found their own niche for different liquid crystal displays. Table 5.1 summarizes the basic performance characteristics of each technology. From Table 5.1, the electron mobility of p-Si and c-Si is, respectively, two and three orders of magnitude higher than that of a-Si. As a result, their pixel sizes can be made much smaller, i.e. the pixel density is higher. Normally, we define pixel density by dots per inch (dpi). A typical pixel size for a-Si TFT LCD is ~80 μm, which corresponds to

Table 5.1 Comparison of amorphous (a), poly (p) and crystalline (c) silicon TFT technologies[6]

Parameters	a-Si	p-Si	c-Si
Electron mobility	1	10^2	10^3
TFT OFF-current (A)	10^{-12}	10^{-12}	10^{-13}
TFT ON-current (A)	10^{-4}	10^{-4}	10^{-3}
Dots per inch	300	1000	2000
Pixel size (μm)	80	25	10
Infrastructure	Good	Fair	Excellent
Panel size	Large and medium	Medium and small	Small

~300 dpi. Owing to the smaller pixel sizes, the p-Si and c-Si LCD panels can reach beyond 1000 and 2000 dpi, respectively.

Cost is always an important factor for all the display devices. Amorphous silicon involves a simple film deposition process so that its fabrication cost for a large display panel is the lowest among the three silicon technologies. Low-temperature p-Si needs an additional laser annealing process. Although its pixel density is increased by ~3×, its cost is also increased. On the other hand, c-Si is particularly attractive for microdisplays where high pixel density is a significant advantage. A smaller image size leads to smaller and lower cost optics. Therefore, for high-resolution large panel direct-view display, α-Si TFT-LCD has its cost advantage. On the other hand, a p-Si TFT-LCD or c-Si backplane LCD is preferred for transmissive and reflective projection displays. The transmissive LC microdisplay we discuss in this chapter is based mainly on high-temperature p-Si TFT-LCD. The low-temperature p-Si TFT is aimed at small to medium-sized direct-view displays. The reflective LC microdisplay described in Chapter 6 is focused on LCOS.

5.3 Projection System

Figure 5.2 shows a simple LCD projector using three transmissive p-Si TFT-LCD panels. More sophisticated projection systems are discussed in Chapter 10. The incoming white light from a lamp is divided into red, green and blue (RGB) channels by the dichroic mirrors. Each beam passes through a monochrome TFT-LCD panel. The X-cube recombines the RGB beams and the projection lens throws the images on to the screen.

In each channel, an LCD panel is sandwiched between two linear polarizers. The 90° twisted-nematic (TN)[7] and vertical-aligned nematic (VAN)[8] LC cells are two common choices. The TN cell offers three advantages: (1) It uses a positive dielectric anisotropy ($\Delta\varepsilon$)liquid crystal material and simple rubbing (or SiO_2 evaporation) technique for aligning LC molecules. The manufacturing yield is high so that the panel cost is relatively low. (2) The transmissive 90°-TN cell is insensitive to wavelength so that the same cell can, in principle, be used for the RGB channels. (3) The TN cell exhibits a relatively high contrast ratio (~500:1) because of the natural phase compensation effect on the cell boundaries. On the other hand, the VAN cell possesses an unprecedented contrast ratio (>1000:1) for normally incident light. A special requirement for the VAN cell is that it uses negative $\Delta\varepsilon$ liquid crystal and homeotropic LC alignment. Because of their small pixel size and small electrode gap, fringing field effects become important. How the fringing field affects the display contrast ratio and optical efficiency will be discussed later.

The light throughput of an LCD projector is determined by the lamp efficacy, etendue of the projection optics compared to the lamp etendue, optical efficiency of the LCD panel, and efficiency of the optical components. High overall system efficiency would enable the use of a low-wattage arc lamp,

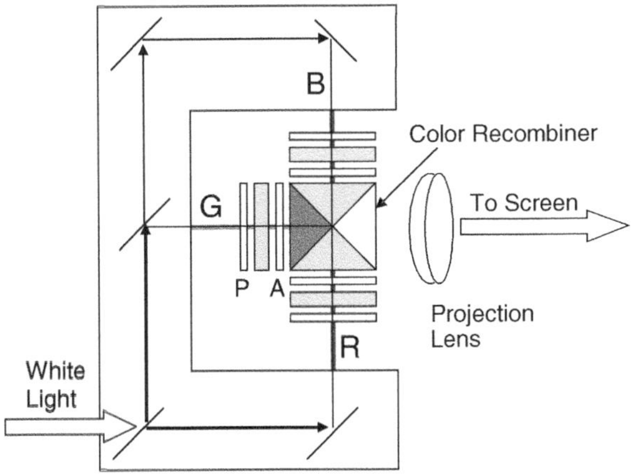

Figure 5.2 Projection system using three transmissive LCD panels

which in turn increases the lamp lifetime and eases the thermal management. Most high-contrast LC devices use linearly polarized light. To convert unpolarized light into linearly polarized light, several polarization conversion schemes have been developed and implemented in projection systems.[9–11] Typical conversion efficiency reaches 70–80%.

The high-power arc lamp usually exhibits two hot-spots near the electrodes. To homogenize the beam, fly's eye[12] and rod integrator[13] are two common approaches. A rod integrator mixes the beam well so that the center-to-corner brightness uniformity is better than 85%. This integrator also helps to define the aspect ratio of the display. For example, the conventional CRT TV set has 4:3 aspect ratio while HDTV has 16:9.

On the LCD panel, TFT aperture ratio plays an important role in determining the light efficiency. For a high-resolution transmissive LC microdisplay, to achieve a large aperture ratio, high-temperature poly-silicon TFT-LCD is commonly used. The pixel size can vary from ~12 to 40 μm, depending on the required device resolution. In such a small pixel, the TFT aperture ratio is usually reduced as the pixel density increases. However, as the technology advances the TFT aperture ratio continues to improve.[14] Another method for enhancing TFT aperture ratio is to use a microlens array.[15] Then the effective TFT aperture ratio is enhanced to >80% for an XGA (1024 × 768) device. Overall a power efficiency of ~10–15 lm/W in the projection system using three transmissive LCD panels has been routinely obtained. Under such a circumstance, a 100 W short-arc lamp would produce 1000–1500 lumens on the screen. The lifetime of the low-wattage lamp reportedly exceeds 10,000 hours. In a palm-sized ultra-portable LCD projector,[16] RGB light emitting diodes (LEDs) replace the arc lamp. The LED lifetime consistently exceeds 50,000 hours, but low lumen output restricts screen size.

5.4 Twisted Nematic Cells

The 90° twisted nematic (TN) cell has been used widely in projection LCDs for its simplicity. In a 90° TN cell, the LC directors are twisted by 90° continuously from the front to the back substrates. The crossed polarizer configuration is preferred as it leads to normally white (NW) display. In a NW display, the dark state can be controlled by the applied voltage. Thus, the contrast ratio is insensitive to the cell gap and operating temperature fluctuations. The front polarizer can be either parallel (called e-mode) or perpendicular (o-mode) to the front LC directors. Their electro-optic effects are rather similar. Let us illustrate the operation principle using the e-mode sketched in Figure 5.3.

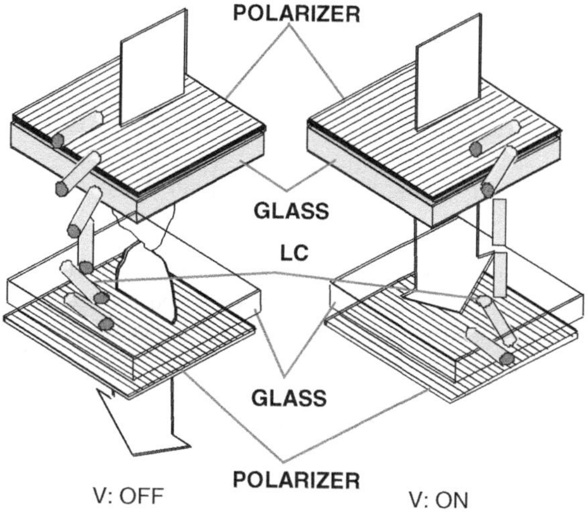

Figure 5.3 Device structure and operation principle of a 90° twist-nematic LC cell. Left: $V = 0$, right: $V = 5\,V_{RMS}$

As shown, the impinging linearly polarized light follows the twist of the LC directors provided that the product of cell gap (d) and birefringence (Δn) is much larger than the wavelength, i.e. $d\Delta n \gg \lambda$. This condition is known as Mauguin's limit.[17] Under such a circumstance, the waveguiding (also known as polarization rotation) effect occurs. The incoming linearly polarized light follows the molecular twist. After traversing through the cell, the polarization remains linear except that the axis is rotated by 90° and is parallel to the transmission axis of the analyzer. Therefore, the analyzer transmits the light leading to a normally white mode. In a voltage-on state ($\sim 5\,V_{RMS}$), the bulk LC directors are reoriented by the electric field except for the boundary layers. As a result, the polarization rotation effect of the TN cell is disrupted. The incoming light no longer follows the molecular twist and, thus, experiences no polarization change as it passes through the LC medium.[18] The outgoing light is absorbed by the crossed analyzer resulting in a dark state. On the other hand, if the analyzer is parallel to the polarizer, then the normally black mode appears. The normally white mode exhibits a larger cell gap tolerance and higher contrast ratio for a broadband light source. Thus, it has been used widely in direct-view and projection displays.

5.4.1 Jones Matrices

The normalized transmittance ($T_\perp$) of a general TN cell can be described by the following Jones matrices as $T_\perp = |M|^2$:[19]

$$M = \begin{vmatrix} \cos\beta & \sin\beta \end{vmatrix} \begin{vmatrix} \cos\phi & -\sin\phi \\ \sin\phi & \cos\phi \end{vmatrix} \begin{vmatrix} \cos X - i\dfrac{\Gamma}{2}\dfrac{\sin X}{X} & \phi\dfrac{\sin X}{X} \\ -\phi\dfrac{\sin X}{X} & \cos X + i\dfrac{\Gamma}{2}\dfrac{\sin X}{X} \end{vmatrix} \begin{vmatrix} -\sin\beta \\ \cos\beta \end{vmatrix} \tag{5.1}$$

where β is the angle between the polarization axis and the front LC director, ϕ is the twist angle, $X = (\phi^2 + (\Gamma/2)^2)^{1/2}$ and $\Gamma = 2\pi d\Delta n/\lambda$ where d is the cell gap. By simple algebraic calculations, the normalized transmittance is derived:

$$T_\perp = \left(\frac{\phi}{X}\cos\phi\sin X - \sin\phi\cos X\right)^2 + \left(\frac{\Gamma}{2}\frac{\sin X}{X}\right)^2 \sin^2(\phi - 2\beta). \tag{5.2}$$

Equation (5.2) is a general formula describing the light transmittance of a TN cell as a function of twist angle (ϕ), β angle, and $d\Delta n/\lambda$.

For the 90° TN cell, we substitute $\phi = \pi/2$ into (5.2) and obtain the normalized light transmittance as follows:

$$T_{\perp} = \cos^2 X + \left(\frac{\Gamma}{2X}\cos 2\beta\right)^2 \sin^2 X \,. \tag{5.3}$$

Equation (5.3) has a trivial solution, which is cos $X = \pm 1$, for the normally white mode where $T_{\perp} = 1$. When cos $X = \pm 1$ (i.e. $X = m\pi$; m = integer), then sin $X = 0$ and the second term in (5.3) vanishes. Therefore, $T_{\perp} = 1$, which is independent of β. Setting $X = m\pi$ and knowing that $\Gamma = 2\pi d\Delta n\lambda$, we derive the Gooch–Tarry condition for the 90° TN cell:[20]

$$\frac{d\Delta n}{\lambda} = \sqrt{m^2 - \frac{1}{4}} \,. \tag{5.4}$$

When $m = 1$, $d\Delta n/\lambda = \sqrt{3}/2$. This is the Gooch–Tarry first minimum condition for 90° TN cells. The Gooch–Tarry analysis shows practical TN cells fall short of the Mauguin limit.

Figure 5.4 plots the normalized light transmittance ($T_{\perp}$) of the 90° TN cell as a function of $d\Delta n/\lambda$ at $\beta = 0$, 15, 30 and 45 degrees. From (5.3) and (5.4), the first $T_{\perp} = 1$ occurs at $d\Delta n/\lambda = \sqrt{3}/2$, independent of β. However, the color dispersion (i.e. the wavelength dependency of the light transmittance) strongly depends on β. At $\beta = 0$, $T_{\perp}$ is insensitive to $d\Delta n/\lambda$ beyond the first minimum. As β increases, the color dispersion increases.

Figure 5.5 plots the voltage-dependent light transmittance (VT) of the 90° TN cell at $\beta = 0$. The cell gap is chosen to be $d = 4.8\,\mu$m and $\Delta n = 0.1$ to satisfy the first minimum condition at $\lambda = 550$ nm. In the figure, the bandwidths of the RGB bands are considered, as indicated in the curves. Within each band, the VT curve is calculated in 10 nm increments and the final result represents the average

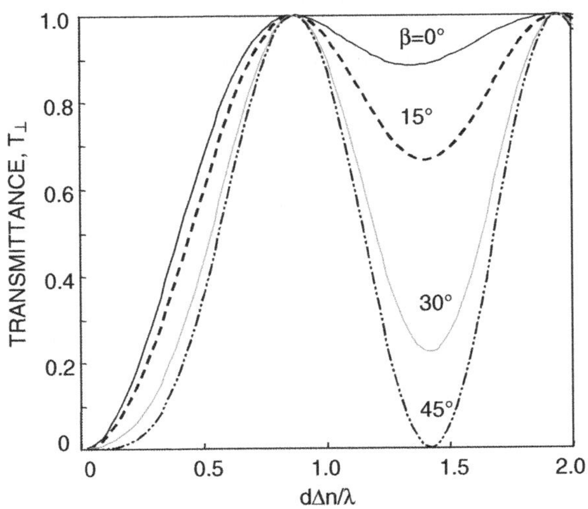

Figure 5.4 Normalized transmittance of a transmissive 90° TN cell as a function of $d\Delta n/\lambda$ and β; polarizers are crossed

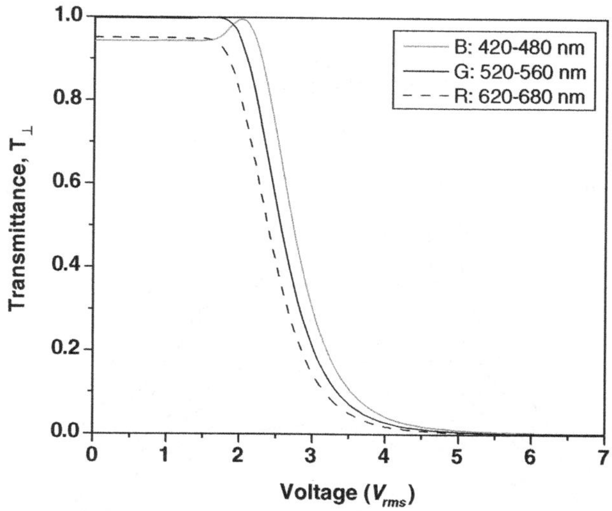

Figure 5.5 Voltage-dependent transmittance of a 90° TN cell. LC: MLC-6694-000; $d\Delta n = 480\,\text{nm}$, $\beta = 0$

transmittance of the given band. From Figure 5.5, the color dispersion of the TN cell is fairly small. Thus, the same panel can be used for RGB channels.

5.4.2 Viewing Angle

Viewing angle is an important characteristic for both direct-view and projection displays. For direct-view LCD, the viewing angle is usually defined by the iso-contrast region with contrast ratio greater than 10:1. Projection display demands a much higher contrast ratio, say >1000:1. If the LCD panel has a wide acceptance angle, then a smaller *f*-number projection lens can be used. The light collection efficiency would be greatly improved.[21]

Figure 5.6 plots the iso-contrast contour of a 90° TN LC normally white display at $\beta = 0$. The LC employed is Merck MLC-6694-000; its parameters are listed as follows: $n_e = 1.5567$, $n_0 = 1.4729$ (at $\lambda = 589\,\text{nm}$ and $T = 20°\text{C}$), $\varepsilon_{\parallel} = 10.1$, $\varepsilon_{\perp} = 3.3$ (at 1 kHz), $K_{11} = 13.1\,\text{pN}$ and $K_{33} = 22.7\,\text{pN}$. The cell gap is $d \sim 5.7\,\mu\text{m}$ so that $d\Delta n = 480\,\text{nm}$. The pretilt angle is ~2°, and the cell gap to pitch length ratio is $d/p = 0.25$. In a TN cell, a small percentage of chiral dopant (whose pitch length is p) is added to the LC mixture to ensure that the LC directors are twisted 90° from top to bottom substrates, with consistent handedness. During simulations, the off-state voltage was set at $0.7\,V_{RMS}$ and on-state voltage at $6.0\,V_{RMS}$. The bandwidth considered for simulation is from $\lambda = 520$ to $560\,\text{nm}$. From Figure 5.6, the 300:1 contour lines look like a butterfly and the 1000:1 contour looks like a boomerang. The horizontal viewing angle is relatively symmetric, but the vertical viewing angle is narrow and asymmetric. The intrinsically narrow viewing angle limits the contrast ratio of the TN LCD projectors to ~500:1, unless employing optical compensation.

5.5 Vertically Aligned Nematic (VAN) Cells

A VAN cell offers an excellent contrast ratio and relatively low operating voltage, although the manufacturing process is more complicated and the negative $\Delta\varepsilon$ LC materials are more difficult to synthesize.

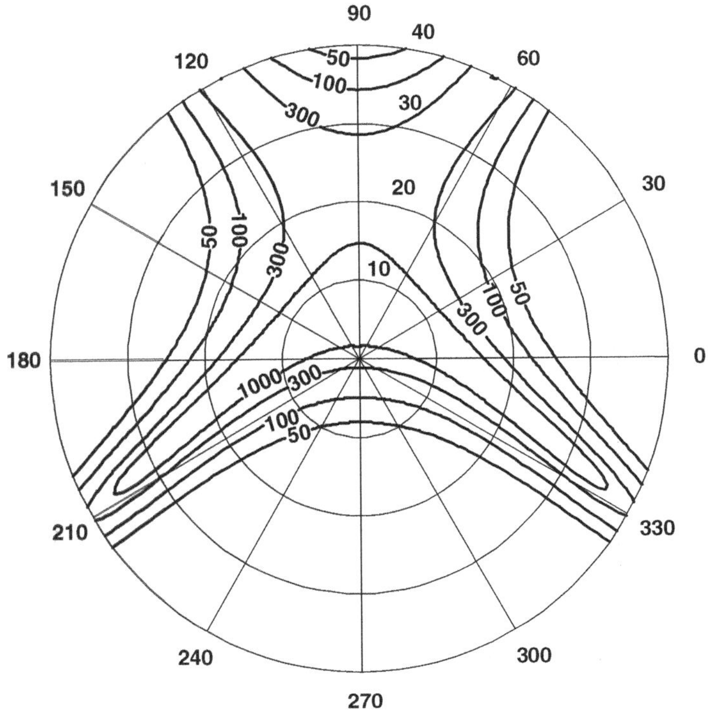

Figure 5.6 Iso-contrast contour of a transmissive 90° TN cell. LC: MLC-6694-000; $d\Delta n = 480\,\text{nm}$, $\beta = 0$, bandwidth 520–560 nm

In the voltage-off state, the LC directors are aligned nearly perpendicular to the substrates, with a small pretilt angle. As the applied voltage exceeds the Freedericksz transition threshold, the LC directors are reoriented by the electric field to be parallel to the substrate surface except for the boundary layers. Since the polarizer's axis is 45° with respect to the LC directors, phase retardation effect occurs. The outgoing light transmits through the crossed polarizer.

The advantages of VAN cells are twofold: (1) its black state is very dark and is insensitive to the cell gap, wavelength, and temperature, and (2) its on-state voltage depends on the $d\Delta n/\lambda$ value and can be made relatively low. The former is due to the nearly vertical alignment and the crossed polarizers while the latter is due to the pure birefringence effect. However, the VAN cells exhibit two major drawbacks: (1) its viewing angle is narrow, and (2) its molecular alignment is more difficult. To widen the viewing zone for accepting a smaller *f*-number projection lens, a negative birefringence C-plate can be considered. In general, the pure birefringence effect of VAN is easier to compensate with retarder plates than TN, facilitating CR > 1000 in VAN projectors.

5.5.1 LC Alignment

Both organic (polyimide) and inorganic (SiO_x) alignment layers have been used for aligning negative $\Delta\varepsilon$ LC mixtures. For large-screen direct-view LCDs, buffed polyimide is a favorable choice because of its low cost. For projection LCDs, SiO_x inorganic alignment layers are more robust than buffed polyimide to high-power arc lamp illumination. High pressure arc lamp radiation consists of rich UV and deep blue components. These wavelengths are harmful to the organic alignment layers and LC materials.[22]

Thus, a long pass filter with a cutoff wavelength above 420 nm should be used in the projection system in order to protect the LCD panels.

SiO_x evaporation

The medium-angle and shallow-angle deposition (MAD/SAD) of SiO_x layers has been used for aligning planar (or homogeneous) LC cells.[23] To extend this MAD/SAD method to produce tilted-homeotropic cells, an extra surfactant layer favoring vertical alignment is needed. In the alcohol treatment method, the ITO glass substrates are first deposited with 150 nm SiO_x layers at oblique angles (30° MAD and 5° SAD) and then thermally exposed to the vapor of long-chain aliphatic alcohol ($C_{18}H_{37}OH$) to bond alkoxy groups to the surface.[24] The LC molecules are aligned nearly perpendicular to the substrate surfaces with a small pretilt angle varying from one to two degrees. Such alcohol-treated homeotropic cells are found to have good temperature and photo stability. Moreover, contrast ratios higher than 1000:1 have been routinely obtained. An imperfection of the MAD/SAD deposition method is that a small splay (similar to open fingers) is observed. That means the LC directors are not exactly parallel to each other. This does not affect the dark state since all the LC directors are perpendicular to the substrate surfaces. However, the on-state transmittance might not reach 100%.

To eliminate the small splay in LC directors and improve manufacturing, a simplified process called moving deposition (MD) has been developed.[25] The MD alignment provides a two-step treatment technique for preparing a substrate that induces a tilt to the LC directors. First, a ~10 nm SiO_2 layer is deposited over the ITO-coated glass by moving the substrate past a magnetron in-line sputtering source. After the deposition of the silica layer, the substrate is contacted with the alcohol vapor at a sufficiently high temperature for the alcohol to react with the hydroxyl groups on the surface of the silica layer. The long-chain alcohol is typically evaporated at 140 °C. The pretilt angle is from 0.5 to 4 degrees, and is of uniform azimuth such that the projections of the directors onto the substrate all lie approximately parallel. Such a MD alignment is stable to temperature variation.

A simplified technique for producing VAN LC alignment by single oblique evaporation of SiO_2 without surfactant has been developed.[26] This single oblique evaporation of SiO_2 alone can produce either homogeneous or homeotropic alignment depending on the dielectric anisotropy of the employed LC materials. Pretilt angle of 2–30° is achieved by varying the deposition thickness and evaporation angle. Since this method does not require alcohol treatment, the process time is shortened and cost reduced. Moreover, without alcohol the ionic contamination is reduced so that the voltage-holding ratio (VHR) is improved. For active matrix LCDs, a high VHR (>95%) is needed in order to avoid image flicker.

In general, the SiO_x evaporation technique produces excellent alignment with good uniformity, high contrast ratio, and sharp threshold. The absence of rubbing marks is important in microdisplays, since they are always subject to magnification; while the elimination of electrostatic charging caused by rubbing benefits all displays. It has been used in some small LCD panels. However, due to the need of vacuum, it is inconvenient and costly for processing large substrates. For large panels, polyimide rubbing is more convenient.

Mechanical buffing

Unidirectional buffing of spin-coated polyimide (PI) film produces sufficient anchoring energy for aligning nematic liquid crystals.[27] Rubbed polyimide surfaces generally exhibit microscopic grooves, which lie parallel to the rubbing direction, together with alignment of polymer chains.[28] If the rubbing direction is along the x-axis of the xy plane, then the benzene ring of the PI film will tilt up a few degrees (for JSR-1, the tilt angle is about 5.9°) along the xz plane.[29] Consequently, the injected LC molecules exhibit a pretilt angle along the xz plane. Due to its simplicity and reliability, mechanical buffing of a polymer-coated surface with fine velvet is the most frequently used alignment technique for large panel LCDs. However, rubbing-induced surface scratches are commonly observed, as shown in Figure 5.7(a).

Some polyimide materials, such as SE-7511L and SE-1211 from Nissan Chemicals and JALS-203 from Japan Synthetic Rubber, are commercially available. To prepare a PI cell, we could spin-coat a thin PI layer, e.g. SE-1211, onto the ITO (indium-tin-oxide)-glass substrates, bake the substrates at

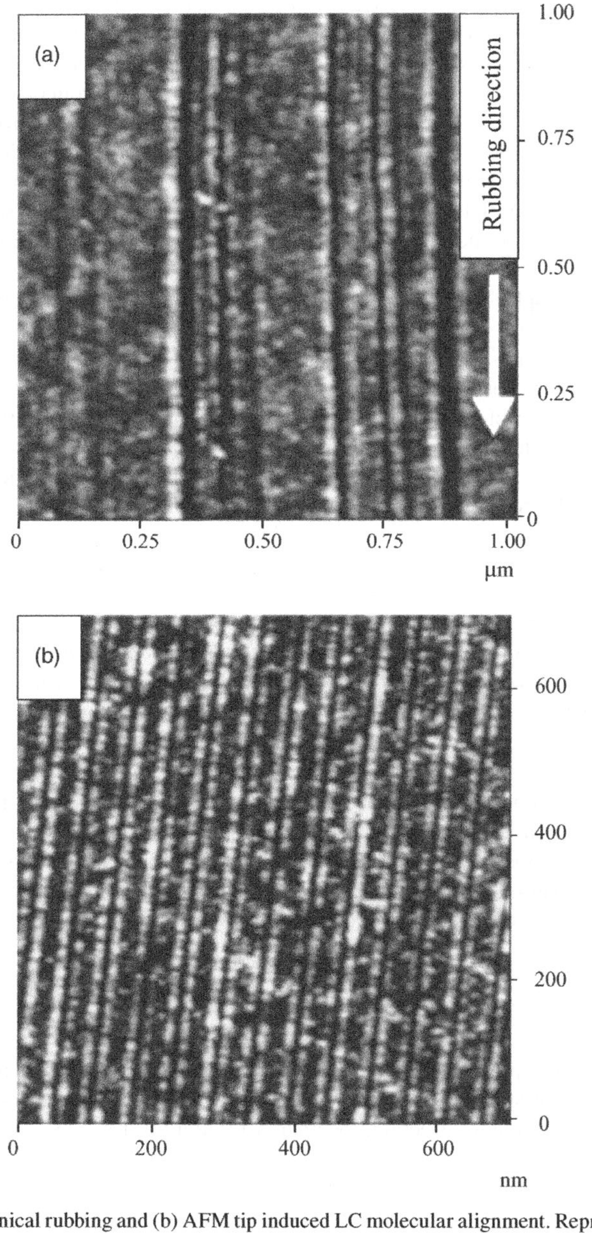

Figure 5.7 (a) Mechanical rubbing and (b) AFM tip induced LC molecular alignment. Reproduced by permission of American Institute of Physics

80 °C for 5 minutes and then at 180 °C for 1 hour. Afterwards, we then gently rub the ITO-glass with cloth in unidirectional parallel motions. The rubbing-induced pretilt angle is about 87° and the polar anchoring energy is $\sim 3 \times 10^{-4}$ J/m^2.[30]

The rubbing process is so simple that it has been commonly utilized for fabricating large LCD panels; however, for projection displays the image is usually magnified by ~50–100×. A small rubbing-induced scratch could be clearly observed on the screen. Rubbing has to be more delicate in microdisplays and is achieved in planar alignment of TN devices, generally with some sacrifice in pretilt magnitude. Rubbed homeotropic alignment is still in the development stage, and it appears that gentle rubbing provides insufficient pretilt, while stronger rubbing generates excessive defects. To reduce mechanical scratches, scanning PI film with an AFM (atomic-force microscope) tip has been investigated.[31] Instead of inducing strain as observed in conventional rubbing, with the AFM scanning the lateral forces exerted by the tip appear to displace the polymer backbone along the scan direction. Since the AFM tip diameter is only 20 nm, the produced surface morphology appears quite smooth, as shown in Figure 5.7(b). However, to make the AFM technique practical, the scanning area has to be increased dramatically.

Ion-beam etching

Ion-beam etching produces high-quality planar LC alignment, which is suitable for projection displays.[32] Unlike mechanical buffing, no scratching is found from ion-beam etching. To etch the PI surface and generate submicron grooves, the incident ion beam is at an oblique angle, say 45°. The pretilt direction is opposite to that of rubbing.[29] A concern for the ion-beam technique is that the process has to be conducted in a vacuum.

5.5.2 Electro-optic Effects

In this section, we show some simulation results to illustrate the basic electro-optic effects of VAN cells. For computer calculations, a Merck high-resistivity mixture MLC-6608 is used. Some physical properties of MLC-6608 are summarized as following: $n_e = 1.558$, $n_0 = 1.476$ (at $\lambda = 589$ nm and $T = 20$°C); clearing point $T_c = 90$°C; dielectric anisotropy $\Delta\varepsilon = -4.2$, and rotational viscosity $\gamma_1 = 186$ mPa at 20 °C. In the VAN cell, the pretilt angle is assumed to be $\theta_p = 88°$ and anchoring energy is assumed to be infinity, unless otherwise specified.

Voltage-dependent transmittance

The normalized transmittance ($T_\perp$) of a homeotropic cell under crossed polarizers is expressed as follows:

$$T_\perp = \sin^2(\pi d \Delta n / \lambda). \tag{5.5}$$

From (5.5), $T_\perp = 1$ occurs at $d\Delta n = \lambda/2$. For the green band at $\lambda = 540$ nm, $d\Delta n \sim 270$ nm. However, this is the minimum $d\Delta n$ value required because under such conditions $T_\perp = 1$ would occur at $V \gg V_{th}$. Due to the finite voltage swing from the TFT, the $d\Delta n$ value needs to be increased. For example, if we limit the on-state voltage to 5 V_{RMS}, then the $d\Delta n$ value needs to be increased to 376 nm. Another important factor to consider is the operating temperature. In a projector, due to the thermal effect of the lamp the TFT-LCD panel temperature can reach 50–60 °C. As the temperature increases, the LC birefringence decreases as

$$\Delta n = (\Delta n)_0 S \tag{5.6}$$

where S is the order parameter and $(\Delta n)_0$ stands for the birefringence at T = 0K. The order parameter can be approximated as[33]

$$S = (1 - T/T_c)^{\alpha} \tag{5.7}$$

Where T_c is the LC clearing temperature and α is a material parameter. It is quite straightforward to determine the parameters $(\Delta n)_0$ and α experimentally. We can measure the temperature-dependent birefringence at a given wavelength and then fit the experimental data with (5.6) and (5.7) using $(\Delta n)_0$ and α as fitting parameters.

For full-color projection displays, refractive index dispersion of the LC employed needs to be taken into consideration. The extended Cauchy equations[34] have been derived and experimentally validated for describing the refractive indices of a multi-component LC mixture:

$$n_e = A_e + \frac{B_e}{\lambda^2} + \frac{C_e}{\lambda^4} \tag{5.8a}$$

$$n_0 = A_0 + \frac{B_0}{\lambda^2} + \frac{C_0}{\lambda^4} \tag{5.8b}$$

Where $A_{e,0}$, $B_{e,0}$ and $C_{e,0}$ are fitting parameters. For the low birefringence ($\Delta n < 0.1$) LC mixtures employed for projection display, the λ^{-4} term can be ignored and only two fitting parameters are required.[35] Figure 5.8 shows such fittings for two low birefringence LC mixtures at $T = 25\,°C$. UCF-280 is a negative $\Delta\varepsilon$ LC mixture while MLC-6241-000 is a Merck positive $\Delta\varepsilon$ TFT LC mixture. The fitting parameters for the n_e and n_0 of UCF-280 and MLC-6241-000 are: $[A_e, B_e] = [1.5443, 0.0083]$ and $[1.5395, 0.0076]$, and $[A_0, B_0] = [1.4689, 0.0053]$ and $[1.4616, 0.0049]$, respectively.

Figure 5.9 plots the voltage-dependent transmittance of the VAN cell using MLC-6608. The bandwidths of the RGB bands considered are as follows: blue band from 420 to 480 nm, green band from 520 to 560 nm, and red band from 620 to 680 nm. The cell gap is $d = 4.5\,\mu m$. To compare the LC mode performance, we only plot the normalized transmittance; the optical losses of the dichroic polarizers and substrate reflections are all ignored. From Figure 5.9, the threshold voltage is about $2\,V_{RMS}$. The

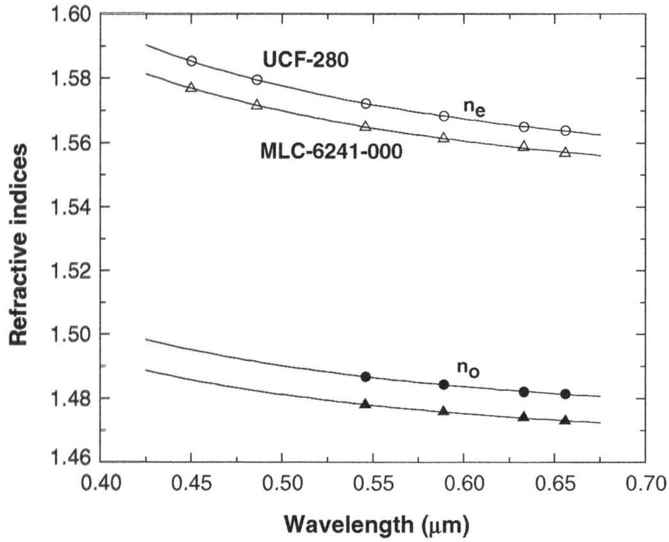

Figure 5.8 Wavelength-dependent refractive indices of UCF-280 ($\Delta\varepsilon < 0$) and MLC-6241-000 ($\Delta\varepsilon > 0$). $T = 25\,°C$. Dots are experimental data and lines are fitted curves using the first two terms of (5.8a) and (5.8b)

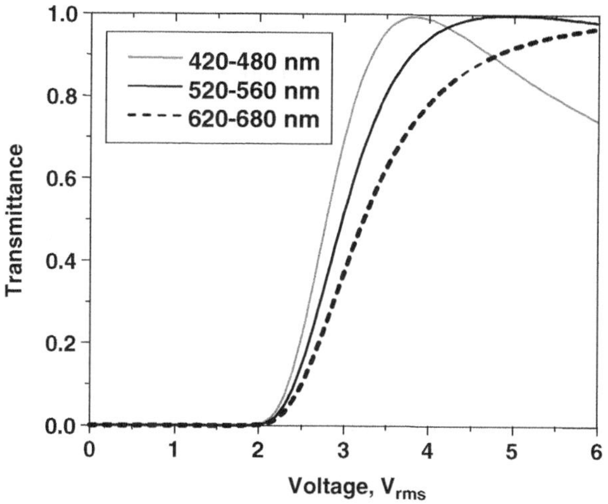

Figure 5.9 Voltage-dependent transmittance of a homeotropic LC cell. MLC-6608, cell gap $d = 4.5\,\mu m$

blue wavelength reaches the maximum at the lowest voltage, followed by green and red because of the combined wavelength and birefringence effects. For the green band centered at $\lambda = 540\,nm$, the on-state voltage is $5\,V_{RMS}$. To achieve a balanced white, lumen content of the green, red, and blue colors should be in the ratio 6:3:1. Thus, the green band plays the most important role among the three primary colors in terms of luminance, due to the spectral sensitivity of the eye. The popular UHP mercury lamp is deficient in red, consequently systems should be designed to optimize the red throughput.

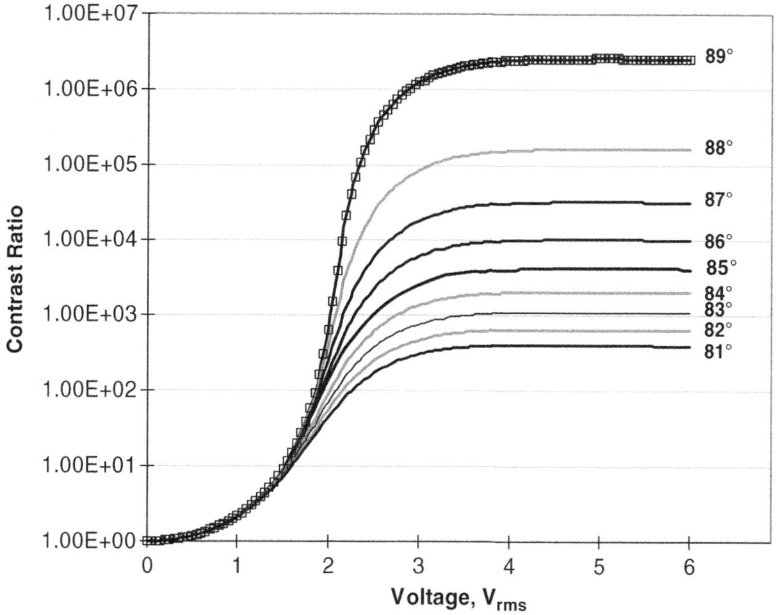

Figure 5.10 Simulated contrast ratio of a homeotropic LC cell at different pretilt angle. MLC-6608, $d = 5.4\,\mu m$

A unique feature of the homeotropic cell is that its dark state is insensitive to the cell gap, wavelength, and birefringence. Using the abovementioned bandwidth in calculation, the voltage-dependent transmittance curves are indeed nearly identical to those using the central wavelength of each band.

Pretilt angle effect

The pretilt angle definition of a VAN cell is somewhat different from that of a TN cell and needs clarification. In a VAN cell, the LC directors are aligned to be nearly perpendicular to the substrates. Thus, when $\theta_p = 90°$ it actually means no pretilt angle. A small pretilt angle away from normal is needed so that the LC molecules know which way to rotate during voltage switching.

Pretilt angle affects the contrast ratio and threshold behavior of a homeotropic cell. Figure 5.10 depicts the simulated contrast ratio of the VAN cell employing MLC-6608 at normal incidence for different pretilt angles. The cell gap used for calculation is 4.5 μm, wavelength $\lambda = 520$–560 nm, and the pretilt angle varies from 89° to 81°. The threshold voltage of MLC-6608 is ~2 V_{RMS}. In theory, sharp threshold behavior exists only when there is no pretilt angle, i.e. $\theta_p = 90°$. Even at $\theta_p = 89°$ (or 1° off normal), the threshold behavior is already not very sharp. Due to the excellent black state of the homeotropic cell, a little transmittance below threshold voltage could result in a noticeable contrast ratio change. During simulation, the polarizers are assumed to have a perfect extinction ratio. As $V > V_{th}$, CR increases sharply. For $\theta_p = 89°$ at 5 V_{RMS}, the simulated CR is greater than 10^6:1. In the voltage-on state, CR decreases as the pretilt angle deviates further from surface normal. From Figure 5.10, in order to obtain CR > 10^4:1, the pretilt angle should be kept above 86°, i.e. less than 4° from normal.

Viewing angle

Figure 5.11(a) plots the iso-contrast contour of the 4.5 μm transmissive VAN cell employing MLC-6608. Although the wavelength used for simulations is $\lambda = 540$ nm, the red and blue panels should have very similar results. From Figure 5.11(a), the homeotropic cell has a rather narrow viewing angle at the ±45° azimuthal angles. In a narrow range along the horizontal and vertical axes, its viewing angles are wide. Overall, the 2000:1 contrast ratio is limited to a ~7° viewing cone.

To widen the viewing angle, a negative birefringence phase compensation film (a negative c-plate) can be considered.[36] The negative c-plate has isotropic refractive indices in the xy plane ($n_x = n_y$), which is parallel to the substrate surface. However, its index in the z-axis is smaller than that in the xy plane ($n_z < n_x = n_y$). Since most LC mixtures have a positive birefringence, i.e. $n_e > n_o$, the negative c-plate would compensate the positive LC birefringence and reduce the light leakage of the LC cell at oblique angles. Therefore, in order to compensate the phase retardation of the VAN cell in the voltage-off state resulting from oblique incidence, an ideal compensation film should possess the following properties: (1) negative birefringence and the same $d\Delta n$ as the LC cell at any incident angle, and (2) similar birefringence dispersion as the employed LC material. The former condition would cancel the phase retardation of the outgoing light after traversing the LC cell over a wide range of incident angles. The latter would result in perfect phase cancellation for all three primary colors intended for a full-color display. By selecting a polycarbonate c-plate with its $d\Delta n = -353$ nm, the viewing cone is improved as shown in Figure 5.11(b). The 2000:1 iso-contrast ratio contour lines extend to beyond the 10° cone. This improved viewing angle would allow a smaller *f*-number projection lens and a larger arc lamp to be used.[37] As a result, the optical efficiency and lamp lifetime are both improved. The major challenge of this c-plate is that it has to withstand the high-power lamp illumination. Thus, thermal-induced stress may cause inhomogeneous phase retardation profile.

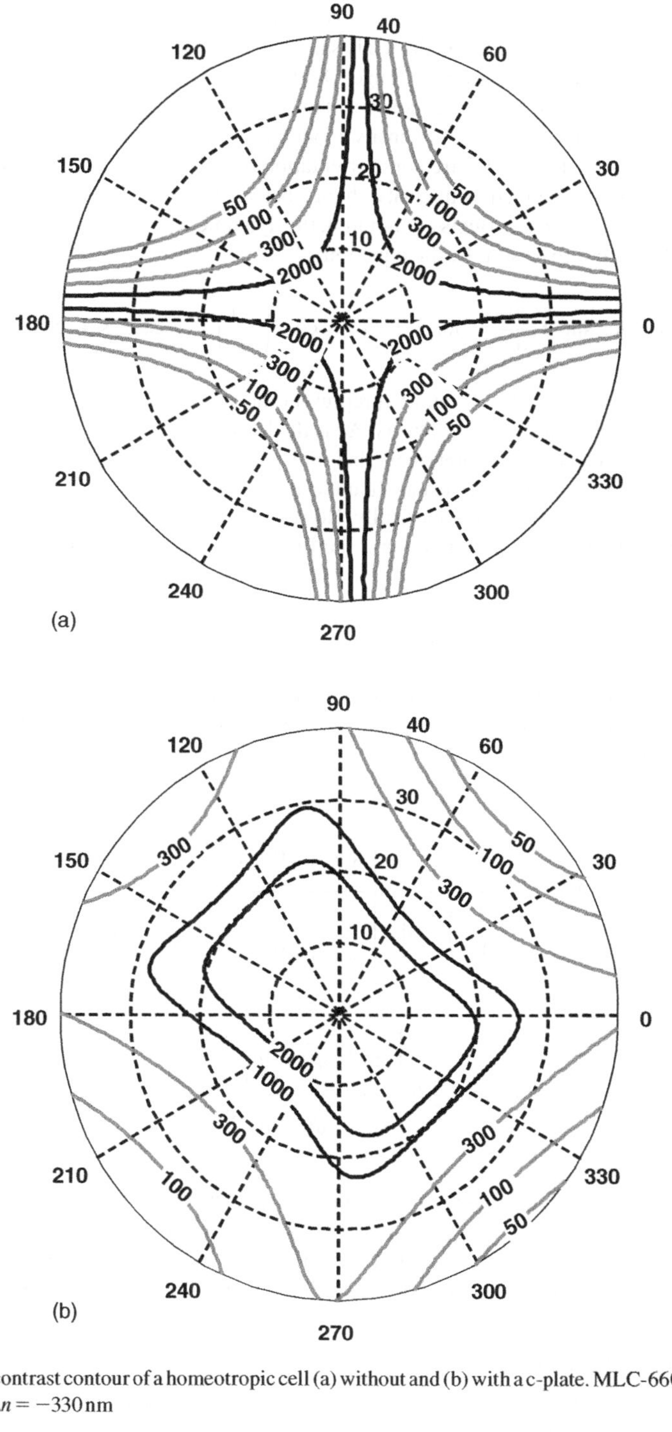

Figure 5.11 Iso-contrast contour of a homeotropic cell (a) without and (b) with a c-plate. MLC-6608, $d\Delta n = 376\,\mathrm{nm}$, negative c-plate $d\Delta n = -330\,\mathrm{nm}$

5.5.3 Response Time

Fast response time is highly desirable for projection displays. A slow LC response would result in image blurring. For display applications, the LC device is sandwiched between two polarizers. The measured response time represents the transmittance change. On the other hand, for a phase-only modulator such as optical phased arrays, the measured response time is the corresponding phase change. There is no doubt that the optical response time for intensity modulation and the phase response time for phase modulation must be related to the LC director reorientation time. To quantify a display device, the rise and decay times are usually defined as intensity change between 10% and 90%.

When the backflow and inertial effects are ignored, the dynamics of the LC director reorientation are described by the following Erickson–Leslie equation: [38,39]

$$\left(K_{11}\cos^2\phi + K_{33}\sin^2\phi\right)\frac{\partial^2\phi}{\partial z^2} + \left(K_{33}-K_{11}\right)\sin\phi\cos\phi\left(\frac{\partial\phi}{\partial z}\right)^2 + \varepsilon_0\Delta\varepsilon E^2\sin\phi\cos\phi = \gamma_1\frac{\partial\phi}{\partial_t} \tag{5.9}$$

where γ_1 is the rotational viscosity, K_{11} and K_{33} represent the splay and bend elastic constants, respectively, $\varepsilon_0\Delta\varepsilon E^2$ is the electric field energy density, $\Delta\varepsilon$ is the LC dielectric anisotropy, and ϕ is the tilt angle of the LC directors. It should be pointed out that in the Erickson–Leslie equation the strong surface anchoring and zero pretilt angle at the surface boundaries are assumed.

In general, (5.9) can only be solved numerically. However, when the tilt angle is small ($\sin\phi \sim \phi$) and $K_{33} \sim K_{11}$ (so-called small angle approximation),[1] the Erickson–Leslie equation is reduced to

$$K_{33}\frac{\partial^2\phi}{\partial z^2} + \varepsilon_0\Delta\varepsilon E^2\phi = \gamma_1\frac{\partial\phi}{\partial t}. \tag{5.10}$$

Under such circumstances, we can derive the analytical solutions for the optical rise (T_{rise}) and decay (T_{decay}) time of the VAN cells.[40] Here, the rise time and decay time are defined when light transmittance changes from 10% to 90%:

$$T_{\text{rise}} = \frac{1}{2}\frac{\tau_0}{\left|\left(\frac{V}{V_{\text{th}}}\right)^2 - 1\right|}\ln\left[\frac{\dfrac{\delta_0/2}{\sin^{-1}\left(\sqrt{0.1}\sin\left(\frac{\delta_0}{2}\right)\right)} - 1}{\dfrac{\delta_0/2}{\sin^{-1}\left(\sqrt{0.9}\sin\left(\frac{\delta_0}{2}\right)\right)} - 1}\right] \tag{5.11}$$

$$T_{\text{decay}} = \frac{\tau_0}{2}\ln\left[\frac{\sin^{-1}\left(\sqrt{0.9}\sin\left(\frac{\delta_0}{2}\right)\right)}{\sin^{-1}\left(\sqrt{0.1}\sin\left(\frac{\delta_0}{2}\right)\right)}\right] \tag{5.12}$$

where τ_0 is the LC director reorientation time ($1 \rightarrow 1/e$). It is related to cell gap d and visco-elastic coefficient (γ_1/K_{33}) as $\tau_0 = \gamma_1 d^2/K_{33}\pi^2$, and δ_0 is the net phase change of the VAN cell. In (5.11) and (5.12), the phase terms inside the brackets do not vary too significantly and can be approximated as a constant. Thus, optical response time is about two times faster than the LC director's reorientation time.

Overdrive and undershoot effects

To improve the response time, a thin cell gap, low-viscosity LC material, elevated temperature operation and driving voltage methods have been considered.[3] To reduce the cell gap, a high birefringence

($\Delta n \sim 0.15$) LC needs to be employed. High birefringence implies a longer molecular conjugation. Usually, the viscosity increases as molecular conjugation increases because of the increased moment of inertia. To lower viscosity, we could take advantage of the built-in thermal effect in a LCD projector. The heat generated by the arc lamp keeps the chassis temperature at about 50–60 °C. For every 15 °C temperature rise, the LC visco-elastic coefficient (γ_1/K_{33}) decreases by ~2×. However, to compensate for the birefringence decrease at elevated temperatures, a slightly thicker cell needs to be used.

The voltage overdrive and undershoot method has been used to reduce LC response time.[41,42] In an LCD, grayscale switching (e.g. from V_1 to V_2; assuming $V_2 > V_1$) is slow, especially in the vicinity of threshold. This is because the voltage difference is small. To shorten the rise time, we could apply a high voltage for a short time and then reduce to a lower level until the optical signal reaches the designated gray level. This is known as the overdrive. During decay from gray level 2 to 1, if we just reduce the voltage from V_2 to V_1 this process will be slow. Instead, we could reduce the voltage to zero for a short period until the transmittance reaches gray level 1 and then apply a holding voltage for it to stabilize. This is known as the undershoot effect, or transient nematic effect. The art of the overdrive and undershoot technique is to foresee the gray level of the input video signals and prepare for the proper voltage and duration of the voltage pulses.

5.6 Fringing Field Effect

In a transmissive microdisplay, the fringing field effect is usually assumed to be negligible because of aperture ratio restriction by the addressing lines, allowing the reverse tilt disclination to be hidden by the black matrix defining the aperture. However, this presumption no longer holds for LC panels using a negative ($\Delta\varepsilon < 0$) LC mixture.

Figure 5.12 shows the fringing field effect on a homeotropic LC cell when the neighboring pixels are off–on–off. The on-pixel is at 5 V_{RMS} and off-pixel at 1 V_{RMS}, which is lower than the threshold voltage $V_{th} = 2\,V_{RMS}$. The LC used for simulation is Merck MLC-6609. The cell parameters are as follows: cell gap 5 μm, pretilt angle 88°, $\lambda = 550$ nm, pixel size 40 μm, and pixel gap 5 μm. Changing the inter-pixel gap from 1 to 5 μm does not affect the fringing field effect noticeably. From Figure 5.12 it can be seen

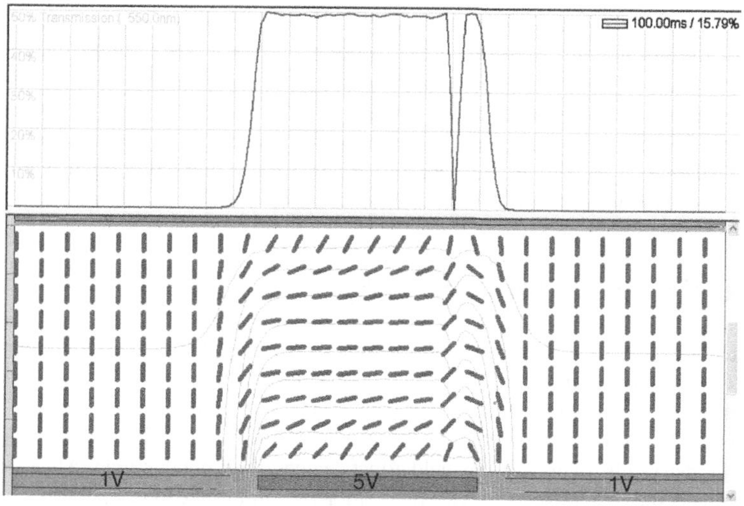

Figure 5.12 Fringing field effect of a transmissive VAN LC cell. The pixel voltages are 1, 5 and 1 V_{RMS}. LC: MLC-6609, $V_{th} \sim 2\,V_{RMS}$, cell gap $d = 5$ μm, pixel size = 40 μm and electrode gap = 5 μm

that the dark pixels remain black. The major effect of the fringing field is on the right edge of the on-pixel. The on-pixel is split into two unequal parts. This not only reduces the optical efficiency, and increases the response time, but also causes the annoying edge effect to the displayed images resolving the pixel.

In HDTV projection applications, the pixel is beyond the limits of visual perception at normal viewing distance. Although the pixel detail is not perceived, the fringe field effect on luminance compromises the limiting resolution. To remain competitive, HTPS microdisplays are squeezing pixel dimensions to 10 μm, while keeping aperture ratio >50%. Resolution is compromised by VAN cell fringe effects, which can no longer be contained by the black matrix aperture.

In order to eliminate the fringing field effect in a transmissive VAN cell, the circularly polarized light is shown to be helpful.[43] Both the image sharpness and the brightness of the panel are dramatically improved. In addition, the dynamic transition time of the device switching from the dark–bright–dark state to the all-bright state is reduced significantly, which sufficiently overcomes the image blurring effect. The idea of using circularly polarized light is more feasible for transmissive LCDs than it is for reflective microdisplay LCDs. In the reflective projection system, if we laminate a quarter-wave film to the exit side of the polarizing beam-splitter, it is difficult to configure a normally black VAN mode because, on double pass, it becomes 1/2 wave making it the bright state. Without using the NB mode the attractive features, such as high contrast ratio and large cell gap tolerance, of VAN mode vanish. This problem can be easily overcome in a transmissive LCD. We could simply replace the linear polarizers with two crossed circular polarizers and the troublesome fringing field effects would be eliminated. However, there is a need to identify circular polarizers with $CR > 2000$ and wide field of view.

By contrast, the transmissive TN cell is less sensitive to fringing field effects. Figure 5.13 shows the simulated 2D transmittance profile of the TN cell. The pixel size is 40 μm and the inter-pixel gap is 5 μm. The $d\Delta n$ is 477.5 nm for the TN cell. As shown in Figure 5.13, the transmittance profile is almost perfect for the TN cell. The calculated optical filled factor is as high as 99% and the dark–bright–dark contrast ratio exceeds 1000:1 at normal incidence. Therefore, the transmissive TN cell is not affected by the fringing field effect. Its low contrast ratio is mainly due to the narrow acceptance angle, as shown in Figure 5.6.

Table 5.2 summarizes the major performance characteristics of the VA and 90° TN cells. The homeotropic cell exhibits an unprecedented contrast ratio within the central 10° viewing cone. If a c-plate is added, its viewing cone is doubled so that a smaller *f*-number projection lens can be used for improving optical efficiency. On the other hand, TN is a mature technology and its production cost is lower. However, its contrast ratio is limited to about 500:1 because its high contrast regime is quite narrow. Despite the manufacturing issues, the homeotropic cell exhibits an annoying edge effect due to the fringing field effects, with high modulation at limiting resolution. The circularly polarized light method has been proposed to eliminate the fringing field effect and thus to improve the display sharpness and brightness. The optical configuration for implementing circular polarization is fairly easy. It can be readily achieved by replacing the linear polarizers by two crossed circular polarizers, having high extinction and wide viewing angle.

5.7 Liquid Crystal Ionic Effects

Ionic impurities are always present in liquid crystals, providing electrical conductivity and charge accumulation effects. Ionic current flow induces the "dynamic scattering effect," employed in the earliest devices, with ionic species introduced into the liquid crystal to optimize performance.[44] Electrolytic effects limited the lifetime of these pioneering devices, prompting research into the behavior of ions and electrodes in liquid crystals. In present-day field effect devices, ionic conduction is a handicap; the manufacture of liquid crystal materials and device fabrication now minimize ionic contamination. However, gradual degradation of the liquid crystal over the device lifetime produces ionic byproducts, and electrical resistance becomes a measure of material stability. Ionic conductivity imposes alternating voltage

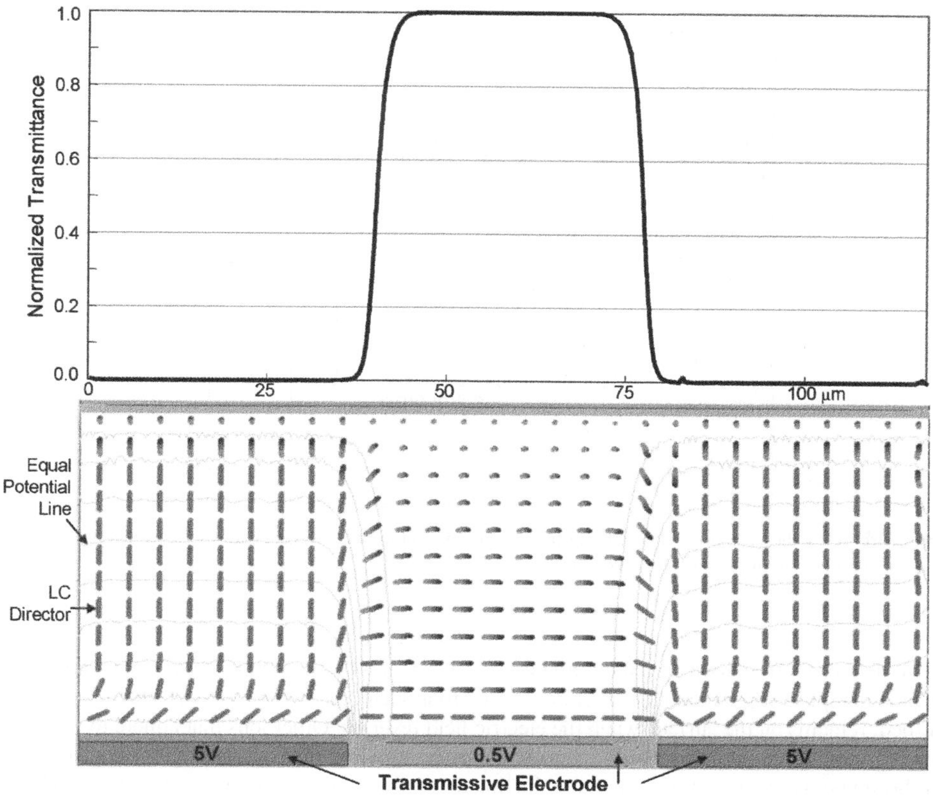

Figure 5.13 Fringing field effect of a transmissive NW 90° TN LC cell. The pixel voltages are 5, 0.5 and 5 V_{RMS}. LC: cell gap = 4.2 μm, pixel size = 40 μm and electrode gap = 5 μm

addressing on liquid crystal devices to minimize ionic charging effects that give rise to image sticking. The general behavior of ions in liquid crystal devices, including electrode behavior, is complex and a subject of continuing study; we present some simple concepts that help elucidate device behavior.

5.7.1 Ionic Conduction

A neutral ionic solution has equal numbers of +ve and −ve ions, which associate by electrostatic attraction to form neutral pairs that do not participate in conduction. Thermal, optical, or applied

Table 5.2 Comparison of VA and 90° TN cells for projection displays

Property	VA	TN
LC $\Delta\varepsilon$	Negative	Positive
Contrast ratio	>1000:1	~500:1
Viewing cone (uncompensated)	~10°	~10° (off center)
Fringing field	Edge effect	Little effect
Circularly polarized light	Yes	Not needed

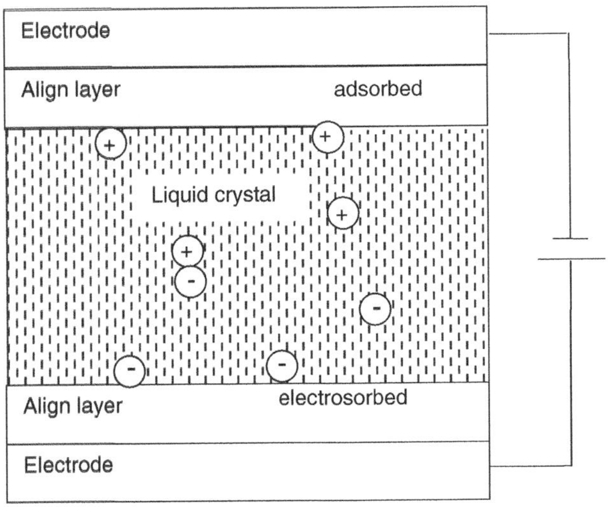

Figure 5.14 Cell showing liquid crystal alignment layers with ions adsorbed on surface, ions electrosorbed, and ions associated or free in the bulk liquid crystal

electric field excitation disrupts the binding energy of ion pairs providing free ions for conduction, as indicated in Figure 5.14. An ion approaching an electrode surface is subject to short-range forces and, if adsorbed, remains on the surface when the electric field reverses. Other ions that adhere to the surface by electrostatic force (electrosorbed) are more easily removed by field reversal. For given conditions, equilibrium between generation and recombination determines the concentration of free ions. The bulk conductivity (σ) is a function of free ion concentration (n), ion charge (q), and ion mobility (μ), and is given by[45–47]

$$\sigma = nq\mu. \tag{5.13}$$

There are generally a number of ionic species with different concentrations, mobility, and possibly multiple charges; for simplicity, we ignore such complication. Ion size and fluid viscosity determine mobility, which always increases with temperature. The typical range of mobility is $\sim 10^{-8}$ to $10^{-10}\,\mathrm{m^2/Vs}$, but much lower mobility (slow ions) exists. The transit time (τ_r) of an ion across the cell depends on cell gap (d) and voltage (V):[46]

$$\tau_r = \frac{d^2}{\mu V}. \tag{5.14}$$

For the above mobility range, the ion transit time across a 3 μm cell gap with 5 V applied is 0.18 to 18 ms, comparable with the video frame time 16.7 ms at 60 frames/sec. Diffusion and multiple species smear the observed ion transits.

5.7.2 Space Charge

Drift, diffusion, generation, recombination, and electrode behavior determine the distribution of ions. Without charge exchange at the electrodes an applied DC electric field separates the ions, with the +ve ions drifting towards the −ve electrode, and vice versa. A space charge is created that opposes the applied electric field, where equilibrium is established when drift and diffusion currents cancel. The total ionic charge per unit area (qnd), compared to the space charge limit (Q_{SCL}), indicates the influence of space charging:[46]

$$Q_{SCL} = \frac{\varepsilon}{d}V \tag{5.15}$$

where ε is the dielectric constant of the liquid crystal. The electrode charge is Q_{SCL} when there are no ions, making it the limiting ionic charge at the electrodes that reduces the bulk liquid crystal field to zero. When $qnd \ll Q_{SCL}$, an applied direct voltage produces a transient current which approaches zero in time comparable to ion transit time (τ_r). The electric field falls in the bulk, but increases near the electrodes.

In active matrix addressing a pixel receives a refresh pulse of duration <1 µs, and is isolated for a frame period ~10 ms. The frame time is comparable to the ion transit time. The liquid crystal capacitance C_{LC} is padded by the addition of storage capacitance C_S, to give a total capacitance $C = C_{LC} + C_S$. A refresh voltage pulse $\pm V$ produces a refresh charge $\pm VC$, the internal field falls with reversal of ionic charge distribution in time comparable to τ_r, and pixel voltage (V_p) approaches.

$$V_p \to V - \frac{2qnd}{VC}. \tag{5.16}$$

The voltage-holding ratio (V_p/V) between refresh pulses should approach unity to maintain grayscale integrity and attenuate flicker.

In the limit $qnd \gg Q_{SCL}$, the bulk ion density remains constant, and ionic surface charges absorb the applied voltage while the bulk field decays exponentially. The (per area) equivalent circuit resistance ($d/qn\mu$) and capacitance (ε/d) are valid. Outside this limit, the equivalent circuit will depend on frequency relative to the ionic transit time.[46,47]

5.7.3 Image Sticking

A constant image maintained for some time may persist as a ghost image, slowly decaying after the addressed image has changed. Ghosting or image sticking is due to ionic charging in liquid crystal devices.[47] Adsorbtion and electrosorbtion at the boundary determine the ion discharge time. If a sustained image with pixel AC addressing RMS voltage (V_{a1}) has a video-signal dependent DC component $V_{a1}S_{DC}$, ionic charge accumulates in the electrode regions producing a comparable reverse screening voltage where coefficient S_{DC} is constant in a linear approximation. A replacement addressed pixel voltage (V_{a2}) is immediately modulated by stored ionic voltage $\sim -V_{a1}S_{DC}$, representing the previous image. The liquid crystal responds to the square of the applied voltage (making RMS voltage an appropriate measure). Adding a small constant DC component V_{DC} (including $V_{a2}S_{DC}$) gives

$$\textit{Pixel mean square } V_p^2 \approx V_{a2}^2 + \left(V_{DC} + V_{a1}S_{DC}\right)^2. \tag{5.17}$$

The ghost voltage term is associated with V_{a1}:

$$\textit{Ghost mean square } V_g^2 = V_p^2 - V_{a2}^2 - V_{DC}^2 = 2V_{DC}V_{a1}S_{DC} + V_{a1}^2 S_{DC}^2. \tag{5.18}$$

V_{DC} is a global voltage controlled by adjustment of the counter-electrode potential, but includes a spurious pixel component. Depending on the signs of S_{DC} and V_{DC}, the terms in (5.18) may cancel somewhat. When V_{DC} is dominant, it amplifies the effect of any charged image-dependent DC level. If the ghost voltage exceeds the minimum perceptible gray-level step, a ghost image is apparent. To attenuate image sticking it is important to minimize spurious DC shift (V_{DC}), as well as eliminating addressing flaws giving rise to S_{DC} terms.

Global DC shift (V_{DC}) corrupts the gray levels and introduces flicker. The gray level RMS voltage error in the limit $V_{DC} \ll V_a$ becomes $V_{DC}^2/2V_a$; e.g. putting $V_a \sim 1$ volt and $V_{DC} \sim 0.1$ volt gives ~ 5 mV grayscale error. Video quality requires the liquid crystal response be comparable to frame rate, consequently $V_{DC} \pm V_a$ introduces flicker modulation amplitude V_{DC} at half the frame-rate frequency. Discharge of the pixel between refresh pulses also contributes to flicker. Lowering flicker amplitude and raising the frequency renders flicker imperceptible.

The high resistance of liquid crystal cells make them susceptible to spurious direct voltages. Galvanoelectric effects have been reported in cells with differing electrode materials.[48] Nematic liquid crystal distortion, inherent to device operation, gives rise to DC electric fields determined by the flexoelectric coefficient, noticeable in hybrid-aligned cells.[49] Asymmetry in the cell structure generates a DC electric field through contact potential or ion trapping effects.[50–52]

5.7.4 Electrode Effects

Consider the electrodes in direct contact with ideal ion-free liquid crystal. The material design and operating conditions inhibit charge transfer at the electrodes because it involves electrochemical degradation. A difference in electrode work function produces a potential difference between the electrodes (contact potential). For example, the contact potential between ITO and aluminum is ~ 0.8 V; however, contact potential is sensitive to molecular surface layers and observed values show substantial variation. In a closed circuit, the contact potentials sum to zero, consistent with zero circulating current. Electrode contact potential produces an electric field in the liquid crystal that influences molecular orientation, and consequently electro-optic response. An applied field adds or subtracts from the internal field according to polarity, making the electro-optic response asymmetric.

Ions included in the liquid crystal form a space charge in response to the contact potential lowering the bulk electric field until cancellation of drift and diffusion currents establish equilibrium. Interactions between the ions and electrode surfaces depend on detailed properties of the materials. Moreover, liquid crystal alignment layers on the electrodes are a basic requirement in a microdisplay. The layers are thin enough to allow ionic diffusion to the electrode surface, complicating electrode surface behavior rather than isolating it. Asymmetry in cell structure generally gives rise to a contact voltage effect in the liquid crystal equivalent to a global V_{DC} term in the addressing voltage. Sensitivity to surface and interface properties generally results in a significant variation of V_{DC} over the cell surface, together with time and temperature dependence. Inherent variation in V_{DC} limits the ability to cancel it by an applied voltage. The best solution is to minimize contact potentials at source by additional coatings; e.g. ITO coating on aluminum, which eliminates contact potential but lowers the reflectivity. Adjustment of the ITO work function to match aluminum, without compromising transmission, is preferable to disturbing the reflecting aluminum pixel structure. Transmission microdisplays have an advantage in ITO/ITO electrode symmetry minimizing contact potential effects.[53]

5.8 References

1 Kim S. S. (2005) The world's largest (82-inch) TFT-LCD. *SID Digest* **36**, 1842.
2 Stupp, E. H. and Brennesholtz, M. S. (1999) *Projection Displays.* John Wiley.
3 Wu, S. T. and Yang, D. K. (2001) *Reflective Liquid Crystal Displays.* John Wiley.
4 Howard, W. E. (1994) Limitations and prospects of a-Si:H TFTs. *Intl Display Research Conference*, 6–10.
5 Ohshima, H. (1994) Status and prospects of poly-Si TFT Technology. *Intl Display Research Conference*, 26–29.
6 Salerno, J. P. (1994) Single-crystal-silicon AMLCDs. *Intl Display Research Conference*, 39–44.
7 Schadt, M. and Helfrich, W. (1971) Voltage-dependent optical activity of a twisted nematic liquid crystal. *Appl. Phys. Lett.* **18**, 127.

8 Schiekel, M. F. and Fahrenschon, K. (1971) Deformation of nematic liquid crystals with vertical orientation in electric fields. *Appl. Phys. Lett.* **19**, 391.
9 Shikama, S., Toide, E. and Kondo, M. (1991) A polarization transforming optics for a high-luminance LCD projector. *Proc. SID* **32**, 301.
10 Nicolas, C., Loiseaux, B. and Huignard, J. P. (1995) Polarized light source for LCD projection. *Displays* **16**, 43 [review].
11 Duelli, M. and Taylor, A. T. (2003) Novel polarization conversion and integration system. *SID Digest* **34**, 766.
12 Itoh, Y., Nakamura, J. I., Yoneno, K. *et al.* (1997) Ultra-high efficiency LC projector using a polarized light illuminating system. *SID Digest* **28**, 993.
13 Duelli, M., McGettigan, T. and Pentico, C. (2001) Integrator rods with polarization recycling functionality. *SID Digest* **32**, 1078.
14 Shirochi, Y., Murakami, K. and Endo, H. (2003) 50/60-inch Hivision LCD rear-projection TV with excellent picture quality. *SID Digest* **34**, 114.
15 Hamada, H., Nakanishi, H., Funada, F. and Awane, K. (1994) A new bright single panel LC-projection system. *Intl Display Research Conference*, 422–3.
16 Willemsen, O. H., Krijn, M. P. C. M. and Salters, B. A. (2005) A handheld mini-projector using LED light sources. *SID Digest* **36**, 1706.
17 Mauguin, M. (1911) *C. Bull. Soc. Fr. Mineral.* **34**, 71.
18 Berreman, D. W. (1974) *Appl. Phys. Lett.* **25**, 12.
19 Wu, S. T. and Wu, C. S. (1999) Mixed-mode twisted-nematic cell for transmissive liquid crystal display. *Displays* **20**, 231.
20 Gooch, C. H. and Tarry, H. A. (1975) The optical properties of twisted nematic liquid crystal structures with twisted angles ≤ 90 degrees. *J. Phys. D8*, 1575.
21 Jepsen, M. L., Ammer, M. J., Bolotski, M. *et al.* (2002) High resolution LCOS microdisplay for single-, double- or triple-panel projection systems. *Displays* **23**, 109.
22 Wen, C. H., Gauza, S. and Wu, S. T. (2005) Photostability of liquid crystals and alignment layers. *J. SID* **13**, 805.
23 Janning, J. L. (1972) Thin film surface orientation for liquid crystals. *Applied Physics Letters* **21**, 173.
24 Lackner, A. M., Margerum, J. D., Miller, L. J. and Smith, W. H. (1990) Photostable tilted-perpendicular alignment of liquid crystals for light valves. *Proc. SID* **31**, 321.
25 Smith, W. H., Garvin, H. L., Robinson, K. and Miller, L. J. (1994) Inducing tilted perpendicular alignment in liquid crystals. US Patent 5,350,498.
26 Lu, M., Yang, K. H., Nakasogi, T. and Chey, S. J. (2000) Homeotropic alignment by single oblique evaporation of SiO_2 and its application to high resolution microdisplays. *SID Digest* **31**, 446.
27 Cognard, J. (1982) Alignment of nematic liquid crystals and their mixtures. *Mol. Cryst. Liq. Cryst. Suppl.* **1**, 1.
28 Geary, J. M., Goodby, J. W., Kmetz, A. R. and Patel, J. S. (1987) The mechanism of polymer alignment of liquid-crystal materials. *J. Appl. Phys.* **62**, 4100.
29 Stohr, J., Samant, M. G., Luning, J. *et al.* (2001) Liquid crystal alignment on carbonaceous surfaces with orientational order. *Science* **292**, 2299.
30 Nie, X., Lin, Y. H., Wu, T. X. *et al* (2005) Polar anchoring energy measurement of vertically aligned liquid-crystal cells. *J. Appl. Phys.* **98**.
31 Rastegan, A., Skarabot, M., Blij, B. and Rasing, Th. (2001) Mechanism of liquid crystal alignment on submicron patterned surfaces. *J. Appl. Phys.* **89**, 960.
32 Stohr, J. *et al.* (1998) *Macromolecules* **31**, 1942.
33 Haller, I. (1975) Thermodynamic and static properties of liquid crystals. *Prog. Solid State Chem.* **10**, 103.
34 Li, J. and Wu, S. T. (2004) Extended Cauchy equations for the refractive indices of liquid crystals. *J. Appl. Phys.* **95**, 896.
35 Li, J. and Wu, S. T. (2004) Two-coefficient Cauchy model for low birefringence liquid crystals. *J. Appl. Phys.* **96**, 170.
36 Wu, S. T. (1994) Film-compensated homeotropic liquid crystal cell for direct-view display. *J. Appl. Phys.* **76**, 5975.

37 Jepsen, M. L., Ammer, M. J., Bolotski, M. *et al.* (2002) High resolution LCOS microdisplay for single-, double- or triple-panel projection systems. *Displays* **23**, 109.
38 Erickson, J. L. (1961) *Trans. Soc. Rheol.* **5**, 23.
39 Leslie, F. M. (1968) *Arch. Ration. Mechan. Anal.* **28**, 265.
40 Wang, H., Wu, T. X., Zhu, X. and Wu, S. T. (2004) Corrections between liquid crystal director orientation and optical response time of a homeotropic cell. *J. Appl. Phys.* **95**, 5502.
41 Wu, S. T. and Wu, C. S. (1989) High speed liquid crystal modulators using transient nematic effect. *J. Appl. Phys.* **65**, 527.
42 Nagata, S., Takeda, E., Nanno, Y. *et al.* (1989) Capacitively coupled driving in TFT-LCD. *SID Digest* **20**, 242.
43 Fan Chiang, K. H., Chen, S. H. and Wu, S. T. (2005) High-definition vertically aligned liquid crystal microdisplay using a circularly polarized light. *Appl. Phys. Lett.* **87**.
44 Bahadur, B. (1990) Dynamic scattering mode LCDs. In B. Bahadur (ed.) *Liquid Crystals Applications and Uses*, Vol. 1. World Scientific.
45 Sussman, A. (1974) Electrochemistry in nematic liquid-crystal solvents. In E. B. Priestly, P. J. Wojtowicz and P. Sheng (eds) *Introduction to Liquid Crystals.* Plenum Press.
46 Maximus, B. and Colpaert, C. (1995) Overview of the conduction regimes in LCDs. *SID Digest* **26**, 609–12.
47 Naemura, S. (2000) Liquid-crystal-material technologies for advanced display applications. *J. SID* **8**, 5–9.
48 Melnik, G. *et al.* (2000) Galvanic activity in reflective LCDs. *SID Digest* **31**, 252–5.
49 Tanaka, Y. *et al.* (1998) Reduction of DC offset voltage of HAN-LCDs. *Intl Display Research Conference*, 1043–6.
50 Lu, M. and Yang, K. H. (2000) Reflective nematic LC devices for LCOS applications. *SID Digest* **31**, 1–7.
51 Huang, H. C. *et al.* (2000) On the minimization of flicker and image retention in silicon light valves. *SID Digest* **31**, 248–51.
52 De Vleeschouwer, H. *et al.* (2001) Asymmetrical properties of reflective LCDs. *Mol. Cryst. Liq. Cryst.* **368**, 643.
53 Cuypers, D., Van Calster, A. and De Smet, H. (2005) V_{com} drift phenomena in VAN LCOS panels. *Intl Display Works*, Japan, 1931–4.

6

Reflective Liquid Crystal Microdisplays

6.1 Introduction

Reflective displays based on liquid-crystal-on-silicon (LCOS) have been developed as the optical engines for projection[1–3] and virtual display[4,5] systems.[6] LCOS is a reflective image transducer that is capable of accepting video signals and converting them into high-brightness, high-contrast, large-screen images. Figure 6.1 depicts the device structure of a silicon backplane where a CMOS transistor and a capacitor are fabricated and connected. The CMOS transistor can support a high data rate owing to the high electron mobility of single-crystal silicon. The electron mobility of c-Si is about two orders of magnitude higher than that of amorphous silicon (a-Si). The reflective aluminum electrode is connected to the drain of the transistor and capacitor through a light-blocking metal layer. After chemical–mechanical polishing, the bare reflectivity of a pixilated aluminum mirror approaches 91%. The LCOS electrode can be smaller than $10\,\mu m \times 10\,\mu m$ and the electrode gap $<0.5\,\mu m$ while maintaining a filling factor $>90\%$. A light-blocking layer is implemented to prevent light leakage from the electrode gaps, which would activate the transistors, smear the device resolution, and degrade the image quality. Based on the light blocking and thermal loading, the JVC light valve is able to tolerate the 15,000 lumens required for electronic cinema.[7]

The major advantages for the LCOS display are threefold: high pixel density, high brightness capability, and potentially low cost. High-definition television (HDTV) needs 1920 (H) × 1080 (V) resolution elements in the 16:9 aspect ratio. If the pixel pitch is $10\,\mu m \times 10\,\mu m$, then the imager diagonal is only 22 mm. The small imager size reduces the optics size and weight. Since optical components contribute a large portion to the projector cost, small optics size helps reduce the overall projector cost. Moreover, large aperture ratio (or filling factor) produces film-like images. This advantage manifests when high resolution and high pixel density are required. Finally, the LCOS is a reflective device, which enables a high-wattage lamp to be used for achieving high display brightness.

Introduction to Microdisplays D. Armitage, I. Underwood and S-T. Wu

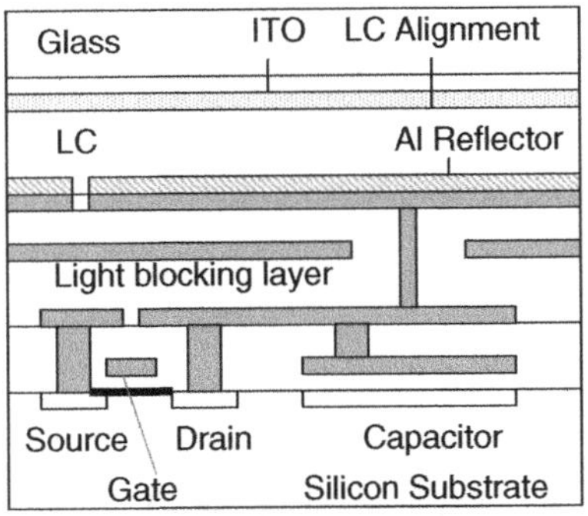

Figure 6.1 Pixel structure of an LCOS microdisplay

Figure 6.2 sketches the projection system using three reflective LCOS panels. The white light coming from the lamp passes through a homogenizer, polarization converter, aperture, UV/IR filters, and is collimated by a lens. The outgoing light is linearly polarized, but not perfectly, because the polarization converter does not have 100% conversion efficiency. Some high-quality polarizing beam-splitters[8] or wire grid polarizers (with cleanup polarizer)[9] exhibit an extinction ratio higher than 2000:1. If a low-contrast polarizing beam-splitter (PBS) is used, then it is necessary to add a pre-polarizer and post-polarizer before and after the PBS in order to achieve high contrast ratio. In Figure 6.2, a Philips prism splits the incoming white light into red, green, and blue LCOS panels. Depending on the LC mode employed, each LCOS panel may be optimized for the specified color band. The modulated beams are reflected by the LCOS, recombined by the prism, and thrown to screen by the projection lens.

Figure 6.3 shows a 58-inch rear-projection TV developed by Brillian Corp.,[10] in which three LCOS panels with 1280×720 resolution elements are used. Since the system employs folded reflective optics, the display depth is only 20 inches.

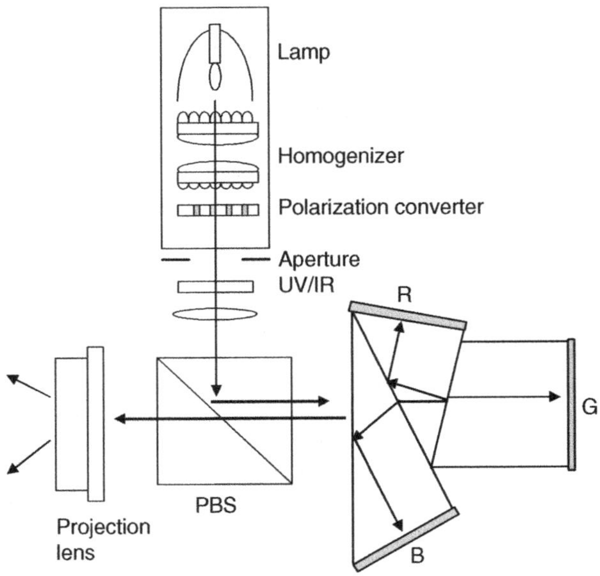

Figure 6.2 Projection system using 3-panel reflective LCOS

Figure 6.3 Prototype of Syntax-Brillian's 58-inch rear-projection TV. Reprinted courtesy of Society for Information Display

In a projection system, the display contrast ratio is determined by several factors, such as the *f*-number of the projection optics, the acceptance angle and extinction ratio of the PBS, the stress bi-refringence of the PBS, the thermal effect of the LCOS panels, and the intrinsic contrast ratio of the LCOS panels. In addition to contrast ratio, the *f*-number also plays an important role in the light throughput.[11] The *f*-number is defined as the ratio of the focal length to the diameter of the projection lens. The light throughput scales approximately with the inverse square of the *f*-number, as depicted in Figure 6.4. Under the same conditions, the light throughput of the *f*-1 system is nearly nine times higher than that of the *f*-3 system. Another major advantage of the small *f*-number LCOS system is that

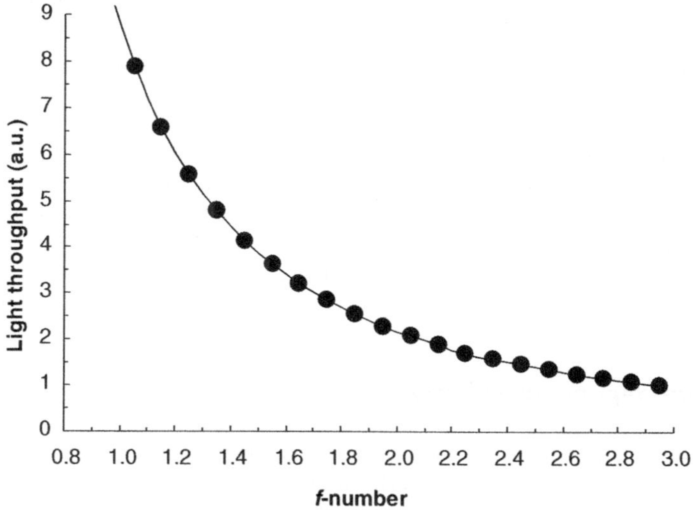

Figure 6.4 Light throughput as a function of *f*-number of the projection lens

it enables a long-arc lamp to be used. The long-arc lamp is less expensive and has a longer lifetime.[12] However, to accommodate the small *f*-number, the PBS should have a high extinction ratio at a large acceptance angle, and the LCOS panel should have a high contrast ratio in a large viewing cone. Normally, the LC viewing angle for projection displays is not a big concern. However, to enable a high contrast ratio, high brightness, and low cost projection display incorporating a small *f*# projection optics and long-arc lamp, the LCOS panels should have at least 10° viewing cone in which the contrast ratio is greater than 2000:1. This is an important criterion when selecting a suitable LC operating mode for LCOS projector.

In the following sections, we analyze the pros and cons of several operating modes developed for LCOS. We first discuss the voltage-dependent reflectance and then the viewing angle for a single-pixel electrode. The fringing field effect resulting from the voltage difference between the adjacent pixels will follow. The LC modes described hereafter include: (1) normally black homeotropic cell, also known as vertically aligned nematic (VAN) cell, (2) normally white film-compensated homogeneous cell, (3) normally white 90° mixed-mode twisted nematic (MTN) cell, (4) normally white film-compensated 63.6° MTN, (5) normally black 63.6° TN, (6) bisector effect of 60° MTN cell, (7) normally white film-compensated 45° MTN cell, (8) normally black 45° TN cell, and (9) finger-on-plane cell.

6.2 Normally Black Homeotropic Cell

Homeotropic cells[13] exhibit an unprecedented contrast ratio when viewed at normal direction. Moreover, the contrast ratio is insensitive to the incident light wavelength, LC layer thickness, and operating temperature. Therefore, they are a strong contender for projection[14,15] and direct-view displays.[16,17]

A technical challenge for the homeotropic cell is the need for high-resistivity LC mixtures with negative dielectric anisotropy. High resistivity is required for active matrix LCDs in order to avoid image flickering, and negative $\Delta\varepsilon$ is required for obtaining a useful electro-optic effect. To obtain negative $\Delta\varepsilon$, the dipoles, in particular the fluoro groups, need to be in the lateral positions. Significant progress in material development has been obtained in the past decade. Nevertheless, the selection of negative $\Delta\varepsilon$ LC compounds is still more limited than the positive ones. Besides, the lateral dipole groups often exhibit a higher viscosity than the axial compounds due to an increased moment of inertia.

A special device configuration that utilizes positive $\Delta\varepsilon$ LC mixture in homeotropic alignment has been demonstrated.[18] The LC directors are reoriented by the lateral fringing field to achieved a wide viewing angle. Such a homeotropic cell exhibits a fast response time, but its operating voltage is too high. The homeotropic cell we discuss in this chapter is one which uses a negative $\Delta\varepsilon$ LC mixture.

For computer simulations, a Merck high-resistivity negative $\Delta\varepsilon$ LC mixture MLC-6608 is used. Some physical properties of MLC-6608 are summarized as follows: $n_e = 1.558$ and $n_0 = 1.476$ (at $\lambda = 589\,\text{nm}$ and $T = 20°\text{C}$); clearing point $T_c = 90°\text{C}$; dielectric anisotropy $\Delta\varepsilon = -4.2$; and rotational viscosity $\gamma_1 = 186\,\text{mPa}$ at 20°C. The generic optical system used for computer simulation is plotted in Figure 6.5, where the LC cell can be a homeotropic or twisted alignment. To evaluate the performance of the LCOS alone, the PBS is assumed to have perfect extinction ratio and to have no loss. In the homeotropic LC cell, the pretilt angle is assumed to be $\theta_p = 88°$, unless otherwise specified.

6.2.1 Voltage-dependent Reflectance

Figure 6.6 plots the voltage-dependent reflectance (VR) of a homeotropic cell for three primary color bands: red (R), green (G), and blue (B). The cell gap is chosen to be $d = 2.3\,\mu\text{m}$ (or $d\Delta n = 192\,\text{nm}$ at $\lambda = 540\,\text{nm}$) so that the on-state voltage for the green band occurs at $4.8\,V_{RMS}$ at room temperature. In a projector, the chassis temperature could go up to 50–60°C due to the thermal effect of the arc lamp.

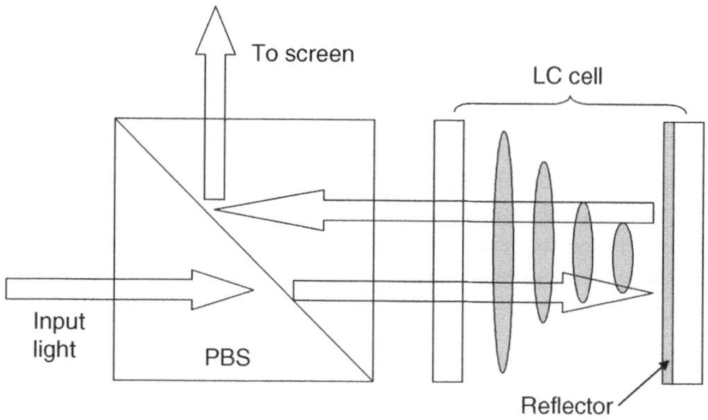

Figure 6.5 Generic projection system used for computer simulations

Thermally induced stress on the optical components could result in contrast ratio degradation.[19] At elevated temperatures, the LC birefringence decreases according to $\Delta n = (\Delta n)_o(1 - T/T_c)^\alpha$; where T_c is the clearing temperature of the LC employed, and $(\Delta n)_o$ and α are material parameters.[20] On the other hand, the threshold voltage will decrease, but at a slower rate. As a result, the on-state voltage will increase slightly.

The blue band used in simulations extends from 420 to 480 nm, the green band from 520 to 560 nm, and the red band from 620 to 680 nm. The curves shown in Figure 6.6 represent the spectral average for each band. For comparison purposes, we only consider the normalized reflectance; the optical losses of the PBS and substrate reflections are all ignored. The threshold voltage of the homeotropic cell is $\sim 2\,V_{RMS}$. The blue wavelength reaches the maximum reflectance at the lowest voltage, followed by green and red because of the combined wavelength and birefringence effects. To achieve a balanced white, the lumen content of the green, red, and blue colors should be in the ratio 6:3:1. Thus, green plays the most important role in luminance among the three primary colors.

A unique feature of the homeotropic cell is that its dark state is insensitive to the cell gap, wavelength, and birefringence. Using the abovementioned bandwidth in calculation, indeed, the voltage-dependent reflectance curves are very similar to those using the central wavelength of each band.

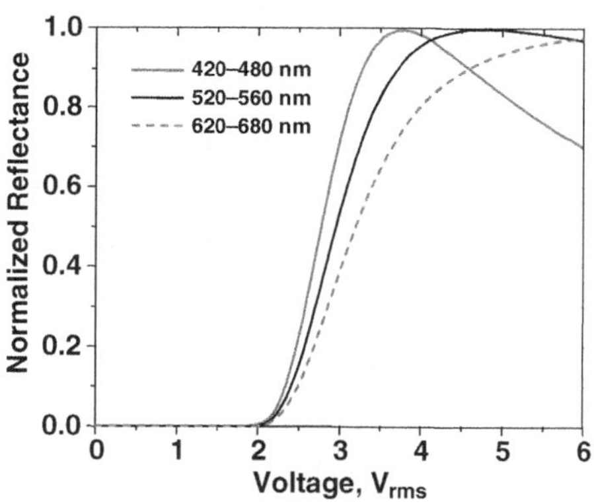

Figure 6.6 Voltage-dependent reflectance of a homeotropic LC cell at the specified RGB bands. MLC-6608, cell gap $d = 2.3\,\mu m$, pretilt angle $= 88°$ and $T = 22\,°C$

6.2.2 Pretilt Angle Effect

Pretilt angle has an important effect on threshold voltage and contrast ratio. A large pretilt angle would smear the threshold behavior, broadening the Freedericksz transition threshold[21]. The contrast ratio of a homeotropic cell is critically dependent on the pretilt angle. A large pretilt angle would generate noticeable phase retardation for the incident light and cause light leakage through the crossed polarizers. Similar to a transmissive homeotropic cell described in Figure 5.10, when the pretilt angle of a reflective homeotropic cell is between 86° and 89°, the contrast ratio maintains above 10^4:1. Throughout this chapter, unless otherwise mentioned, the pretilt angle used for the simulations is $\theta_p = 88°$, or 2° off normal.

6.2.3 Viewing Cone

The viewing cone of the LCOS panel determines what *f*-number projection lens can be used. A small *f*-number projection lens collects more light and improves the optical efficiency. To accommodate a small *f*# lens while keeping a high contrast ratio, the LCOS panel should have a large viewing cone. Figure 6.7 plots the iso-contrast contour curves of a reflective homeotropic cell. The LC mixture employed is Merck MLC-6608 and cell gap $d = 2.3\,\mu m$ (or $d\Delta n = 192\,nm$ at $\lambda = 540\,nm$). Although the green band centered at $\lambda = 540\,nm$ is used for simulations, the red and blue panels should have very similar results.

From Figure 6.7, the homeotropic cell exhibits a relatively narrow viewing angle. A wide viewing angle exists only along the horizontal and vertical axes. Overall, the 2000:1 contrast ratio is limited to the central 8° viewing cone. By comparing the results shown in Figure 5.11, the reflective homeotropic cell exhibits a wider viewing angle than its transmissive counterpart. This is because of the mirror image effect.[22] The viewing angle of a reflective homeotropic cell is equivalent to a two-domain transmissive cell.

To widen the viewing angle, a negative birefringence phase compensation film (also known as a negative c-plate) is added.[23] A negative c-plate has isotropic refractive indices in the xy plane

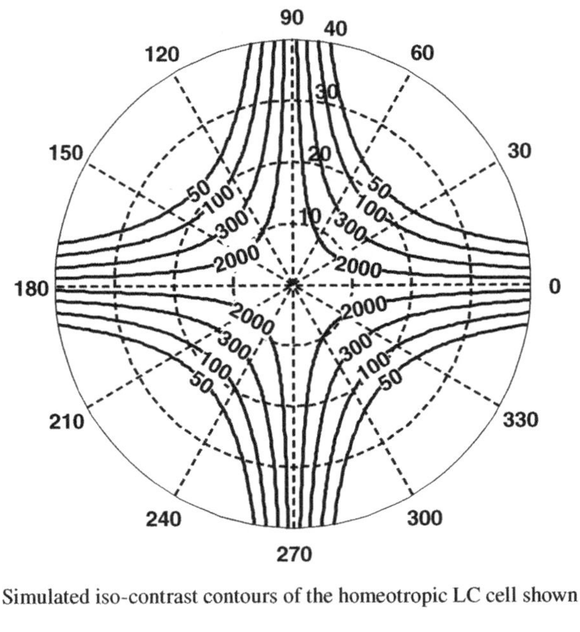

Figure 6.7 Simulated iso-contrast contours of the homeotropic LC cell shown in Figure 6.6

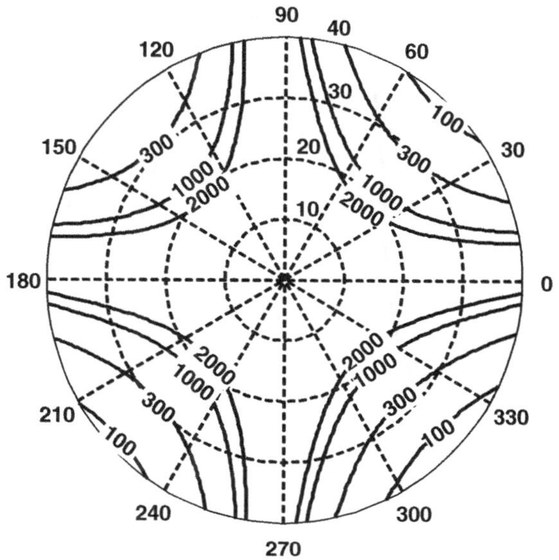

Figure 6.8 Same as Figure 6.7 except with a c-plate compensation: $d\Delta n = -183\,\text{nm}$

($n_x = n_y$), which is parallel to the substrate surface. However, its index in the z-axis is smaller than that in the xy plane (i.e. $n_z < n_x = n_y$). Since most LC mixtures have a positive birefringence (i.e. $n_e > n_o$) the negative c-plate would compensate for the positive LC birefringence and reduce the light leakage of the LC cell at oblique angles. Therefore, in order to compensate the phase retardation of the homeotropic cell in the voltage-off state resulting from oblique incidence, an ideal compensation film should possess the following properties: (1) negative birefringence and the same $d\Delta n$ as the LC cell at any incident angle, and (2) similar birefringence dispersion as the employed LC material. The first condition would cancel the phase retardation of the outgoing light after traversing through the LC cell over a wide range of incident angle. The second would result in a perfect phase cancellation for all three primary colors intended for a full-color display. By selecting a polycarbonate c-plate with $d\Delta n = -183\,\text{nm}$, the viewing cone is nearly doubled as shown in Figure 6.8. The 2000:1 iso-contrast ratio contour lines are extended to $\sim 18°$ viewing cone. This improved viewing angle permits a smaller f-number projection lens and a longer arc lamp to be used. As a result, the light throughput and lamp lifetime are both improved.

The price paid to widen the viewing cone is the addition of a negative birefringence compensation film. The c-plate can be laminated to the PBS cube or to the glass side of the LCOS panel. Since both locations are near the imaging plane, any defect in the film or lamination process would be magnified. Moreover, the thermal stress of the film is another concern.

6.2.4 Fringing Field Effect

In the LC modes discussed above, the voltage-dependent reflectance is calculated only for a single pixel; no fringing field from the neighboring pixels is considered. In reality, each pixel may be operated at different voltages in order to achieve grayscales. Thus, the fringing field effect has to be taken into account.[24,25] As the pixel density increases, the fringing field effect becomes more important.[26] Moreover, the LCOS pixel size is around 10 μm so that the diffraction effect for the visible light becomes significant in reducing light throughput efficiency.

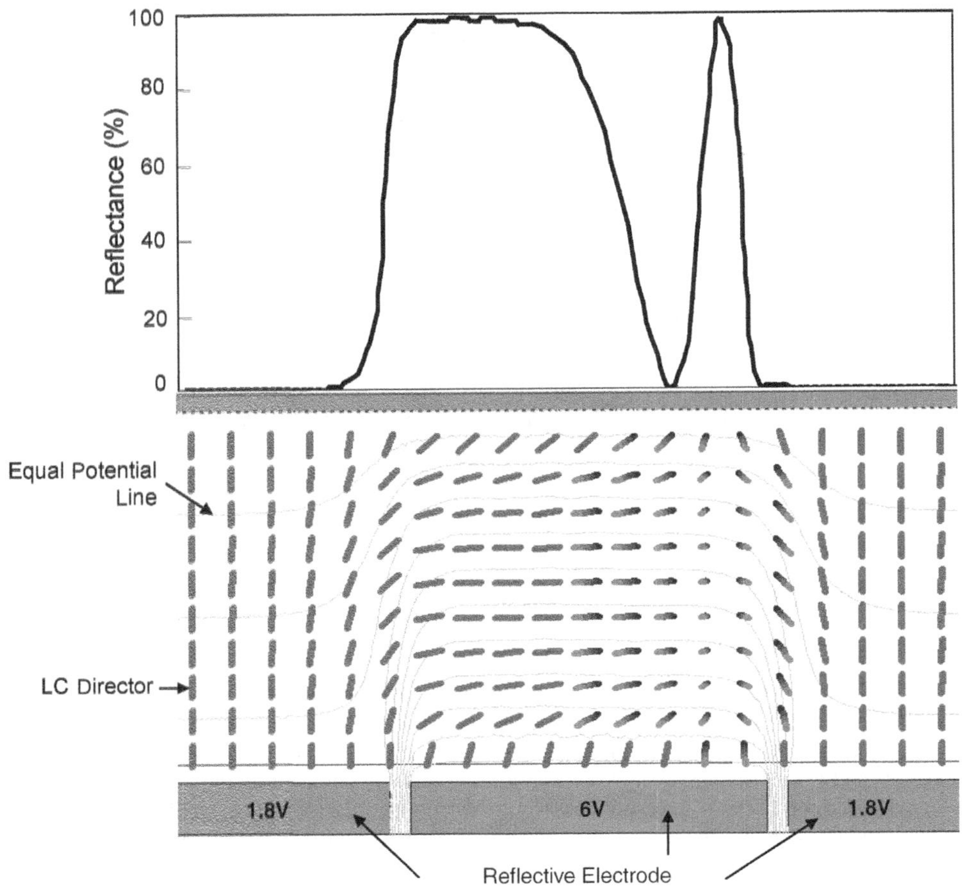

Figure 6.9 Fringing field effect on the reflective homeotropic LC cell with pixels off-on-off. MLC-6609, d = 2.0 μm, $d\Delta n$ = 156 nm. Pixel size = 15 μm and electrode gap = 0.9 μm. Top: optical signals, bottom: electric field equal potential lines and LC directors

Figure 6.9 shows the simulated fringing field effect of a homeotropic cell between crossed polarizers. The negative $\Delta\varepsilon$ LC mixture MLC-6609 is used for calculations. The LC parameters are $\Delta n = 0.0777$ (at $\lambda = 589$ nm and $T = 20°$C), $\varepsilon_{\perp} = 7.1$, $\varepsilon_{//} = 3.4$, $K_{11} = 17.2$ pN, $K_{22} = 8.4$ pN, and $K_{33} = 17.9$ pN. The fringing field effect becomes stronger for a thicker cell gap. For minimizing the fringing field effect, the cell gap is taken to be $d = 2$ μm. The pixels are at off-on-off conditions. The bright pixel is at 6 V_{RMS} and the adjacent dark pixels are at 1.8 V_{RMS}, which is slightly below the threshold voltage. In a VA cell, the dark state outside the pixels remains excellent and is not affected by the fringing field. However, the fringing field splits the bright pixel into two unequal areas. This causes edge effect near the right edge of the bright pixels. Of course, if the pretilt direction is reversed, the edge effect will change sides. By integrating the on-pixel area from 15.9 to 30.9 μm, the overall optical loss due to this split pixel is ~30%.

Figure 6.10 shows the displayed Chinese character (eagle) using a homeotropic LCOS panel.[27] The disclination lines are clearly observed. In this example, the pixel pitch was 15.5 μm, the inter-pixel gap 0.7 μm and the LC $d\Delta n$ 206 nm. This split pixel effect is a concern for the LCOS projectors employing VA cells, at limiting resolution with high modulation.

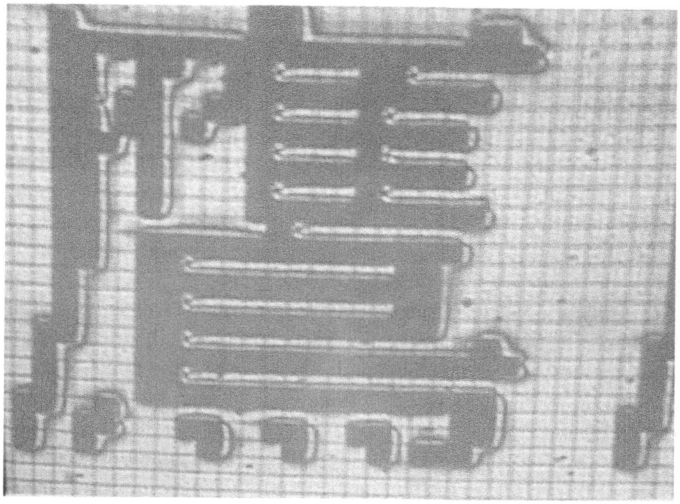

Figure 6.10 A photo (Chinese eagle character) showing the fringing field effect on the reflective homeotropic LCOS. Pixel size = 15.5 μm and electrode gap = 0.7 μm. Reproduced with permission of the Japan Society of Applied Physics

Changing the electrode slope has been studied for minimizing the fringing field effect.[28,29] For the VA cell, the observed dark line is not influenced by the electrode slope. In the flat field condition, in which all the pixels are at the same voltage, the disclination line disappears. As expected, this stationary state leads to a very high contrast ratio for the VA cell. However, if the adjacent pixels differ substantially in voltage, then the disclination lines shown in Figure 6.9 appear, and the local contrast is compromised.

To reduce diffraction loss, JVC has developed a reflection-enhanced D-ILA device with level inter-pixel gaps.[30] A few layers of dielectric mirror are deposited on top of the aluminum reflector. This dielectric layer not only enhances the reflectivity of the aluminum reflector to ~95%, but also improves the reflectivity (R_g) of the inter-pixel gaps, enhancing effective fill-factor and reducing diffraction loss.

In summary, the major advantages of the VA cell are threefold: high contrast ratio, large cell gap tolerance, and no need for a phase compensation film although the c-plate will widen the viewing cone. The drawbacks are fringing field effects, less selection on the negative $\Delta\varepsilon$ LC materials, and more complicated LC alignment procedures. The latter is discounted by the stability of inorganic alignment in projector conditions.

6.2.5 Effect of Field Fringing on Image Quality

Modulation transfer function (MTF) characterizes the resolution of imaging components such as the projection lens. A sinusoidal luminance pattern input to the lens emerges with attenuated modulation (contrast) described by Equation (1.2) in Chapter 1. MTF describes that output modulation as a function of input spatial frequency. At low spatial frequency the projection lens MTF approaches unity and at very high spatial frequency the MTF approaches zero; details of the characteristic follow from optical aberrations, scattering and diffraction. The product of component MTFs gives the overall MTF of a system. Image quality benefits from high MTF over the range of spatial frequencies comprising the image.

The video signal addressing a microdisplay identifies the luminance modulation at each pixel, suggesting a device MTF relating actual luminance to ideal addressed luminance. The luminance observed depends on aperture ratio, addressing details and field fringing. We restrict our attention to field fringing, taking the VAN cell as an example of severe fringing. When addressing all pixels to the same luminance level, corresponding to zero spatial frequency, inter-pixel gaps impose field fringing that lowers the luminance at the pixel boundaries. However, the effect is negligible in LCOS with pixel gaps less than 0.5 μm. At the highest spatial frequency addressing the display with full modulation, adjacent pixels have extremes of dark and bright; exhibiting worst-case field fringing. Under such conditions, a typical LCOS VAN microdisplay with 10 μm pixel pitch suffers luminance loss as high as 50% in the on-state pixel, making the effective MTF 50%. Reducing the addressed depth of modulation lowers the voltage difference between pixels, attenuating the fringe field and raising the effective MTF to approaching 100% as the transverse voltage vanishes.

Addressing a sinusoidal luminance distribution of arbitrary spatial frequency gives rise to transverse pixel voltage and fringe field effects. At low spatial frequency, or low modulation depth, the transverse voltage is small, making the MTF attenuation small. By analyzing the effect of transverse voltage on pixel luminance, over a range of luminance and transverse voltage, the fringing attenuation of MTF over the upper range of spatial frequency could be assessed. However, nonlinearity limits the value of an MTF concept in general; the weaker MTF of high spatial frequency imposes on lower spatial frequencies when both are present, due to the nonlinearity introduced by the fringe field effect.

In HDTV, the pixel pitch is below, or at the limit of, visual perception at normal viewing distance; consequently the pixel detail is not significant. Any attenuation in high spatial frequency content of the image will degrade image quality, but quantitative assessment is difficult. The success of VAN LCOS in HDTV applications suggests fringe field effects on image quality are minor. Moreover, the quest for optimum image quality is driving the cell gap below 2 μm, providing reduced fringing and enhanced response speed.[31]

With sufficient magnification the pixel becomes visible, revealing the fringe field effect, making severe disclination effects a distraction. A rear-projection computer monitor viewed at reading distance may reveal pixel structure. However, direct-view displays are preferred in monitor applications.

Fringe field disclination introduces a trailing tail in fast moving images due to the recovery time of the disclination region.[32] Reducing the pretilt angle to less than 84 degrees increases the strength of the directional tilt, reducing the recovery time to attenuate tailing below perception.

6.2.6 Cell Gap

There are several considerations regarding cell gaps in general. A small cell gap is desirable for fast response and weak fringe fields. Cells requiring a precise cell gap are undesirable because of low manufacturing yield, and the need to tune gap or liquid crystal for different color bands. A near achromatic black state is desirable to reduce color variation due to cell gap nonuniformity.[33] The latter factors essentially rule out normally black cells, except the VAN cell. Gap uniformity compromises yield, even for the most favorable cell design. Manufacturers resort to adjusting pixel voltage look-up tables to compensate for cell gap nonuniformity. Adjustment is made at the final test procedure, where other defects in color management can be corrected along with the cell gap. Cell gap tolerance remains tight at 3% even with pixel voltage compensation adjustment.

6.3 Normally White Homogeneous Cell

The homogeneous cell is generally ruled out for transmissive display because of its narrow viewing angle.[34] However, for reflective projection displays, the film-compensated homogeneous cell can be considered because of the improved viewing angle originating from the mirror image effect.[35] For the same material and the same LC layer thickness involved, the homogeneous cell exhibits a faster response time than any twisted nematic cell because the splay elastic constant K_{11} is about two times

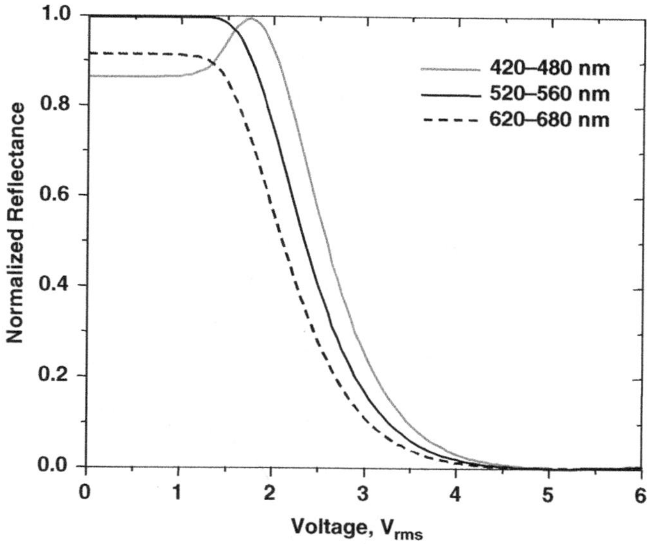

Figure 6.11 Voltage-dependent reflectance of a NW film-compensated homogeneous cell at the specified RGB bands. MLC-6694-000, $d\Delta n = 165\,\text{nm}$, $d = 1.95\,\mu\text{m}$, $\beta = 45°$ and uniaxial film $d\Delta n = 20\,\text{nm}$

larger than the twist elastic constant K_{22}. Moreover, the required $d\Delta n$ for a homogeneous cell is smaller than that of any TN cell due to the pure birefringence effect. This advantage becomes obvious when the cell gap is reduced to $d = 1\,\mu\text{m}$ in order to achieve fast response time. The small $d\Delta n$ requirement enables a low-birefringence and high-resistivity commercially available LC mixture to be employed.

6.3.1 Voltage-dependent Reflectance

Figure 6.11 depicts the voltage-dependent reflectance of a film-compensated homogeneous cell with $d\Delta n = 165\,\text{nm}$, $\beta = 45°$, pretilt angle $\sim 2°$, and uniaxial compensation film $d\Delta n = 20\,\text{nm}$. For the phase compensation effect to occur, the axis of the compensation film should be oriented at 90° with respect to the LC directors. From Figure 6.11, the voltage-off state color dispersion is not too large. The dark state occurs at $V \sim 5.2\,V_{RMS}$ for the employed MLC-6694-000 cell with cell gap $d = 1.95\,\mu\text{m}$. This dark-state voltage can be lowered by increasing the $d\Delta n$ value of the compensation film.

6.3.2 Viewing Cone

Figure 6.12 plots the iso-contrast contour of the film-compensated homogeneous cell. The viewing cone for the CR = 1000:1 exceeds 10°. Therefore, the film-compensated homogeneous LC layer works equally well for projection display as for other normally white MTN modes discussed in this chapter. Two factors that affect the viewing cone are $d\Delta n$ value and beta angle, which represents the birefringence effect. A smaller $d\Delta n$ value and lower birefringence effect would lead to a wider viewing cone. The film-compensated homogeneous cell has a smaller $d\Delta n$, but its birefringence effect is larger. On the other hand, the MTN cell has a smaller birefringence effect, but it requires a larger $d\Delta n$ value. Thus, their 1000:1 contour boundaries are comparable.

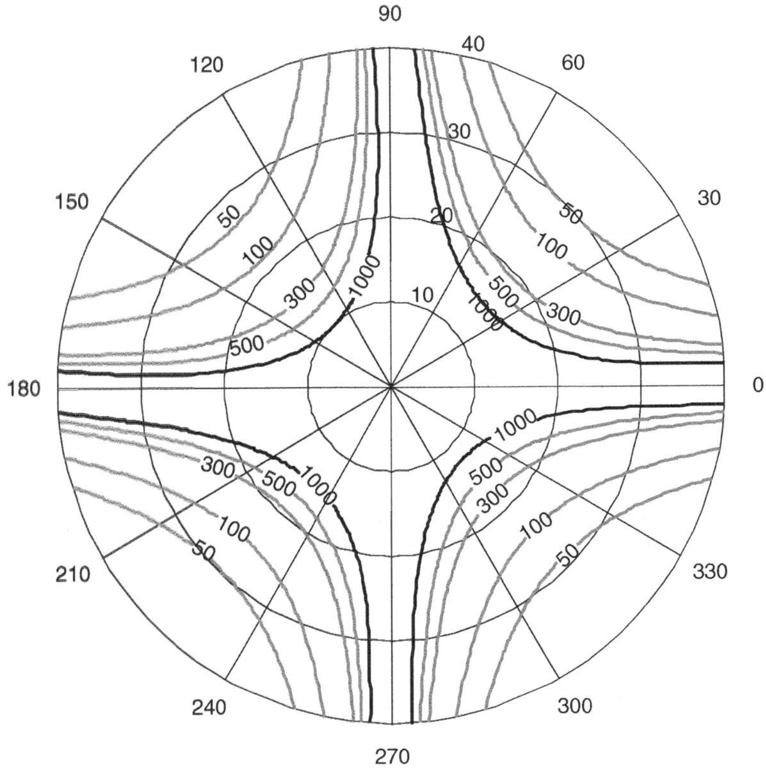

Figure 6.12 Simulated iso-contrast contours for the NW film-compensated homogeneous cell shown in Figure 6.11

6.3.3 Fringing Field Effect

Figure 6.13 shows the fringing field effect on the film-compensated homogeneous cell. Three checkerboard pixels with dark–bright–dark configuration are evaluated. The on-pixel voltage is 4 V_{RMS} and off pixels are 1.5 V_{RMS}, which is slightly below V_{th}. The pixel size is 15 μm and the electrode gap is 0.9 μm. From Figure 6.13, the left dark pixel exhibits a good dark state and the on-pixel reaches 100% reflectance. However, the right dark pixel exhibits a relatively large spike light leakage, which would degrade the local contrast ratio. If the pretilt direction is reversed, then the spike will occur at the left dark pixel.

The advantages of the film-compensation homogeneous cell are high optical efficiency and fast response time. However, the need of phase compensation film, gap and temperature sensitivity, limit its widespread application. Input polarization 45 degrees to the nematic director makes it impossible to index match the nematic refraction and compensator refraction. Reflections from such anisotropic material alter polarization, compromising CR.[36,46]

6.4 Reflective TN Cells

Unlike the homeotropic cell, the reflective TN cells use positive $\Delta\varepsilon$ LC materials. The optical system is the same as that sketched in Figure 6.5. Here, the PBS functions as a pair of crossed polarizers. In this

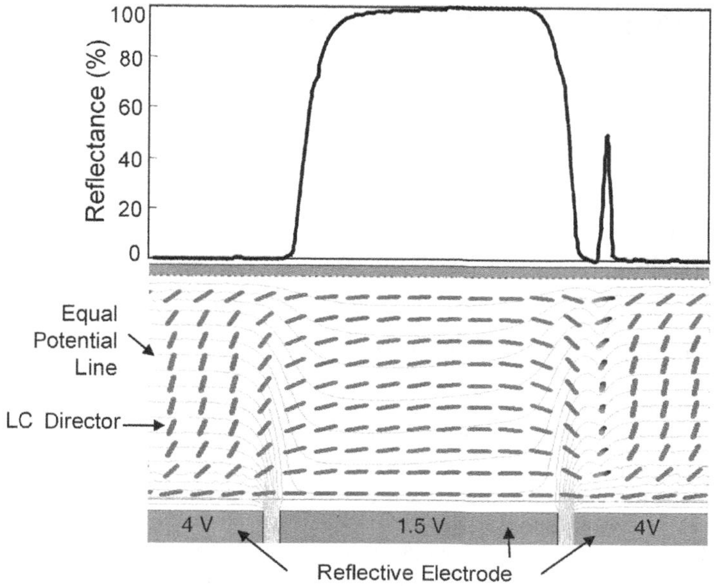

Figure 6.13 Fringing field effect on the reflective film-compensated homogeneous LC cell shown in Figure 6.11. Pixel size = 15 μm, electrode gap = 0.9 μm. Top: optical signals; bottom: electric field equal potential lines and LC directors

chapter, we illustrate some normally white (NW) displays using mixed-mode twisted nematic (MTN) cells or film-compensated homogeneous cells, and two normally black (NB) LC modes using TN cells. The reflective MTN cells exhibit a much smaller $d\Delta n$ value than the Gooch-Tarry first minimum so that they are more suitable for NW displays. On the other hand, the TN cells with twist angles ranging from 45° to 65° are good candidates for NB displays. Each mode has its own merits and demerits. For instance, the normally white mode usually exhibits a larger cell gap tolerance while the normally black mode has a lower on-state voltage.

For a reflective TN cell with twist angle ϕ, the normalized reflectance ($R_{\perp}$) can be solved by the Jones matrix method. The analytical solution is [37]

$$R_{\perp} = \left(\Gamma \frac{\sin X}{X}\right)^2 \left(\sin 2\beta \cos X - \frac{\phi}{X} \cos 2\beta \sin X\right)^2. \qquad (6.1)$$

Here, β is the angle between the polarization axis and the front LC director, ϕ is the twist angle in π radians, $X = [\phi^2 + (\Gamma/2)^2]^{1/2}$ and $\Gamma = 2\pi d\Delta n/\lambda$, where d is the cell gap. For a normally white display, we need to search for $R_{\perp} = 1$. On the other hand, for a normally black display, we simply solve for $R_{\perp} = 0$. From (6.1), if $\sin X = 0$ (i.e. $X = m\pi$ and m is an integer), then $R_{\perp} = 0$ independent of β. This special condition leads to

$$\phi^2 + (\Gamma/2)^2 = m^2. \qquad (6.2)$$

In the following sections, some MTN and TN cells are chosen as examples to illustrate the design principles. Both single-electrode (1D) and multiple-electrode configurations (2D fringing field effect) will be considered.

6.5 Normally White 90° MTN Cell

The 90° twisted nematic cell has orthogonal boundary layers; as a result, its dark-state voltage is low and contrast ratio is high. To illustrate the design principles, we use Equation (6.1) to plot the normalized reflectance ($R_\perp$) as a function of $d\Delta n/\lambda$ and β and then calculate the voltage-dependent light reflectance. The term MTN refers to the mixed-mode twisted nematic cell, where there is a proper mixing between the polarization rotation and birefringence effects.[38] The MTN cell has $d\Delta n$ value below the Gooch–Tarry first minimum. Generally speaking, the MTN has a large cell gap tolerance and is useful for reflective direct-view and projection displays.

6.5.1 Voltage-dependent Reflectance

Figure 6.14 depicts the computer simulation results of $R_\perp$ as a function of $d\Delta n/\lambda$ at $\beta = 0, 20$, and 40°. At $\beta = 0°$, the peak reflectance only reaches ~70%. As β increases to 20°, the maximum reflectance ($R_\perp = 88\%$) occurs at $d\Delta n \sim 0.45\lambda$. As β continues to increase, the maximum reflectance gradually declines. At $\beta = 40°$, the maximum reflectance drops to 68%. Therefore, the 90° MTN cell has optimal performance at $d\Delta n \sim 240$ nm and $\beta = 20°$.

Figure 6.15 shows the voltage-dependent reflectance of a 90° MTN cell with $d\Delta n = 240$ nm and $\beta =$ 20° for the RGB bands. The LC employed is Merck MLC-6694-000; its parameters are listed as follows: $n_e = 1.5567$, $n_o = 1.4729$ (at $\lambda = 589$ nm and $T = 20°C$), $\varepsilon_\parallel = 10.1$, $\varepsilon_\perp = 3.3$ (at 1 kHz), $K_{11} =$ 13.1 pN, $K_{22} = 7.0$ pN, and $K_{33} = 22.7$ pN. In the visible spectral region, the LC birefringence decreases as the wavelength increases. At $\lambda = 540$ nm, $\Delta n \sim 0.0857$. Thus, the required cell gap is $d = 2.8\,\mu$m. Unless otherwise specified, the pretilt angle of the LC cell is assumed to be $\theta_p \sim 2°$. Within each color band, the voltage-dependent reflectance curve is calculated every 10 nm. The final results represent the average for the specified band. The blue band covers from 420 to 480 nm, the green band from 520 to 560 nm, and the red band from 620 to 680 nm.

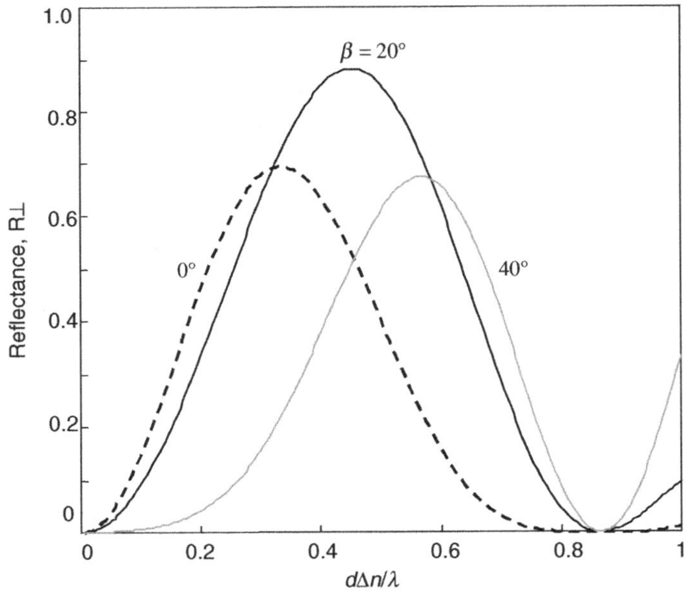

Figure 6.14 Normalized reflectance as a function of $d\Delta n/\lambda$ at $\beta = 0, 20$ and 40° for the 90° MTN cell

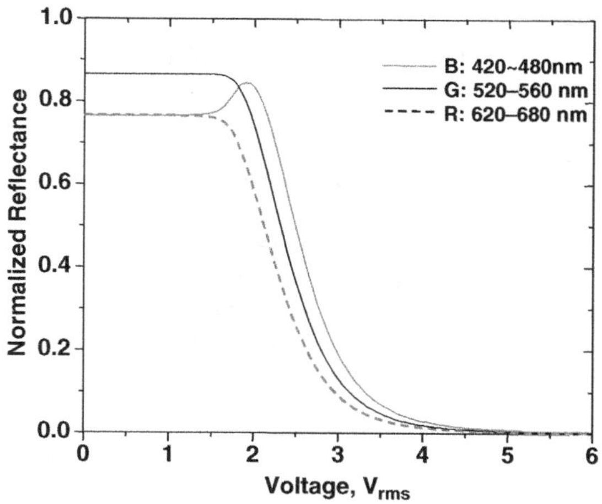

Figure 6.15 Voltage-dependent reflectance of a 90° MTN cell at the specified RGB bands. MLC-6694-000, $d\Delta n = 240\,\text{nm}$ and $\beta = 20°$

In the voltage-off state, the color dispersion is weak and thus the display appears white. As the voltage exceeds the optical threshold, molecular reorientation takes place and reflectance declines sharply. At 5 V_{RMS}, a good common dark state is obtained for the RGB bands studied. High contrast ratio and low operation voltage are two built-in natures of the 90° MTN cell. The boundary layers of such a cell are orthogonal to each other so that their residual phase retardations compensate for each other. Consequently, its dark state takes place at a relatively low voltage.

6.5.2 Viewing Cone

Figure 6.16 plots the iso-contrast contour of the 90° MTN cell for the green band. A contrast ratio greater than 2000:1 can be obtained within the central 10° viewing cone. This result is comparable to that using a VA LC cell without compensation film, except that the VA cell has a higher CR at the normal incidence. For projection displays employing a high extinction ratio polarizing beam-splitter (PBS), a 2000:1 contrast ratio should be achievable using the 90° MTN cell. The major drawback of the 90° MTN cell is that its maximum reflectance is only 88%.

6.5.3 Fringing Field Effect

Although the maximum reflectance of the 90° MTN cell is only 88%, it does not require any phase compensation film to achieve high contrast ratio. Figure 6.17 shows the fringing field effect on the 90° MTN cell with $d\Delta n = 240\,\text{nm}$ and $\beta = 20°$. Since this is a normally white mode, high voltage leads to a dark state. In Figure 6.17, the applied voltages are 6, 1.6, and 6 V_{RMS}, respectively. A small spike is observed only at the right dark pixel. In comparison to other LC modes reported in this chapter, the 90° MTN cell is least sensitive to the fringing field.

As shown in Figures 6.7 and 6.16, the VA and MTN cells exhibit a comparable viewing cone for the CR = 1000:1 contour lines without using any compensation film. Figure 6.10 shows that the split pixel, which is inherent to the VA cell, reduces the local optical efficiency by ~30% and causes the undesirable edge effect. Thus, after considering the fringing field effect, MTN possesses a slightly (~10%)

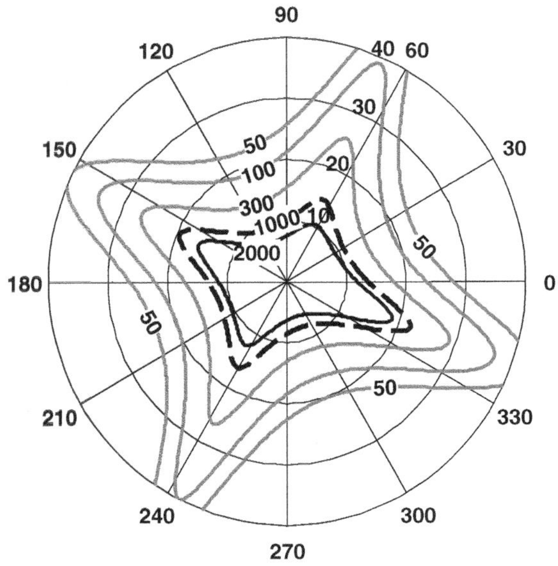

Figure 6.16 Simulated iso-contrast contours for the 90° MTN cell shown in Figure 6.15

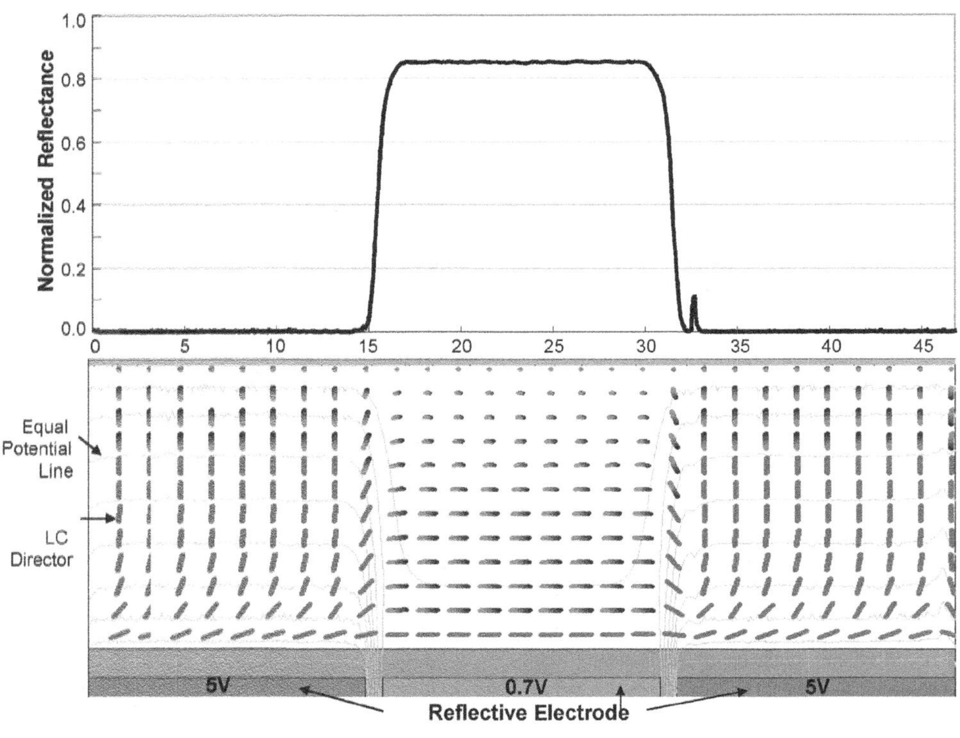

Figure 6.17 Fringing field effect of the NW 90° MTN LC cell with pixels off-on-off. Pixel size = 15 μm, electrode gap = 0.9 μm. LC: $d = 1\,\mu m$, $\Delta n = 0.24$, $\beta = 20°$

higher local optical efficiency than the VA cell. However, the VA cell has an unprecedented contrast ratio. Another difference is that the VA needs a negative $\Delta\varepsilon$ LC while MTN uses a positive $\Delta\varepsilon$ LC. Therefore, these two modes are strong contenders for the LCOS industry. However, a projector designer will struggle to avoid throughput loss in any component.

6.6 Normally White 63.6° MTN Cell

When $\beta = 0$, Equation (6.2) has a special solution for $R_\perp = 1$ which leads to $\phi = \sqrt{2}\pi/4$ (= 63.6°) and $d\Delta n/\lambda = \sqrt{2}/4$.[39] As the twist angle departs from 90°, the natural phase compensation for obtaining a good dark state disappears so that a phase compensation film has to be used. The uncompensated 63.6° MTN cell ($d\Delta n \sim 192$ nm) exhibits a poor contrast ratio (~25:1) at V = 5 V_{RMS}, which is too low for projection display applications.

Due to the special twist angle involved, it is necessary to define the coordinates of the film-compensated 63.6° MTN cell. The rubbing direction on the top glass substrate is chosen to be the horizontal axis and the molecular twist is counterclockwise. In the uncompensated scheme, the polarizer's axis should be parallel to the front LC rubbing direction, i.e. $\beta = 0$. However, with compensation film the optimal beta angle is found to be $\beta = 4°$ and the LC $d\Delta n$ value needs to be increased in order to obtain $R_\perp \sim 1$ at $V = 0$. The axis of the uniaxial film is oriented at 136° with respect to the horizontal axis.

6.6.1 Voltage-dependent Reflectance

Figure 6.18 depicts the voltage-dependent reflectance of the uniaxial film-compensated 63.6° MTN cell with $d = 2.5\,\mu$m, $\Delta n = 0.0857$, $\beta = 4°$ and film $d\Delta n = 15$ nm. In the voltage-off state, the color dispersion is weak. A good black state for the RGB bands is achieved at $V \sim 5\,V_{RMS}$. If a larger film $d\Delta n$ is selected, then the LC $d\Delta n$ has to be increased accordingly in order to maintain $R_\perp \sim 1$ at $V = 0$.

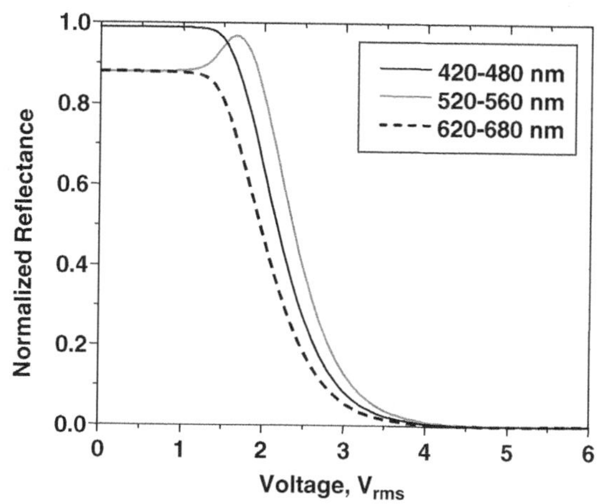

Figure 6.18 Voltage-dependent reflectance of a film-compensated 63.6° MTN cell at the specified RGB bands. MLC-6694-000, $d = 2.5\,\mu$m, $\Delta n = 0.0857$, $\beta = 4°$. Film: $d\Delta n = 24$ nm and its optical axis is at 136° with respect to the horizontal axis

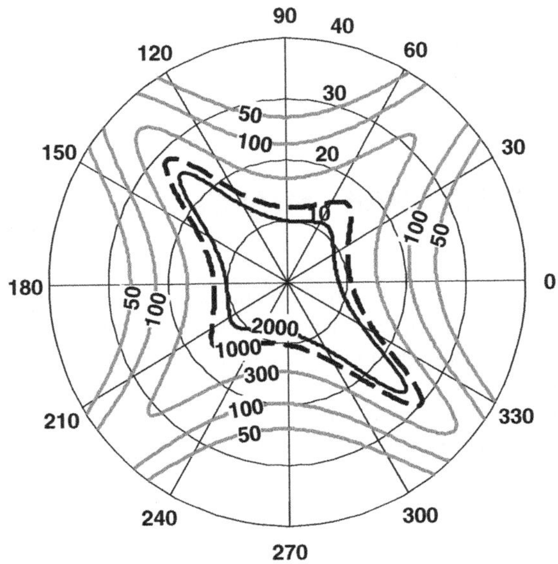

Figure 6.19 Simulated iso-contrast contours for the film-compensated NW 63.6° MTN cell

6.6.2 Viewing Cone

Figure 6.19 plots the iso-contrast ratio contour for the film-compensated 63.6° MTN cell. Within the central ~10° viewing cone, contrast ratio greater than 2000:1 can be realized. The viewing angle of the film-compensated 63.6° MTN cell is slightly wider than that of the 90° MTN cell because of its smaller $d\Delta n$ value. However, the tradeoff is on the requirement of a uniaxial compensation film. The film with $d\Delta n \sim 15\,\text{nm}$ is not easy to fabricate. Moreover, the thermal stress[40] and survivability of the film under high-power arc lamp is a concern.

6.6.3 Fringing Field Effect

Figure 6.20 depicts the fringing field effect of the film-compensated NW 63.6° MTN cell. The checkerboard pattern with pixels in off–on–off configuration are used for simulations. The bright state is at $V = 0.7\,V_{RMS}$ and dark state at $5\,V_{RMS}$. The fringing field effect depends on the cell gap; the thicker the cell gap, the larger the fringing field effect. In this simulation, the cell gap is taken to be $d = 2.5\,\mu\text{m}$. On the left side of the dark state, no light leakage is observed. The bright pixel exhibits nearly 100% reflectance. However, in the right dark pixel there is a spike, which leads to light leakage in the dark pixel and degrades the local contrast ratio significantly.

6.7 Normally Black 63.6° TN Cell

For normally black mode operation, we use Equation (6.2) to find the relationship between the twist angle and its corresponding $d\Delta n/\lambda$ value. For instance, if $\phi = \sqrt{2}\pi/4$ (= 63.6°), then its dark state (at $V = 0$) occurs at $d\Delta n/\lambda = \sqrt{7}/8 \sim 0.9355$, independent of β. At $\beta = 0$, the TN cell has the largest $d\Delta n/\lambda$ tolerance. This is the preferred operation mode. For the green band centered at $\lambda = 540\,\text{nm}$, the required $d\Delta n = 505\,\text{nm}$. Similarly, if $\phi = 50°$, then the dark state occurs at $d\Delta n/\lambda \sim 0.9606$. As a matter of fact,

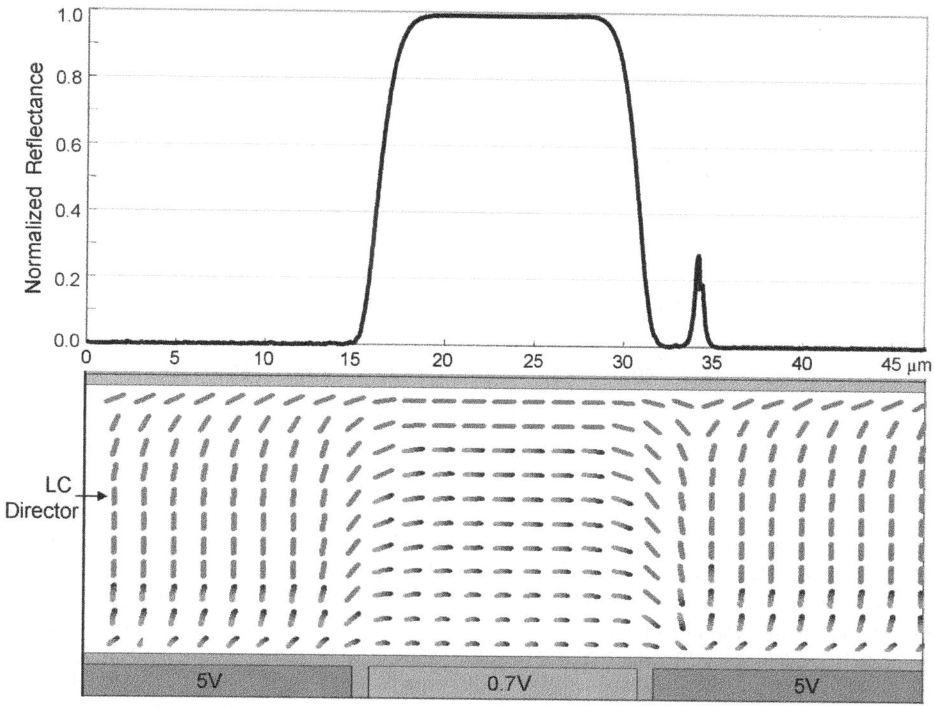

Figure 6.20 Fringing field effect of the NW 63.6° MTN cell with pixels off-on-off. Pixel size = 15 μm and electrode gap = 0.9 μm. LC: $d = 2.5\,\mu m$, $\Delta n = 0.0857$ and uniaxial film $d\Delta n = 24\,nm$

the electro-optic performance and viewing angle for twist angle ranging from 65° to 40° are very similar. Thus, we choose $\phi = 63.6°$ as an example to illustrate the normally black mode operation.

6.7.1 Optimal d/p Ratio

The normally black mode derived from Equation (6.2) does not guarantee that its bright state reflectance can actually reach $R_\perp = 100\%$. A common method to enhance reflectance is to adjust the d/p ratio; here d represents the LC cell gap and p the pitch length of the chiral dopant.[41] In an ideal case, $d/p = \phi/2\pi$; ϕ is the twist angle.

Figure 6.21 plots the optimal d/p ratio for each twist angle in order to achieve $R_\perp = 100\%$ at $\lambda = 540\,nm$ and $\beta = 0$. For the 52° TN cell, $R_\perp = 100\%$ occurs at $d/p = 0$, i.e. no chiral dopant is required. When $\phi \sim 56°$, the ideal condition $d/p = \phi/2\pi$ (= 0.156) is satisfied. For each twist angle, there exists an optimal d/p ratio so that $R_\perp = 100\%$ can be obtained. For example, for the NB 63.6° TN cell studied here, its optimal d/p ratio is ~0.6. The negative sign shown in Figure 6.21 implies that the twist of the chiral dopant is in the opposite direction to the TN cell.

6.7.2 Voltage-dependent Reflectance

Figure 6.22 plots the voltage-dependent normalized reflectance of the 63.6° TN cell at $\beta = 0$, $d\Delta n = 512\,nm$, and $d/p = 0.6$. Since this cell is optimized for the green band, its $R_\perp$ reaches 100% at $V \sim 3.2\,V_{RMS}$. For

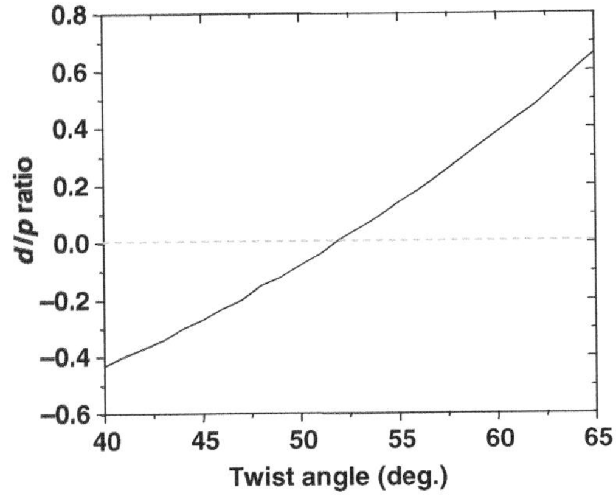

Figure 6.21 The optimal d/p ratio for each TN cell to achieve $R_\perp = 100\%$

the red and blue bands, the $d\Delta n$ values need to increase and decrease respectively, in order to compensate for the wavelength difference. The optimized panels will exhibit a dark state at $V = 0$.

Cell gap tolerance is a major concern for the normally black mode. Unlike a normally white display, whose dark state can always be obtained by controlling the applied voltage, the dark state of a normally black mode is predetermined by the initial $d\Delta n$ value. Thus, it is crucial to design a NB mode that can withstand a large cell gap variation, gap nonuniformity, broad bandwidth of each color filter, and operating temperature fluctuation. Figure 6.23 depicts the cell gap tolerance of the 63.6° TN cell. With ±5% variation in $d\Delta n$ value, the black state remains reasonably dark except that the on-state voltage is slightly affected.

6.7.3 Viewing Cone

Figure 6.24 plots the iso-contrast ratio contours for the NB 63.6° TN cell at $\beta = 0$. Within the central 15° viewing cone, contrast ratio greater than 1000:1 can be realized. Such a NB mode has a wider

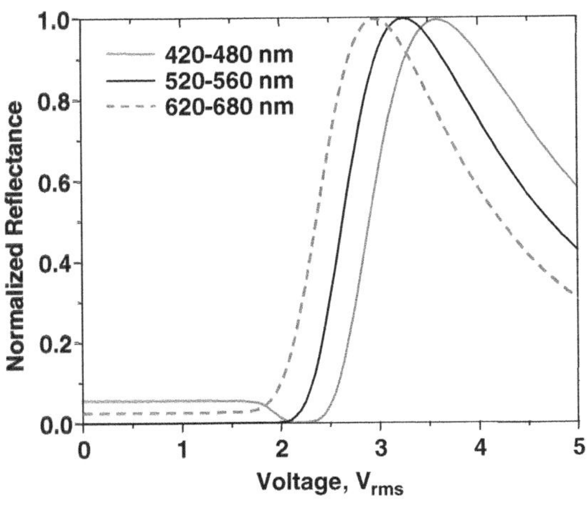

Figure 6.22 Voltage-dependent reflectance of a NB 63.6° TN cell at the specified RGB bands. MLC-6694-000, $d\Delta n = 512\,\text{nm}$, $d/p = 0.6$ and $\beta = 0°$

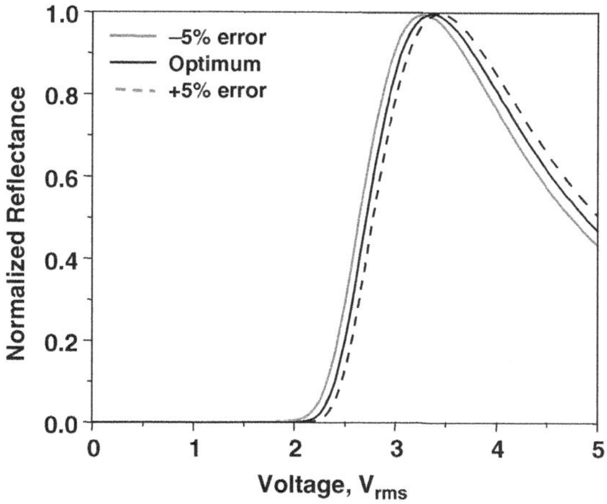

Figure 6.23 Cell gap tolerance of the NB 63.6° TN cell for the green band. Optimum: $d\Delta n = 512\,\text{nm}$

viewing cone than the homeotropic cell and NW MTN modes even though its $d\Delta n$ value is much larger. The narrow viewing cone of the NB homeotropic cell is attributed to the birefringence effect, which is sensitive to the incident angle.

6.7.4 Fringing Field Effect

Figure 6.25 depicts the fringing field effect of the NB 63.6° TN cell. The checkerboard pattern with pixels in off–on–off configuration are used for simulations. The dark-state voltage is at $V = 2\,V_{RMS}$

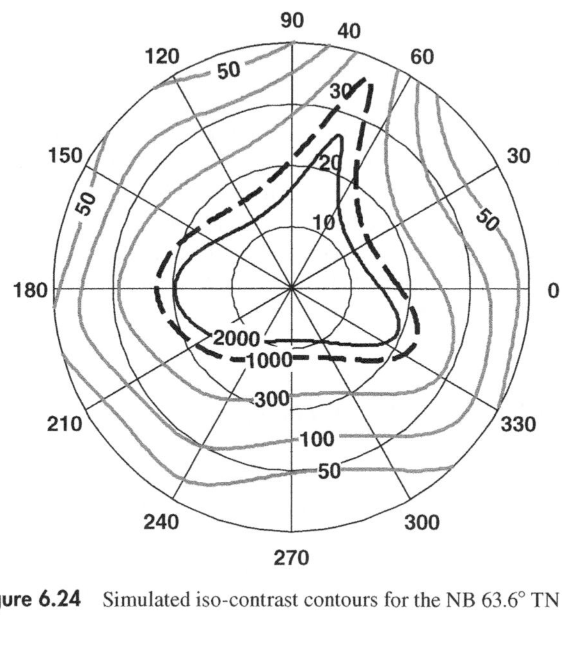

Figure 6.24 Simulated iso-contrast contours for the NB 63.6° TN cell

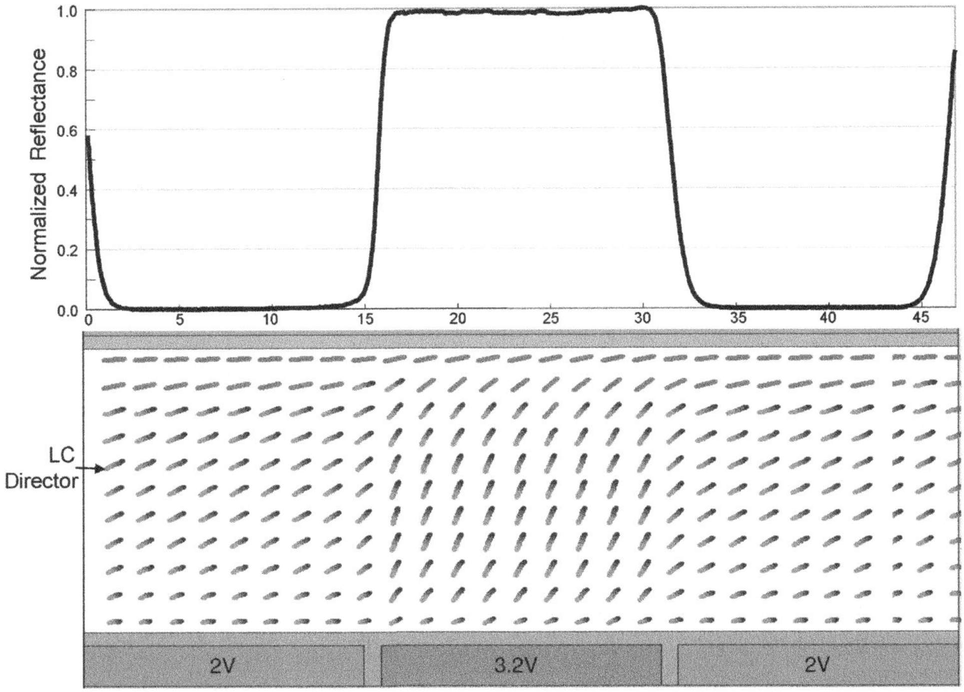

Figure 6.25 Fringing field effect of the NB 63.6° TN cell. LC cell gap $d = 2.1\,\mu m$, $\Delta n = 0.24$ and $\beta = 0$

and bright state at 3.2 V_{RMS}. As mentioned earlier, the fringing field effect depends on the cell gap; the thicker the cell gap, the larger the fringing field effect. In this simulation, the cell gap is taken to be $d = 2.1\,\mu m$ and a high birefringence LC with $\Delta n = 0.24$ is employed. At first glance, no side lobes due to fringing field are observed. However, the middle bright pixel is supposedly to cover from ~15.9 μm to 30.9 μm. From Figure 6.25, the bright state extends to ~33 μm, converting part of the right side dark pixel. The image looks continuous, but the dark pixel has actually been shrunk in area by ~10%.

In summary, the NB 63.6° TN cell exhibits several advantages, such as a large cell gap for assuring high manufacturing yield, a large cell gap tolerance, a low operating voltage, and a wide viewing cone for using a small *f*-number projection lens. The required $d\Delta n \sim 505$ nm is comparable to that of a transmissive 90° TN cell. To eliminate the spike-like light leakage that is observed in other TN cells, a thin cell gap approach is promising. However, to maintain a cell gap in the 2–3 μm range for achieving fast response time while smoothening fringing field effect, high birefringence ($\Delta n \sim 0.17$–0.25) and good stability LC mixtures need to be developed.

6.8 Normally White 60° MTN Cell

Although the 60° MTN cell is very close to the 63.6° MTN cell discussed above, it has unique phenomena that deserve special mention. Figure 6.26 shows the $R_\perp$ vs. $d\Delta n/\lambda$ plot of the 60° twisted cell at $\beta = 0$, 15, 30, and 45°. The reflectance has double peaks: one at $\beta = 0$ and $d\Delta n/\lambda \sim 350$ nm and another at $\beta = 30°$ and $d\Delta n/\lambda \sim 625$ nm. The case for $\beta = 0$ is no surprise because its twist angle is so close to 63.6°. Its performance is expected to be very similar to that of the 63.6° MTN cell. The case of $\beta = 30°$ is special because this is the bisector effect.[42,43] In this section, we will focus on the bisector effect.

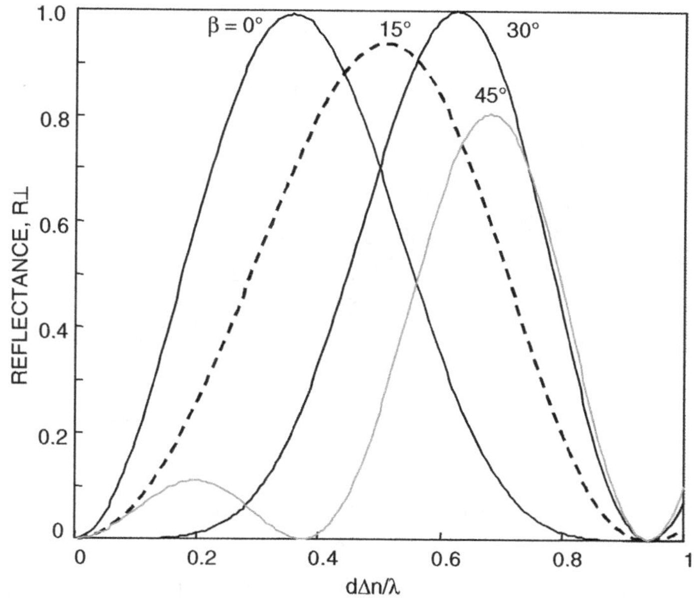

Figure 6.26 Normalized reflectance as a function of $d\Delta n/\lambda$ at $\beta = 0$, 15, 30 and 45° for the 60° TN cell

6.8.1 Bisector Effect

Figure 6.27 shows the voltage-dependent reflectance ($R_{\perp}$) of the 60° MTN cell with $d\Delta n = 344\,\text{nm}$ and $\beta = 30°$. As expected, a good dark state takes place at a relatively low voltage. Benefiting from the bisector effect, its dark-state voltage is lower than that of the 90° MTN cell.

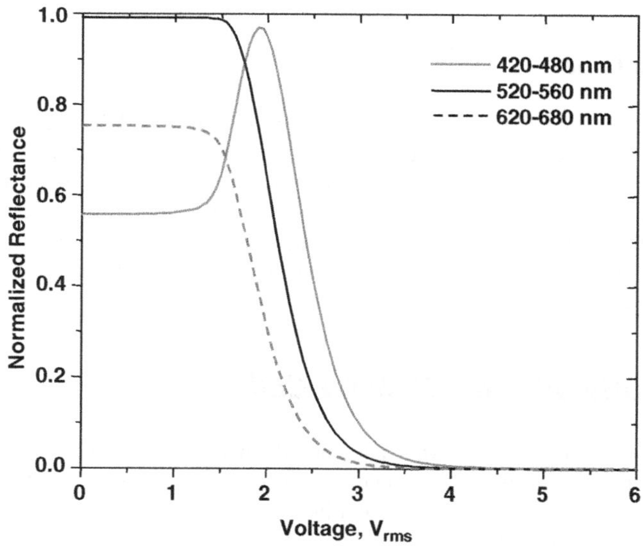

Figure 6.27 Voltage-dependent reflectance of a 60° MTN cell at specified RGB bands. MLC-6694-000, $d\Delta n = 344\,\text{nm}$ and $\beta = 30°$

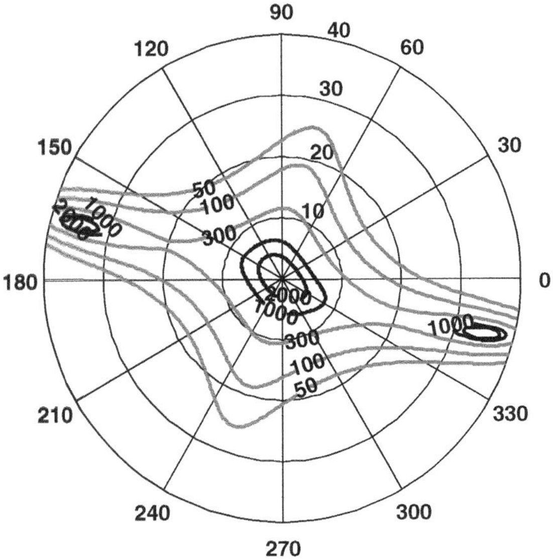

Figure 6.28 Simulated iso-contrast contours for the NW 60° MTN cell. The off-state voltage is 0.7 V_{RMS} and on-state voltage is 5 V_{RMS}

The major problem of the 60° MTN cell is its relatively large color dispersion.[44] From Figure 6.27, in the voltage-off state the reflectance difference between the green and the blue bands is as large as 60%. This implies that the off-state is not a balanced white. Fortunately, for projection displays using three panels, each panel can be filled with a different LC mixture. One could use a higher Δn LC for the red channel and a lower Δn LC for the blue channel such that the $d\Delta n/\lambda$ for the RGB panels is nearly identical. Under these circumstances, the red and blue curves shown in Figure 6.27 will overlap with the green. High reflectance, low operation voltage, and zero color dispersion can be obtained simultaneously.

6.8.2 Viewing Cone

Figure 6.28 plots the iso-contrast contour for the green band of the 60° MTN cell at $\beta = 30°$. A large β implies a large birefringence effect. The birefringence effect has narrower viewing characteristics than the polarization rotation effect. Moreover, the large $d\Delta n$ also makes a negative contribution to viewing angle. Figure 6.28 shows that the viewing cone of the bisector 60° MTN cell is less than 5° for the 1000:1 iso-contrast contour lines. Moreover, in order to achieve 1000:1 contrast ratio, the required on-state voltage is at least 5 V_{RMS}. The narrow viewing cone would limit the application of the bisector effect for high-contrast projection displays.

6.9 Normally White 45° MTN Cell

For the 45° twisted cell, both normally white and normally black modes can be realized, depending on their $d\Delta n/\lambda$ values. Figure 6.29 plots the normalized reflectance as a function of $d\Delta n/\lambda$ for the 45° TN cell at different β angles. Two modes are particularly useful for projection displays: the normally black mode at $\beta = 0$ and the normally white mode at $\beta = -15°$. Let us differentiate the former as NB 45° TN cell and the latter as NW 45° MTN cell.

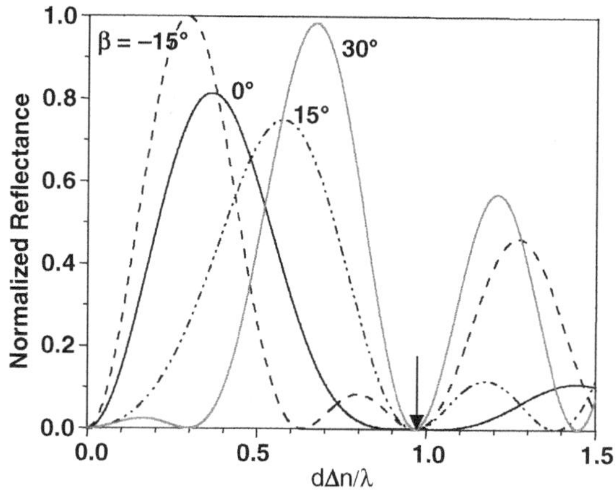

Figure 6.29 Normalized reflectance of a 45° TN cell as a function of $d\Delta n/\lambda$ at $\beta = -15$, 0, 15 and 30°

Let us first discuss the normally white 45° MTN cell at $\beta = -15°$. In the voltage-off state, the reflectance $R_\perp = 1$ occurs at $d\Delta n/\lambda \sim 0.3$. In the high-voltage state, almost all of the bulk LC directors are reoriented by the electric field except for the boundary layers. Since these boundary layers are at 45° with respect to each other, they do not compensate each other like the 90° TN cell. Thus, a phase compensation film is required in order to achieve a high contrast ratio.

6.9.1 Voltage-dependent Reflectance

In the uncompensated scheme, i.e. $\beta = -15°$, although $R_\perp = 1$ is reached at $V = 0$ the dark state is poor. To add a uniaxial compensation film the orientation angles of the polarizer, LC cell, and uniaxial film need to be optimized. Let us assume the rubbing direction of the top glass substrate is along the horizontal axis and the LC directors are twisted counterclockwise. To obtain $R_\perp \sim 1$ at $V = 0$, the following parameters are found: LC $d\Delta n = 197\,\text{nm}$, film with $d\Delta n = 27\,\text{nm}$, the polarizer's axis is oriented at $\beta = 78°$ (equivalent to $-12°$) and the uniaxial film is oriented at 110° with respect to the horizontal axis. Under these circumstances, a good dark state occurs at $V = 5\,V_{RMS}$.

Figure 6.30 shows the voltage-dependent light reflectance of a film-compensated 45° MTN cell for the RGB bands. This normally white mode exhibits a high reflectance, high contrast ratio, and weak color dispersion. To further lower the operation voltage, a film with a larger $d\Delta n$ value could be considered. The major attraction of this operating mode is the small $d\Delta n$ requirement. In the color sequential or scrolling color projection display,[45] the LC response time needs to be less than 1 ms. To achieve such a fast response time, using a thin cell gap is the simplest approach. Philips has successfully fabricated 1 μm spacer posts in the LCOS devices.[46] With 1 μm cell gap, the rise time (white to black) is ~0.2 ms and decay time is ~1 ms. A black pre-write at the color-scrolling boundary eliminates crosstalk between successive colors.

6.9.2 Viewing Cone

Figure 6.31 plots the iso-contrast ratio contour for the film-compensated NW 45° MTN cell. The operating voltages used for these simulations are $V_{ON} = 5\,V_{RMS}$ and $V_{OFF} = 0.7\,V_{RMS}$. The 1000:1 contour lines are somewhat asymmetric. Although the birefringence effect originated from $\beta = -12°$

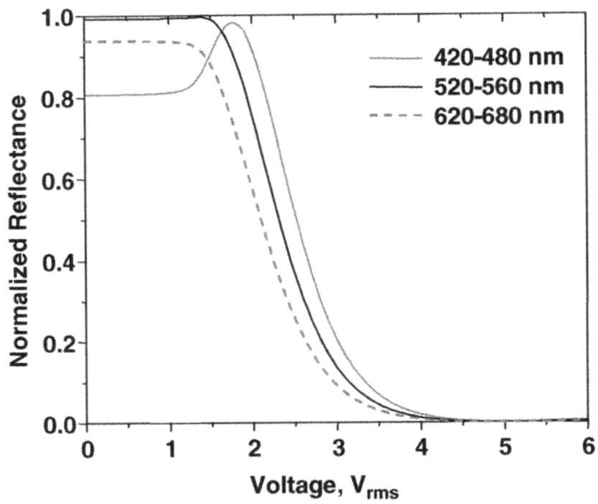

Figure 6.30 Voltage-dependent reflectance of a NW film-compensated 45° MTN cell at the specified RGB bands. MLC-6694-000, $d = 2.3\,\mu m$, $\Delta n = 0.0857$, $\beta = -12°$ and film $d\Delta n = 27\,nm$ and its optical axis is at 110° with respect to the horizontal axis

is unfavorable to the viewing angle, the small $d\Delta n$ value counterbalances this undesirable effect. As a result, within the central 10° viewing cone the calculated contrast ratio is still greater than 2000:1.

The viewing angle of the film-compensated NW 45° MTN cell is close to that of the film-compensated NW 63.6° MTN cell. The cell gap difference is also negligible. A common problem for both approaches is the need for a small $d\Delta n$ uniaxial film and the survivability of the film under intense arc lamp illumination.

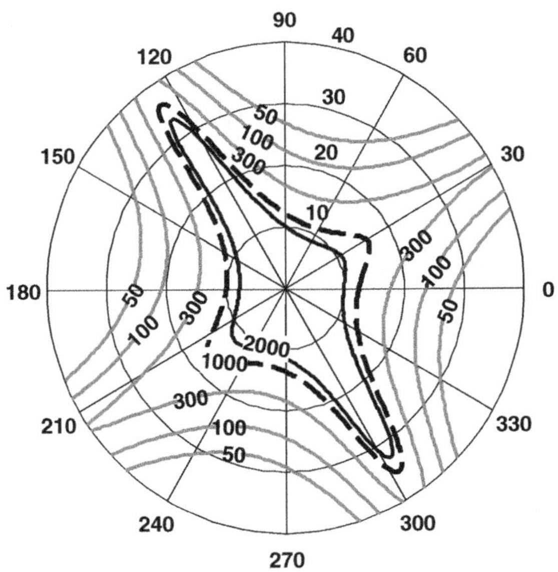

Figure 6.31 Simulated iso-contrast contours for the NW film-compensated 45° MTN cell for the green band from 520 to 560 nm

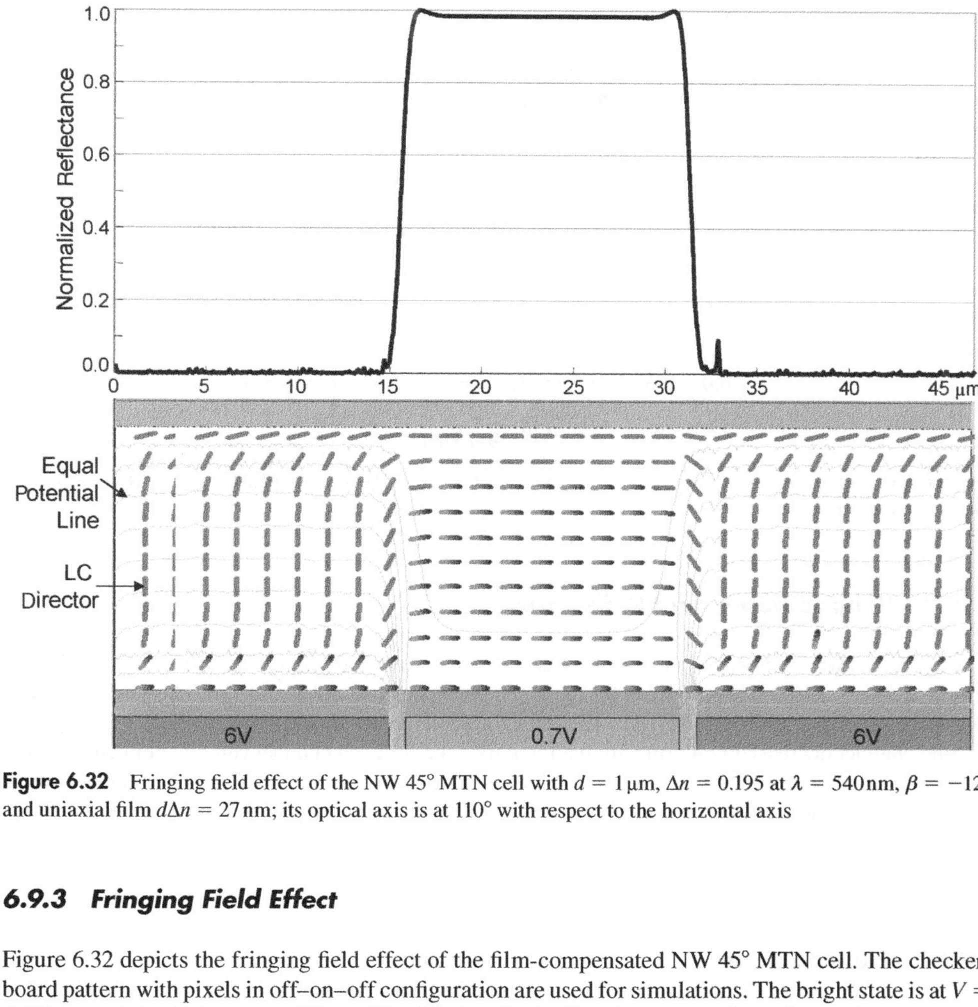

Figure 6.32 Fringing field effect of the NW 45° MTN cell with $d = 1\,\mu m$, $\Delta n = 0.195$ at $\lambda = 540\,nm$, $\beta = -12°$ and uniaxial film $d\Delta n = 27\,nm$; its optical axis is at 110° with respect to the horizontal axis

6.9.3 Fringing Field Effect

Figure 6.32 depicts the fringing field effect of the film-compensated NW 45° MTN cell. The checkerboard pattern with pixels in off–on–off configuration are used for simulations. The bright state is at $V = 0.7\,V_{RMS}$ and the dark state at $6\,V_{RMS}$. The fringing field effect depends on the cell gap; the thinner the cell gap, the smaller the fringing field effect. In this simulation, the cell gap is chosen to be $d = 2.5\,\mu m$. On the left side of the dark state, no light leakage is observed. The bright pixel exhibits nearly 100% reflectance. However, in the right dark pixel there is a small spike. This light leakage degrades the local contrast ratio. If the LC cell gap is reduced to $1\,\mu m$, then the spike peak is reduced. As a result, the device contrast ratio is improved.

6.10 Normally Black 45° TN

The normally black 45° TN mode was first implemented in Hughes photoactivated liquid crystal light valve as published in 1975.[47] After three decades, it is still a strong contender for LCOS due to its large cell gap and low operating voltage. As shown in Equation (6.2), the black state occurs at $d\Delta n/\lambda = \sqrt{15}/4 \sim 0.968$ for the 45° TN cell. Under this circumstance, $R_\perp = 0$ is independent of β. Figure 6.29 shows that the curve with $\beta = 0$ is least sensitive to the $d\Delta n/\lambda$ variation. As β approaches 30°, birefringence effect is pronounced and the reflectance becomes very sensitive to $d\Delta n/\lambda$. Under such a circumstance, the cell gap tolerance is greatly reduced.

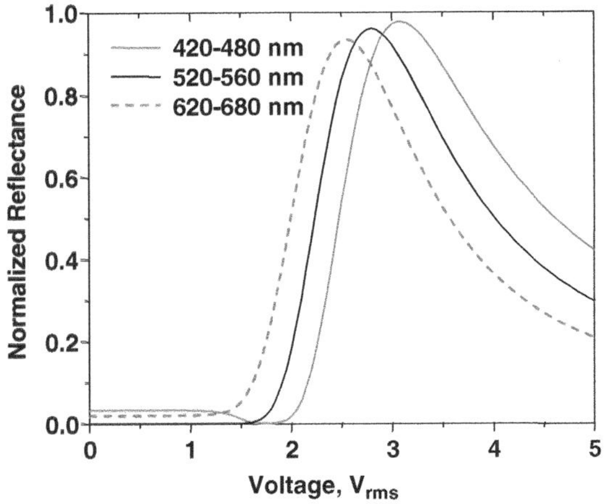

Figure 6.33 Voltage-dependent reflectance of a NB 45° TN cell at the specified RGB bands. MLC-6694-000, $d\Delta n = 523$ nm and $\beta = 0°$

6.10.1 Voltage-dependent Reflectance

Figure 6.33 shows the voltage-dependent reflectance of the normally black 45° TN cell with $d\Delta n = 523$ nm at $\beta = 0$. The cell is optimized for the green band centered at $\lambda = 540$ nm. Thus, there is some light leakage for the red band. The $d\Delta n$ value for the red channel should be increased to 629 nm and the blue channel decreased to 436 nm so that the dark state for each channel can be obtained simultaneously. Also noticed from Figure 6.33 is that the on-state voltage is $\sim 3\,V_{RMS}$. Low-voltage operation is particularly attractive for the high pixel density LCOS. As the pixel size continues to decrease, the maximum silicon backplane output voltage would decrease to $2.5\,V_{RMS}$. Higher voltage requires more expensive hybrid circuitry.

6.10.2 Viewing Cone

Figure 6.34 plots the iso-contrast contour of the NB 45° TN cell for the green band. In the computer simulations, the dark-state voltage is $V = 0.7\,V_{RMS}$ and the on-state voltage is at $2.8\,V_{RMS}$ for the MLC-6694-000 mixture employed. A CR > 1000:1 can be obtained within the central 10° viewing cone.

In Figure 6.33, the maximum reflectance only reaches 95%. By changing β to $-12°$ and reducing $d\Delta n$ to 434 nm, the reflectance for RGB can reach 100%. However, its viewing angle is much narrower due to the increased birefringence effect. The 1000:1 iso-contrast contour lines are shrunk to the central ~5° viewing cone. This tradeoff is not worthwhile. Another solution is to add a chiral dopant with $d/p \sim -0.27$ as shown in Figure 6.21. However, the viewing cone for the 1000:1 contour lines is reduced to ~8°.

6.10.3 Fringing Field Effect

Figure 6.35(a) shows the simulated fringing field effect of the normally black 45° TN mode with $d\Delta n = 523$ nm and $\beta = 0°$. No chiral dopant is added to enhance the reflectance. The on-state pixel is at ~2.8 V and the adjacent dark pixels are at $0.7\,V_{RMS}$, which is below threshold. The fringing field on the left side of the bright pixel causes about 15% optical loss. Moreover, the shallower rising slope contributes to

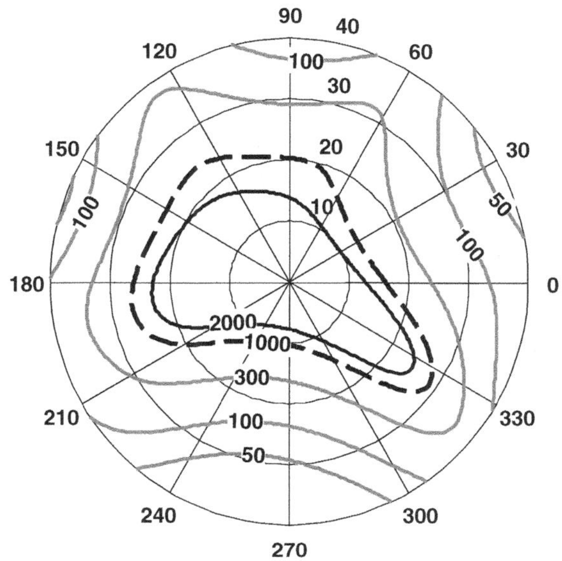

Figure 6.34 Simulated iso-contrast contours for the NB 45° TN cell

an additional optical loss. By integrating the bright pixel area from 15.9 to 30.9 µm, the overall optical efficiency only reaches ~70%. Moreover, there is an appreciable amount of light leakage on both sides of the dark pixels. This light leakage will undoubtedly degrade the local contrast ratio. Several LCOS projectors using the 45° TN panels have been produced, with contrast ratio usually limited to ~500:1.

To reduce the fringing field effect, a thinner cell gap could be considered. As the cell gap is reduced, the reflectance at the left side of the bright pixel is gradually increased and the light leakage at the dark pixels is decreased. Figure 6.35(b) plots the simulated fringing field effect of the 45° TN cell using a high birefringence LC ($\Delta n = 0.24$) and a thin cell gap ($d = 2.2\,\mu m$). Indeed, the optical efficiency in the bright pixel is improved and the light leakage near the left edge of the bright pixel is suppressed. The side lobe in the right edge remains except that its amplitude is reduced. In the 15 µm pixel size and 0.9 µm electrode gap configuration, the critical cell gap for eliminating the left side lobe is $d < 2.5\,\mu m$. Another advantage of using a thinner cell is the faster response time. The major technical challenge for the thin cell approach is that high Δn, high resistivity and good stability LC materials remain to be developed.

6.11 Finger-on-Plane Structure

To reduce the fringing field effect between adjacent pixels, a finger-on-plane (FOP) common electrode has been developed. Figure 6.36 shows the cross-sectional view of the cell structure. A periodic structure of common electrodes is fabricated on the insulation layer along the column direction. The major electrodes are on the top of each inter-pixel gap with a few equally spaced minor electrodes interposed between the adjacent major electrodes. Both the common electrode and the pixel electrode are made of aluminum.

The displayed image, indeed, shows no pixel-splitting phenomenon. However, diffraction due to the tiny periodic electrodes as illustrated in Figure 6.36 still occurs. The optical efficiency of the FOP mode is about 15% lower than that of the VA mode at $d\Delta n = 550\,nm$ and this difference is further enlarged in the blue and red regions. The poor optical efficiency limits the FOP mode from widespread applications.

(a)

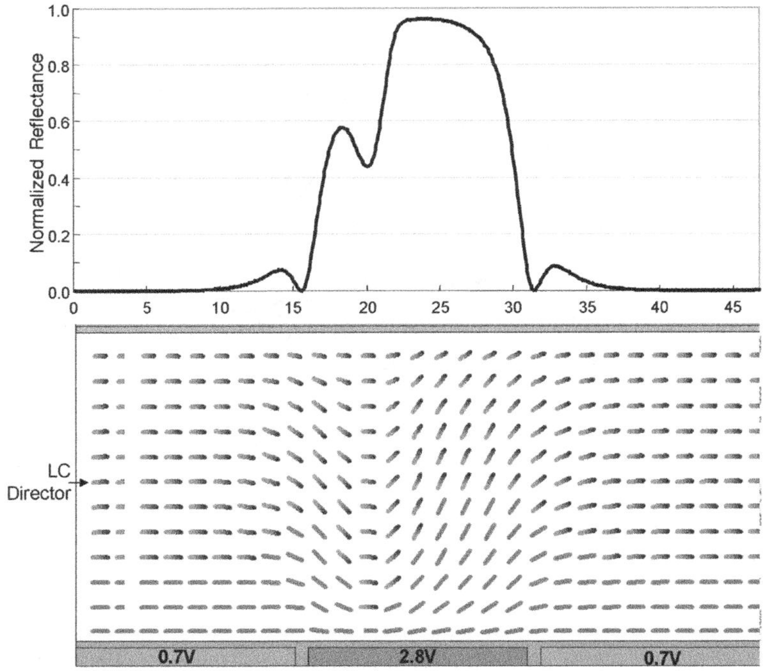

(b)

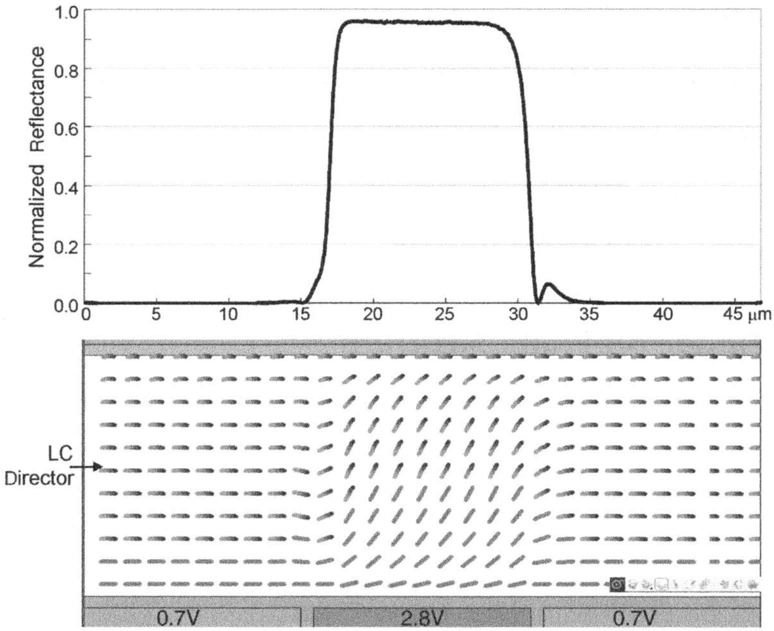

Figure 6.35 Fringing field effect on the NB 45° TN LC cell with pixels off-on-off. Pixel size = 15 μm, electrode gap = 0.9 μm. LC: $d\Delta n$ = 523 nm, $\beta = 0$. (a) Cell gap d = 4.8 μm, Δn = 0.11. (b) d = 2.2 μm, Δn = 0.24

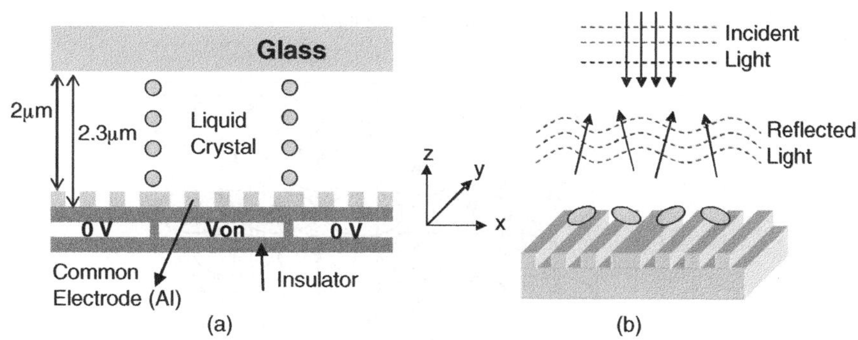

Figure 6.36 Device structure of the finger-on-plane LC mode

6.12 Scattering and Diffractive Microdisplays

A controllable optical scattering effect can generate an image. The first liquid crystal devices employed the dynamic scattering effect produced by an electric current in NLC, but were quickly superseded by electric field effect devices based on polarizing optics. Liquid crystals are fundamentally phase modulators, readily applied to the formation of phase-modulated gratings favored by diffraction-modulated projectors. Schlieren optic projectors employing phase gratings approach 90% throughput efficiency, an attraction that has sustained interest in this approach for many years.

6.12.1 Polymer Dispersed Nematics

Interest in scattering revived when a dispersal of NLC droplets in a solid matrix demonstrated switching from a scattering to non-scattering state with application of an electric field. Figure 6.37 illustrates the structure and compares the off- and on-states. In the on-state, an applied electric field aligns the optical axis of NLC droplets in the direction of the applied field, and optical propagation in that direction samples the NLC ordinary optical refractive index (n_o). The polymer matrix refractive index (n_p) matches the NLC, i.e. $n_p \rightarrow n_o$, giving minimum optical scattering in the on-state. The NLC unified orientation is lost when the electric field is switched off and the effective optical index ($n_o < n < n_e$) increases towards the extraordinary index n_e, forming an optical scattering off-state. The details of droplet orientation in the off-state are determined by interface orientation effects and droplet shape. The strength of the scattering varies with applied electric field, providing analog addressing capability.

Scattering devices do not require polarized light, eliminating the losses, cost, and complexity of polarization. In a projection system employing a scattering microdisplay, the readout light is scattered out of the projection aperture, to modulate the image intensity on the output screen, as indicated in Figure 6.38 (see Chapter 10). The scattering is more effective in modulating to dark levels for small projection aperture (high *f*-number). Consequently, high contrast ratio requires high projection *f*-number, but high throughput requires low *f*-number, implying an inherent tradeoff in CR against throughput.

The first devices were prepared from an aqueous emulsion of NLC and polymer, coated onto the substrate and dried to form a thin layer of NLC droplets separated by polymer membranes. The structure prompted the description NCAP (nematic curvilinear aligned phase). Further work showed a solution of liquid crystal molecules and precursor monomers, exposed to polymerizing UV light, forms liquid crystal droplets in a polymer matrix due to decreased solubility with polymerization. The resulting structure is described as polymer-dispersed liquid crystal (PDLC). Both structures in general can be described as liquid crystal dispersions.[48]

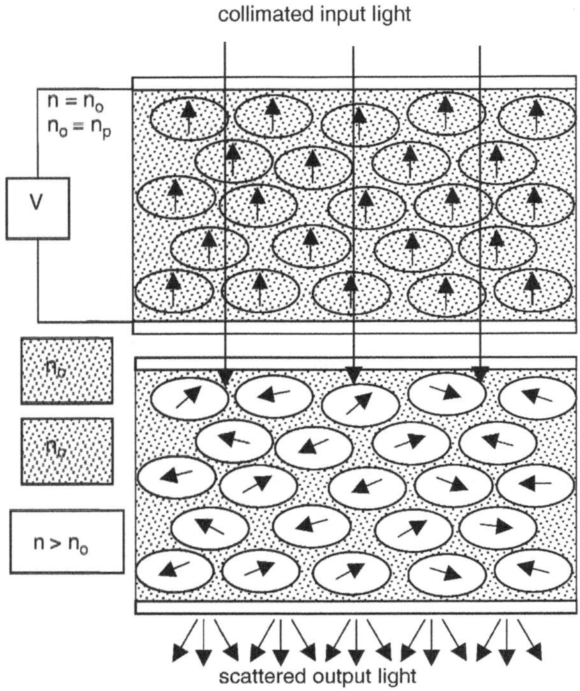

Figure 6.37 Nematic droplets in polymer matrix. The applied voltage orients optic axis of nematic to match polymer refractive index $n_p = n = n_o$. Removal of voltage gives mismatched scattering orientation with index n ($n_o = n_p < n < n_e$)

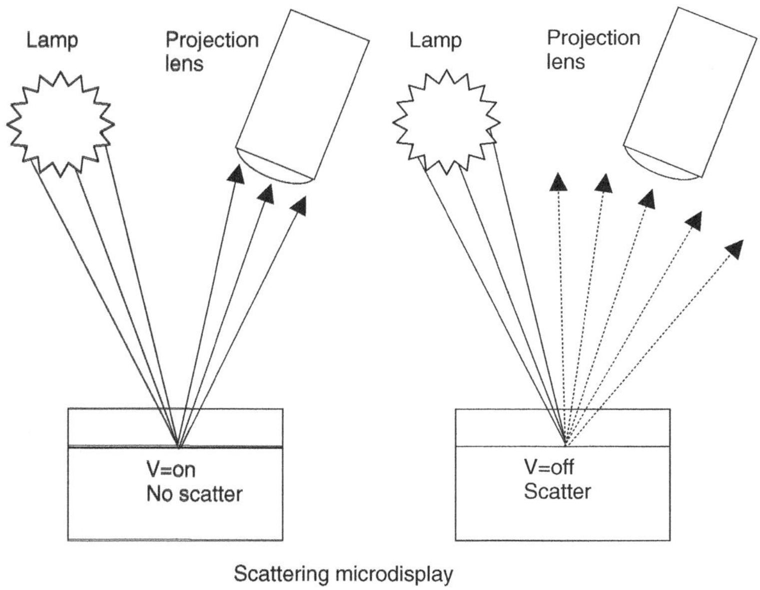

Figure 6.38 Primitive arrangement of scattering microdisplay projector indicating dark-state pixel produced by scattering outside projection aperture

The voltage needed to drive NCAP or PDLC devices increases with layer thickness (cell gap) and reduction in droplet size. The scattering effect increases with layer thickness and favors an optimum droplet size (~1 μm) depending on wavelength. Response time quickens with decreasing droplet size. Reflective devices have an advantage in effectively doubling the layer thickness for the same voltage. The limited voltage available from microdisplay addressing methods requires the material be optimized for voltage, response time, and scattering properties.

A transmission projection display using polysilicon matrix addressing with 12 μm thick PDLC readout was demonstrated, but at a 2.8-inch diagonal hardly qualifies as a microdisplay.[49] The large display area compensated for the high *f*-number needed to raise the CR while maintaining adequate throughput, in an etendue limited projection system.

A LCOS type microdisplay employing NCAP was developed to prototype level, together with a projection system.[50,51] Upper voltage level 7 V and NLC droplet size ~2 μm gave response time <50 ms. Projection throughput 600 lumen was achieved with a 100 W short arc lamp, and CR > 40.

Work on dispersed nematic scattering microdisplays has declined for a number of reasons. The tradeoff in throughput against CR is a disadvantage, compounded by the development of polarizing conversion systems for birefringent microdisplays that erode the scattering advantage. Hysteresis remains a problem, and response time is slow by current pure nematic standards. Silicon technology has advanced, along with microdisplay fabrication and assembly methods that result in high-yield manufacturing of non-scattering devices. Development work continues in PDLC for various applications, but is unlikely to benefit microdisplays, except in special cases. Polymer network structures that influence and stabilize liquid crystal geometries are a related field that may have applications in microdisplays employing birefringence effects.[48]

6.12.2 Diffraction

Diffractive systems are similar to scattering and require Schlieren readout. However, diffraction controls the distribution of light in the output plane, enabling the design of a Schlieren filter to optimize the bright and dark states (see Chapter 10). A phase grating modulated over the range 0 and $\lambda/2$ controls the Schlieren transmission over the range 0 and 90%. It is a challenge in liquid crystals to design a phase grating that is independent of polarization; only achieved by alternating the surface alignment at the grating period.[24] At least one cycle of grating is required per pixel, making the alignment difficult to pattern. Progress in assessing the CR and thoughput achievable in a practical system has not been forthcoming.

6.13 Ferroelectric Liquid Crystals

Nematic liquid crystal responds to a dipole moment induced by the applied electric field, making nematic director reorientation proportional to the square of the electric field, and consequently independent of polarity. Nematics are electrically driven in one direction and rely on elasticity to restore the original configuration on lowering the applied voltage. Ferroelectric liquid crystals (FLCs) possess a spontaneous dipole moment P_s, providing electric torque $E \times P_s$ on molecular orientation, the torque changing sign when electric field E reverses. Therefore, the FLC device has the advantage that it can be driven to both the on-state and off-state.

Ferroelectricity imposes symmetry restrictions, such that the simplest FLC is the chiral smectic-C phase (S_C*). Smectic liquid crystals have layered structures, where molecular diffusion within the layer is much faster than interlayer diffusion. Figure 6.39 illustrates the structure of the S_C* FLC phase. The spontaneous polarization vector P_s lies in the smectic layer plane, and is orthogonal to the local molecular director n. Relative to the smectic layers, n is constrained to a constant polar tilt angle θ, and variable azimuth angle ϕ. For successive smectic layers, ϕ increases and n spirals with helical pitch p

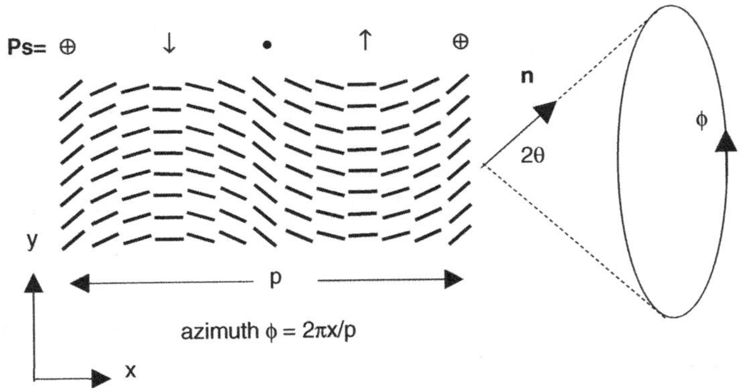

Figure 6.39 Ferroelectric S_C* director orientation $n(\theta,\phi)$ has constant cone angle θ, while ϕ advances with successive molecular layers. The polarization vector P_s is orthogonal to n. Director orientation is shown projected on xy plane, $n_x = \cos\theta$, $n_y = \sin\theta \sin\phi$

around a cone angle 2θ. The polarization follows the rotation, where the helical P_s pattern results in zero macroscopic polarization (helioelectric).[52]

6.13.1 Surface-Stabilized FLC

In the surface-stabilized ferroelectric liquid crystal (SSFLC) device, the FLC cell alignment surfaces play a critical role in stabilizing the FLC configuration, while allowing a rapid switching effect.[53] Figure 6.40 illustrates the structure of an SSFLC cell. Parallel surface alignment promotes the "bookshelf geometry," where the smectic layers are perpendicular to the electrode surfaces, suppressing helical twist of the FLC when the cell spacing is significantly less than the pitch.

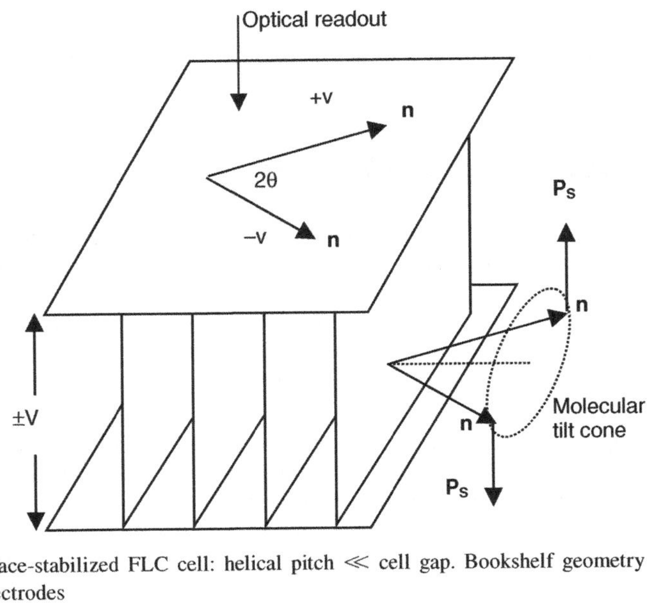

Figure 6.40 Surface-stabilized FLC cell: helical pitch ≪ cell gap. Bookshelf geometry has smectic layers perpendicular to electrodes

In the untwisted state, P_s is perpendicular to the cell surface and the polarity can be switched according to the polarity of an applied voltage. As P_s reverses, the molecular director n rotates (around the cone angle) through twice the S_C* tilt angle (2θ), while the smectic layer structure remains intact. Through crossed polarizers, the normalized reflectivity can be written

$$R = \sin^2(4\theta)\sin^2\left(2\pi d\Delta n/\lambda\right). \tag{6.3}$$

For optimum black/white switching the FLC tilt angle θ = 22.5 degrees and cell thickness $d = \lambda/4\Delta n \sim 1\,\mu\text{m}$. The reflective cell then behaves as a half-wave retarder rotating through 45 degrees, consequently imposing a 90-degree rotation on the input polarization. The small cell gap is an advantage in raising the electric field and reducing field fringing, but makes cell gap uniformity more difficult to achieve.

The SSFLC has bistable on/off states, directly applicable to switching or computing applications. Display applications require time or space division to achieve gray levels; the latter sacrifices resolution, while the former introduces pulse-width-modulation artifacts and demands high frame rate. Residual ionic conductivity imposes periodic reversal of applied voltage, reducing the effective duration of the on-time and the off-time. The cell bistability can be exploited to minimize the sacrifice in operating time;[54] if needed, the effect of alternating voltage can be cancelled by an additional transmission cell driven in antiphase.[55]

Direct-view and microdisplay SSFLC devices have been demonstrated for many years and microdisplay devices are now available in commercial quantities. The binary response speed is substantially faster than that achieved by NLC at comparable cell gap, temperature, and voltage. The switching speed is inadequate for binary pulse code modulation similar to the DMD. However, in NTE applications FLC pulse width modulation subdivides by pulse modulation of the LEDs providing the readout light. A field-sequential-color SSFLC microdisplay serves the camera viewfinder and head-mounted market at standard QVGA resolution and SXVGA resolution.[54, 56] In the dark state, the input polarization is along the birefringent axis, maintaining the dark state over a wide field of view, enhanced by the minimal $d\Delta n$. Cell gap $<1\,\mu\text{m}$ suppresses fringe field effects. Drawbacks to the SSFLC are temperature sensitivity and alternating voltage imposed by ionic conduction, which halves the optical duty cycle.

6.13.2 Other FLC Modes

The shortcomings of the SSFLC microdisplay would be overcome by a structure that achieves analog modulation, without sacrificing response speed. The deformed helix FLC device structure (DHFLC) exploits helical pitch smaller than the optical wavelength. Optical propagation is described by a refractive index averaged over the local director configuration. An applied electric field deforms the helix and modifies the configuration average, rotating the effective optic axis to give a continuous grayscale.[52,57] Uniformity and contrast ratio issues in the DHFLC mode handicap microdisplay development.

An antiferroelectric liquid crystal (AFLC) variant of chiral smectic-C (S_{C^*a}) has P_s reversing polarity at successive smectic planes. Application of an electric field disturbs the balance of $\pm P_s$, which is coupled to the director orientation and hence the birefringence. In an SSFLC configuration, analog modulation in the form of a V-shaped characteristic is observed as indicated in Figure 6.41. Gray-level independent of voltage polarity and potential for fast response has attracted the interest of microdisplay developers.[58] However, hysteresis effects and material development have limited progress. A similar V-shaped characteristic is observed in FLC, under appropriate operating conditions. The general theory of V-shaped switching in FLC is evolving and further development may result in high-performance analog microdisplays.

Some of the issues in the design of FLC devices can be mitigated by applying polymer dispersal methods developed for NLC to form polymer stabilized FLC. The development of FLC materials and devices is an active area of research that may eventually challenge the ascendancy of NLC

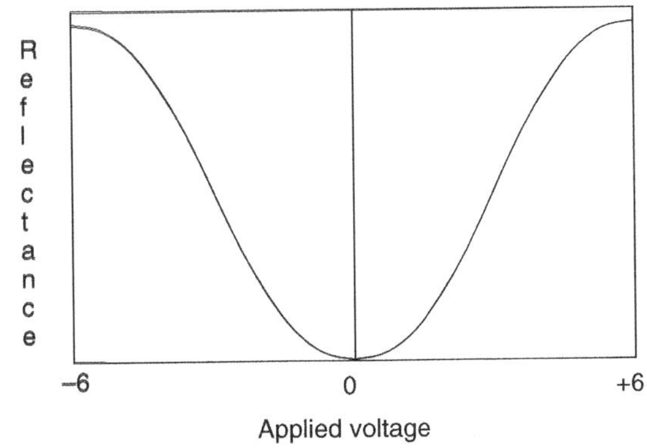

Figure 6.41 V-shaped switching characteristic

in microdisplay applications.[59] However, FLC device performance will be measured against NLC achievements currently providing a driven response of 0.2 ms and a relaxation of 1 ms to 90% completion, while retaining CR ~ 800.[60]

The electroclinic effect in smectic-A arises from the soft-mode behavior of the tilt angle as the phase transition to S_{C^*} is approached. In the "bookshelf" geometry of S_A, rotation of the optic axis is linearly dependent on applied electric field, providing analog modulation.[52] The effect is significant over a narrow temperature range close to the transition temperature, where the response speed is comparable to SSFLC devices. Studies of polysiloxane materials have shown that the effect can be larger over a wider temperature range. Development of polysiloxanes may result in an electroclinical response of value in microdisplay applications.[61]

6.14 References

1 Alt, P. M. (1997) Single crystal silicon for high resolution displays. *Intl Display Research Conference*, M19–28.
2 Sato, F., Yagi, Y. and Hanihara, K. (1997) High resolution and bright LCD projector with reflective LCD panels. *SID Digest* **28**, 997–1000.
3 Kurogane, H., Doi, K., Nishihata, T. *et al.* (1998) Reflective AMLCD for projection displays. *SID Digest* **29**, 33.
4 Bergstrom, N., Chuang, C. L., Curley, M. *et al.* (2000) Ergonomic wearable personal display. *SID Digest* **31**, 1138.
5 Schott, D. J. (1999) Reflective LCOS light valves and their application to virtual displays. *Intl Display Research Conference*, 485.
6 Melcher, R. L. (2000) LCoS-microdisplay technology and applications. *Information Display* **16**, 1, 20–23.
7 Sterling, R. D. and Bleha, W. P. (2000) D-ILA technology for electronic cinema. *SID Digest* **31**, 310 (2000).
8 Bruzzone, C. L., Ma, J., Aastuen, D. J. W. and Eckhardt, S. K. (2003) High-performance LCOS optical engine using Cartesian polarizer technology. *SID Digest* **34**, 126.
9 Gardner, E. and Hansen, D. (2003) An image quality wire-grid polarizing beam splitter. *SID Digest* **34**, 62.
10 Melcher, R. L. (2003) LCoS for high performance displays, *IEEE LEOS Annual Meeting Conference Proceedings* **2**, 812–13.
11 Jepsen, M. L., Ammer, M. J., Bolotski, M. *et al.* (2002 High resolution LCOS microdisplay for single-, double- or triple-panel projection systems. *Displays* **23**, 109.

12 Moench, H., Giese, H., Hechtfischer, U. *et al.* (2003) UHP lamps with increased efficiency. *SID Digest* **34**, 758.
13 Schiekel, M. F. and Fahrenschon, K. (1971) Deformation of nematic liquid crystals with vertical orientation in electric fields. *Applied Physics Letters* **19**, 391.
14 Grinberg, J., Bleha, W. P., Jacobson, A. D. *et al.* (1975) Photoactivated birefringence liquid crystal light valve for color symbology display. *IEEE Trans. Electron Devices* **22**, 775.
15 Sterling, R. D. and Bleha, W. P. (2000) D-ILA technology for electronic cinema. *SID Digest* **31**, 310.
16 Takeda, A., Kataoka, S., Sasaki, T. *et al.* (1997) A super-high-image-quality multi-domain vertical alignment LCD by new rubbing-less technology. *SID Digest* **29**, 1077.
17 Oh-e, M., Yoneya, M. and Kondo, K. (1997) Switching of negative and positive dielectric anisotropic liquid crystals by in-plane electric fields. *J. Appl. Phys.* **82**, 528.
18 Kim, K. H., Park, S. B., Shim, J. U. *et al.* (1998) New LCD modes for wide-viewing-angle applications. *SID Digest* **29**, 1085.
19 Cline, R., Duelli, M. and Greenberg, M. (2002) Thermal stress birefringence in LCOS projection displays. *Displays* **23**, 151.
20 Haller, I. (1975) Thermodynamic and static properties of liquid crystals. *Prog. Solid State Chem.* **10**, 103.
21 Freedericksz, V. and Zolina, V. (1933) Forces causing the orientation of an anisotropic liquid. *Trans. Faraday Soc.* **29**, 919.
22 Kuo, C. L., Wei, C. K., Wu, S. T. and Wu, C. S. (1997) Reflective display using mixed-mode twisted nematic liquid crystal cell. *Jpn. J. Appl. Phys.* **36**, 1077.
23 Wu, S. T. (1994) Film-compensated homeotorpic liquid crystal cell for direct-view display. *J. Appl. Phys.* **76**, 5975.
24 Yang, K. H. and Lu, M. (1998) Nematic LC modes and LC phase gratings for reflective spatial light modulators. *IBM J. Res. Dev.* **42**, 401–10.
25 Ji, Y., Gandhi, J. and Stefanov, M. (1999) Fringe-field effects in reflective CMOS LCD design optimization. *SID Digest* **30**, 750.
26 De Smet, H., Van den Steen, J. and Van Calster, A. (2001) Microdisplays with high pixel counts. *SID Digest* **32**, 968.
27 Chou, W. Y., Hsu, C. H., Chang, S. W. *et al.* (2002) A novel design to eliminate fringe field effects for liquid crystal on silicon. *Jpn J. Appl. Phys.* **41**, 7386.
28 Fan-Chiang, K. C., Wu, S. T. and Chen, S. H. (2002) Fringe field effect on liquid-crystal-on-silicon devices. *Jpn J. Appl. Phys.* **41**, 4577.
29 Zhang, Y. L., Bos, P. J. and Chung, D. B. (2003) Modeling of the effect of electrode topology and interpixel gap on LCOS devices. *SID Digest* **34**, 1382.
30 Kozakai, T., Hori, Y., Sugihara, M. *et al.* (2001) Reflection improvement for D-ILA. *SID Digest* **32**, 910.
31 Hashimoto, S. *et al.* (2005) SXRD (silicon X-tal reflective display): a new display device for projection displays. *SID Digest* **36**, 1362–5.
32 Peng, J., Wang, S. Y., Wang, C. J. *et al.* (2005) Improvement on fringe field in VA mode LCOS panel. *SID Digest* **36**, 1294–7.
33 Skaife, J. J., Flynn, M. and Bone, M. (2004) Optimization of a MTN mode for all digital LCoS imager for projection applications. *SID Digest* **35**, 986–9.
34 Wu, S. T. and Wu, C. S. (1998) Optimization of film-compensated homogeneous cells for liquid crystal displays. *Liquid Crystals* **24**, 811.
35 Wu, S. T. and Wu, C. S. (1998) A biaxial film-compensated thin homogeneous cells for reflective liquid crystal displays. *J. Appl. Phys.* **83**, 4096.
36 Chen, J., Robinson, M. G. and Sharp, G. D. (2005) High contrast MacNeille PBS based LCOS projection systems. *Proc. SPIE*, **5740**, 76–87.
37 Wu, S. T. and Wu, C. S. (1998) Bisector effect on the twisted nematic cells. *Jpn J. Appl. Phys.* **37**, L1497.
38 Wu, S. T. and Yang, D. K. (2001) *Reflective Liquid Crystal Displays.* Wiley–SID, Ch. 4.
39 Sonehara, T. and Okumura, O. (1990) A new twisted nematic ECB mode for a reflective light valve. *Proc. SID* **31**, 145.
40 Ou, C. R., Lee, W. T., Chung, S. C. *et al.* (2001) Thermal-optical simulation on colorQuad for image quality. *SID Digest* **32**, 922.

41 Huang, Y-P. *et al.* (2004), Full-color transflective Ch-LCD with image-enhanced reflector. *SID Digest* **35**, 882–5.
42 Van Sprang, H. A. (1996) Reflective liquid crystal display device with twist angle between 50 and 68 degrees and the polarizer at the bisectrix. US Patent 5,490,003.
43 Yang, K. H. (1996) A self-compensated twisted-nematic liquid crystal mode for reflective light valves. *Intl Display Research Conference*, 449.
44 Yang, K. H. and Lu, M. (1999) Nematic liquid crystal modes for Si wafer-based reflective spatial light modulators. *Displays* **20**, 211.
45 Janssen, P. (1993) A novel single light valve high brightness HD color projector. *Intl Display Research Conference*, 249–52.
46 Janssen, P., Shimizu, J. A., Dean, J. and Albu, R. (2002) Design aspects of a scrolling color LCOS display. *Displays* **23**, 99–108.
47 Grinberg, J., Jacobson, A., Bleha, W. *et al.* (1975) A new real-time noncoherent to coherent image converter: the hybrid field effect liquid crystal light valve. *Optics Eng.* **14**, 217.
48 Drzaic, P. S. (1995) *Liquid Crystal Dispersions.* World Scientific Publishing.
49 Nakanishi, K. *et al.* (1994) A 45-inch high-brightness rear projector using a 350K-pixel PDLC panel. *SID Digest* **25**, 749–52.
50 Havens, J. *et al.* (1997) NCAP projection displays. *Proc. SPIE*, **3013**, 185–92.
51 Nagae, Y. *et al.* (1995) Compact liquid-crystal projectors with high optical efficiency. *SID Digest* **26**, 223–6.
52 Beresnev, L. A. *et al.* (1988) Ferroelectric liquid crystals. *Mol. Cryst. Liq. Cryst.* **158A**, 3–150.
53 Clark, N. A. and Lagerwall, S. T. (1980) Submicrosecond bistable electro-optic switching in liquid crystals. *Applied Physics Letters* **36**, 899.
54 Birch, M. *et al.* (2002) SXGA resolution FLC microdisplays. *SID Digest* **33**, 954–7.
55 Clark, N. A. *et al.* (2000) FLC microdisplays, *Ferroelectrics* **246**, 1003–16.
56 Displaytech, Colorado: www.displaytech.com
57 Funfschilling, J. and Schadt, M. (1989) Fast responding and highly multiplexible distorted helix ferroelectric liquid-crystal displays. *J. Appl. Phys.* **66**, 3877–82.
58 Chan, L. K. M. *et al.* (2000) Antiferroelectric microdisplays on silicon. *Intl Display Research Conference*, 261–4.
59 Kobayashi, S. *et al.* (2004) Fabrication and electro-optic characteristics of polymer-stabilized V-mode ferroelectric liquid crystal displays and intrinsic H-V-mode ferroelectric liquid crystal displays: their application to field sequential fullcolor active matrix liquid crystal displays. *Optics Eng.* **43**, 290–7.
60 Sloof, W. A. and Brennersholtz, M. S. (2003) A high-resolution LCOS imager for HDTV systems. *Intl Display Workshop*, Japan, 1577–80.
61 Pivnenko, M. N. *et al.* (2004) Novel electroclinic organosiloxane materials for optoelectronic devices. *Proc. SPIE*, **5289**, 16–24.

7

LCD Assembly and Testing

7.1 Background

The demand for flat panel displays fueled the rapid development of liquid crystal display manufacturing technology. Weaker demand for microdisplays handicapped growth of their manufacturing technique until recent years. Projection TV is a potentially huge market for competitively priced microdisplays. Attention to manufacturing issues determines which microdisplay achieves the best combination of price and performance.

The assembly and test of LC microdisplays follows the same general principles as direct-view LCDs, but modified by the smaller area and magnification requirement of microdisplays.[1–4] Magnification inherent to microdisplays reveals small blemishes that would not be visible in direct-view displays, placing stringent requirements on materials and assembly procedures, implying semiconductor cleanroom standards.[5,6] Small microdisplay area benefits from established semiconductor packaging methods, and allows more attention to optical details such as interface reflections and ITO transmission losses. Figure 7.1 indicates the basic elements of an assembled LCOS microdisplay, and Figure 7.2 illustrates a finished device.

The structure must withstand the high optical flux (~1 W) and operating temperature of a projector (>40 °C). Optical and electronic power dissipation is higher per unit area than in direct-view devices, requiring attention to heat sinking.[7] Color uniformity requires LC cell gap uniformity, achieved in direct-view displays by a random distribution of particle spacers throughout the cell area. A similar approach in microdisplays produces a noticeable degradation in magnified image quality. Substrates containing many devices are wafer-scale assembled, simplifying the initial assembly step and minimizing particle contamination. Finally, the assembled cell must be mounted in its package, without cell gap distortion and secure against significant movement over lifetime.

The assembly details mainly describe LCOS, where more information is available. However, transmission microdisplays have similar assembly problems, particularly cell gap uniformity, and follow similar assembly processes, with a transmissive active matrix substrate replacing the silicon wafer. Transmission devices differ in package geometry, restricting support to the peripheral region; however,

Introduction to Microdisplays D. Armitage, I. Underwood and S-T. Wu

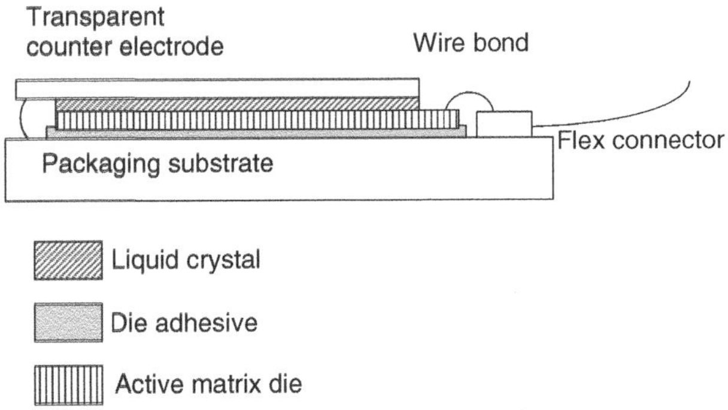

Figure 7.1 Basic elements of assembled liquid crystal microdisplay

the support structure is similar to LCOS. In general, the issues are similar, prompting similar solutions. Assembly and test issues of MOEMS are discussed in the chapter devoted to micromechanical devices; likewise, similar issues in OLEDs are included in the emissive chapter. Yield quantifies the efficiency of the fabrication and assembly process; less than the 100% ideal yield is inevitable, while profit margin determines the lowest acceptable yield.

7.2 Back-end Processing

Back-end processing is a term used in silicon technology to describe any additional processing applied to wafers received from the foundry. We employ the term to describe any further work on the substrates prior to assembly. The wafers are typically 200 mm diameter, by 0.8 mm thick, and contain tens to hundreds of die (device elements) depending on die size. The active surface has arrays

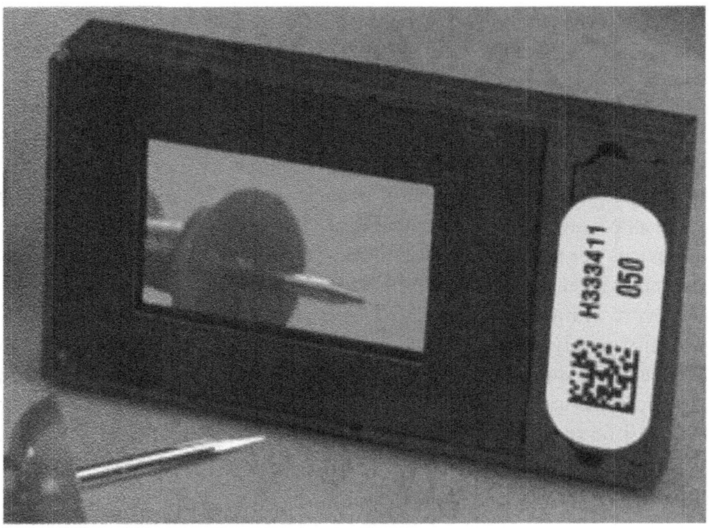

Figure 7.2 LCOS microdisplay with detachable flex connector. Reprinted courtesy of Syntax–Brillian Corp.

of highly reflecting aluminum pixel electrodes, with the opposite surface ground flat. The 1 mm thick counter-electrode glass substrate closely matches the silicon expansion coefficient, is coated with an ITO transparent electrode, and preferably is of circular form to match the silicon wafer; having the same form simplifies automatic handling. Optical inspection detects surface quality, flatness, and any discoloration or staining of the active matrix wafers and counter-electrode substrates, with consequent discarding of substandard wafers and ITO substrates. Faulty die, identified by optical inspection or standard electrical test methods applied to the wafers, exit the assembly process at the singulation step. A variety of automated test equipment supports microdisplay manufacturing.

7.2.1 Dielectric Mirror

Processing at the foundry levels the pixel surface and inter-pixel gap to provide a smooth interface to the liquid crystal and high aluminum reflectivity. The reflectivity (R) of aluminum is calculated from its complex refractive index $n_1 = 0.895 - i6.67$ at $\lambda = 550$ nm: $R = (n_1 - n_2)(n_1 - n_2)^*/(n_1 + n_2)(n_1 + n_2)^*$, n_1 and n_2 are the refractive indices at the interface. For air $n_2 = 1$ and $R = 92\%$, for nematic liquid crystal $n_2 \sim 1.5$ and $R = 83\%$. There is considerable loss of metal reflectivity due to the liquid crystal interface. Further loss is introduced by the aperture ratio, diffraction, and scattering losses. Deposition of a dielectric mirror over the surface reduces all of these losses.[8]

A dielectric mirror appropriate to a photo-addressed microdisplay contains 20 or more alternating layers, such as silicon dioxide and titanium dioxide, to provide high reflection ~99% over the required color band, and help shield the photoconductor from readout light. However, a thick dielectric mirror sacrifices considerable pixel voltage in capacitive coupling to the liquid crystal. A few enhancement layers, incurring a voltage loss of approximately 0.1 V, achieve the substantial gain shown in Figure 7.3. Complications arise from dielectric layers between the pixel electrode and liquid crystal; enhanced ionic charging effects increase susceptibility to ghost imaging, aided by unstable DC levels related to contact potential. Sufficient electrical leakage is required in the final layer to attenuate ionic charging and match interface potentials.

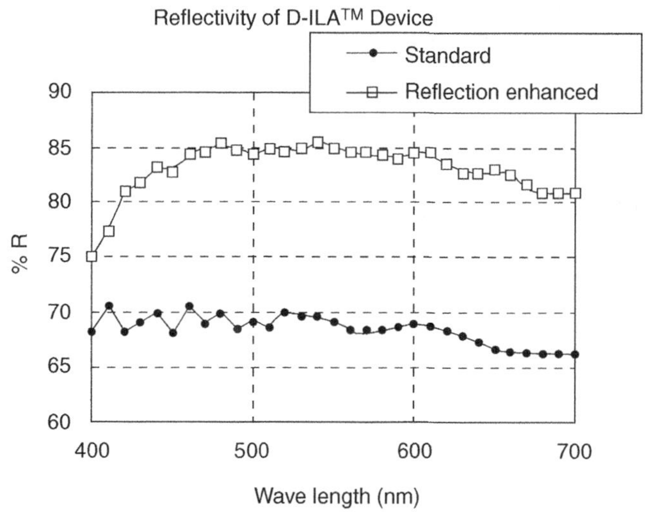

Figure 7.3 Dielectric enhanced reflection of microdisplay. Reprinted courtesy of Society for Information Display

7.2.2 Liquid Crystal Alignment Layer

Surfaces impose a favored alignment direction on liquid crystal that propagates into the bulk material, controlling the orientation distribution throughout the cell in conjunction with the applied voltage. The substrates form the liquid crystal boundaries, and have surface alignment layers appropriate to the liquid crystal cell design. Standard buffed polyimide methods developed for direct-view displays achieve tilted parallel alignment suited to twisted nematic cells; however, a polyimide is required that can tolerate projector operating conditions, particularly the short-wavelength region of the readout light.[9,10] Nematic cells require alignment layers on both surfaces of the cell, while a single surface alignment layer is often favored by FLC cells.

Polyimide layers ~50 nm thickness are formed by spin coating a precursor solution, followed by curing at elevated temperature <200 °C. The buffing process has evolved empirically, supported by the development of special machinery, polyimides, and buffing cloths. Microdisplay alignment favors a gentle buffing process, minimizing the blemishes inherent to rubbing. Liquid cleaning removes particulate matter produced by buffing.[11]

Buffed alignment has proved ineffective in providing tilted vertical alignment for VAN, due to the difficulty of achieving adequate pretilt without excessive rubbing defects. Alignment relies on anisotropic inorganic coatings deposited by oblique evaporation or sputtering methods.[12,13] Superior alignment results, without the scratches, dust and electrostatic charging problems of buffing; moreover, inorganic alignments, such as silicon oxide, are insensitive to optical damage.[10] A typical alignment process deposits a few hundred ångströms of silicon oxide at an oblique incidence angle to provide VAN alignment with zero-voltage pretilt >6 degrees from vertical, tilted in the desired direction. Oblique evaporation gives reliable alignment, but the angular alignment direction splays over the area of the wafer, due to the finite distance of the evaporation source. Sputtering provides constant alignment direction by control of the angular range of the sputtered beam.[13]

Exposure to short-wave blue light in projectors compromises the lifetime of buffed polymer alignment, and results in differential aging in multiple-microdisplay projectors, giving rise to color distortion. Optimum choice of polyimide[9] demonstrates ~20,000 hours of life, but security favors inorganic alignment with indefinite lifetime. Desirable features of VAN cells reinforce the choice of inorganic alignment achieved with a single evaporation per surface. Twisted nematic cells also align with inorganic oblique deposition, but require two depositions.[14] The 90-degree twist cell performs well in transmission, however, VAN transmission cells gain in popularity. VAN cells completely dominate reflective microdisplays in color parallel projectors. Parallel-aligned cells provide the fastest response speed, essential to field-sequential-color operation. The liquid crystal industry in general is developing improved alignment methods for both parallel and vertically aligned cells.

7.3 Assembly Components

Assembly of microdisplay components draws attention to compatibility requirements and structural integrity. Experimental devices assembled on an individual basis, without regard to long-term behavior or optimum performance, contrast with manufacturing processes optimized for cost, performance, and life. Wafer-scale assembly has a number of advantages in manufacturing. Clearly, it reduces the handling problem by fabricating many devices in one operation on much larger substrates. Moreover, forming the empty liquid crystal cells before breaking the wafer into individual devices (singulation) avoids ongoing particle contamination of the cells from edges damaged by singulation.

Tight tolerance on cell gap uniformity is essential for color uniformity in liquid crystal displays. Cell gap uniformity ultimately has the greatest impact on yield, and demands exceptional flatness in

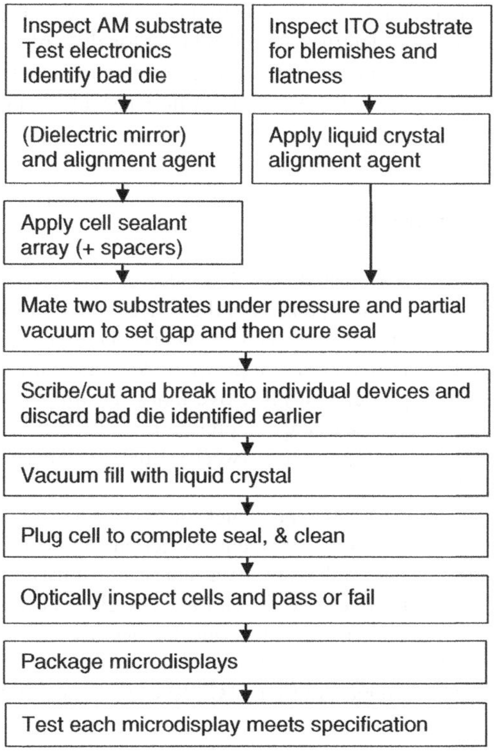

Figure 7.4 Assembly process flow with detailed testing

the component substrates. The tolerance can be relaxed somewhat by adjusting the look-up table (LUT) determining the pixel voltage to take into account cell gap nonuniformity.[15] Figure 7.4 indicates the assembly process flow.

7.3.1 Active Matrix Substrate

The active matrix addressing circuit, fabricated on a substrate ~1 mm thick, forms the electrically active surface and half the optical–mechanical structure of the liquid crystal cell. For a transmission cell the basic substrate is a transparent glass, quartz, or sapphire wafer, polished to high optical quality and flatness, with the TFT active matrix array on the inner surface. The outer surface of the substrate is anti-reflective coated to improve efficiency and reduce multi-reflections. Index matching of the transparent conducting pixel to the liquid crystal index is an advantage. Index matching is a term used to describe additional interface optical layers designed to reduce overall reflection when contacted with material (liquid crystal) of the design index.

In reflective LCOS, the silicon wafer forms the substrate. Stringent flatness requirements impose conditions on circuit layout details and fabrication procedures.[16,17] Metal pixels and underlying metal shields protect LCOS from readout light. Dielectric coatings may be added to enhance the pixel reflectivity. Wafer size influences assembly methods and production economics. Fully depreciated foundries offer lower fabrication cost, but inferior technology and wafer diameter 200 mm, or less. Production is moving to 300 mm wafers offering smaller higher-speed circuitry, improved flatness, and greater rigidity of thicker silicon. Optical inspection of incoming active matrix substrates eliminates wafers with excess flaws and identifies flawed regions for discard at a later stage. Electronic testing at the wafer stage identifies faulty circuits, subject to rejection from the assembly chain at the earliest opportunity. Data on the received state of the wafers provides rapid unambiguous feedback to the foundry.

7.3.2 Transparent Counter-Electrode

A glass substrate ~1 mm thick, with deposited ITO electrode, forms the transparent counter-electrode. The glass is a close match to the thermal expansion of the active matrix substrate to minimize cell gap distortion with temperature, and carefully polished to achieve flatness and surface quality. Flatness and surface quality are important in maintaining high yield. An optical layer under the ITO electrode achieves index matching to the liquid crystal refractive index, minimizing total reflection from the interface boundary. VAN cells have an advantage in presenting only their ordinary index to the index-matching plane; planar aligned cells compromise between ordinary and extraordinary index, or impose input polarization conditions.[18,19]

The cell gap is comparable with the coherence length of the readout light, causing interference of pixel reflection and ITO reflection, imposing stringent restrictions on ITO reflection to keep intensity spatially uniform. Absence of a highly reflecting pixel makes transmission cells less sensitive to interference effects. Standard anti-reflective coatings on the outer glass surface minimize reflections to improve optical efficiency and attenuate multi-reflections that degrade contrast ratio. Multi-reflections support multiple passes through the liquid crystal; accumulating optical retardation to the detriment of CR. Short electrical path accommodates lower ITO electrical conductivity, consistent with lower transmission loss.

7.3.3 Cell Gap Spacers

Cell gap uniformity determines optical modulation uniformity over the display area. The eye is more sensitive to color variation than intensity variation, making color uniformity the criterion of gap uniformity. Parallel color projection from multiple microdisplays, each dedicated to one primary color, places stringent requirement on their cell gap uniformity. Random variations in the cell gap of each microdisplay implies similar variation in primary luminance, causing color nonuniformity. Tight gap uniformity tolerance is required. Color-sequential projection from one microdisplay has residual color nonuniformity due to gap nonuniformity, since modulation is nonlinearly dependent on wavelength and gap; however, gap tolerance is increased. Polishing produces a smooth surface with gradual undulations, imposed on gap uniformity; sudden changes in gap are associated with surface damage. Gentle variation in gap produces gradual color change, making it harder to detect. To get around the stringent gap requirement, the LUT determining the pixel voltage can extend to take into account gap variation over the cell. The LUT cannot accommodate gap correction for every pixel; each LUT setting operates over a local region of the active area.[20,21] Continuity between local regions sets a limit on overall gap nonuniformity, typically ~3%, that can be accommodated by LUT settings. Interpolating values between tabulated points provides most effective use of the LUT, keeping intensity steps below visual perception.

A random distribution of spacer particles sets and maintains the cell gap in direct-view LCD, with the liquid crystal sealed at lower pressure than atmosphere to maintain a compressive force on the cell. Magnification inherent to microdisplays makes random particle distribution undesirable due to optical noise appearing as haze. Particle spacers are effective in controlling cell gap when confined to the boundary seal by mixing with the cement prior to dispensing. Boundary gap control works best with small cells, in keeping with other economic incentives to reduce cell diameter.

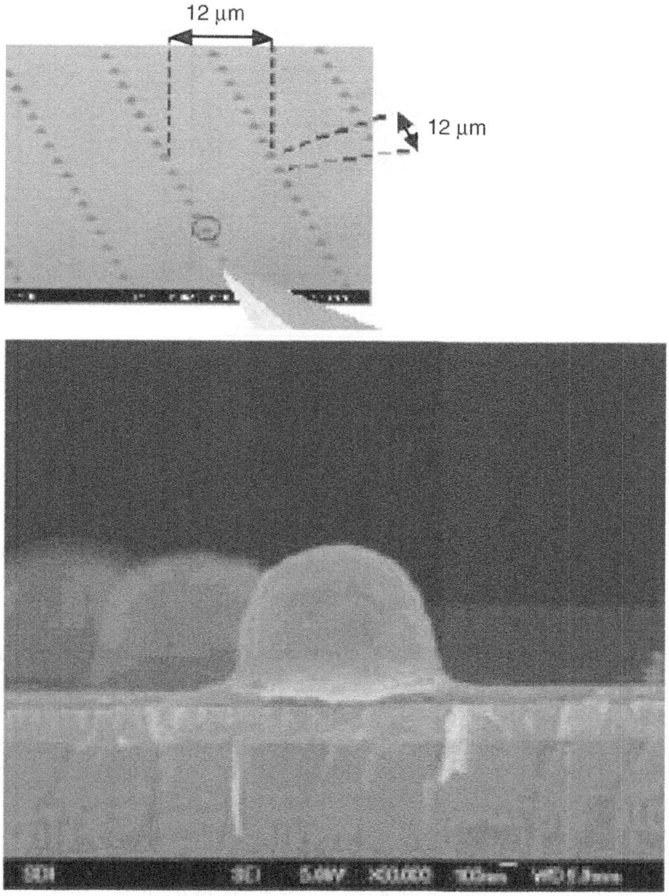

Figure 7.5 ITO substrate with buffed polyimide alignment layer with subsequent 0.9 μm high spacers formed from negative photoresist. Reprinted courtesy of Society for Information Display

Distortion of the liquid crystal in contact with internal spacer particles amplifies their effect, lowering the contrast ratio of normally black cells (darkest state at lowest voltage). The normally white state (brightest state at lowest voltage) minimizes contrast degradation, since the driven black state removes some of the nematic distortion due to the spacer. Pillar spacers built into the silicon die by lithography serve the thinnest cells $<2\,\mu m$ used in reflection, as discussed in Chapter 3. However, such spacers disturb the liquid crystal alignment layer process, making them unsuited to cell gaps $>2\,\mu m$. The spacers shown in Figure 7.5 are formed from negative photoresist and can be fabricated after the buffed polyimide alignment process.[22] Spacers tied to the pixel periodicity do not degrade the image quality, since quality is inherently limited by pixel period.

7.3.4 Liquid Crystal Seal

The cell boundary seal prevents the liquid crystal leaking out and blocks ingress of contaminants such as oxygen and water that would compromise lifetime. UV-cure polymeric cements suffice to hold the substrates together and seal the cell, while minimizing assembly forces causing warping. The sealant

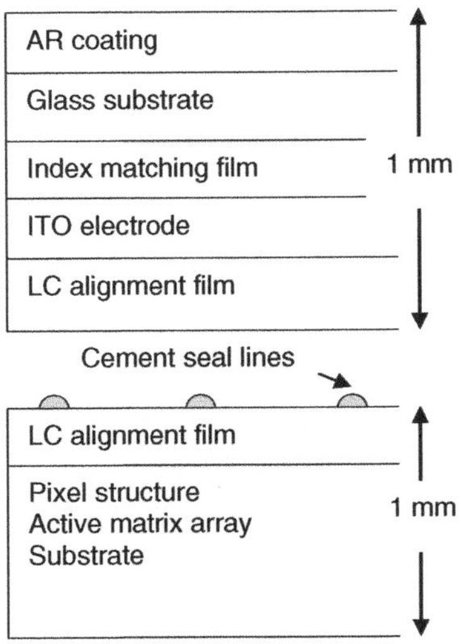

Figure 7.6 Mating of two substrates under partial vacuum in wafer scale assembly

makes contact with the liquid crystal, restricting the choice to chemically inert materials.[23–25] Mixing spacer balls in the cement sets the cell gap at the boundary seal; substrate flatness maintains the gap in the active area of the cell. A thin bead of cement dispensed or printed on one of the substrates forms the cell boundary, while leaving a break for liquid crystal filling at a later stage.[26] An automated machine can dispense all the cell boundaries over the surface of a wafer, before mating with the opposing substrate to form many cells in one operation. Figure 7.6 shows the substrates prepared for mating under partial vacuum, where applied compression sets the gap according to spacer balls, followed by UV cure. The seal is <1 mm wide and a few microns thick, depending on cell gap. Silicon area is expensive, imposing minimum seal area consistent with a reliable seal. Subsequent singulation reveals the circuit connection pads and the ITO contact surface of each cell.

A plug of UV curing cement, known as end-seal, completes the cell seal after liquid crystal fill. The chance of contamination at end-seal is significant, since the liquid crystal is in contact with the liquid cement before the cement is UV-cured.

7.3.5 Liquid Crystal

Specialized chemical companies manufacture liquid crystals for specific display applications. Large customers have liquid crystal mixtures optimized for their particular application. Nematic liquid crystals for displays are organic compounds with flow viscosity about an order of magnitude greater than water. Surface tension is comparable to that of water, and wetting ensures capillary filling of a cell. To prevent crystallization at the lower range of storage temperature, it is essential to employ mixtures of compounds. Mixtures also facilitate fine-tuning of properties to achieve desired optical and electrical requirements, with favorable temperature dependence.

Liquid crystals engineered for displays have high chemical stability and essentially 100% transparency to visible light. At shorter wavelengths, liquid crystal compounds absorb the photon energy responsible for degradation.[27] Liquid crystals are shielded from UV light where possible. Degradation is also caused by high temperatures; however, the upper limits of the nematic phase in displays are typically <100 °C, assuring operation safely below incipient degradation. Operating well below the nematic/isotropic phase transition confers stability of physical properties for modest temperature change. Purification of liquid crystal compounds removes contaminants that may influence display performance and life; ionic purification is particularly important in reducing the electrical conductivity to very low values, essential for active matrix addressing. Prevention of ionic contamination dictates the cleanliness of the assembly process.

Nematic cells fill in the nematic-liquid-crystal phase, although some advantage in alignment accrues from filling at the higher temperature isotropic phase, but thermal distortion and higher vapor pressure outweigh such considerations. FLC only exists in the viscous smectic-liquid-crystal phase, requiring cells be filled in the higher temperature nematic or isotropic phase.

7.4 Assembly Methods

Assembly methods are the key to high yield. Details are proprietary, but some general principles hold.

7.4.1 Wafer-Scale Assembly

Wafer-scale assembly has several advantages. Dicing the wafer to separate individual die reveals inferior die flatness in isolated form, due to stress relaxation. Superior flatness in wafer-scale assembly improves cell gap uniformity. Eventual singulation will relax stress somewhat, with some distortion of the assembled cell; however, the result can be no worse than assembly after singulation, and is generally better. The edges of singulated die are rich in loose particles that can migrate to the surface during cell assembly, destroying cell gap uniformity. Wafer scale essentially seals the cell before singulation, eliminating edge particle contamination.

Handling is easier at wafer scale, including dispensing the boundary seal with minimum sacrifice of silicon area. Evaporated silicon oxide forming the liquid crystal alignment layer has a splayed directional distribution due to wafer width and finite distance from the evaporation source. The liquid crystal tilt alignment direction varies systematically over the wafer surface according to the evaporation direction; alignment variation over the cell dimension is not significant, but direction cannot be optimum for all cells. The angular dispersion of alignment can be corrected by substrate positioning when the oxide deposits on individual die, an advantage to singular assembly.

Figure 7.7 shows regions of 8-inch wafer-scale assembled substrates before singulation.[15] The cells on the right have active diagonal 0.78 inches, with cell gap $<2\,\mu m$ and inorganic alignment layers. The gap is determined by spacers embedded in the cell boundary seal, leaving the active area clear.

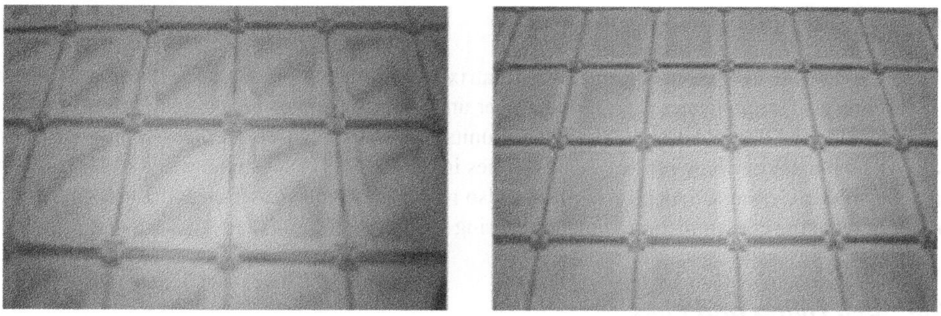

Figure 7.7 Wafer assembled microdisplays before scribe and break. Left: without wafer planarization, interference fringes reveal poor cell gap uniformity. Right: with wafer planarization, cell gap uniformity within 3% eliminates fringes. Reprinted, courtesy of Society for Information Display

The small cell gap promotes strong interference effects in the empty cells, easily seen under ordinary fluorescent lights such as the orange light illuminating the cells. Older wafers lacking planarization produce assembled cells with poor uniformity betrayed by the obvious interference fringes (Newton rings). Wafers processed with CMP planarization produce assembled cells with gap uniformity better than 3%. At 600 nm wavelength, one fringe represents a cell gap change of 300 nm, i.e. 15% of the 2000 nm cell gap, consistent with the absence of fringe patterns in the planarized cells. An improvement in cell-gap uniformity of devices followed from wafer-scale assembly.

7.4.2 Seal and Spacer Application

The favorite assembly method embeds particle spacers in the cell seal, employing spherical particles of glass or polymer, having a narrow distribution of diameter about the desired value. Sufficient particle density per unit area ensures little variation of average particle diameter over the seal area. Pressure applied to spread the cement seal deforms the particles and surface contact region, making the cell gap somewhat dependent on applied pressure.

Cell seal favors a UV-curing polymeric cement with low enough viscosity to squeeze to the required gap value at modest pressure. The seal protects the cell from atmospheric contaminants, such as water and oxygen. A polymeric seal allows some exchange with atmosphere, but many years of experience and experimental testing prove its reliability in liquid crystal devices. Choice of a chemically inert material avoids chemical reaction of the seal and liquid crystal. The liquid crystal industry has developed a variety of polymeric seals with established reliability.

The spacers are mixed with cement in appropriate proportions, ready for dispensing. A computer controlled dispensing machine writes a thin bead of cement defining the seal for each cell of the wafer. A gap left in the seal allows for liquid crystal fill. Alternatively, printing the seal on the cell is an advantage in assembly throughput speed for small-area microdisplays with die count ~100/wafer.[26]

7.4.3 Assembly Pressure

The substrates forming the liquid crystal cell are brought together with proper orientation under uniform pressure. A gas pressure bag is often used to apply uniform pressure and works well with internal spacers. With boundary spacers, gas pressure induces dishing deformation of the cell center, barring it from use; moreover, with sufficient pressure, cell surfaces press together, disturbing the alignment structure. Pressure applied by a mechanical form directing force to the seal regions avoids dishing effects. UV-cure proceeds with seal flow complete and pressure maintained.

7.4.4 Singulation

The extension of the ITO electrode and active matrix bond pads at opposite ends of the cell, as indicated in Figure 7.1, complicates singulation; upper and lower substrates break along offset lines. The assembly cannot be cut into simple rectangular units. Crystal silicon discourages simple scribe and break. An automatic machine scribes cleavage lines in the glass/ITO substrate, while partial saw-cuts define the silicon cleavage. Sapphire substrates also require saw-cuts. Laser cleaving tools may prove superior in reducing contamination as manufacturing progresses.

7.4.5 Cell Filling and Plug

Individual cells have a small opening for liquid crystal filling. Discarding cells that failed the initial electrical test or optical test, the remaining cells fill by capillary action on immersing the opening in a liquid crystal reservoir under modest vacuum limited by the liquid crystal vapor pressure. Quick fill inhibits loss of the most volatile components of the liquid crystal mixture that would alter the composition. Elevating the temperature lowers the viscosity, increasing the flow rate, but also increases the vapor pressure. Fill time is reduced by raising the pressure after the initial entry of the liquid crystal, to increase the differential pressure driving the flow. However, excessive pressure differential will collapse the cell without internal spacers, damaging the alignment layers. Annealing the cells for several hours after filling ensures uniform alignment, and relaxes cell stress, stabilizing the long-term structure.

Complete seal of the cell requires plugging the fill hole. The plug material is usually the same cement as the boundary seal. The fill hole is cleaned and a small drop of cement applied. Capillarity of the cement sucks it into the gap, or the release of slight cell pressure provides suction. UV exposure quickly cures the cement plug to seal the cell. For a short period, the liquid crystal is in contact with uncured cement fluid, a source of contamination. Thoroughly cleaning of the cell exterior follows end-seal, in preparation for optical testing and packaging.

7.4.6 One-Drop Filling

Large-area direct-view displays fill by the drop-on-demand method, where a number of precisely measured liquid crystal droplets placed on the substrate ensure cell formation and filling when the substrates are mated and exposed to UV.[24] The small volume of a microdisplay makes the method difficult to apply, but may be introduced as manufacturing evolves. Figure 7.8 lists an experimental one-drop-filling process for microdisplays.[28] Figure 7.9 shows the silicon wafer with seal dispensed and liquid crystal drop placed in the center of each seal region, and Fig. 7.10 illustrates the vacuum chamber and assembly platform.

The boundary seal and liquid crystal are in contact prior to cure, requiring careful choice of material to avoid mixing and reaction. The seal material has viscosity ~200,000 cP and is dispensed to provide a seal width 1 mm at 2 µm cell gap, controlled by particle spacers included in the seal. A liquid crystal area of 153 mm^2 implies 0.306 mm^3 volume weighing ~0.3 mg. Adjustment of vacuum pressure during assembly influences the cell gap uniformity, achieving uniformity within 1.5 interference fringes, i.e. 12% uniformity with 70% yield. First-class imaging requires cell gap uniformity of 3%, followed by LUT voltage correction of gap distortion at final projection setup.

Dispense seal on bottom substrate (Si wafer)
Dispense liquid crystal drop at center of each seal region
Transport Si wafer and ITO substrate to assembly chamber
Position rotation angle, translation, and parallelism
Evacuate chamber and apply assembly pressure
UV cure seal while screening LC area
Vent vacuum
Transport to oven, bake at 120 deg C for 1 hr to complete cure
Slow cool to room temperature to align liquid crystal

Figure 7.8 One drop filling process[28]

Figure 7.9 Seal and liquid crystal drop dispensed on 8-inch wafer ready for assembly. Reprinted courtesy of Society for Information Display

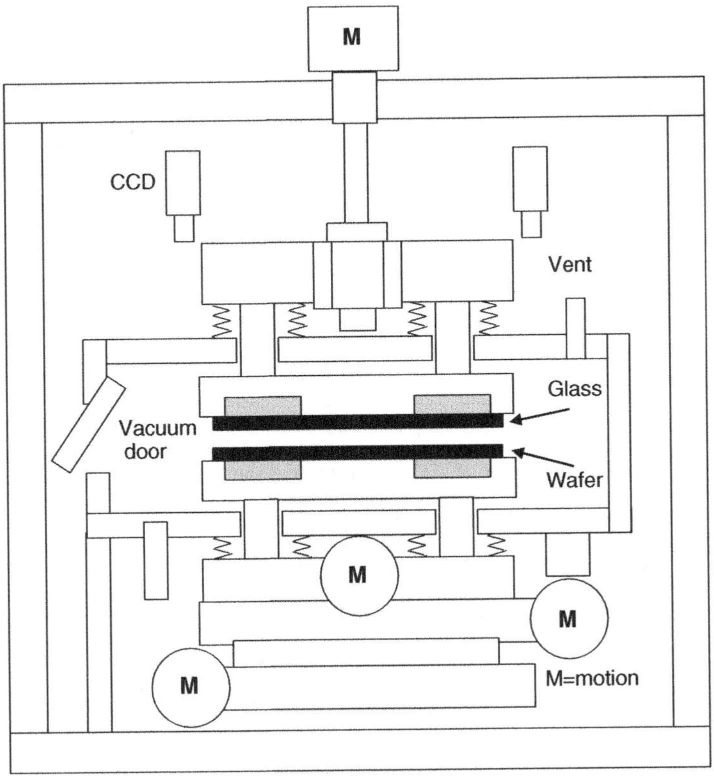

Figure 7.10 Vacuum chamber and assembly platform. Reprinted courtesy of Society for Information Display

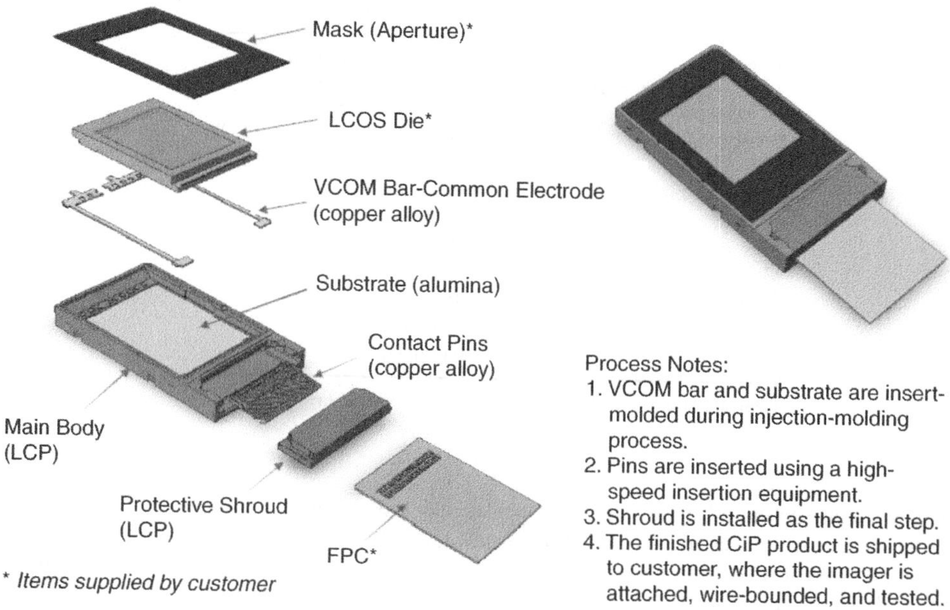

Figure 7.11 Package and assembly.[29] Reprinted courtesy of Quantum Leap Packaging Inc.

7.4.7 Packaging

Packaging serves several purposes: it protects the cell from handling damage and contamination, while providing an optical mounting structure and serving as a heat sink.[25 29] Figure 7.11 illustrates a package structure with detachable flex connector. The adhesive securing the cell to the package substrate should have low thermal resistance, while being compliant enough to preserve uniform cell gap, but preventing movement during the microdisplay lifetime. Mechanical stability is required to maintain optical convergence of two or more microdisplays in a color projector; stability criteria are relaxed for microdisplays destined for color sequential applications. The packaging substrate should have adequate mechanical strength and thermal conductivity, e.g. metal or ceramic, and approach the linear expansion coefficient of silicon to minimize thermal stress. The substrate also provides an anchor for the flex cable carrying the electrical connections wire bonded to the die pads. Encapsulation protects the wire bond region. A black metal frame delineates the optical active area, shielding and protecting the rest of the device. Wire bonding is labor intensive, implying replacement by an automatic method (TAB) as manufacturing progresses.

A semiconductor packaging company has studied the requirements of microdisplay packaging and designed a standard package that can be adapted to a range of LCOS microdisplays. Table 7.1 lists package issues and solutions. The housing is injection molded from liquid crystal polymer (LCP) Vectra-150i, having high stability, very low expansion coefficient, yet flows very easily into the fine product features. This facilitates insertion molding of various components or materials such as lead-frames and metal or ceramic substrates attenuating differential expansion between die and substrate.

The substrate favors materials matching the low expansion coefficient of silicon, with thermal conductivity adequate to heat sinking. Experience in semiconductor packaging provides the electrical connector details. Figure 7.12 shows strip and matrix packaging formats employed in manufacturing.

Table 7.1 Package specification[29]

Property	Specification
Housing material	Vectra LCP E150i
Substrate	Ceramic, Alloy 42, Kovar LCP or other stable material
Connector pin	Copper alloy
Plating	Selective gold over nickel
Max voltage	250 V
Max current	1.0 A/pin
Contact resistance	10 mΩ max
Insertion force	29 grams/pin
Normal force	60 grams/pin

Reprinted courtesy of Society for Information Display

The die attaches to the substrate by pressure-sensitive or thermal curing adhesives.[25,30] Experiments show that softer and thicker adhesive layers introduce less cell distortion. A thick adhesive layer will compromise the heat sinking property of the substrate, implying a compromise between thermal transfer and cell distortion.

Transmission microdisplays require peripheral support and rely mainly on air-cooling rather than heat sinking by the package as previously illustrated in Figure 4.9. The early devices were embedded in plastic packaging; however, higher optical and electrical power levels and improved cell uniformity demand metallic packaging. Adhesives secure position, and low differential expansion minimizes cell distortion.

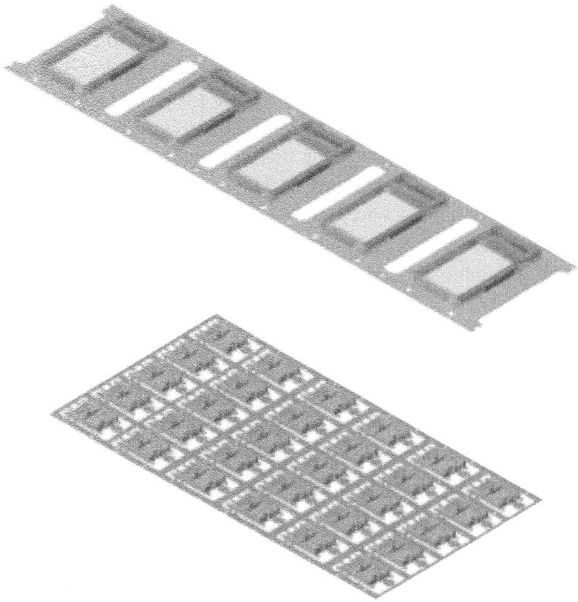

Figure 7.12 Strip and matrix package formats.[29] Reprinted courtesy of Society for Information Display

Table 7.2 Tests

Stage	Test
Received wafers	Optical quality and electric
Mirror and alignment	Repeat optical and electric
Received ITO substrates	Optical quality, conduction
Assembled wafers	Gap uniformity
Singulation	Gap uniformity
Ready for package	Optical quality, LCD clearing temperature
Packaged	Full specifications
Temperature cycling	Stress test
Temperature and humidity	Stress test
Vibration	Stress test

7.5 Testing

There are a number of reasons for testing at various stages of the assembly process. Intensive testing is necessary when starting a new production line, where optimum conditions at each stage are evolving. More testing is required in the early manufacturing of a new microdisplay design to assess that specifications are met. A well-established line with high yield may reduce cost by just testing the final microdisplay product, checking it meets specification. Early tests can identify faulty units, removed from further processing at the first opportunity, but the cost saving in assembly may not cover the cost of testing. Arguably,[31] testing prior to packaging is not cost effective in some cases for yields above 60%. However, sampled testing (quality control) is important in providing data identifying the stage responsible for any fall in yield, and monitoring known critical stages.

Stress tests determine the microdisplay reliability when subject to a range of temperature, humidity, and vibration. Accelerated life-testing requires prolonged exposure to humidity, elevated temperature, and temperature cycling; severe radiation at the shortest blue wavelength may also limit life. Table 7.2 lists assembly and packaged tests.

7.5.1 Assembly Tests

The effectiveness of each stage in the assembly is tested. Several companies supply sophisticated test equipment designed for inclusion in production lines, with minimum impact on process flow. It is important to examine the received substrates; defects in the starting substrates may dominate the overall yield.[6] Test circuitry designed into the active matrix substrates enhances the electrical test detail gathered from probing the substrate. The test data identifies some faulty die for discard at the earliest opportunity; this e-sort is also valuable feedback to the silicon foundry. Comprehensive electronic testing is not feasible without electro-optic observation; after e-sort, the remaining die may have faults that show up when the packaged microdisplay is tested.

Optical quality expresses surface flatness as worst-case peak-to-valley error over specific surface area, and RMS values averaged over surface area. Peak/valley $\lambda/10$ over the active device area is a desirable target. A polished glass surface is marred by scratches and pits (digs), where mechanical quality is given a scratch/dig number; scratch $= 10\times$(width of scratch in microns), and dig $= 10\times$(depth of dig in microns) Scratch/dig numbers 10/5 are desirable.[6] ITO substrates are also examined for coating defects.

The silicon wafers are re-examined, following any dielectric depositions and LC alignment layers to see if further damage has occurred. However, all handling introduces damage, and the alignment coatings are fragile. Sample testing is sufficient.

Comparison of gap uniformity before and after singulation indicates the stress level in assembly. The filled and end-sealed cells, cleaned in preparation for packaging, are checked for optical quality and gap uniformity. The nematic-to-isotropic phase-transition temperature observed in the filled cells is a sensitive monitor of composition integrity and non-ionic contamination, when compared with standard value. Transition temperature varying over the device surface is a sign of boundary or surface contamination.

The assembly tests are extensive when setting up a new production line. They relax to sample testing when production is satisfactory.

7.5.2 Specifications

A standard set of specifications established for flat panel displays and projectors facilitates comparison between different products. Microdisplay specifications support design and computation of display specifications for any system incorporating the microdisplay. Reliability and lifetime data are valuable. Table 7.3 lists the desired microdisplay specification.

Microdisplays normally include an ASIC providing electronic control circuitry, temperature control, etc. The ASIC design determines gray levels, frame rate, operating voltages, etc. Achievable frame rate is limited by the liquid crystal response time. Properties most sensitive to processing variables appear at the lower end of Table 7.3. Production tests set the temperature at normal operating temperature for the test duration. Tests include a range of wavelengths appropriate to the microdisplay design.

Table 7.3 Specifications

Property	Specification
Optical	Transmission/reflection
Liquid crystal	Nematic or FLC
Nematic cell	VAN, TN, other
Characteristic	Grayscale vs. voltage and temperature stability
Geometry	Basic dimensions
Weight	Microdisplay only
Pixel pitch	Microns
Active area	Array size and diagonal
Aperture ratio	Percent
Temperature	Operating and storage
Power	Electrical and optical
Thermal	Heat sinking
Lifetime	Predicted
Interconnect	Cable
Electronics	ASIC
Grayscale	ASIC levels
Frame rate	Hertz
Response time	rise and fall (at T deg C)
Flicker	Frequency and level
Defective pixels	Number and location
Local defects	Number and location
Uniformity	9 point test
Throughput	Efficiency
CR sequential	f/# dependence, with and without compensation

7.5.3 Specification Tests

Measurement standards are established for flat panel displays and projectors.[32, 33] A microdisplay can be subjected to standard tests by installing it in a high-quality projector and testing the projector. The projector test examination is a selling point for some manufacturers who guarantee every microdisplay they offer has passed such a visual test. Direct testing of microdisplays requires automated optical inspection, followed by sophisticated data processing to reveal sub-pixel defects.[34]

Every microdisplay is tested for defective pixels. Dead pixels remain in the on or off state, irrespective of the addressing signal. Partial pixels operate over a substandard dynamic range. Production-line test equipment raises the temperature of completed microdisplays as they approach the test head. Positioning mechanisms place the microdisplay for microscopic examination, while making electrical connection. Electronic test patterns are input, and optical output recorded on an electronic camera resolving pixel detail. Computational analysis of the output data identifies the locations of faulty pixels. Particulate defects and small blemishes are included in pixel defects at this stage. With improvement in technology, many displays now offer zero pixel defects, although some defects are tolerable.[35]

The human eye determines if a display imperfection is objectionable, but is slower and more expensive than test machinery. The initial electronic testing at wafer level (e-sort) identifies faults such as broken lines, but does not evaluate the pixel electrode response. Pattern recognition algorithms can detect most of the optical flaws in microdisplays that the human eye perceives;[36] however, high-quality projection microdisplays are subject to human inspection in a projection system, to catch faults missed by machine tests. In a severe machine test, some acceptable devices suffer rejection, while a less stringent test passes some devices that are not acceptable, making it difficult to set the pass/fail bar. Figures 7.13, 7.14 and 7.15 show optical defects causing rejection of microdisplays that passed the starting electrical and optical tests.

The uniformity test, applied on a multiple-pixel scale, detects contaminated regions affecting the optical transmission or reflection, e.g. residue from dried water droplets; yet larger scale reveals cell gap variation. The ANSI flat panel test divides area into 3×3 regular regions, and compares the intensity of the nine regions for constant gray level. Automated optical inspection can perform uniformity test on a pixel scale if required, and as a function of gray level. Uniformity measured in mid-graylevels is more sensitive to cell gap variations.

Total throughput tests are easily performed and not normally sensitive to process variations. Contrast ratio is becoming harder to test because of the very high values >1000 achieved in some devices. Microdisplays requiring optical compensation test with and without optical compensation retarder. Off-state darkness dominates contrast ratio; measurement is essentially a test of the dark-state liquid crystal alignment. The sequential on/off CR measured on-axis at high *f*-number gives the highest value, checking with and without compensating retarder. CR as a function of *f*-number, with compensating retarder set to optimum, indicates the performance in an optical system. The ANSI contrast ratio test employing a checkerboard pattern is a test of the optical system rather than just the microdisplay performance.

Response time is not sensitive to process; slow response suggests a drastic change in nematic alignment strength or cell gap. However, response is easy to measure within the general electro-optical test.

7.5.4 Stress Tests

Stress tests establish the robustness of the microdisplay to extreme working conditions and storage. Semiconducting devices are routinely subject to tests of temperature cycling, humidity, and vibration. These tests have evolved to establish the reliability of electronic components, and apply directly to microdisplays.

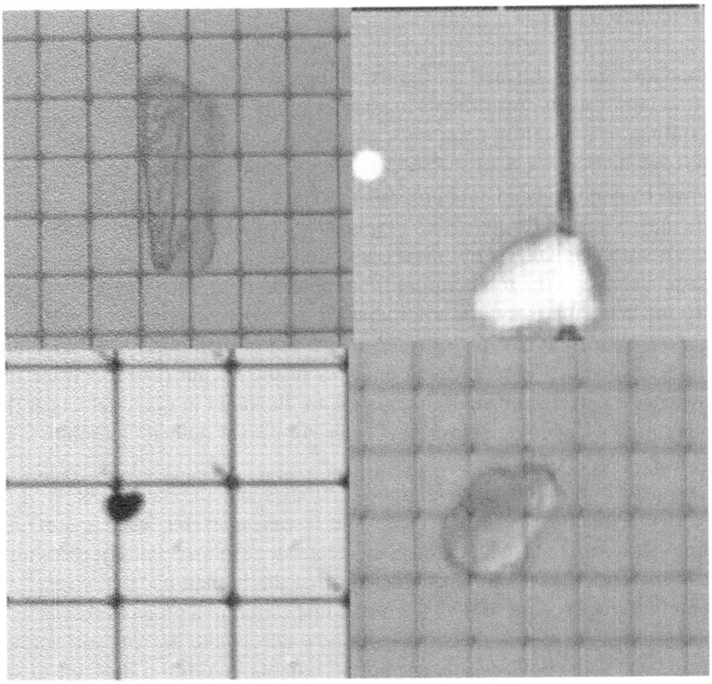

Figure 7.13 Examples of wafer optical defects observed in production. Reprinted courtesy of Society for Information Display

Accelerated life-testing is ongoing and conducted at different stress levels to refine estimated time to failure. Storage at elevated temperature and humidity reveal material limits that compromise lifetime. In projection applications, it is difficult to filter out near-UV light without compromising the lower limit of the blue spectrum. The sensitivity of the microdisplay to intense blue light and near-UV predicts lifetime limits in this respect.

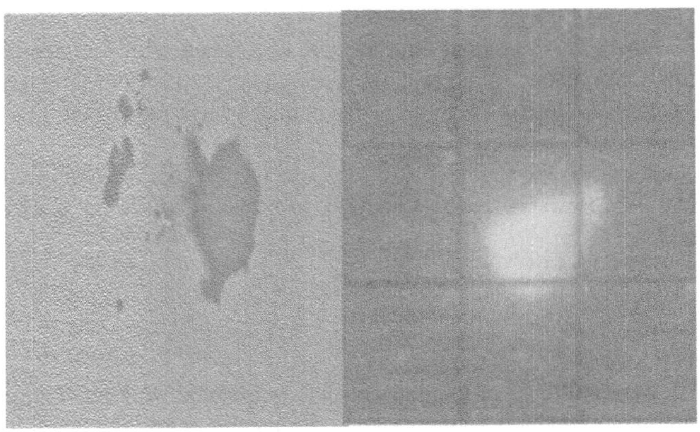

Figure 7.14 Microdisplay rejected for (left) ITO coating defect.[6] Reprinted courtesy of Society for Information Display

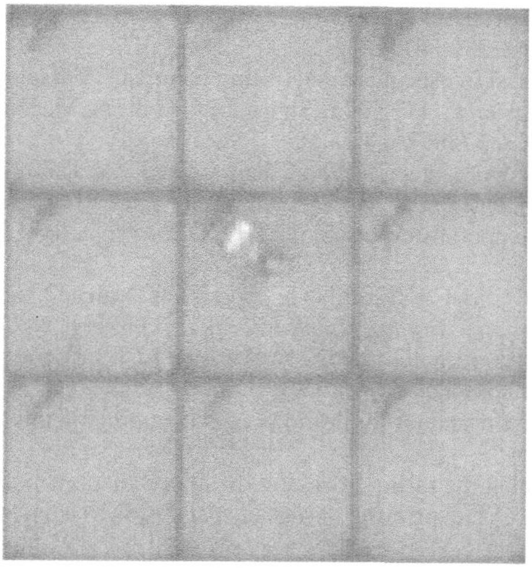

Figure 7.15 Microdisplay rejected due to particle on ITO surface.[6] Reprinted courtesy of Society for Information Display

7.6 References

1 Chen, W. S. and Melcher, R. (2001) Silicon foundry fabricates microdisplay with liquid crystal on silicon. *Solid State Technology*, May, 81–8.
2 Stefanov, M. E. *et al.* (2000) LCoS microdisplay manufacturing technology. *Intl Display Research Conference*, 17–21.
3 Stauss, H. C. (2004) The resurgence of LCOS displays. *SID Information Display*, **20/11**, 16–20.
4 Derrington, C. E., Smith, P. A. and Stauss, H. C. (2000) High-volume microdisplay manufacturing. *SID Digest* **31**, 371–3.
5 Boyce, M. and Wilson, M. D. (2002) Automation for production microdisplay testing. *Microdisplay*, 67–9.
6 Worthington, S., Hanna, M. and Uyeda, S. (2005) LCoS microdisplay manufacturing issues and glass. *SID Digest* **36**, 1186–9.
7 Cheong, K-B. *et al.* (2002) Thermal analysis and design for cooling system on LCoS imager. *Intl Display Research Conference*, 547–9.
8 Kozakai, T. *et al.* (2001) Reflection improvement for D-ILA™. *SID Digest* **32**, 911–3.
9 Yakovenko, S., Konovalov, V. and Brennesholtz, M. (2004) Lifetime of single panel LCOS imagers. *SID Digest* **35**, 64–7.
10 Wen, C-H., Gauza, S. and Wu, S-T. (2005) Ultraviolet stability of liquid crystal alignment layers and mixtures. *Proc. SPIE*, **5740**, 67–75.
11 Kim, S. G., Park, S. M. and Cho, C. S. (2002) A study on post-buffer washing processes for LCOS microdisplays. *Intl Display Research Conference*, 539–41.
12 Lu, M. *et al.* (2000) Homeotropic alignment by single oblique evaporation of SiO_2 and its application to high resolution microdisplays. *SID Digest* **31**, 446–9.
13 Murauski, A. *et al.* (2004) Ion beam sputter deposition process of inorganic vertical alignment layers. *SID Digest* **35**, 574–7.
14 Kahn, F. J. (1977) Capacitive analysis of liquid crystal displays. *Mol. Cryst. Liq. Cryst.* **39**, 109–23.
15 Hashimoto, S. *et al.* (2005) SXRD (silicon X-tal reflective display): a new display device for projection displays. *SID Digest* **36**, 1363–5.
16 Miremont, C. *et al.* (2002) Improving the flatness of silicon backplanes for high quality FLCOS microdisplays. *Displays* **23**, 115–19.

17 Hung, Y. C. *et al.* (2003) Study of flatness improvement of LCoS wafers using post-treatment. *Intl Display Workshop*, Japan, 1739–40.
18 Janssen, P. *et al.*(2002) Design aspects of a scrolling color LCoS display. *Displays* **23**, 99–108.
19 Chen, J. M. G. and Robinson, G. D. (2005) Sharp, high contrast MacNeille PBS-based LCOS projection system. *Proc. SPIE*, **5740**, 76–87.
20 De Moerloose, W., Van Doorselaer, G. and Van Calster, A. (2002) Reducing greyvalue non-uniformities in microdisplays. *SID Digest* **33**, 635–7.
21 Aruga, S. (2002) High temperature poly-Si TFT LCD for projection. *Intl Display Manufacturing Conference*, 551–6.
22 Kim, S-H., Kim, J-H. and Cho, C-S. (2002) The method of 1-μm cell gap in one panel LCOS imager for HD TV projection systems. *Intl Display Research Conference*, 543–6.
23 Cuypers, D. *et al.* (2002) Assembly of an XGA 0.9-inch LCoS display using inorganic alignment. *Intl Display Research Conference*, 551–4.
24 Kojima, K. (2003) The main sealant for liquid crystal dropping method. *Intl Display Works*, Japan, 605–7.
25 Palmateer, L. (2002) Packaging of liquid crystal on silicon display modules. *Microdisplay*, 46–8.
26 Davis, D. J. *et al.* (2000) Offset printing of gasket seals for microdisplay applications. *Intl Display Manf. Conference.*
27 Oepts, W. *et al.* (2002) Degradation of liquid crystal alignment upon high intensity illumination in microdisplays. *Intl Display Research Conference*, 201–4.
28 Wang, J. M. *et al.* (2005) The study of one-drop-filling process on LCOS panel manufacturing. *Intl Display Works*, Japan, 1947–50.
29 Lowe, D. *et al.* (2001) Liquid crystal on silicon (LCoS) micro-display packaging standardization. *Microdisplay*, 44–7.
30 Kim, S-H. *et al.* (2002) Reliability of WXGA LCOS imager packages. *Microdisplay*, 49–52.
31 Wilson, M. D. (2001) In-line production microdisplay test system. *Microdisplay Tech. Digest*, 63–6.
32 Downen, P. (2006) A closer look at flat-panel-display measurement standards and trends. *Information Display* **22**, 1, 16–21.
33 Collier, L. (2006) Optics, illumination, and processing tradeoffs for display inspection. *Information Display* **22**, 1, 22–7.
34 Collier, L. (2004) LCoS testing turns five. *Information Display* **20**, 5, 40–3.
35 Swinkels, S., Heynderickx, I. and Langendijk, E. (2005) How visible are defective pixels in high-resolution displays? *SID Digest* **36**, 386–9.
36 Pratt, W. K. R, Pendergrass, A. and Sawkar, S. S. (1999) Defect detection in reflective liquid-crystal microdisplays. *SID Digest* **30**, 468–71.

8

Micromechanical Devices

8.1 Background

Micro-optical-electromechanical (MOEM) devices have a long history.[1] The Eidophor (image bearer) was invented in 1939 and remains active in high-power projection applications.[2,3] It is based on optical phase gratings formed by electrostatic distortion of an oil film, where phase modulation is converted to intensity modulation by Schlieren optical readout (see Chapter 10). A scanning electron beam, modulated according to the desired image, charges the oil film. Technical and financial issues curtailed the development of an elastomer to replace the oil film in the Eidophor. However, elastomers were pursued in photoconductor addressed and electrode addressed diffraction devices. The Ruticon (wrinkle image) was a photoconductor addressed diffractive light valve developed up to a prototype stage.[4,5] An active matrix addressed deformable elastomer used in a Schlieren optical system (AMADEUS) is illustrated in Figure 8.1. The electrostatic force between the mirror electrode and addressing electrodes compresses the elastomer, forming a phase grating on the mirror surface. A pixel is resolved by one or more cycles of modulation, implying two or more electrodes per pixel. Interest in AMADEUS has also waned.[6–8]

Membranes can form deformable mirror structures; an example using electron-beam addressing is shown in Figure 8.2. The electron beam transmitted by the 200 nm thick metal alloy membrane charges the local glass surface under the membrane. The resultant electrostatic force deforms the membrane mirror, facilitating Schlieren optical readout.[9] The electrical conductivity of the glass is sufficient for discharge at TV frame rates. Photo-addressed or electrode addressed deformable metal membrane structures have also been demonstrated.[1] A deformable membrane device may be fabricated from a polymeric membrane coated with a metal film mirror. A "charge transfer plate" allows the membrane to be mounted outside the vacuum chamber when electron-beam addressing is used.[10] Active matrix addressing of a polymer membrane has also been demonstrated.[11] In general, it is difficult to achieve sufficient lifetime in elastomer or membrane systems, particularly with metallic coatings. Moreover, diffractive readout expands the etendue (see Chapter 10), with consequent lower throughput in compact systems.[12]

Introduction to Microdisplays D. Armitage, I. Underwood and S-T. Wu

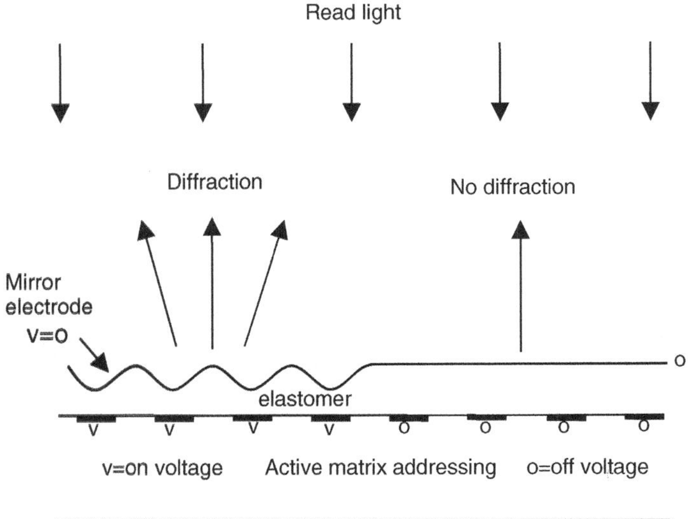

Figure 8.1 Active matrix addressed elastomer-mirror showing diffracted readout light in on-state

Progress in deformable micromirror devices required materials that resisted fatigue and were amenable to micro-fabrication methods. Amorphous silicon dioxide based micromirrors formed by semiconductor fabrication techniques are fatigue resistant and appear to have unlimited lifetime.[13–15] An electron-beam addressed device employing silicon dioxide micromirrors, known as the mirror-matrix tube, was developed to prototype level.[13] The mirror-matrix tube was a significant step in the development of practical devices. The micromirrors were fabricated from a silicon-on-sapphire disc, which ultimately formed the tube faceplate. The silicon surface was oxidized and the mirror geometry defined in the oxide layer by lithography. Careful etching of the silicon from under the oxide left an oxide micromirror supported by a remnant silicon post. A final aluminum deposition provided high reflection for the mirror and an electrode structure under the mirror gaps, as shown in Figure 8.3. Silicon micromechanical technology has developed to make oxide or silicon mirror structures easier to manufacture, raising the possibility of a revival of interest in some of the early devices.

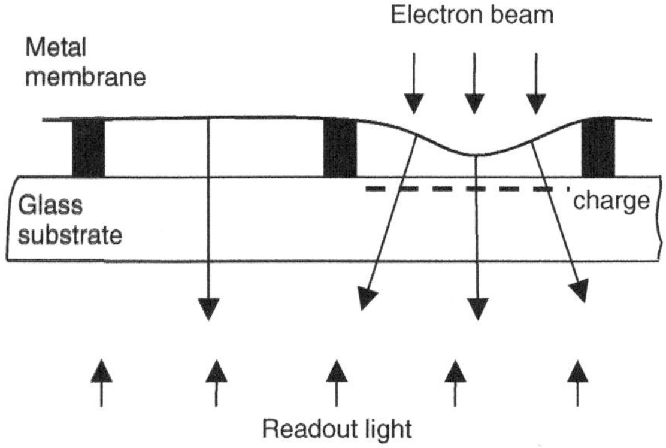

Figure 8.2 Electron-beam addressed membrane-mirror showing deflected readout light in on-state

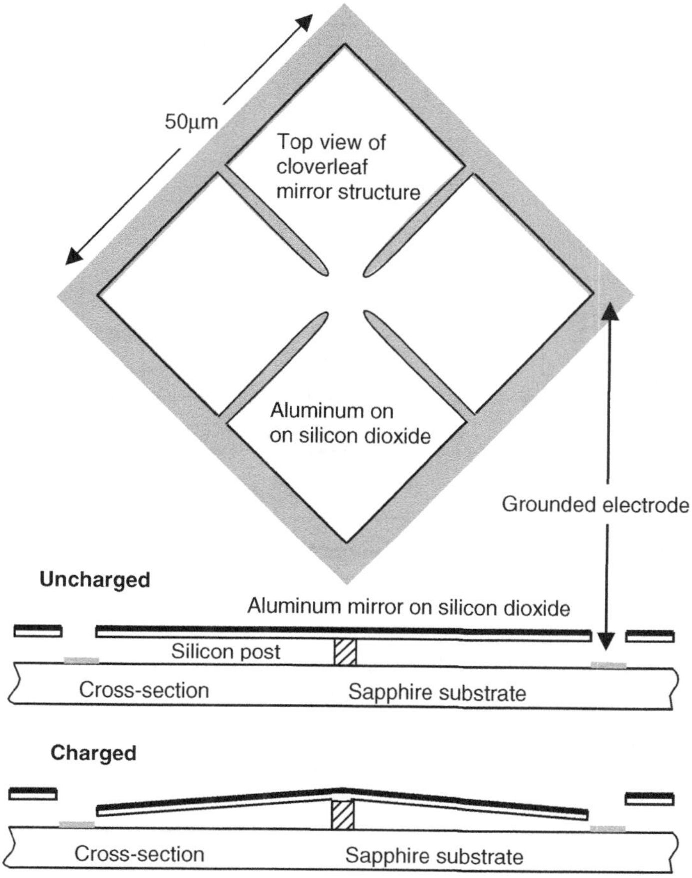

Figure 8.3 Pixel structure of mirror-matrix tube illustrating electrostatic deflection of four leaves

8.1.1 Electrostatic Deflection

A square mirror supported at one of its corners, shown in cross-section in Figure 8.4, illustrates the physics of a micromirror deflected by electrostatic force. Mechanical deflection ϕ is dependent on voltage V applied between the conducting mirror and drive electrode. The optical readout light is steered towards the output aperture by the mirror deflection.

For narrow electrode, small d, and small ϕ, as shown in Figure 8.4, the energy stored in the device capacitance approximates

$$E = \frac{1}{2}CV^2 = \frac{A\varepsilon_0 V^2}{2(d - r\phi)} \tag{8.1}$$

where ε_0 is free space permittivity, and A is an area term. The electrical torque (T_e) on the micromirror becomes

$$T_e = \frac{dE}{d\phi} = \frac{Ar\varepsilon_0 V^2}{2(d - r\phi)^2}. \tag{8.2}$$

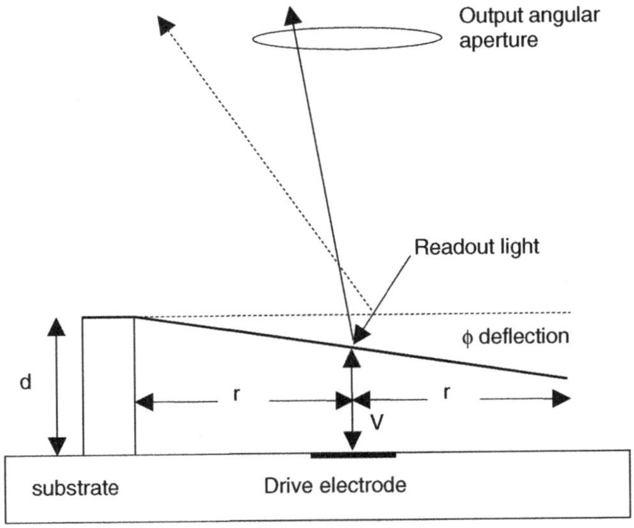

Figure 8.4 Cross-section of square mirror MOEM device supported at corner

T_e increases with V and ϕ in a highly nonlinear manner, making analog control difficult. Adding the elastic restoring torque $k\phi$ gives the total torque T on the mirror

$$T = \frac{Ar\varepsilon_0 V^2}{2(d - r\phi)^2} - k\phi. \tag{8.3}$$

For a given voltage the mirror deflection ϕ is given by the torque balance $T = 0$, where (8.3) can be rewritten in simpler form with constants c and h:

$$T = 0,\ h = \frac{r}{d},\ c = \frac{2kd}{A\varepsilon_0},\ V^2 = \frac{c}{h}\phi(1 - h\phi)^2. \tag{8.4}$$

A critical V is reached at $\phi = \phi_s = 1/3h$, the stable deflection limit. Further increase in V makes $dT/d\phi > 0$, where the mirror becomes unstable and ϕ increases to a maximum value $\phi = \beta = 1/2h$ when the mirror corner contacts the substrate. The ratio of stable dynamic range to maximum range is $\phi_s/\beta = 2/3$. The stable voltage limit is given by putting $\phi = \phi_s = 1/3h$ in (8.4):

$$V_s^2 = \frac{4c}{27h^2}. \tag{8.5}$$

Moving the drive electrode closer to the hinge increases ϕ_s, but also increases V_s in the same proportion. In practice, the area of the drive electrode is comparable to the mirror area; typically, analog addressing limited to ϕ_s sacrifices ~33% of full dynamic range β. Moreover, the highly nonlinear characteristic makes it difficult to achieve uniformity over a micromirror array. A mechanically limited mirror tilt $\beta = 12°$ implies $\phi_s = 8°$ and consequently $h = 1/3\,\phi_s = 2.387$. Putting the constant $c/h = 3400$ gives critical potential $V_s = 14.53$ V for the limit of stable operation. A 1% safety margin limiting the drive potential to 99% of the critical value gives $V = 14.38$ V, corresponding to $\phi = 6.73$ degrees, a substantial sacrifice compared to maximum dynamic range $\phi_s = 8$ degrees.

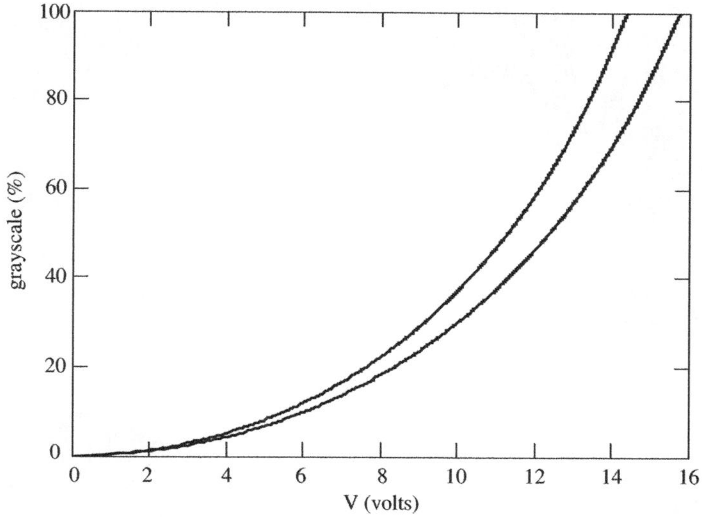

Figure 8.5 Grayscale response shift for material tolerance ±10%

An optical system with numerical aperture sin(6.73) = 0.117, i.e. *f*-number 4.3, allows all of the read light to be deflected in or out of the projection lens aperture by mirror tilt of 6.73 degrees (ignoring diffraction), which produces optical deflection of twice that angle. Output optical flux is determined by the overlap of output light cone and output aperture as a function of mirror tilt angle, assuming the read light intensity is independent of angular distribution. Output flux, or grayscale, plotted against voltage in Figure 8.5, also shows the effect of a 20% increase in elastic constant k. A 2% decrease in elastic response would bring the instability voltage below the 14.38-volt drive potential, which is not tolerable in an analog system. A greater stability margin requires a reduction in maximum drive voltage and throughput, where the plots in Figure 8.5 can be interpreted as ±10% tolerance in mechanical uniformity. Material properties and geometry, particularly the hinge region, influence the uniformity of response. It is difficult to achieve adequate grayscale uniformity, at numerical apertures required for sufficient throughput, due to limited uniformity in mechanical and geometric factors.

Although limited in accuracy, the simple model demonstrates the salient features of analog electrostatic micromirrors, and applies to the DMD. The same general features result from more sophisticated electromechanical models developed to account for detailed experimental observations.[14,16]

8.2 Digital Mirror Device

8.2.1 Background

DMD originally meant deformable mirror device and covered a range of structures from foils to various pixel mirror geometries, driven by silicon wafer CMOS. Development was funded by optical processing and computing interests from 1977 onwards, producing various prototypes demonstrating analog amplitude and/or phase modulation. The problems of analog devices prompted work on digital devices based on binary electromechanical states. The binary device blossomed and DMD became the digital mirror device in 1987, eliminating the earlier devices. Analog operation is restricted in dynamic range by the need to avoid mirror bottoming; digital operation removes that restriction and allows a wider dynamic

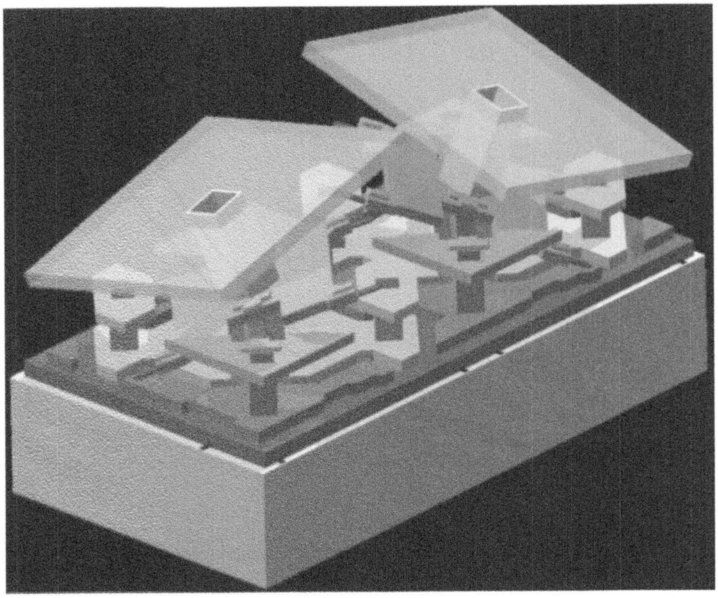

Figure 8.6 Illustration of first-generation DMD showing on-state and off-state pixels. Reprinted courtesy of Texas Instruments

range for essentially the same voltage level. Binary switching provides the benefits of higher throughput and contrast ratio. Further development applied to projection systems culminated in the launch of Digital Light Processing^{TM} technology (DLP 1996), based on the DMD illustrated in Figure 8.6.

The digital mirror device (DMD) furnishes each pixel with a square mirror hinged at opposing corners, supporting rotation about the diagonal with electrostatic drive. Diagonal rotation minimizes off-state diffraction into the projection aperture, improving contrast ratio. Early devices sacrificed considerable pixel aperture accommodating the torsion hinges supported on posts. The hidden hinge design achieves maximum aperture ratio by mounting the mirror on a hinged yoke structure completely hidden beneath the mirror.

Figure 8.6 shows pixels in alternate binary states. The yoke, suspended by thin torsion hinges attached to posts, supports the mirror superstructure. Drive electrode lower region acts on the yoke and is supplemented by a raised portion acting on the mirror. Binary drive electrodes and yoke/mirror potential control the mirror position. Yoke extensions impact electrodes (at yoke potential), imposing the binary stop positions.

First-generation pixel pitch 17 μm, with 0.7 μm gaps and $\pm$ 10° rotation, has given way to 14.5 μm pitch, 0.7 μm gap and $\pm$12° rotation in current products. Greater rotation accommodates higher numerical aperture, enhancing throughput in projector applications. Smaller pixel pitch shrinks the area, with cost benefit to microdisplay and optical system. The DMD has a clear advantage in color field sequential projection, dominating lightweight portable applications, and competing strongly in consumer rear-projection TV.

8.2.2 Structure

In the second-generation DMD shown in Figures 8.7 to 8.10, mirror deflection is restricted to on/off binary states by mechanical stops, and grayscale is achieved by pulse width modulation (PWM) of

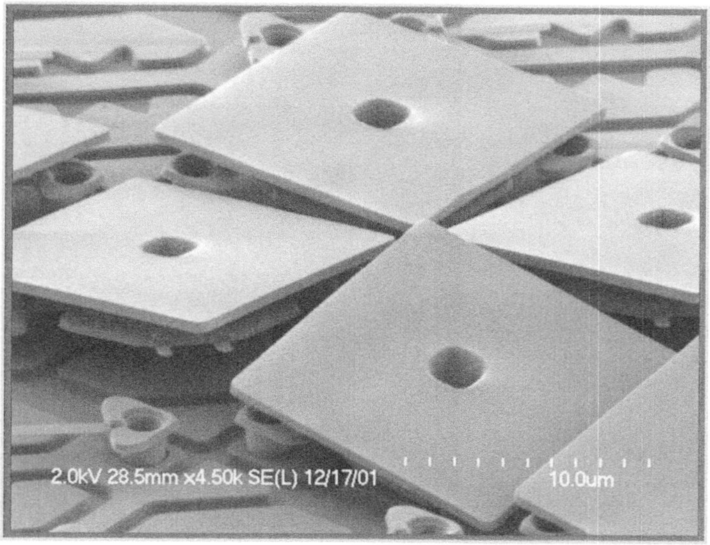

Figure 8.7 SEM picture of on-state and off-state pixels of second-generation DMD with material removed to show electrode structure. Reprinted courtesy of Texas Instruments

the on-state.[17–20] Analog uncertainty is replaced by mechanical contact problems that have proved easier to solve. In Figure 8.7, removal of some of the pixel mirrors reveals support structure, drive electrodes and rotation mechanism. One pixel mirror tilts towards the viewer, two tilt away, and one is flat. The details of the yoke structure appear in Figure 8.8, showing the torsion hinge, and spring tips free of the mechanical stop, and contacting the stop in Figure 8.9. A drive electrode

Figure 8.8 SEM of yoke and pixel structure of DMD revealing torsion hinge and spring tips. Reprinted courtesy of Texas Instruments

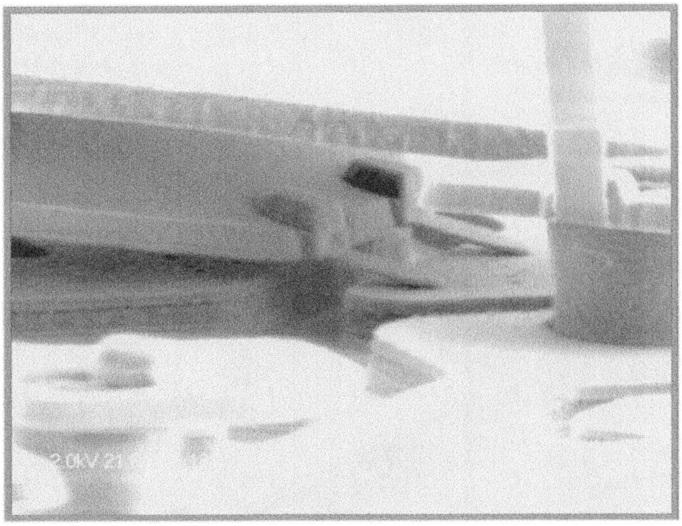

Figure 8.9 SEM picture of yoke with spring tips contacting stop. Reprinted courtesy of Texas Instruments

appears under the yoke, isolated from the mechanical stop electrode connected to the yoke. The second-generation drive electrode is planar and electrostatic interaction is essentially restricted to the yoke. Hiding the mechanism beneath the mirror (hidden hinge) maximizes the mirror aperture ratio. Electron microscopy reveals some of the underlying structure in Figure 8.10, not visible in optical readout.

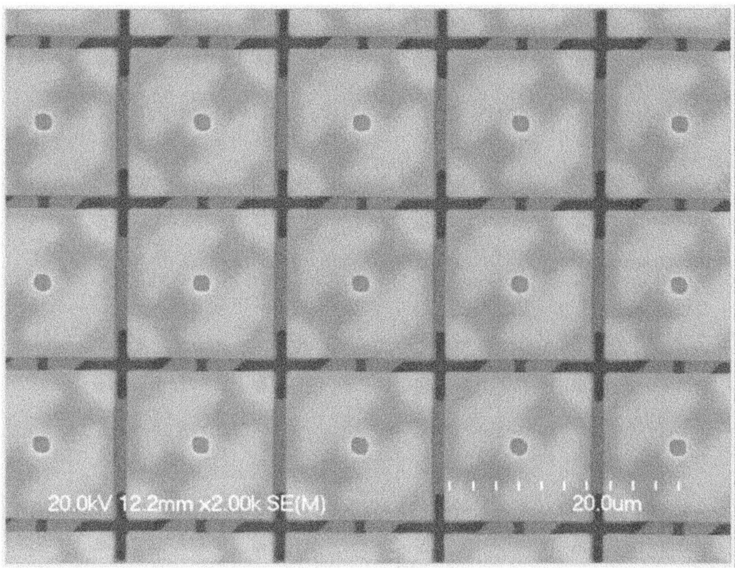

Figure 8.10 SEM picture of DMD pixels showing underlying yoke structure. Reprinted courtesy of Texas Instruments

The mirror, driven electrically to either binary state, need not rely on strong torque from the mirror support hinge for fast response speed. Consequently, the hinge is made very thin (~60 nm) to inhibit fatigue failure. Moreover, the hinge thickness is comparable to the grain size of the hinge material, circumventing accumulation of dislocations at grain boundaries that instigate fatigue failure. An electrode at the same potential as the mirror, to eliminate charging effects at the contact, forms the mechanical stop. A passivation coating applied to lower the surface energy attenuates sticking at the mechanical stops. The coating may be a silane surfactant applied by vapor phase methods. Spring tips at the contact points are a further precaution against sticking.[18]

8.2.3 Fabrication

The fabrication of MEMS follows lithographic methods developed by the microelectronics industry. The two-dimensional etched patterns forming the layered structure of an integrated circuit are extended into the third dimension for MEMS devices. Three-dimensional lithography requires the use of sacrificial layers and etch-stop techniques. Passivation glass layers seal a typical silicon circuit chip before separation from its wafer. High-class clean rooms are not needed for singulation and packaging of an ordinary chip.

A MEMS device such as the DMD must be protected from particle and other contamination for its entire lifetime. Consequently, high-quality cleanliness is maintained through to the final sealed package. Capillary water on the control surfaces would degrade performance and must be excluded by hermetic sealing. Ceramic packaging with welded seals provides the best protection for long life. Efforts to develop inexpensive plastic seals have not succeeded to date. Figure 8.11 shows a ceramic packaged device with rear pin connections. The flow chart in Figure 8.12 illustrates the fabrication process.

The addressing circuitry is fabricated in CMOS and then a silicon oxide layer is deposited on the wafer to facilitate chemical–mechanical surface polishing, providing a flat surface for the DMD

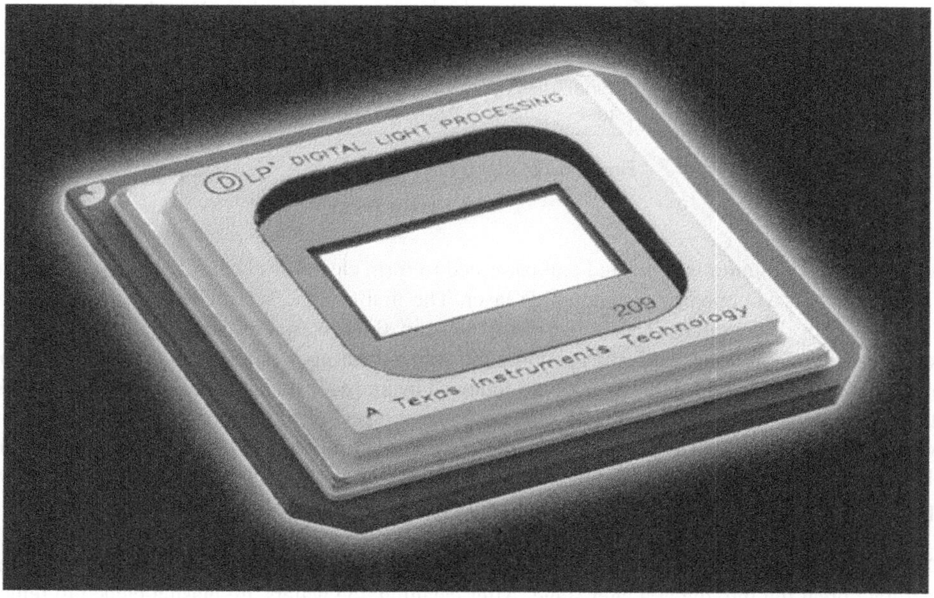

Figure 8.11 Ceramic packaged 720 p DMD with rear connection pins. Reprinted courtesy of Texas Instruments

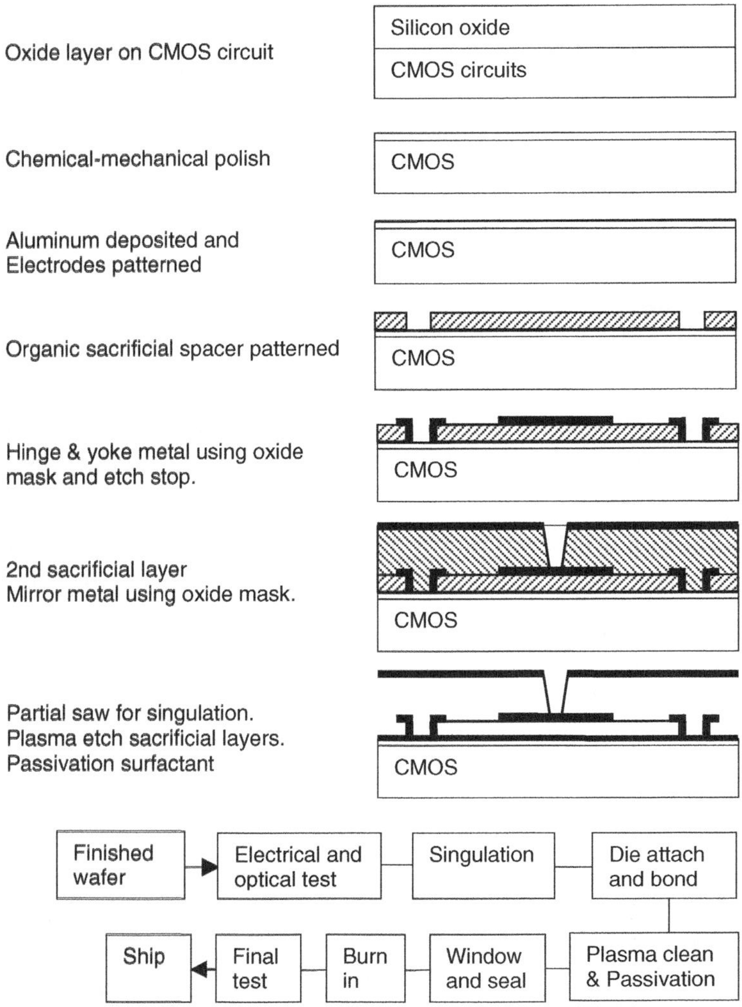

Figure 8.12 DMD fabrication steps

superstructure. Aluminum is deposited and patterned to form electrodes connected to the underlying addressing circuits via openings in the oxide layer. The first organic sacrificial layer is spin-coated and patterned by lithography. The hinge metal ~60 nm thick is sputter-deposited, followed by a plasma-deposited SiO_2 layer that is then patterned to define the hinges. A much thicker layer of aluminum ~600 nm is sputtered for the yoke, again followed by an oxide masking layer to define the yoke. Metal accumulating on the via surfaces forms the support posts for the torsion hinges. The yoke and hinge are patterned in a single plasma etch, leaving intact the hinge metal under the yoke to enhance the hinge to yoke bond. A second organic layer is spin-coated and patterned to accommodate the mirror support post. The final aluminum layer is sputtered-deposited, followed by an oxide layer, patterned to define the mirror.

To minimize contamination and damage, the wafers are partially sawed along the singulation lines to facilitate breaking into die at a later stage. After cleaning, the wafers are plasma-etched to remove the sacrificial material and a thin self-limiting passivation layer (surfactant) is deposited. Passivation lowers the surface energy, which inhibits sticking of the micromechanics and any particles that might make contact. Faulty die are identified by electrical and optical testing before singulation. Good die are mounted in the package and wire bonded to the interface terminals. Final plasma cleaning and repeat of the anti-stick passivation is followed immediately by a welded seal of the package, using a lid containing the optical window. A small component of solid passivation material included in the sealed package sublimates a "lubricating" atmosphere repairing any damage to the surface passivation layer. Any MEMS defects usually show up early in the operating life, revealed by multiple burn-in periods and additional testing before the DMD devices are qualified.

8.2.4 Operation and Throughput Efficiency

The DMD deflects light towards or away from the projection aperture of the optical system, as shown in Figure 8.13. The cone angle (2β) of the readout light is identified with the f-number = $1/(2\sin\beta)$ of the

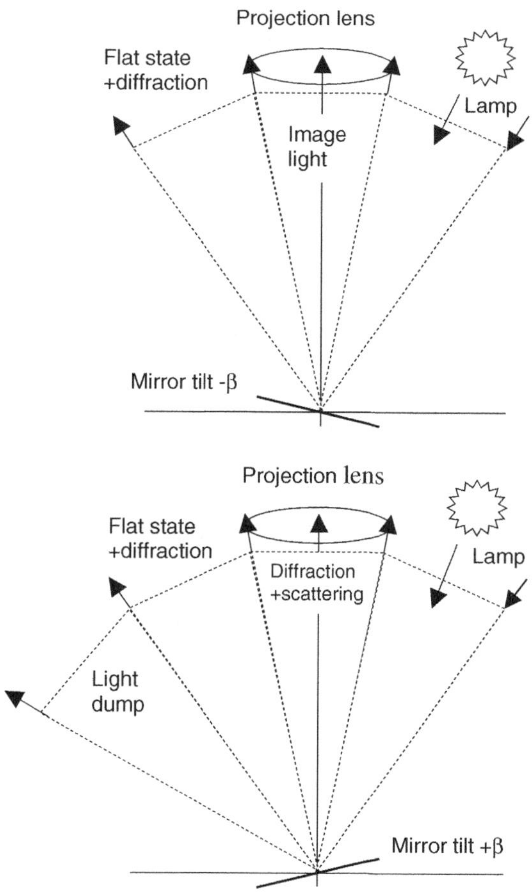

Figure 8.13 DMD optical configurations for on-state and off-state

system, and limited by the tilt angle β of the micromirror. Exclusion of the flat-state eliminates window reflections, transient mirror response, and any faulty pixels that settle at zero tilt appear dark. When a micromirror rotates, the reflected light rotates through twice the mirror rotation angle. All the directly reflected light moves out of the projection aperture for mirror position $+\beta$, but diffracted light and scattered light are collected by the projection lens. An inspection of the diffraction pattern of an array of square apertures reveals that the diffraction is weakest along the diagonal directions of the array. The DMD micromirrors rotate in the diagonal direction to enhance the dark state by exploiting the weaker diffraction towards the projection aperture. Scattering from the micromirror edges is also deflected away from the output aperture.

The device throughput efficiency is the fraction of device input light flux collected by the projection aperture. It is limited by several factors,[21] including mirror (pixel) aperture ratio s, the ratio of active pixel area to total pixel area. Aluminum alloy reflectivity R can reach 88% in microdisplay structures. The projection aperture collects light over a limited angular range, diffraction beyond that angle is lost, giving rise to a diffraction efficiency term DE. In pulse width modulation, electronic addressing limitations introduce an on-time factor that reduces throughput.[22]

$$\textit{Throughput efficiency} \quad \eta = (\mathrm{R})(\mathrm{s})(\mathrm{DE})(\mathrm{timing}). \tag{8.6}$$

Lithography and the processes involved in fabricating micromechanical structures limit the aperture ratio. The current DMD has a pixel pitch $p = 14.5\,\mu\text{m}$ and inter-mirror gap $g = 0.7\,\mu\text{m}$, giving mirror width $13.8\,\mu\text{m}$. The mirror support via of area $<4\,\mu\text{m}^2$, illustrated in Figure 8.10, makes little contribution to image light (specular reflection), giving $s = (13.8^2 - 4)/14.5^2 = 88.7\%$.

Aluminum alloys developed by the microelectronics industry are optimized for electrical conductivity and stability in microelectronics applications. The best available materials achieve a reflectivity of 88% when the surface is optically smooth. Further materials development to optimize optical properties could result in a small improvement in reflectivity. Addition of appropriate dielectric layers would enhance the aluminum reflectivity, but stresses introduced by the deposition methods and differential expansion would degrade mirror flatness, resulting in an overall throughput loss and deterioration in contrast ratio.

The reported diffraction efficiency is 85% for the $17\,\mu\text{m}$ pitch DMD, having $16\,\mu\text{m}$ mirrors, $s = 89.9\%$ and timing factor 92%.[21]

$$\textit{Throughput efficiency} \quad \eta = (88\%)(89.9\%)(85\%)(92\%) = 62\%. \tag{8.7}$$

A phased-array addressing scheme has greatly improved the timing factor.[22] Diffraction efficiency will fall as the pixel size is reduced, for given projection aperture.

8.2.5 Diffraction Efficiency and Contrast Ratio

Figure 8.14 illustrates the diffraction of plane waves from a one-dimensional array of mirrors rotating $\pm\beta$ about their centers. Mirror periodicity restricts the diffraction intensity to specific directions given by

$$\sin\theta_i + \sin\theta_r = \frac{n\lambda}{p} \tag{8.8}$$

where θ_i is the incident angle of a plane wave on the array, θ_r the reflection angle, n the diffraction order, λ wavelength, and p the mirror pitch. Angles measured in the anticlockwise direction are positive, opposite directions are negative. When $\beta = 0$, the strongest diffraction is the zero-order

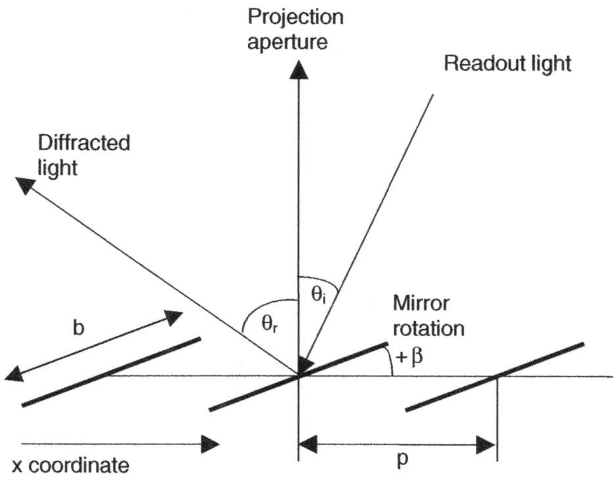

Figure 8.14 Array of one-dimensional micromirrors

$n=0$, but for $\beta \neq 0$, the shift in phase over the mirror width alters the diffraction amplitudes. The maximum diffraction peak is shifted from $n=0$ to $n \sim 2\beta p/\lambda$. Only for integer values of $2\beta p/\lambda$ is the peak intensity comparable to the zero-order case, demonstrating the enhancement in wavelength discrimination achieved by a "blazed grating." For non-integer values of $2\beta p/\lambda$, the diffraction intensities of order n close to $2\beta p/\lambda$ are enhanced so that the sum of the diffraction intensities is preserved.

Adequate details of the diffractive behavior can be calculated from scalar diffraction theory in the far-field approximation.[23,24] The angular distribution of diffraction intensity from an individual one-dimensional micromirror illuminated by a plane wave at angle θ_i is written in normalized form[25]

$$I = \left[\frac{\sin\left[\frac{\pi}{\lambda} b\{\sin(\theta_r - \beta) + \sin(\theta_i - \beta)\}\right]}{\frac{\pi}{\lambda} b\{\sin(\theta_r - \beta) + \sin(\theta_i - \beta)\}} \right]^2 . \tag{8.9}$$

A plot of the intensity distribution over projection aperture angular range ±12 degrees is shown in Figure 8.15 for $\beta = -12$ degrees, and readout angles $\theta_i = 2\beta$ and $\theta_i = 2\beta - 6$ degrees, with $\lambda = 550\,\text{nm}$, $p = 10\,\mu\text{m}$, and $b = 9\,\mu\text{m}$. The vertical lines in Figure 8.15 show the positions and relative intensity of the diffraction peaks for an array of mirrors according to Equation (8.8). The first diffraction minima occur at $\theta = \pm(\lambda/b)$, in a small angle approximation, showing how the diffraction broadens with increase in λ/b.

The throughput efficiency $\eta = Rs(DE)$ of a single micromirror can be determined by integrating the diffraction intensity over the projection angular aperture. Readout light is represented by an angular spectrum of unit-amplitude incoherent plane waves. Equation (8.9) is derived from a Fourier transform of the mirror optical amplitude with respect to spatial frequency $\lambda^{-1}\sin[(\theta_r - \beta) + (\theta_i - \beta)]$; therefore integration with respect to $d\{\lambda^{-1}\sin[(\theta_r - \beta) + (\theta_i - \beta)]\}$ refers normalized throughput to the mirror intensity/b, introducing a term $(b/\lambda)\cos(\theta_r - \beta)d\theta_r$ in the integral for DE. The input angular spectrum requires a normalizing factor $1/2\beta$. Mirror aperture loss is given by the aperture ratio $s = b/p$, and angular dependence of aperture ratio s is accounted for by the $\cos(\theta_i - \beta)/\cos\theta_i$ term. The micromirror

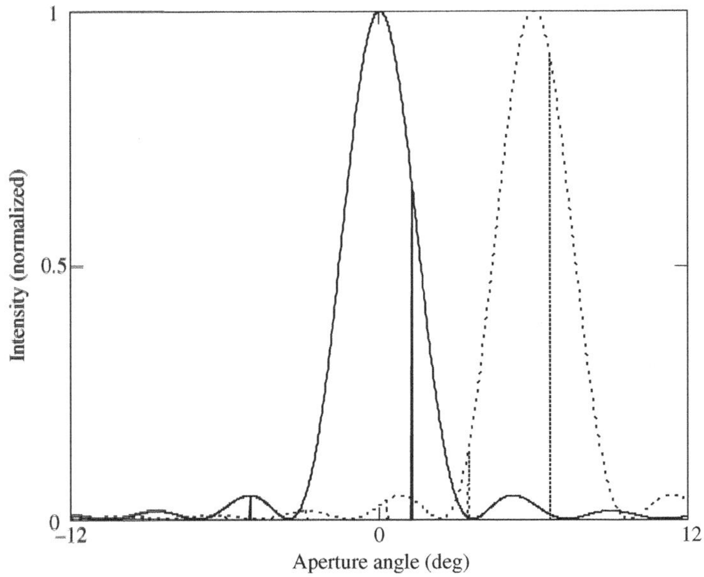

Figure 8.15 Diffraction intensity distribution for on-state individual 1-dimensional micromirror width $b=9\,\mu m$, with $\beta=-12$ degrees, readout angles $\theta_i=2\beta$, $\theta_i=2\beta-6$ degrees shown dotted, $=550\,nm$. Vertical lines indicate the relative intensity and mirror array diffraction angles for periodicity $10\,\mu m$

throughput efficiency is expressed:

$$\eta=\frac{Rsb}{\lambda}\int_{-|3\beta|}^{-|\beta|}\int_{-\beta}^{\beta} I(\theta_r,\theta_i)\frac{\cos(\theta_r-\beta)}{2\beta}\frac{\cos(\theta_i-\beta)}{\cos\theta_i}d\theta_r d\theta_i\,. \tag{8.10}$$

All regions of the micromirror diffraction pattern are sampled with equal probability by the angular selectivity introduced by a micromirror array, when the angular range and waveband of the readout light are taken into account.[25] Therefore, the throughput efficiency of a blazed micromirror array is essentially that of an individual micromirror, even when the aperture ratio approaches unity. It is an advantage in an imaging system for the *DE* to be independent of array size, since image luminance is then independent of resolution.[24] For the geometry described earlier, Equation (8.10) gives ideal ($R=1$) throughput efficiency of 88.6%, implying $DE=98.4\%$, for $s=90\%$. We observe that angular dependence of aperture ratio makes separation of diffraction efficiency and aperture ratio a matter of convention. Self-shadowing of the mirror array is avoided for the parameters quoted. In the non-blazed case $\beta=0$, the throughput efficiency of an array approaches 100% when the aperture ratio approaches unity. In diffraction terms, the angular selections of the array approach the zeros of the micromirror diffraction, excepting the zero-order peak.

Reversing the mirror tilt angle to $+\beta$ gives the diffraction pattern shown in Figure 8.16, for $\theta_i=-2\beta$ and $\theta_i=-2\beta-2$ degrees. Integrating over aperture angle and input angular range as before gives $\eta=0.185\%$, and contrast ratio $88.6/0.185=479$.

The diffraction calculation can be extended to a two-dimensional (x, y) array of micromirrors tilted β along the diagonal axis by small angle approximation $\beta_x=\beta/\surd 2$, $\beta_y=\beta/\surd 2$, giving I_x and I_y components of diffraction, and integrating over the square aperture $\pm\beta$. Changing the integration limits of θ_{rx} and θ_{ix} to respectively, $\surd[(\beta^2+\theta_y{}^2]$, and $/\surd 2\beta\pm\surd[\beta^2+(\theta_{yi}-\surd 2/\beta/)^2]/$, while normalizing to $\pi\beta^2$, accommodates a circular aperture with *f*-number $1/2\sin\beta$. The term I_{xy} accounts for the angular dependence of aperture ratio. The two-dimensional aperture ratio $s=(b/p)^2$.

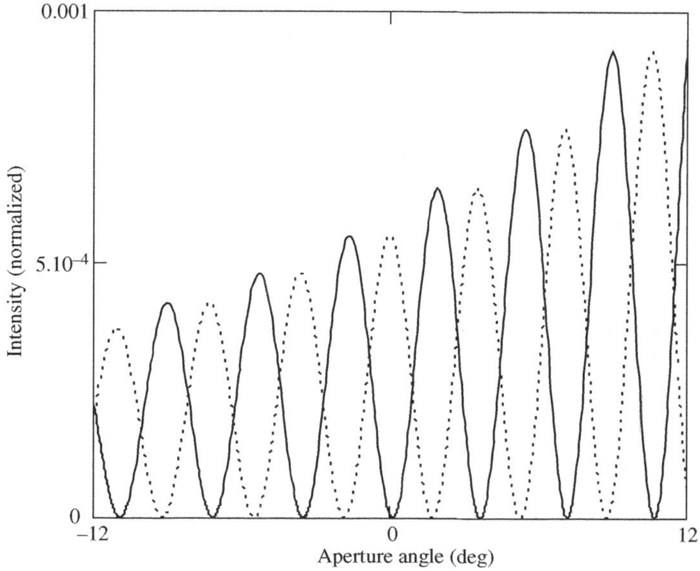

Figure 8.16 Diffraction intensity distribution for off-state individual 1-dimensional micromirror width $b = 9\,\mu\text{m}$, with $\beta = +12$ degrees, readout angles $\theta_i = -2\beta$, and $\theta_i = -2\beta - 2$ degrees shown dotted, $\lambda = 550\,\text{nm}$

$$I_x = \left[\frac{\sin\left[\frac{\pi}{\lambda} b\{\sin(\theta_{xr} - \beta_x) + \sin(\theta_{xi} - \beta_x)\}\right]}{\frac{\pi}{\lambda} b\{\sin(\theta_{xr} - \beta_x) + \sin(\theta_{xi} - \beta_x)\}}\right]^2 \cos(\theta_{xr} - \beta_x) \rightarrow \left[\frac{\sin\left[\frac{\pi}{\lambda} b\{\theta_{xr} - 2\beta_x + \theta_{xi}\}\right]}{\frac{\pi}{\lambda} b\{\theta_{xr} - 2\beta_x + \theta_{xi}\}}\right]^2 = i_x$$

$$I_y = \left[\frac{\sin\left[\frac{\pi}{\lambda} b\{\sin(\theta_{yr} - \beta_y) + \sin(\theta_{yi} - \beta_y)\}\right]}{\frac{\pi}{\lambda} b\{\sin(\theta_{yr} - \beta_x) + \sin(\theta_{yi} - \beta_y)\}}\right]^2 \cos(\theta_{yr} - \beta_y) \rightarrow \left[\frac{\sin\left[\frac{\pi}{\lambda} b\{\theta_{yr} - 2\beta_y + \theta_{yi}\}\right]}{\frac{\pi}{\lambda} b\{\theta_{yr} - 2\beta_y + \theta_{yi}\}}\right]^2 = i_y$$

$$I_{xy} = \frac{\cos\left[\left\{(\theta_{xi} - \beta_x)^2 + (\theta_{yi} - \beta_y)^2\right\}^{1/2}\right]}{\cos\left\{(\theta_{xi}^2 + \theta_{yi}^2)^{1/2}\right\}}$$

$$\eta = Rs\left(\frac{\mathrm{b}}{2\beta\lambda}\right)^2 \int_{-|\sqrt{2}\beta-\beta|}^{-|\sqrt{2}\beta+\beta|} \int_{-|\sqrt{2}\beta-\beta|}^{-|\sqrt{2}\beta+\beta|} \int_{-\beta}^{\beta} \int_{-\beta\eta}^{\beta} I_x I_y I_{xy}\, d\theta_{xr}\, d\theta_{yr}\, d\theta_{xi}\, d\theta_{yi}$$

$$\eta \rightarrow \frac{\cos(\beta + |2\beta|)}{\cos(2\beta)} Rs\left(\frac{b}{2\beta\lambda}\right)^2 \int_{-|\sqrt{2}\beta-\beta|}^{-|\sqrt{2}\beta+\beta|} \int_{-|\sqrt{2}\beta-\beta|}^{-|\sqrt{2}\beta+\beta|} \int_{-\beta}^{\beta} \int_{-\beta\eta}^{\beta} i_x i_y\, d\theta_{xr}\, d\theta_{yr}\, d\theta_{xi}\, d\theta_{yi} \qquad (8.11)$$

Table 8.1 Calculated DMD efficiencies at $\lambda = 550$ nm

b (μm)	p (μm)	s^a (%)	f/#	DE (%)	CR ($\times 10^4$)	Time (%)	R (%)	η (%)	Exp (%)
16	17	83	3	92.4	12^b	92	88	62.0	61.9[21]
16.3	17	89.9	3	92.4	13^b	98	88	71.6	
13.8	14.5	87.7	2.4	94.5	14^b	98	88	71.5	68[27]
13.8	14.5		2.4		0.104[26]	100		72[26]	71[26]
13.8	14.5		2.4		0.1[28]				65[28]
9.3	10	80.5	2.4	90.0	6^b	98	88	62.5	
9.7	10	88.1	2.4	90.6	6.5^b	98	88	68.8	

[a]Includes via loss, [b]Diffraction only.

The integral can be approximated by the small angle form shown in (8.11) to give within +0.3% error in thoughput and +8% error in CR. The approximate form reduces the time required to compute the numerical integration, particularly if an extra integration over wavelength is performed.

Table 8.1 gives calculated values of device throughput efficiency that are comparable with DMD performance reports.[21,26–28] Experimental values depend on optical system details; e.g. throughput efficiency depends on the angular dependency of readout intensity, which we have assumed independent of angle. Distortions in real DMD structures will lower the throughput. Existing technology could shrink the pixel pitch to 10 μm,[29] but unless the mirror gap $g = p - b$ is reduced there is a significant fall in throughput. A small amount of self-shadowing of the micromirror corner occurs at large illumination angles and negative β, but the effect on throughput is negligible. An on-time factor of 98% has been assumed for typical projection systems. The aperture ratio takes into account the mirror via loss $s = (b^2 - v)/p^2$, where $v = 4$ or 15 is the mirror via area in microns squared. Table 8.1 gives the sequential on/off contrast ratio (CR) for micromirror diffraction alone, and is the limiting contrast ratio that can be achieved for the said operating conditions. In practice, the CR is much lower.

Laser experiments have shown that mirror surface scattering is not significant in degrading CR.[30] The mirror central support via scatters light towards the projection aperture, and considerable CR improvement was achieved by shrinking it to 2×2 μm and designing the via orientation to duplicate that of the micromirror. Further reduction in via size might compromise the mechanical strength of the mirror support. Light that passes through the mirror gaps is reflected by the underlying electrode structure and emerges from the gaps to degrade the CR. Reduction in mirror gap would raise both the throughput and CR. Reducing the reflectivity of the structure below the mirror (dark metal DMD) boosts the CR $>$ 1000.[31] A standard foundry process of depositing titanium nitride can substantially reduce the reflectivity. Optimization of the shape of the readout aperture can enhance CR for a small sacrifice in throughput.[31] Every refresh addressing operation of the micromirror causes a bounce in angular position, which reduces CR and throughput, but not significantly. The scattering from the mirror via and pixel edges has been modeled by scattering functions fitted to experimental measurements. The scattering model facilitates the computation of throughput and CR as a function of output aperture, in good agreement with experimental values.[26] Dominant scattering is generated by the corner regions of the pixels;[31] smaller pixels should give rise to more scattering because of the increase in number of pixel corners for given microdisplay area, with consequent decrease in CR.

8.2.6 Addressing

Active matrix addressing takes the form of a static random access memory (SRAM) element at each pixel, implemented in 5-volt CMOS. Binary data is distributed by (x,y) word and bit lines and stored in the pixel SRAM. The binary value determines the routing of the 5-volt SRAM level to either driving electrode. A binary-1 generates $V_a = 5$ and $V_{\bar{a}} = 0$ in Figure 8.17, and binary-0 switches $V_a = 0$ and $V_{\bar{a}} = 5$.

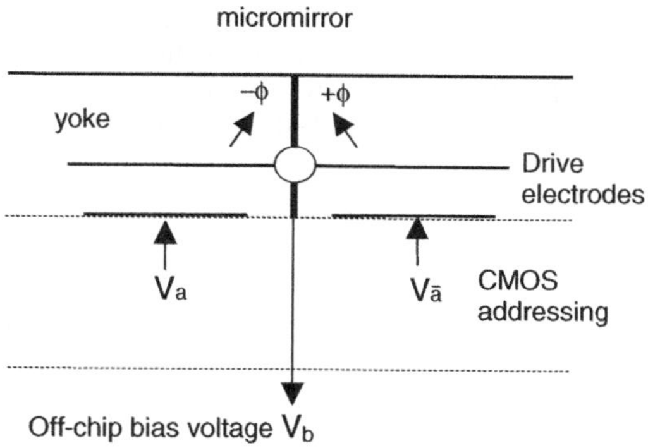

Figure 8.17 DMD pixel addressing electrodes, voltages, and rotation $\pm\phi$

A bias potential $V_b \sim 24$ volts input to the chip and applied to the mirror/yoke, enhances the action of the addressing voltage, and maintains the mirror state during the SRAM refresh cycle. The bias voltage also pulls the yoke towards the substrate, stressing the hinges to produce a stronger and more uniform hinge torque. When the yoke and one addressed electrode are both driven to positive potential, the torque is voltage biased in the direction of rotation towards the zero potential electrode, due to greater yoke/electrode voltage difference. The general form of the torque equation holds from section 8.1.2, although the DMD hinge structure differs. Clockwise mirror rotation is identified by negative values of ϕ, and vice versa. Voltage levels and nonlinearities in the torque equation determine switching and latching behavior:

$$T = \frac{kh}{c}\left[\frac{V_b - V_a}{1 - h\phi}\right]^2 - \frac{kh}{c}\left[\frac{V_b - V_{\bar{a}}}{1 + h\phi}\right]^2 - k\phi\,. \tag{8.12}$$

The addressing voltage acting alone to produce binary deflection $\phi = \beta$, is given by Equation (8.5) with $V_a = V_s$, $V_b = 0$, and $V_{\bar{a}} = 0$. The c and h values used in section 8.1.2 were chosen to approximate the DMD values, and substitution in (8.5) gives $V_a = V_s = 14.53$ volts, almost three times the available voltage. Equation (8.12) can be solved for the relations between addressing voltage, bias voltage and mirror tilt, by using the balance condition $T = 0$ and instability condition $dT/d\phi = 0$; however, the algebra is clumsy, and our interest is better served by numerical scanned solutions. Limited addressing potential $V_a = 5$ volts requires a bias potential $V_b > 15.6$ volts to achieve an unstable deflection to the $-\beta$ limit. To maintain the given $-\beta$ state when the addressing voltage switches to the alternate electrode ($V_a = 0$, $V_{\bar{a}} = 5$) requires bias $V_b > 19.9$ volts, which is the operational lower bias limit. For zero addressing, $V_a = V_{\bar{a}} = 0$, and bias potential $V_b > \sqrt{c/2h} = 18.9$ volts, the mirror moves from unstable equilibrium at $\phi = 0$ to either of the $\pm\beta$ limits.

The DMD is operated at bias voltage $V_b \sim 24$ volts, with a lower limit ~20 volts. The constants in the primitive model were set to give a bias voltage comparable with the DMD value. Equation (8.12) provides a simple approximation to torque variation throughout the DMD cycle, providing a guide to DMD operation. More sophisticated finite element modeling is required for quantitative analysis.[19,32,33] A detailed model should include the electrode structure, nonlinear elastic behavior, adhesion, electromechanical circuit effects, and resistive damping. Texas Instruments has developed a sophisticated digital mirror device pixel simulation tool (DMDPST) to predict the performance and aid design.[33]

To activate the new addressed state the bias voltage must be lowered. However, stiction effects need to be taken into account. It is found that pulsing the bias voltage induces a vibration in the micromirror structure that can overcome stiction effects and promote binary operation in pixels that otherwise fail to switch. Spring tips, added to the yoke extremities that contact the landing sites defining the binary positions, enhance the vibration effect.[18] A bipolar reset voltage pulse, together with raising addressing voltage to 7.5 V during the switching transient, improves reliability and stability.[32,33]

The switching transient for a 17 μm pitch device with 10° deflection angle is shown in Figure 8.18, along with electrostatic torques computed by the DMDPST developed by Texas Instruments. The

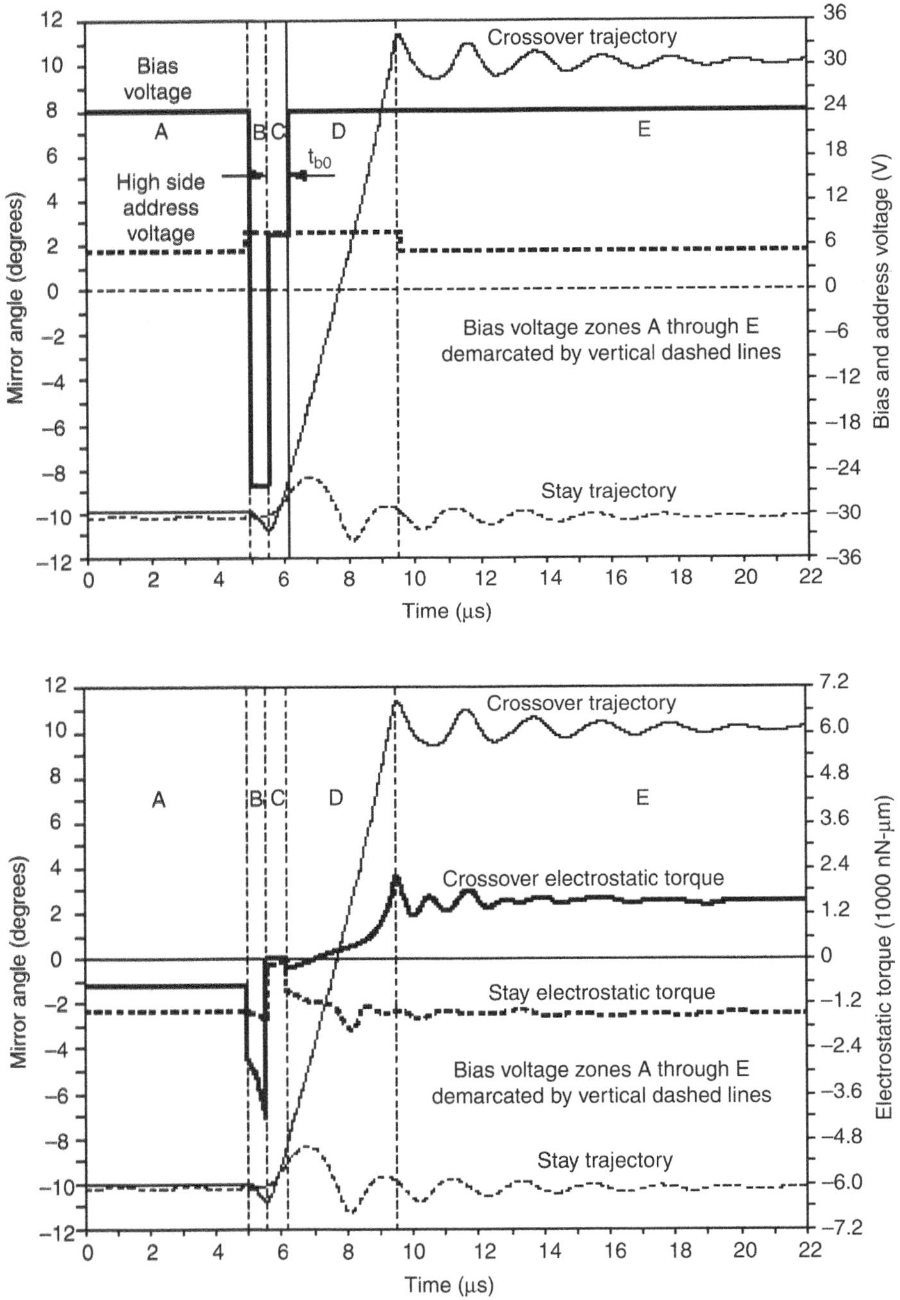

Figure 8.18 First-generation DMD switching transient for ±10 degrees deflection. Reprinted courtesy of Texas Instruments

Table 8.2 DMD switching transient

Time	Bias	Stay-state				Crossover state			
t	V_b	ϕ	V_a	$V_{\bar{a}}$	T	ϕ	V_a	$V_{\bar{a}}$	T
(μs)	(v)	(°)	(v)	(v)	(arb)	(°)	(v)	(v)	(arb)
−1	+24	−12	+5	0	−1431	−12	0	+5	−476
0	−26	−12	+7.5	0	−1493	−12	0	+7.5	−3476
0.7	−26	−12	+7.5	0	−1493	−12.5	0	+7.5	−3854
0.71	7.5	−12	+7.5	0	+487	−12.5	0	+7.5	+766
1.4	7.5	−11	+7.5	0	+461	−10	0	+7.5	+621
1.41	+24	−11	+7.5	0	−1182	−10	0	+7.5	+80
3.0	+24	−10	+7.5	0	−964	−2	0	+7.5	+285
5.0	+24	−12	+7.5	0	−1471	+12	0	+7.5	+1471
5.1	+24	−12	+5	0	−1431	+12	0	+5	+1431
*	+24	+4.5	+7.5	0	−263	−4.5	0	+7.5	+263
*	+24	0	+7.5	0	−304	0	0	+7.5	+304
*	+24	11	+7.5	0	+4.3	−11	0	+7.5	−4.3
*	+24	+7.8	+5	0	+1.4	−7.8	0	+5	−1.4

switching transient is divided into five regions labeled A to E, specified in Figure 8.18. During period A the address electrodes have been set to new values, but the mirrors do not change state, consistent with bias of +24 V and normal address electrode potential of +5 V. A mirror refreshed to switch is identified as *crossover*, while a mirror refreshed to the same state is identified by *stay*. The states are distinguished by substantially differing electrostatic torques reinforcing the current state, where a *crossover* state has lower voltage across the smaller gap. A reset pulse drives the bias voltage to −26 V and the address electrode voltage is stepped to +7.5 V during period B, greatly enhancing the torque on a *crossover* mirror, to boost stored elastic energy in the structure. The bias is then pulsed to +7.5 V during period C, providing weak electrostatic retaining torque on the *stay* mirror, and weak electrostatic release torque on the *crossover* mirror, allowing the mirrors to rebound from their mechanical stops. The higher rebound momentum and lower torque of a *crossover* mirror carries it through the torque reversal region to complete the transition during period D, when the bias returns to +24 V. The lower rebound momentum and higher torque of a *stay* mirror limit its bounce and it returns to its original position. The mirrors settle to their addressed positions during period E, where the electrode address potential steps down to 5 V, and both retaining torques have the same magnitude. The periods B and C ~0.7 μs are tuned to the mechanical resonance of the micromirror structure, optimizing the distinction of *crossover* from *stay* events. The voltage levels are also carefully optimized. Operating a 5-volt CMOS at 7.5 V generally reduces the lifetime, but the duty cycle at 7.5 V is low enough to have negligible effect on life.

The relative value of total torque (electrostatic + elastic) follows from Equation (8.12) for any deflection angle and voltages. Table 8.2 lists total torque at various angles through the switching cycle up to 12.5 degrees maximum, ignoring inertial and spring tip forces. Particular angles of interest have (*) in the time column. According to (8.12), the electrostatic torque is zero at $\phi = 4.5$ degrees, in agreement with Figure 8.18, leaving only the elastic torque component. The elastic torque is zero at $\phi = 0$, leaving only the electrostatic torque component. The torque goes through zero and reverses sign at $\phi = 11$, with 7.5-volt addressing, providing a low margin of safety for a 12-degree device. Stepping the addressing potential down to 5 V reduces the angle to $\phi = 7.8$ degrees, improving the safety margin.

Figure 8.18 and Table 8.2 indicate the high stability of an ideal DMD. The DMDPST can investigate the response of real DMDs to assess manufacturing tolerances, stiction, and aging behavior. Fluctuations in tilt during reset have negligible effect on throughput and contrast ratio. Similar dynamic response is reported for the second-generation DMD.[28]

Inertia and frictional damping limit the response speed. In a vacuum, the DMD is under-damped and considerable ringing occurs during the switching transient. It is essential to seal the device in an inert gas atmosphere so that viscous damping attenuates oscillations. The switching time from initiation to peak response is $<5\,\mu s$. However, reliable refresh must wait until the switching transients have subsided. The overall switching time from bias voltage reset initiation to steady state response is $\sim 15\,\mu s$. The output optical cone sweeps through the projection aperture over the angular range 0 to -10 degrees, making the effective optical switching transient $2\,\mu s$. The damped resonant period $\sim 2.5\,\mu s$, implies a resonant frequency $\sim 400\,kHz$. The high resonant frequency makes the DMD immune to normal mechanical shock effects.

Field-sequential-color projection requires at least 180 color-frames/sec, and grayscale resolution of 8 bits/color field implies a least-significant-bit (LSB) time of $22\,\mu s$. The DMD has the minimum required response speed; however, the addressing time of the entire array (load time) is longer than the LSB time (and other low-level bits), requiring all mirrors be switched to the off-state while the addressing cycle is completed. Circuitry was included to achieve switch-off of all the mirrors (full clear) in a time interval less than the LSB. The early DMDs sacrificed 8% of throughput due to dark-mirror time in the global reset method. Development of a phased reset scheme minimized the reset loss.[22]

The DMD array is partitioned into enough sequential groups for the group load time to be less than the LSB time, and each group has its own reset pulse. Each group is loaded and reset independently from other groups; hence, the term "phased reset." The reset operation in a group may be delayed relative to load completion time of that group. A phased reset PWM sequence controller facilitates the addressing.

Pulse width modulation can introduce local intensity errors in the image (PWM artifacts) due to eye motion missing part of the code, during the pulse code period. A bit-splitting method that distributes the light more uniformly over the frame time attenuates PWM artifacts (discussed in Chapter 10). The phased reset DMD architecture facilitates more bit splitting to provide an improved image, without a throughput penalty.

The luminance response of the eye is logarithmic, consequently a perceptual uniform grayscale is nonlinear in luminance steps. The nonlinear steps require greater bit depth to achieve the desired addressing levels within the linear binary PWM scale. Image contouring effects will be apparent if the minimum gray step is excessive. Image processing techniques such as dithering and error diffusion resolve the grayscale issue to some extent.[34] For further discussion of PWM grayscale issues see Chapters 2 and 10.

8.2.7 Lifetime

The DMD was the most complex MEMS device in production when volume manufacturing began in 1996. Its commercial development raised questions of manufacturing yield and reliability. A thorough program of testing and analysis was developed to ensure the commercial success of the DMD. Standard semiconductor tests, such as temperature cycling, humidity and vibration, are supplemented by tests and analysis of the MEMS structure. The current device must pass a stringent series of tests to qualify for service.

Information gathered over the past 10 years or more, together with accelerated life testing, has established the fabrication techniques and operating conditions to ensure long lifetime. The torsion hinge is fabricated from special alloys developed to extend lifetime. Hinge fracture is a very rare event, even with prolonged operation at elevated temperature. However, *permanent set* is a fault that occurs in the hinge due to slow mechanical deformation under stress, known as creep. It occurs at temperatures beyond 60°C, with prolonged operation when subject to duty factor favoring one of the binary states over the other. Permanent set (also known as hinge memory) is apparent when a mirror fails to return to the flat zero-deflection state at zero voltage. The binary addressing system is designed to tolerate a few degrees of permanent set. However, excessive hinge distortion inhibits reliable switching between the binary states, and so determines an upper operating temperature of 60 °C for the DMD structure. Improvement of hinge alloys remains an active area of research.[35,36]

An important test of pixel switching tolerance applies a standard bipolar reset pulse of −26 V and 0.7 μs duration, but sweeps the bias voltage over a range +18 to +32 volts and the bias off-time (when bias = 7.5 V) over a range 0.33 to 2.0 μs. For each set of conditions, all the pixels are subject to a series of stay and crossover commands. A DMD "solution space map" is generated by counting the number of nonworking pixels for the two-dimensional scan of bias voltage and bias off time.[32,33,35,37] The area of working pixels identifies the boundary values of the operating conditions. During accelerated life testing, the working area will shrink, indicating the long-term favorable operating conditions. For example, the optimum operating bias is 24 V with an initial tolerance of ±20%. Accelerated aging by high temperature, high duty cycle exposure, leaves the optimum value at 24 V, but shrinks the tolerance.

In projectors it was found that residual UV light illuminating the DMD was associated with a small statistical increase in pixel sticking. Analysis did not reveal a physical mechanism for the sticking, but UV damage to the surfactant that inhibits sticking may be involved. Enhancing the UV filter to restrict radiation beyond 400 nm to less than 0.7 mW/cm^2 on the DMD eliminates the effect, and a protective UV filter is now incorporated in the DMD window. High-power projector illumination in the visible has no effect on lifetime, other than that due to temperature rise.[35] Recent developments have improved the UV resistance to allow UV modulation in applications outside the display field.[28]

The only life limiting failure mechanism appears to be temperature-accelerated hinge memory. The projected lifetime of the DMD is greater than 100,000 hours for operational temperature below 50 °C, based on test results and modeling data developed for the DMD.

8.2.8 Fast-Track Pixel Design

The latest development known as fast-track pixel (FTP) provides a design path to smaller pixel size and higher speed.[29,38] Having a technology that can scale provides an important advantage in high-resolution applications. Figure 8.19 illustrates the structure implemented in 13.7 μm and 10.8 μm pixel pitch. Trimming the yoke structure supporting the mirror reduces inertia, promoting

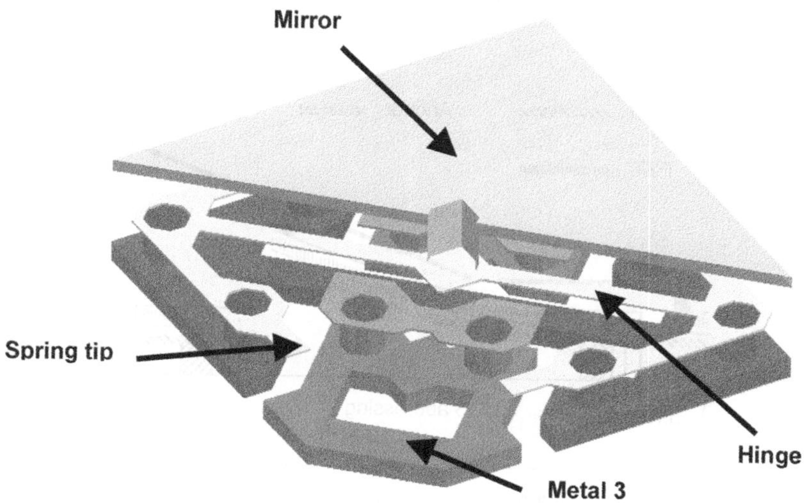

Figure 8.19 Fast track pixel design illustrated with cutaway mirror. Reprinted courtesy of Texas Instruments

faster response, while the drive electrostatic force acts directly on the mirror. Reduced mirror thickness also lowers inertia. The static structure now supports the spring release mechanism, which includes a torsional component in addition to the old cantilever response. Moreover, smaller spring contact area attenuates stiction, reducing the possibility of mirror sticking. A longer torsion hinge supporting the mirror allows improved process control over hinge width and thickness, enhancing uniformity in pixel response over the die area; further improvement followed the elimination of processes that disturb the hinge material structure. Mirror flatness depends on the deposition surface, which improves in flatness with the density of structure imposed by reduced pixel pitch; mirror flatness gains from smaller pixel pitch. Higher resolution lithography and smaller vertical spacing enable smaller inter-mirror gaps and mirror vias, maintaining optical efficiency with pixel scaling.

The CMOS process resolution has improved from 1.2 μm minimum metal line width for the 13.7 μm pitch, to 0.7 μm line width for the 10.8 μm pitch, accommodating a smaller SRAM cell. The voltage level associated with high-speed electronics falls below the electrostatic drive voltage required at the pixel. An architecture including both low- and high-voltage transistors achieves optimum performance. Low-voltage circuitry performs the high-speed addressing functions, located on the die periphery. High-voltage pixel circuitry occupies the central region of the die.

8.3 Piezoelectric Micromirror

8.3.1 Structure and Operation

Deformable mirrors controlled by piezoelectric activators have been in service for many years as phase correction devices in optical systems. The development of micromachining techniques has enabled the fabrication of micromirror arrays employing piezoelectric actuators.

The thin-film micromirror array (TMA) is a piezoelectric actuated device that has been developed to prototype stage for high-power projectors.[39,40] A cross-section of the device is shown in Figure 8.20. A piezoelectric ceramic based on lead, zirconate, and titanate (PZT) is sandwiched between electrodes, and held by a cantilever support anchored to the substrate. An applied addressing voltage induces longitudinal contraction and transverse expansion in the PZT, resulting in an upward bending of the cantilever. The micromirror attached to the actuator rises and tilts as shown.

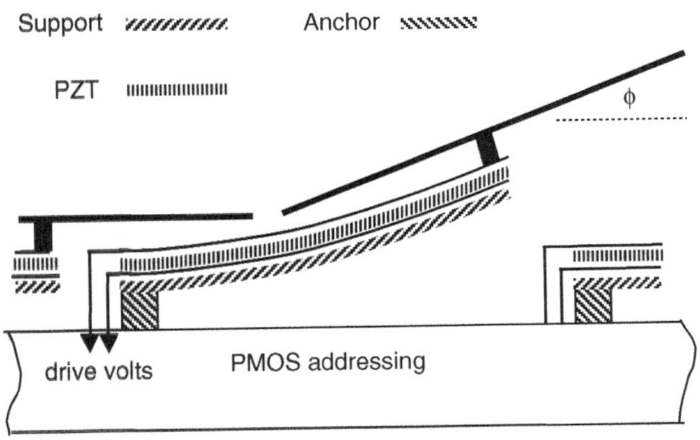

Figure 8.20 Piezoelectric micromirror array

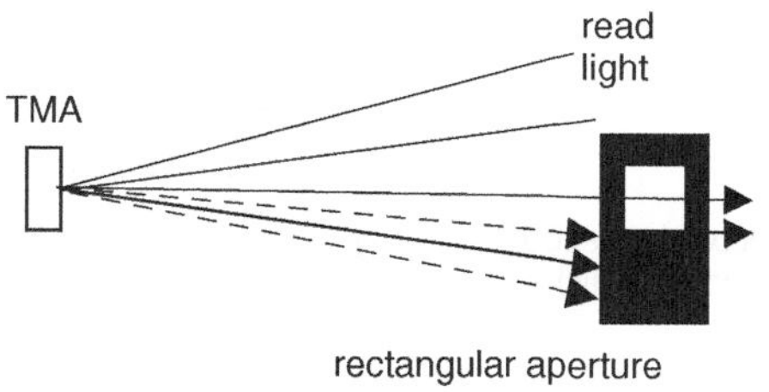

Figure 8.21 TMA projection through rectangular aperture, showing dotted dark state and 50% gray level

In a projector the mirror tilt deflects light into the projection aperture, similar to the DMD operation, but the TMA is designed as an analog modulator. The TMA angular deflection is linear with applied voltage, and the grayscale is linear with voltage when a rectangular projection aperture is used, as indicated in Figure 8.21.

High throughput and adequate CR is achieved at VGA resolution by 97 μm micromirrors with 100 μm pitch, giving 95% aperture ratio. The device active area $48 \times 64\,\mathrm{mm}^2$ makes it rather large for a microdisplay. It will remain large for the targeted 41 μm pitch at SXGA resolution. An addressing potential of 10 volts produces 4 degrees of micromirror deflection, implying an 8-degree light beam deflection. The optical deflection can accommodate a light cone $\pm 4°$ corresponding to projection optics *f*-number = 7.17. Increasing the *f*-number improves the contrast ratio at the expense of throughput, as in all Schlieren type systems. In the TMA, large device area compensates modest deflection angle to achieve sufficient etendue for high throughput efficiency.

The cost of the microdisplay and associated optics increases with device area. To be cost competitive, even in the high-power market, the pixel size will have to shrink considerably, and the micromirror deflection angle increase. Length in the PZT arm enhances the deflection angle, implying that the TMA cannot readily scale to smaller pixel dimensions. TMA response speed at less than 25 μs accommodates field sequential color projection. Piezoelectric deflection does not suffer the instability of electrostatic deflection.

8.3.2 Fabrication

Figure 8.22 illustrates the fabrication steps. It begins with an PMOS active matrix die (or wafer) having a tungsten metallization process designed to withstand the high temperature needed to form the PZT ceramic. The die is sealed with a passivation layer, on which is deposited (LPCVD) a sacrificial polysilicon layer. Chemical–mechanical polishing of the polysilicon provides a flat surface, which improves uniformity. Standard lithography is used to open anchor regions in the polysilicon. Design of the anchor structure is critical in achieving uniform response over the device area.[41] Silicon-rich silicon nitride deposited by LPCVD forms the support and counter balance to the PZT layer. The bottom electrode of platinum on tantalum is DC magnetron sputtered. The piezoelectric layer is formed by the sol-gel method. The sol material is spun onto the bottom electrode, and then heat treated. A rapid thermal annealing process crystallizes the PZT into the desired perovskite structure. The top electrode of platinum is magnetron sputtered, and the lithography steps to define the actuator structure are completed, leaving the polysilicon intact. Via holes are formed to connect the electrodes to the addressing circuit. A polymeric fluid is spin-coated and hardened to form the second sacrificial layer. A hole is opened to form the mirror support post and an aluminum layer is sputter deposited, then patterned into the micromirror format. The polysilicon is removed through the mirror gaps by plasma etching, followed by extraction of the polymer spacer by XeF_2 vapor

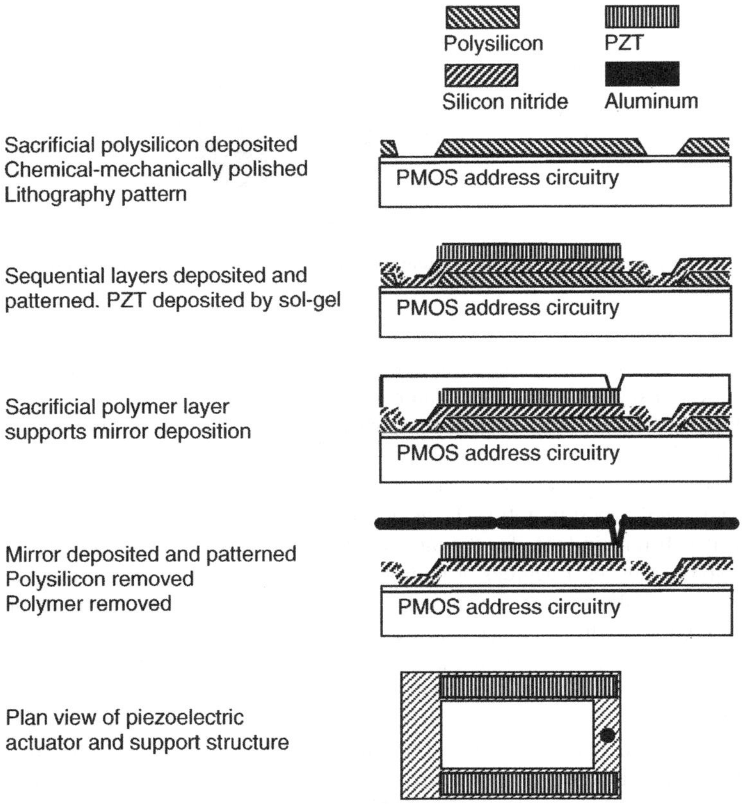

Figure 8.22 Fabrication of thin-film micromirror array

etching. The surface quality of the aluminum mirror is critical in obtaining maximum reflectivity and contrast ratio.[42,43] Accelerated life tests predict no degradation in performance over 10 years of projector operation.[41]

8.4 Grating Light Valve

8.4.1 Operation and Performance

The grating light valve (GLV) or deformable grating device (DGD) is a diffractive device, as illustrated in Figures 8.23 and 8.24.[44] The grating structure is ribbon like, and often described as ribbons in the literature. Each pixel is shown with three moveable gratings and three fixed gratings. A minimum of two gratings are required to resolve a pixel. In the off-state the gratings are not deflected and the device is weakly diffracting according to the grating periodicity p and aperture ratio $s = (p-2g)/p$. Alternate gratings are deflected $\lambda_o/4$ by an applied electric field in the fully on-state, to give strong diffraction at wavelengths near λ_o. A deflection $\lambda_o/4$ gives a phase retardation $\phi = \pi\lambda_o/\lambda$ on reflection at λ wavelength.

Sufficient tensile stress built into the silicon nitride grating during fabrication ensures rapid return of the gratings to the off-state when the applied field drops. The top surface of the grating is aluminum

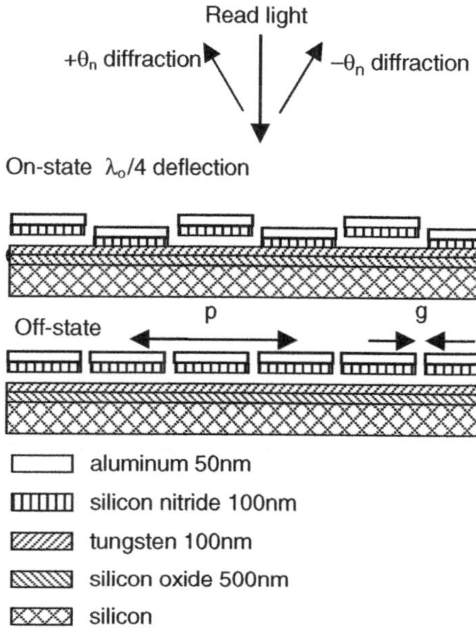

Figure 8.23 Grating light valve cross-section

coated for high reflectivity and serves as an electrode. Tungsten allows high-temperature processing and provides durability in the lower electrode. The maximum grating deflection 0.15 μm occurs for red light, and such small deflection enables very fast on/off switching times ~10 ns for applied ~10 V.[45]

Figure 8.24 Grating light valve pixel. Reprinted courtesy of Silicon Light Machines, now Cypress Semiconductor

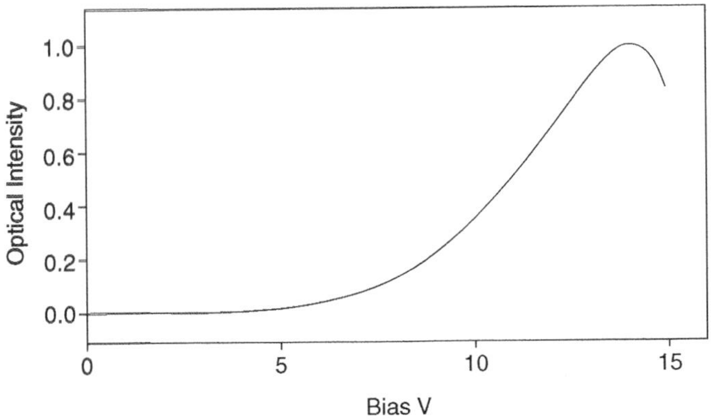

Figure 8.25 Grating light valve analog response. Reprinted courtesy of Silicon Light Machines, now Cypress Semiconductor

The ideal output of the GLV, for unit input intensity, is the sum of diffraction terms derived from a one-dimensional Fourier expansion of the grating:

$$\sum \eta_n = \frac{4}{\pi^2}\sum \frac{1}{n^2}\left[\cos^2\phi \sin^2\frac{n\pi s}{2}\cos^2\frac{n\pi}{2} + \sin^2\phi \sin^2\frac{n\pi s}{2}\sin^2\frac{n\pi}{2}\right] \tag{8.13}$$

$$\textit{Diffraction angle}\ \theta_n = \sin^{-1}\frac{n\lambda}{p} \qquad n = 0, \pm 1, \pm 2, \ldots \tag{8.14}$$

Figure 8.25 shows the analog response of the GLV, where a Schlieren filter (Chapter 10) selects the significant odd-order diffractions as ϕ increases with applied voltage. With $\phi = \pi\lambda_o/\lambda$, averaging $\cos^2\phi$, or $\sin^2\phi$ over a waveband about λ_o gives the maximum throughput efficiencies listed in Table 8.3. A primary color band is represented by a range $\lambda_o/\lambda = 1 \pm 8\%$, which expands to $\lambda_o/\lambda = 1 \pm 27\%$ for white light. The throughput efficiency $2(\eta_1 + \eta_3)$ at λ_o, approaches 90.06%, when $s \to 1$. Under similar conditions, a primary color approaches 89.9% and white light 84.6%. When reflectivity and aperture ratio are taken into account the center band throughput is 70% at $s = 0.95$.[46] The potential high throughput should be reconciled with the expansion in etendue due to diffraction. Efficient blocking of the zero and even orders of diffraction gives low off-state throughput, however some sacrifice in maximum throughput is required to achieve high contrast ratio.[12]

The fast response speed easily accommodates pulse width modulated gray levels, similar to the DMD. The hysteresis in on/off voltage associated with electrostatic forces is sufficient for passive

Table 8.3 GLV diffraction terms

State 100% on/off	$\Delta(\lambda_o/\lambda)$ (%)	η_0 (%)	η_1 (%)	η_2 (%)	η_3 (%)	η_4 (%)
Off	All	$100s^2$	0	$10.1\sin^2(\pi s)$	0	$2.5\sin^2(2\pi s)$
On	0	0	$40.5\sin^2(\pi s/2)$	0	$4.5\sin^2(3\pi s/2)$	0
On	±8	$0.53\,s^2$	$40.3\sin^2(\pi s/2)$	<0.1	$4.5\sin^2(3\pi s/2)$	<0.1
On	±27	$6.0\,s^2$	$38.1\sin^2(\pi s/2)$	$0.6\sin^2(\pi s)$	$4.2\sin^2(3\pi s/2)$	$0.2\sin^2(2\pi s)$

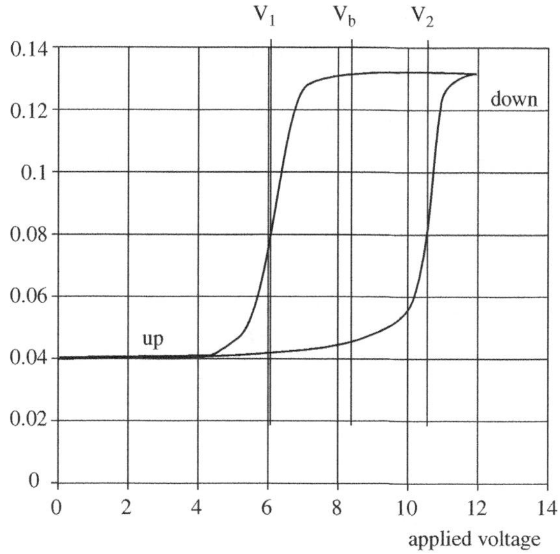

Figure 8.26 Grating valve hysterisis: V_b = bias voltage, V_2 = on threshold, and V_1 = off threshold. Reprinted courtesy of Silicon Light machines, now in Cypress Semiconductor

matrix addressing, as indicated in Figure 8.26. Freedom from active matrix addressing enhances manufacturing yield and power handling.

Efforts to develop a two-dimensional light modulator for projection applications were handicapped by stiction and aperture ratio issues. Figure 8.24 illustrates the difficulty of achieving high pixel aperture ratio in the long direction of the grating. The grating supports take up space, and grating curvature close to the support sacrifices throughput. Increasing the drive voltage reduces the curvature loss, but higher voltage exacerbates the stiction problem.

Display development is now concentrated on GLV laser scanning projection displays. A one-dimensional GLV can write a complete vertical line of the display, while a conventional scanning mirror handles the relatively slow horizontal scan. The one-dimensional modulation of the GLV accommodates poor laser beam quality in the vertical coordinate. Stiction issues are eliminated by avoiding contact in the on-state, since grating curvature is not significant when the readout laser line is concentrated along the center of the gratings. In tailoring the GLV to analog scanned laser displays, the response speed is lowered to ~1 μs, but since an entire line is written in that time, the response is more than adequate. The GLV drivers provide pixel refresh rates 250 kHz per drive channel, which supports a 1920 × 1080 scanned display at 96 Hz frame rate. Grayscale is programmable to 8-bits, which can be extended to 10-bits by dithering over multiple frame refreshes. An auto calibration routine that monitors the grating deflection characteristics maintains the analog grayscale precision.[47,48]

The one-dimension GLV has pixel pitch 25 μm, giving an array length of 27 mm. Developments foresee integrating the driver with the GLV to provide an inexpensive device when manufactured in high volume. When full-color laser diodes have achieved sufficient power, lifetime and low cost, the scanning GLV should be a strong competitor in consumer projection applications. GLV development has broad base support in printing applications and telecommunication switches and controls.

8.4.2 Fabrication and Testing

The GLV is fabricated on a silicon wafer utilizing standard semiconductor processes and equipment. Grating fabrication is a two-mask process, but more steps are required to complete the device, raising

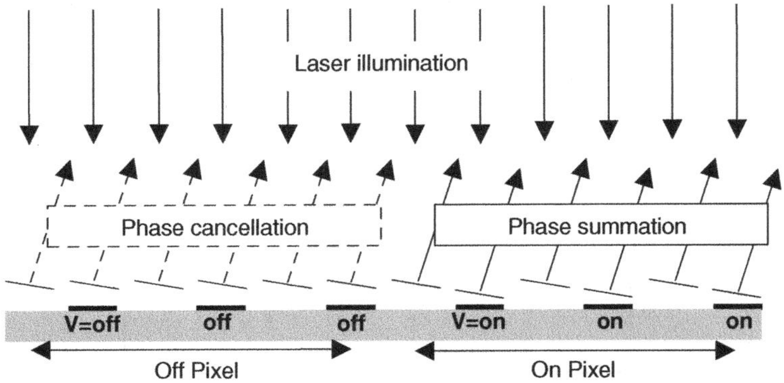

Figure 8.27 Blazed-grating light valve comparing on and off pixels

the mask count to seven. A sacrificial layer defines the air gap under the grating. The silicon nitride layer (LPCVD) is aluminum sputter coated, and the gratings patterned. Removal of the sacrificial layer leaves the gratings suspended over the air space, and held flat by tensile contraction forces. Each grating structure is sealed under nitrogen by a glass lid while still at the wafer stage, to protect the MEMS structure from contamination with subsequent processing. The simple fabrication process ensures high yield.

Silicon nitride is a hard amorphous ceramic chosen for durability and tensile strength. In a fully activated grating, the tensile stress is less than 10% of its fracture level. Change in the mechanical resonance frequency is a sensitive test of change in the grating stress level. Constancy of stress determines the stability of the grating electro-optic characteristic. Resonant frequency is monitored during life tests as an indicator of grating stress. Accelerated life tests show that there is no significant change in grating performance over 3.3×10^{12} switching cycles, in keeping with an expected life of 10^{13} to 10^{14} switching cycles. A thermal cycling test over temperatures in the range 18–100 and 30–150°C showed some change over the first four cycles, but after that initial burn-in the resonant frequency was stable over 110 thermal cycles. The test has been extended to −15 °C without observable degradation.[47]

The GLV has no difficulty in handling the high optical power density required by the laser scanning system. In tests simulating 30 W laser power, the induced temperature rise lowers the resonant frequency due to thermal expansion, while gas damping increases. However, the resonant frequency is stable over time.

A recent development in GLV structure introduces tilt to provide a blazed grating, as shown in Figure 8.27 The first-order diffraction energy is concentrated in a single order, simplifying the Schlieren filter and raising the throughput efficiency to >70%. A 0.25 μm fabrication process gives grating pitch 4.25 μm and >90% aperture ratio. Employing 6 gratings per pixel gives CR = 10,000 in a 5000 screen lumen system.[49]

8.5 Interference Modulation

Interference modulation (IMod) is a recent development in display technology devices.[50] Figure 8.28 indicates the IMod structure and operation.[51,52] An optical resonant cavity formed by a reflecting thin-film stack and a metal membrane reflector modulates the readout light. The cavity is designed to produce strong reflection over a chosen waveband. A voltage applied between the stack and metal reflector closes the air gap, altering the cavity tuning to give weak reflection. Reduction in voltage allows elastic tension to return the membrane to its original position, restoring strong reflectivity to the cavity.

Switching speeds of order 10 μs have been demonstrated, with predicted upper limit ~1 μs. White/black contrast ratios >10:1 are reported, together with 35% on-state reflectivity. Nonlinearity in

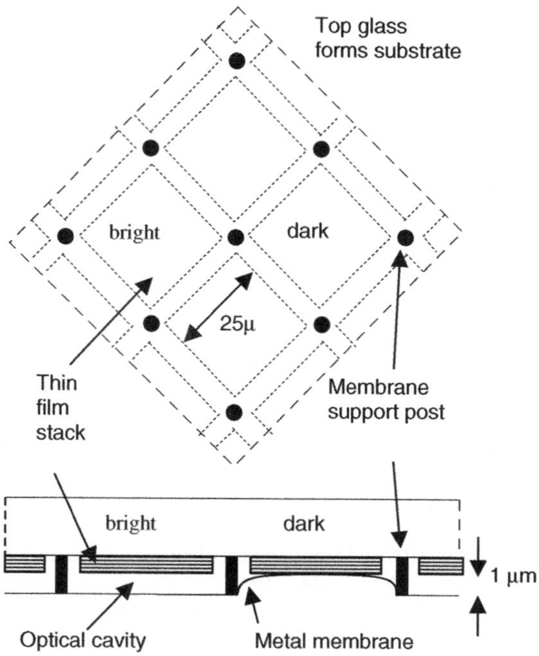

Figure 8.28 IMod top view and cross-section showing bright state and dark state

electrostatic force accommodates passive matrix addressing at a modest drive level of 5 V. The optical cavity air gap controls the reflected color. Forming adjacent cavities with appropriate air gaps generates red green and blue pixels for color imaging. Color is not sensitive to viewing angle.

Fabrication begins with the formation of the thin-film stack on the glass substrate, by conventional deposition methods, to provide broadband reflection. Sufficient electrical conductivity must be included in the stack for addressing purposes. A dielectric layer terminates the stack to maintain insulation when contacted by the metal membrane in the dark state. The stack is patterned according to addressing requirements. A sacrificial layer ~1 μm is deposited to form the cavity air gap, and patterned to accommodate support posts. The support posts and metal membrane are deposited, where nickel is favored for resistance to fatigue failure. Addressing requirements are patterned in the membrane. The sacrificial layer is removed by conventional etching methods.

The Imod is being developed for direct-view portable applications, where strengths such as low power, high reflection, wide viewing angle, and low cost are incentives to adopt the new technology. Spatial dithering or pulse width modulation achieves grayscale. Lifetime experiments on devices under development indicate some changes in the voltage response characteristic with aging. However, the stability of actuation voltage is consistent with long life.

The application of IMod to projection will require much higher contrast ratio and reduction in pixel size, while preserving high throughput. Near-eye applications favor color field sequential, but can tolerate less throughput efficiency. LED readout in near-eye applications should give higher contrast ratios. Evolution of IMod technology will determine future applications.

8.6 Further Development

Advances in microelectronic technology, and movement up the learning curve of existing devices, will enhance performance and lower cost. The MOEMS structure can be fabricated separately from the active matrix addressing electronics, and then assembled on a wafer scale.[53,54] Standard 5-volt CMOS cannot support the high-temperature processing of micromirrors fabricated from silicon nitride or similar strong materials. Separating the micromirror structure from the addressing electronics allows a wider choice of foundry for the 5-volt CMOS wafer, and high-temperature processing of MOEMS on a quartz wafer substrate by a specialized foundry. Precise assembly is required to locate the addressing electrodes relative to the micromirrors, but can be achieved with existing technology. Competing processes among MOEMS manufacturers will improve performance and lower cost.

8.7 References

1 Preston, K. (1972) *Coherent Optical Computers*. McGraw-Hill.
2 Tepe, R. (1985) Analysis of fluid light valve control layers for high-definition television picture projection. *J. Appl. Phys.* **57**, 2355–60.
3 Schulz-Hennig, J. F. (1998) Exciting story of the high-end television projection systems and the novel compact Eidophor AE12. *Proc. SPIE*, **3296**, 72–83.
4 Sheridon, N. K. (1972) The Rutican family of erasable image recording devices. *IEEE Trans. Electron Devices* **19**, 1003–10.
5 Lakatos, A. I. and Bergen, R. F. (1977) TV projection display using an amorphous-Se-type Ruticon light valve. *IEEE Trans. Electron Devices* **24**, 930.
6 Glenn, W. E. (1987) Solid-state-driven deformable television light modulator. *SID Digest* **19**, 72–4.
7 Tepe, R., Gerhard-Multhaupt, R., Brinker, W. and Molzow, W. (1989) Viscoelastic spatial light modulator with active matrix addressing. *Applied Optics* **28**, 4826–34.
8 Gerhard-Multhaupt, R., Brinker, W., Ehrke, H-J. *et al.* (1992) Active-matrix-addressed deformable-elastomer-layer light-valve projection: control layer and optics. *SID Digest* **23**, 451–4.
9 Van Realt, J. A. (1970) A new Schlieren light valve for television projection. *Applied Optics* **9**, 2225–9.
10 Horsky, T. N., Schiller, C. M., Genetti, G. J. *et al.* (1992) Electron-beam-addressed membrane mirror light modulator for projection display. *Applied Optics* **31**, 3980–90.
11 Hornbeck, L. J. (1983) 128 × 128 deformable mirror device. *IEEE Trans. Electron Devices* **30**, 539.
12 Armitage, D. (1997) Diffractive projection displays. *Proc. SPIE*, **3013**, 112–25.
13 Thomas, R. N., Guldberg, J., Nathanson, H. C. and Malmberg, P. R. (1975) The mirror-matrix tube: a novel light valve for projection displays. *IEEE Trans. Electron Devices* **22**, 765–74.
14 Peterson, K. E. (1978) Dynamic micromechanics on silicon: techniques and devices. *IEEE Trans. Electron Devices* **25**, 1241–9.
15 Brooks, R. E. (1984) Micromechanical light modulators for data transfer and processing. *Proc. SPIE*, **465**, 46–54.
16 Wetsel, G. C. and Strozewski, K. J. (1993) Dynamic model of microscale electromechanical spatial light modulator. *J. Appl. Phys.* **73**, 7120–4.
17 Hornbeck, L. J. (1989) Deformable-mirror spatial light modulators. *SPIE Critical Reviews*, **1150**, 86–102.
18 Hornbeck, L. J. (1996) Digital light processing and MEMS: reflecting the digital display needs of the networked society. *Proc. SPIE*, **2783**, 2–13.
19 Van Kessel, P. F. *et al.* (1998) MEMS-based projection display. *Proc. IEEE* **86**, 1687–704.
20 Hornbeck, L. J. (1999) Digital light processingTM update status and future applications. *Proc. SPIE*, **3634**, 158–70.
21 Florence, J. M. and Yoder, L. A. (1996) Display system architectures for digital micromirror device (DMD)-based projectors. *Proc. SPIE*, **2650**, 193–208.
22 Doherty, D. and Hewlett, G. (1998) Phased reset timing for improved digital micromirror device (DMD) brightness. *SID Digest* **29**, 125–8.
23 Armitage, D. (1999) Resolution issues in reflective microdisplays. *Proc. SPIE*, **3634**, 10–19.
24 Verweire, B. and Defever, R. (1999) Limitations of resolution of LCOS-based projection displays by diffraction effects. *Proc IDRC*, 489–92.
25 Jenkins, F. A. and White, H. E. (1957) *Fundamentals of Optics*. McGraw-Hill.

26 Dewald, D. S., Bartlett, T. and Iyengar, A. (2003) Optical model of DMD device in DLP™ projection system. *Intl Display Works*, Japan, 1557–60.
27 Dewald, D. S., Penn, S. M. and Davis, M. (2001) Sequential color recapture and dynamic filtering: a method of scrolling color. *SID Digest* **32**, 1076–9.
28 Dudley, D., Duncan, W. and Slaughter, J. (2003) Emerging digital micromirror device (DMD) applications. *Proc. SPIE*, **4985**.
29 Grimmett, J. and Huffman, J. (2005) Advancements in DLP® technology: the new 10.8-μm pixel and beyond. *Intl Display Works*, Japan, 1879–82.
30 Kearney, K. J. and Ninkov, Z. (1998) Characterization of a digital micromirror device for use as an optical mask in imaging and spectroscopy. *Proc. SPIE*, **3292**, 81–92.
31 Dewald, D. S., Segler, D. J. and Penn, S. M. (2003) Advances in contrast enhancement for DLP projection displays. *J. SID* **11**, 177–81.
32 Knipe, R. L. (1996) Challenge of a Digital Micromirror Device™: modeling and design. *Proc. SPIE*, **2783**, 135–45.
33 Meier, R. E. DMD pixel mechanics simulation. *TI Technical Journal* **15**, 64–74.
34 Watanabe, S. *et al.* (2002) Advantages of 2nd generation home entertainment DLPTM technology. *Intl Display Works*, Japan, 1245–8.
35 Douglass, M. R. (1998) Lifetime estimates and unique failure mechanisms of the digital micromirror device (DMD). *Proc. Intl Reliability Physics Symposium*, 9–16.
36 Douglass, M. R. (2003) DMD reliability: a MEMS success story. *Proc. SPIE*, **4980**, 1–11.
37 Chu, H. *et al.* (1998) DMD superstructure characterizations. *TI Technical Journal* **15**, 75–86.
38 Gong, C. (2005) Digital micromirror device (DMD) pixel shrink for high resolution application. *Proc. TEXMEMS VII*, 103–6.
39 Um, G. (1992) A new display projection system. *SID Digest* **23**, 455–58.
40 Kim, S-G., Hwang, K-H. and Koo, M. K. (1999) Thin-film micromirror array (TMA) for cost-competitive information-display systems. *SID Digest* **30**, 982–85.
41 Kim, S. G. *et al.* (1999) Thin-film micromirror array (TMA) for high luminance and cost-competitive information-display systems. *Proc. SPIE*, **3634**, 207–16.
42 Kim, S. G. *et al.* (1997) Micromachined thin-film mirror array for reflective light modulation. *Ann. CIRP* **46**, 455–8.
43 Yo, Y. S. *et al.* (2002) Optical performance evaluation of thin-film micromirror array in projection display. *Optics Eng.* **41**,409–15.
44 Bloom, D. (1997) The grating light valve: revolutionizing display technology. *Proc. SPIE*, **3013**, 165–71.
45 Amm, D. T. and Corrigan, R. W. (1998) Grating light valve technology. *SID Digest* **29**, 29–32.
46 Corrigan, R. W. *et al.* (1999) Calibration of a scanned linear grating light valve projection system. *SID Digest* **31**, 220–3.
47 Amm, D. T. and Corrigan, R. W. (1999) Optical performance of the grating light valve technology. *Proc. SPIE*, **3634**, 71–8.
48 Corrigan, R. W., Amm, D. T. and Gudeman, C. S. (1998) Grating Light Valve™ technology for projection displays. *Intl Display Works*, Japan, 757–60.
49 Ito, Y. (2006) High-performance blazed GxL™ device for large-area laser projector. *Proc. SPIE*, **6114**.
50 Miles, M. W. (1997) A new reflective FPD technology using interferometric modulation. *SID Digest* **29**, 71–4.
51 Miles, M. W. (2000) Digital paper™: reflective display using interferomentic modulation. *SID Digest* **31**, 32–5.
52 Miles, M. W. *et al.* (2003) Digital paper™ for reflective displays. *J. SID* **11**, 209–15.
53 Huibers, A. G. (1998) Reflective spatial light modulator with encapsulated micromechanical elements. US Patent 5,835,256.
54 A. Huibers, P. W. Richards, R. Grasser, (2006) New Micromirror Device for Projection Applications, SID Digest, **37**, 1618–21.

9

Emissive Microdisplays

9.1 Introduction

At the time of writing, in 2006, the organic light emitting diode (OLED) microdisplay is a rising star in the field of emissive microdisplays. OLED, in the two commercially dominant forms of small-molecule or SMOLED and polymer- or P-OLED (and in various other forms or variants), is a major area of potential and development for displays in general and the developments are beginning to feed into the field of microdisplays. OLED microdisplays form the main theme of the chapter. Some other emissive microdisplay technologies are discussed briefly in section 9.7.

In the context of this book overall, it should be recognized that, again at the time of writing, the field of OLED microdisplays is much less mature and the technology less well developed with fewer active players than microdisplay technologies described in most of the other chapters of this book so that information, in general, is less freely available and less comprehensive. As is the case for many early stage technologies, progress is very rapid so it is likely that specific information offered herein, such as performance parameters, will become out-of-date commensurately quickly.

In this chapter we present a brief overview of OLEDs in general. The intention is not to provide a comprehensive overview, but rather to provide sufficient information to allow the reader to appreciate OLEDs in the context of microdisplays. Designing and manufacturing an OLED microdisplay involves different challenges to those of a conventional OLED display or an LC microdisplay and the end product offers unique advantages for specific applications areas. We look a little wider than what is current. For example, phosphorescent systems are mentioned; there are no microdisplays that currently employ phosphorescent materials but there is a likelihood that in the future that will change.

9.1.1 Background to Organic Electronics and Displays

Since their discovery in the late 1980s, organic heterojunction devices, also known as organic light emitting diode (OLED) devices, have been the subject of intense research and development. Transistors,[1]

Introduction to Microdisplays D. Armitage, I. Underwood and S-T. Wu

Table 9.1 Main attributes of OLED displays contrasted with those of LC displays

OLED technology	LCD technology
Emissive	Reflective or transmissive
No external light source required	Requires backlight or frontlight
Near-Lambertian emission	Limited viewing angle
Emits unpolarized light	Output is typically linearly polarized light
All solid state	Not all solid state
Very fast switching	Slow switching
Colors may be achieved by doping or using different emissive materials	Colors derived from external light source or color filters
Only lit pixels consume power	Backlight power constant
Responds to current	Responds to applied voltage
Current drive circuit	Voltage drive circuit
Relatively short lifetime	Relatively long lifetime
Differential aging	Not an issue
Patterning techniques still maturing	Not an issue

memories,[2] photocells[3] and photonic circuits[4] have all been proposed whilst the most prolifically pursued application has been displays and, in particular, color displays.[5] This intense activity has resulted in the introduction of the first generation of products to utilize the new technology early in the new century. Whilst the long-term goal of much of the activity centers around the development of curved and flexible displays built on plastic substrates, early products have been less ambitious ranging from monochrome character and dot-matrix displays to passive matrix color displays, mostly so far for use in consumer products such as MP3 players. Perhaps surprisingly, among the earliest products to be offered are full-color CMOS active matrix microdisplays, as described in section 9.6, based upon both molecular (SMOLED) and polymer (P-OLED) organic emissive materials.

Some of the general attributes of OLED displays are listed (and compared with those of LC technology) in Table 9.1.

It is fair to say that the some of the main long-term aims of organic emissive displays research are as follows.

- *Large, flat, thin screens.* These will compete with plasma screens, LC displays and rear-projection displays.
- *Non-flat or conformal displays.* It is expected that the manufacturing techniques for organic emissive displays will allow such displays to be manufactured on curved substrates.
- *Flexible displays.* There is a further expectation that the manufacturing techniques for organic emissive displays will allow such displays to be manufactured on flexible substrates thus facilitating one route to electronic paper that can be laid flat to view and rolled or folded up for convenience when not in use.

En route to the above the technology is expected to pass through many stages of development. In the short-term OLED displays are moving from monochrome to area-color to RGB color pixels with the main focus being small rigid displays for consumer electronics products such as MP3 players and cell phones (both primary displays and secondary displays).

9.1.2 Basic Operation

Electrical conduction

Electrical conduction requires mobile charge carriers. In crystalline materials, a quasi-continuum of electron states, called an energy band, extending throughout the crystal lattice facilitates mobility.

When the entire range of states in a band is occupied (full band), there are no available electron states to accommodate charge movement. A few electrons excited from a full valence band into an empty conduction band become mobile; they leave behind empty states in the valence band interpreted as mobile positive charges (holes). Semiconductors have an energy gap between a full valence band and an empty conduction band such that, at room temperature, some electrons are thermally excited across the energy gap. In an intrinsic (undoped) semiconductor, equal numbers of electrons and holes contribute to electrical conductivity. Very small concentrations of impurity atoms (dopants) provide isolated states within the energy gap, controlling the conductivity type. Dopant states close to the conduction band and donating electrons establish dominant electron or n-type conductivity. Conversely, dopant states close to the valance band and accepting electrons generate holes giving dominant hole or p-type conductivity. The interplay of n-type and p-type conductivity, i.e. p–n junctions, determines device function.

The density of states as a function of energy (effective mass) and scattering mechanisms (relaxation time) determine charge carrier mobility. High-mobility semiconductors such as GaAs are therefore sensitive to crystal defect structures and surface conditions. Low mobility is associated with narrow energy bands, high effective mass, complex lattice interactions (polarons), trapping, and hopping conductivity. Crystal defects and surface states are important aspects of inorganic semiconductors, limiting device performance and lifetime. Sensitivity to structural defects is a severe handicap in fabricating large-area inorganic structures.

In disordered semiconductors such as amorphous silicon, the energy band edges become blurred, and defect-state density increases, making doping ineffective in determining carrier type. Alloying hydrogen with amorphous silicon (α-Si) reduces the defect density by terminating broken bonds, enabling p-type and n-type doping. Transistors formed from α-Si sacrifice carrier mobility, but absence of crystallinity enables the fabrication of large-area active matrix addressing circuitry.

Molecules applicable to OLEDs form amorphous solids, retaining detailed structure within the molecule, and some intermolecular short-range order. Delocalized molecular orbitals accommodate the charge carriers, providing hopping conduction between molecules.[6] Appropriate molecular orbitals are associated with sp^2 and π conjugation (double bonds separated by single bond), requiring all organic conductors to be highly conjugated materials. For organics, the lowest unoccupied molecular orbitals (LUMO) act as the "conduction band," and the highest occupied molecular orbitals (HOMO) provide the "valence band" having hole-type conductivity. The ground state energy of the π-bond (S_0) forms the HOMO, with the excited state π^* providing the LUMO at band gap ~2.5 eV, as illustrated in Figure 9.1.

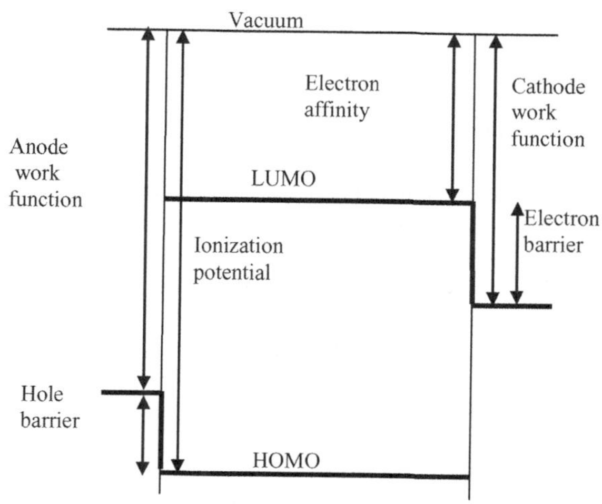

Figure 9.1 OLED interface energy levels referred to vacuum level

Molecular disorder involves some molecular bond distortion, but the chemical bonds remain intact, limiting the generation of defect states.

The electrical conductivity of organics, expressed in semiconductor form, is $ne\mu$, where n is the carrier concentration, e electronic charge, and μ the mobility. This gives the very low mobility ($\mu < 10^{-4}$cm^2/Vs) expected of narrow bands and hopping conduction. Mobility declines in general with loss of order in amorphous materials, associated with carrier localization effects and hopping. Conduction is non-ohmic and usually space charge and trap limited, implying mobility increase with electric field and temperature. Thin film organic materials are often found to be either hole transporting layers (HTL) or electron transporting layers (ETL), where dopants have little effect, implying the influence of structural details. Intrinsic organics have been developed that can be doped to give n-type or p-type conduction.

Carrier injection

Hole or electron injection describes conduction across an interface. The energy levels at the interface (electromotive properties) determine the injection type. The work function of a material is the minimum energy required to remove an electron from the material in a vacuum environment. Electron affinity is the converse term to describe the empty conduction band relative to vacuum level.

Figure 9.1 is an energy level diagram referred to vacuum level and shows how the material work function influences the interface energetics. Hole energy increases in the downward direction, so that a downward step represents a barrier at the hole injection contact.

In a simple organic diode structure, the electron-injecting material identifies the cathode and favors low work function materials such as the alkali metals (e.g. lithium). The hole-injecting material identifies the anode and favors higher work functions (e.g. ITO). For barriers larger than ~0.4 eV, the current is largely determined by the efficiency of carrier injection at the electrodes, and is known as the injection-limited regime.

Intimate contact between two materials creates a molecular diffuse interface state differing from the bulk properties on either side. Compared to inorganic materials, continuity in structure is not critical in organics because the defects are not generally associated with broken or distorted bonds.[7] Consequently, describing the interface in terms of bulk properties of the components is a simple approximation giving reasonable results. The injection characteristics of the electrode can be tailored to the organic by choosing an electrode with appropriate work function. Surface preparation or introduction of an interface material can dramatically change the injection characteristic. Bulk conductivity and injection characteristics determine the current for a given voltage. When the injection rate exceeds the bulk transport rate, charge accumulation gives rise to space charge limited current (SCLC). At higher voltage, the trapping levels fill, and the effective mobility increases.

Electron–hole recombination

Light emission from electron–hole recombination is fundamental to organic displays. The π-bonds determine the energy levels of the LUMO and the HOMO, and therefore an upper limit to the energy available for light emission in electron–hole recombination. When electrons and holes flow into the same region (recombination zone), electron–hole capture takes place giving rise to an electron–hole exciton (excited state). The recombination zone width is determined by the diffusion length of carriers and excitons ~10 nm. The heterojunction structure shown in Figure 9.2 provides barriers to electron and hole transport beyond the junction, confining excitons to the interface region. The binding energy involved in the capture process lowers the combined energy of the electron/hole pair. The exciton binding energy is higher in organics, due to lower dielectric constant, making the exciton stable at room temperature. The greater stability of the exciton results in a lower absorption of band-gap light, promoting higher device efficiency.

In the ground state (S_0) π-bond the two electrons have spin states +½ and −½ to satisfy the exclusion principle, and form an antisymmetric wavefunction. Hole–electron excitons with such antisymmetric

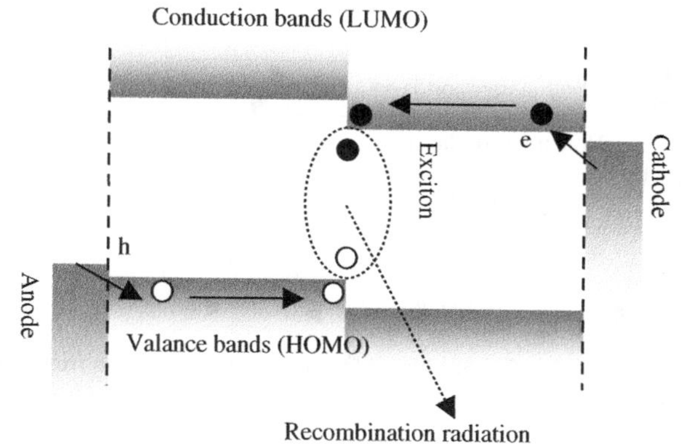

Figure 9.2 Heterojunction OLED showing electron and hole injection, transport, exciton formation, and recombination radiation

spin states have a high probability of radiative decay. Conversely, excitons with a symmetric spin state are long lived and have a low probability of radiative decay. The symmetric state is triply degenerate (triplet), implying only 25% of the excitons are in the singlet radiative state if electron–hole capture is spin independent. If only 25% of the electrons and holes can achieve singlet states, the radiation efficiency is similarly limited to 25%. Figure 9.3 illustrates the energy levels of singlet and triplet excitons. The 25% limit is disputed for fluorescent polymers, where recent experiments support substantially higher limits.[8]

Fluorescent dyes doping the recombination zone absorb singlet energy, controlling the emission color and improving radiation efficiency by lowering self-quenching losses. Doping with phosphorescent dyes (PHOLED) improves the radiation efficiency by providing a radiant path for triplet states. The host material can transfer its triplet exciton energy to the guest dye with subsequent radiation.[9] The dye chosen for high quantum efficiency also controls the color. Doping with a phosphorescent organo-metallic complex such as tri(2-phenylpyridine)iridium, introduces spin-orbit coupling facilitating radiative decay of both singlet and triplet excitons. Any separation of transport and recombination properties allows greater flexibility in material engineering.[7]

Organic light emitting diodes

OLEDs are thin-film solid-state devices that emit light from the plane of the thin film, as illustrated by the primitive device shown in Figure 9.4. Below a certain threshold, no significant current flows in

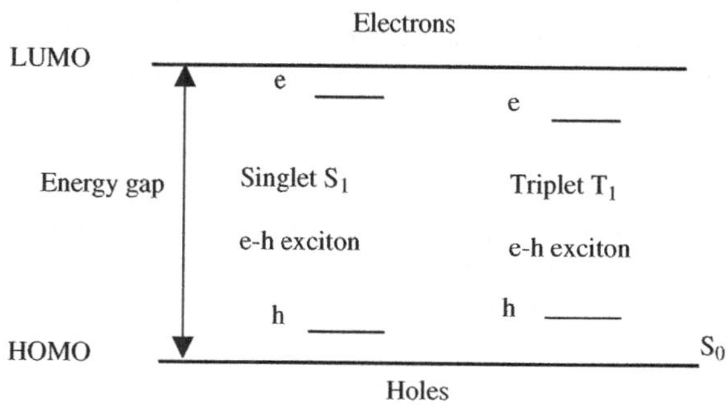

Figure 9.3 Energy levels of singlet and triplet excitons

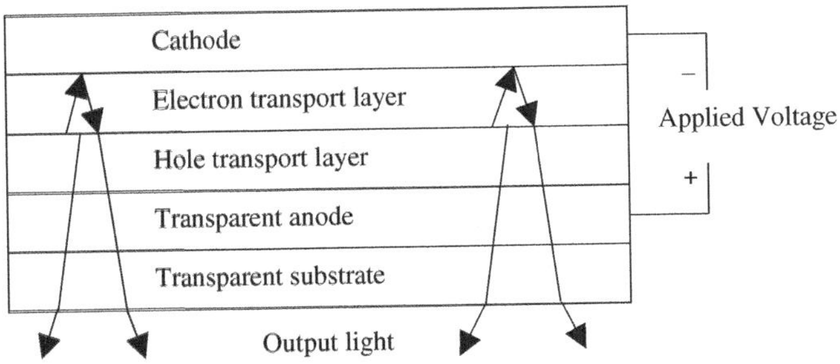

Figure 9.4 Primitive bottom-emitting forward-biased OLED with light emission from interface of electron/hole transport layer region (and reflected from cathode)

forward bias and no light is emitted. Above the threshold, sufficient flow of electrons and holes generates light. The light level rises with increase in current. A thin cathode transmits substantial light, otherwise upward propagating light is reflected or absorbed by the cathode. In reverse bias, the current is low and emitted light absent.

In a practical device, thin layers <100 nm mitigate SCLC reducing drive voltage, and interface layers are included to improve the carrier injection and radiant recombination. The anode and cathode layers are critical in achieving performance and lifetime. Attenuation of the electrical component of optical field close to the electrode inhibits emission; resonant properties of the optical cavity formed by the structure should be taken into account. Outcoupling efficiency accounts for Lambertian emission limited by total internal reflection and absorption losses. Encapsulation is essential in preventing ingress of atmospheric contaminants such as moisture and oxygen that react with OLED materials, particularly the alkali cathode, thereby reducing shelf-lifetime. Detailed discussion of evolving materials and devices follow.

9.2 Organic Emissive Materials

9.2.1 Classification

Classification by chemistry. Organic emissive materials are classified into small molecules (SMOLED) and polymers (P-OLED, POLED, PLED, LEP etc.). The term OLED is sometimes loosely used to cover all of the above, but more correctly describes the pioneering small-molecule variety and tends to retain that flavor. In order to avoid ambiguity we shall use SMOLED and P-OLED as appropriate. SMOLED materials and devices have a slight lead in development over P-OLED, due to a longer record of research and development. Polymers exploit background chemistry adept at structural modification by side chain or backbone components to satisfy multiple requirements.[i] SMOLED materials, on the other hand, offer a wider range of chemical complexity and are easier to purify.

Classification by method of deposition. The most important distinction in practice between molecular and polymeric materials is the method of deposition and patterning. Molecular materials generally require expensive vacuum deposition, while light emitting polymers can be solution processed. That is to say they can be dissolved in solution, deposited whilst in solution and once the solvent has evaporated the solid polymer remains. Deposition techniques include spin-coating and inkjet-printing. In conventional SMOLED displays, patterning of the organic layers is achieved by shadow-masking; in P-OLED

[i]Later in the chapter, for the sake of brevity, points may be illustrated using only one or the other, SMOLED or P-OLED, as examples.

Table 9.2 Classification of organic emissive materials

	Molecular	Polymer
Fluorescent	Organic light emitting diode (OLED) or small-molecule OLED (SMOLED)	Light emitting polymer (LEP, PLED or P-OLED)
Phosphorescent	Phosphorescent OLED (PHOLED)	Dendrimer

displays it is achieved by inkjet-printing. Neither technique is capable of patterning at the small dimensions typical of microdisplays (around 15 μm pitch for a full color pixel).

Classification by mechanism of light emission. Fluorescence refers to light emitted directly in very fast response to a stimulus, essentially ceasing when the stimulus terminates. Lifetimes of order nanoseconds characterize fluorescent states. In comparison, phosphorescence may be slow to respond to stimulus, and continue to emit light for a substantial period on removal of the stimulus. The terms applied to OLED emission identify light emitted from the decay of a singlet exciton as fluorescent, distinguished from the interaction of a triplet exciton with a phosphorescent dye to emit phosphorescent light. The phosphorescent time delay in OLEDs is smaller than time intervals of interest to display work, relegating the optical distinction to an emitting material marker. The classification appears in Table 9.2.

Addition of a phosphorescent dye to the emission region of an OLED can raise the internal quantum efficiency to approach 100%. Both molecular[10] and polymeric[11] phosphorescent materials demonstrate increased brightness and show great long-term promise. However, currently most commercially available displays do not use phosphorescence, and at the time of writing there are no reports of microdisplays utilizing phosphorescent materials.

9.2.2 Small-Molecule Materials

Organic electroluminescence has been known for over forty years[12] with early devices using anthracene crystals as the emitting layer. Serious exploitation of the phenomenon in practical devices really began in earnest after Tang and Van Slyke[13] introduced into the device structure a hole transport layer that significantly reduced the drive voltage and increased the quantum efficiency.

Existing work on the photoconductive properties of organics contributed to the elucidation of emissive properties and development of improved materials and diode structures. The prospect of commercial OLED products engaged the research laboratories of many large chemical companies and appealed to academic workers alike. A continuing flow of new materials and fabrication methods feed the new technology. Desirable charge transport and emissive properties in a material must be accompanied by electrochemical stability, and morphological stability conferred by high glass-transition temperature.

Small molecules dissolved in a solution that coats the substrate, followed by thermal or vacuum evaporation of the solvent, generally form polycrystalline films. Poor homogeneity and grain boundary resistivity make polycrystallinity undesirable in OLEDs. Preparing successive layers without dissolving previous layers compounds the difficulty of casting films from solution. Combining the electro-optic properties of small molecules with solution processing proved a severe challenge to molecular chemistry. Vacuum deposition requires the molecular thermal degradation temperature be substantially higher than the favored sublimation temperature. Chemistry found it much easier to cope with vacuum deposition; consequently, SMOLED materials have carried the burden of vacuum processing from the beginning. However, vacuum deposition enables the preparation of a complex sequence of films to optimize device performance.

Stability of the SMOLED amorphous film favors a high glass transition temperature >100 °C., however, effective sublimation requires heating beyond the glass transition temperature. Figure 9.5 shows

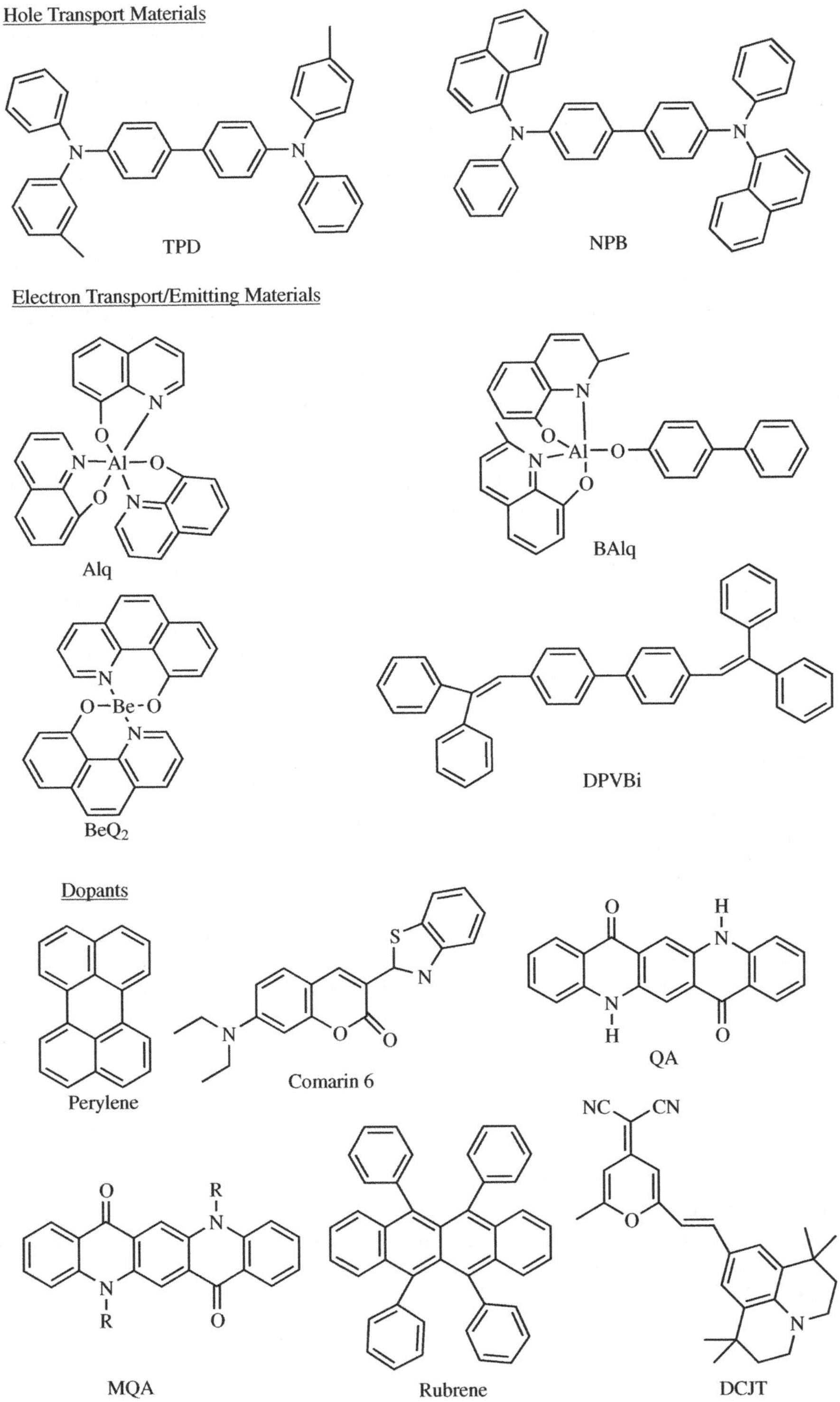

Figure 9.5 Some early SMOLED materials. Reprinted courtesy of Society for Information Display

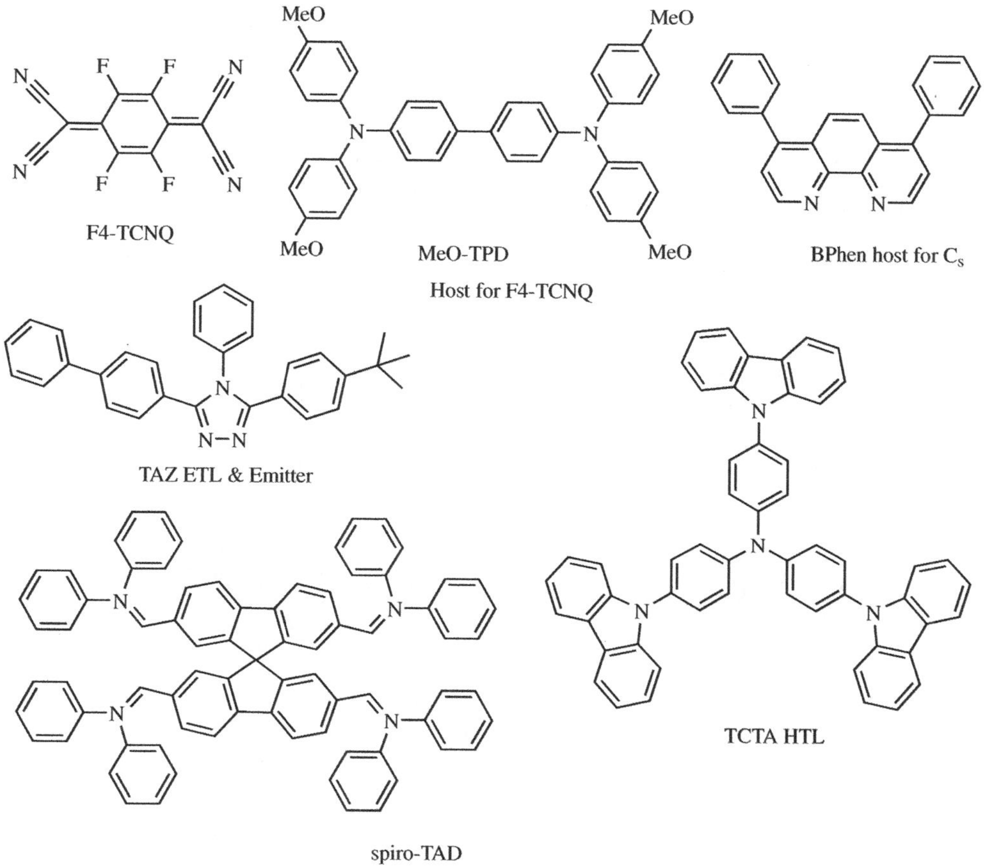

Figure 9.6 SMOLED materials used in PIN OLED. Reprinted courtesy of Society for Information Display

a selection of early SMOLED materials, including the original Tang and Van Slyke diode components; Alq_3 and TPD.[14] The dopants are fluorescent molecules that enhance emission and control color. Some current materials used in a PIN OLED [15] appear in Figure 9.6.

9.2.3 Polymer Materials

Burroughes *et al.*[16] first reported the emission of visible light from a conjugated polymer in 1990. They used the term "light emitting polymer" (LEP); other terms used since include "polymer organic light emitting diode" (PLED, POLED or P-OLED).[ii] They and others have developed conjugated polymers based upon poly(phenylene vinylene) (PPV) and its derivatives while a further family of materials has been developed based upon poly(fluorene) (PF) and its derivatives. Some examples of polymeric materials are shown in Figure 9.7.

Polymer chemistry offers the advantage of incorporating a number of separate molecular features into a single macromolecule, e.g. light emitting factors and solubility factors. A key feature of polymer

[ii]Today the term LEP is commonly used for materials and the other terms for devices or displays.

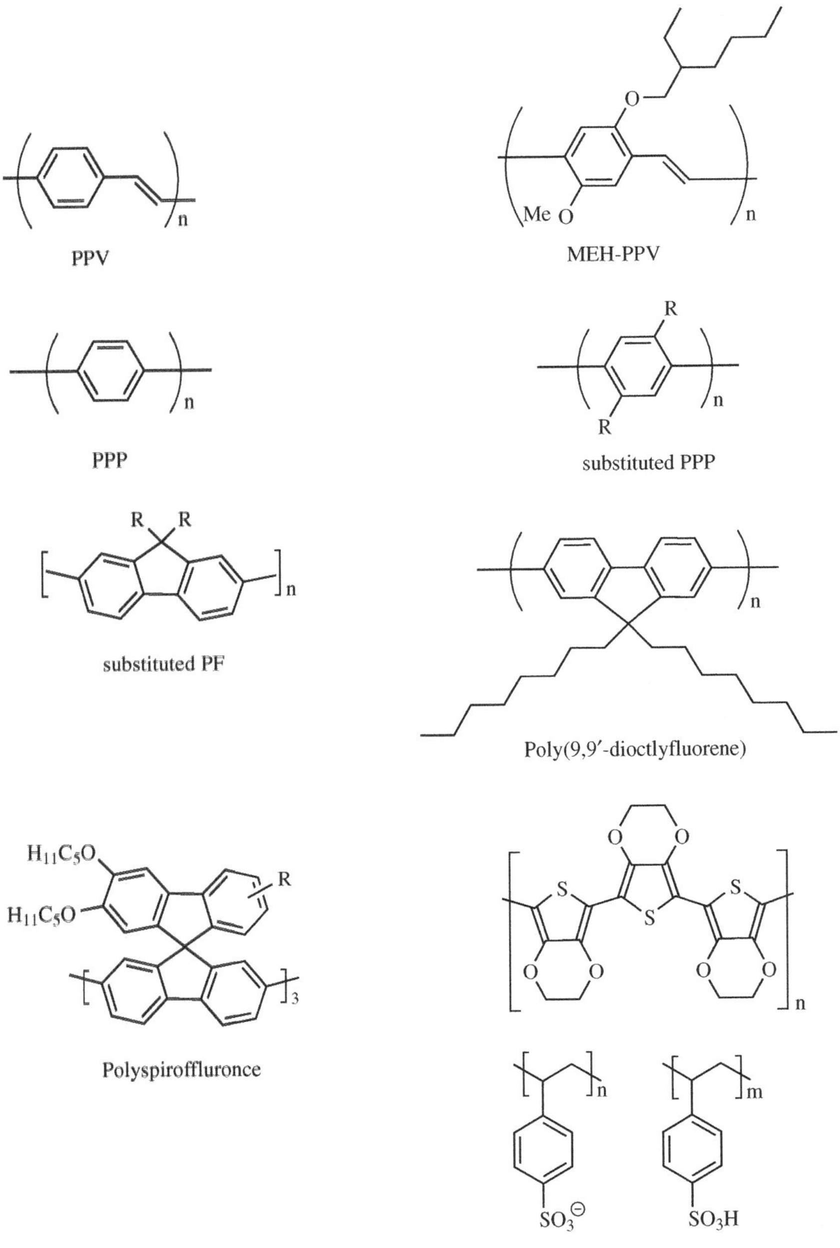

Figure 9.7 Schematic representations of example polymeric materials

OLED is that it can be solution-processed. Thus, manufacturing is possible by many techniques including spin coating, spray coating and inkjet printing.[17] The last of these, in particular, offers the highest long-term likelihood of low-cost roll-to-roll processed large and/or flexible displays. However, the resolution available from inkjet printing in 2006 falls far short of that required for microdisplays.

In comparison with SMOLED, the design of polymeric materials also takes into account the defects in molecular structure and the solid-state conformation of the polymer chain. Molecular weight and purity vary with fabrication methods, giving rise to some inconsistency of reported data. Polymer structure that resists crystallization and has a high glass transition temperature is desirable. Cross-linking by thermal or UV exposure provides an advantage.

9.2.4 Phosphorescent OLED

Phosphorescent emission from a SMOLED (PHOLED) device follows from the addition of a phosphorescent dopant. Some PHOLED materials are shown in Figure 9.8. The high efficiency potential of PHOLED provides driving force to develop appropriate materials with adequate lifetime. The host materials must generate excitons at desirable energy levels, allowing the dopant to transfer exciton energy into radiation of the desired color. Phosphorescent radiation below the host exciton energy level prevents host quenching of the radiation.

The transition metal complexes such as $Ir(ppy)_3$ have strong spin-orbit coupling, providing an intersystem crossing path allowing radiation from both singlet and triplet excitons. Internal quantum efficiencies approaching 100% are achieved. Polymer hosts seek to reconcile high exciton energy with low barrier for charge injection. The copolymer host shown in Figure 9.8 includes an oxadiazole component to enhance electron transport, matching hole transport for efficient recombination.[18]

9.2.5 Dendrimers

Long-chain linear-polymers forming the original P-OLEDs adopt a conformation determined by local interactions in the amorphous solid state. The necessary components for optimum performance incorporated in the polymer fail to harmonize due to uncontrolled structural details. The design of dendrimer polymers optimizes intra- and inter-polymer structure.

The name dendrimer comes from the three-dimensional tree-like branching structure shown in Figure 9.9. The first light emitting dendrimers were reported by the group of Ifor Samuel and Paul Burn in 1999 by Halim *et al.*[19] and in subsequent publications including Lo *et al.*[20] (2002) and Lo *et al.*[21] (2005). Charge transport structure surrounds the light-emitting core containing a phosphorescent dye, isolating it from the cores of adjacent dendrimers. The dendrimer boundary terminates in chemical groups providing solubility. Terminal groups have little influence on light emission from the core, allowing dendrimers to be optimized for solution processing. By combining all of the components within a dendrimer macromolecule, phase separation is limited, preventing aggregation of emitters associated with exciton quenching. In summary, dendrimers optimize the desirable combination of phosphorescent light emission and solution processing.[22,23]

9.2.6 Example Performance Data

Reported state-of-the-art performance data, in late 2005 through early 2006, of various materials and configurations is summarized in Tables 9.3 to 9.6. Note that these performance data are those released by material or device companies and have probably been measured, in each case, in some optimal configuration. The configurations will be different from one to the other making it inappropriate to consider direct cross comparisons. Furthermore, it has been noted already that development in this field is currently very rapid, as illustrated for example in the evolution of blue lifetime shown in Figure 9.10, further confounding attempts at cross comparisons. It is suggested that the reader consult

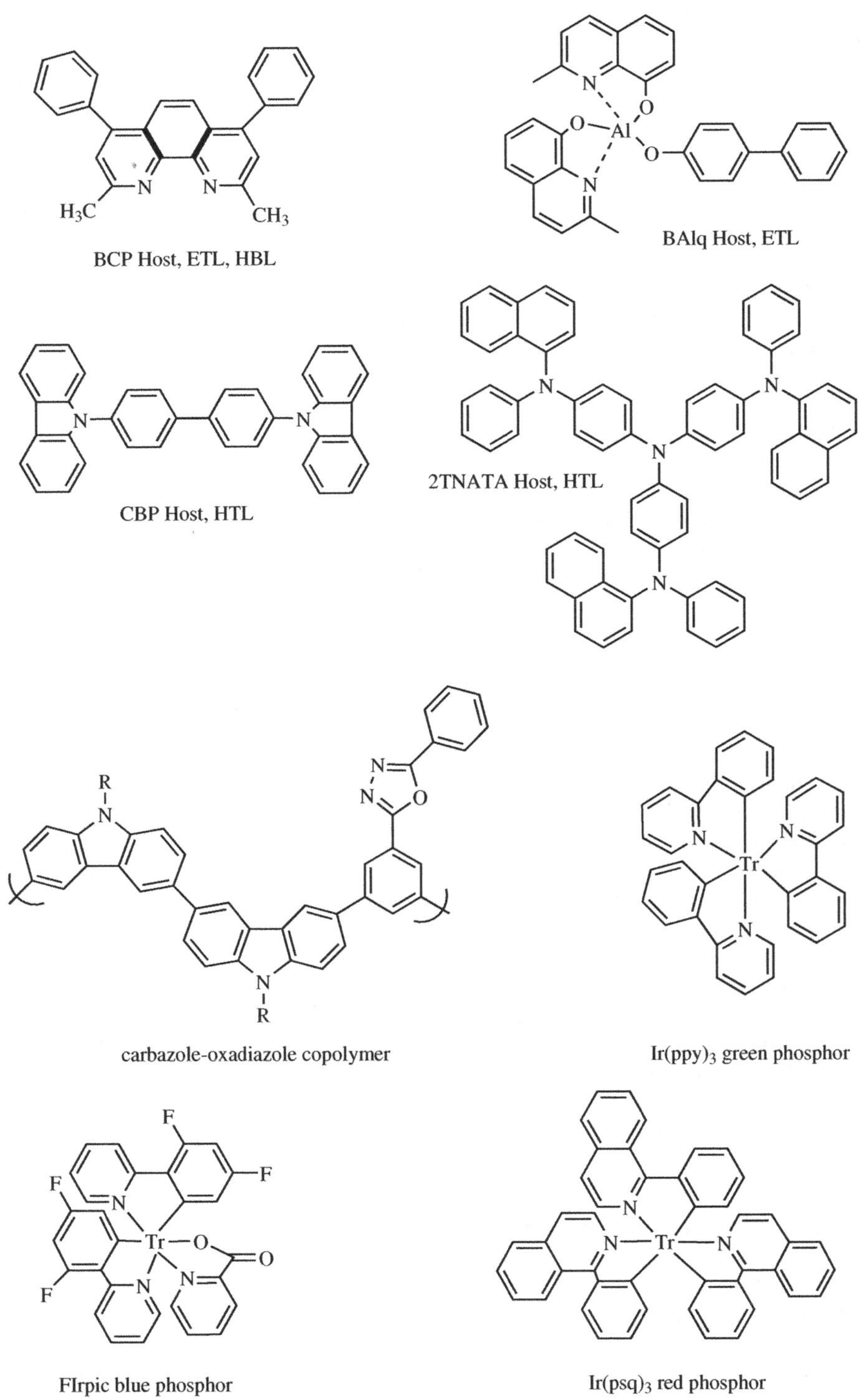

Figure 9.8 Schematic representations of some PHOLED materials

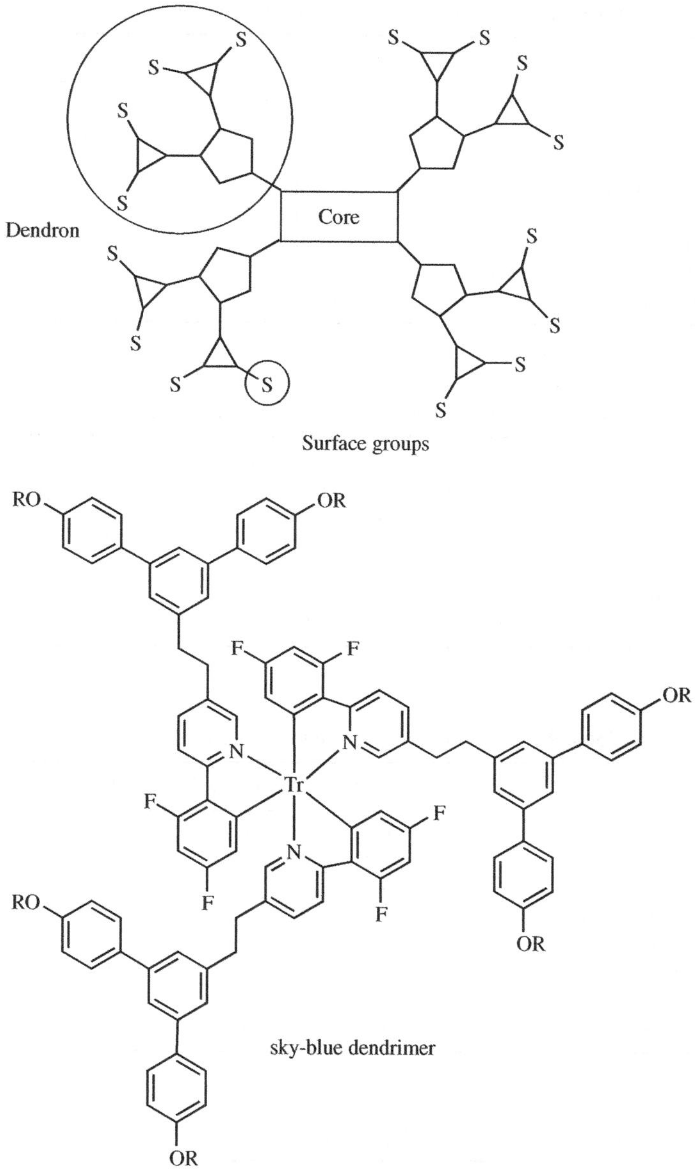

Figure 9.9 Schematic representation of generic dendrimer and example of sky-blue emitting dendrimer. Generic dendrimer reprinted courtesy of Professor Ifor Samuel of University of St Andrews, UK; sky-blue dendrimer. Reprinted courtesy of Society for Information Display

an appropriate source or sources for up-to-date parameter values and that great care is taken in interpreting such data or extrapolating it into a particular situation. For example, the particular constraints of microdisplays, such as the CMOS substrate, top-emitting configuration, etc. mean that such figures may possibly not be achievable in a microdisplay.

Table 9.3 Properties of typical SMOLED materials in 2005[24]

	Red	Green	Blue	White
CIE X	0.66	0.27	0.14	0.32
CIE Y	0.34	0.65	0.14	0.34
Voltage (V)	5–7	5–7	5–7	5–7
Max efficiency (cd/A)	>7	>17	>4	>13
Lifetime (k hours) at 1000 cd/m^2	>25	>15	>10	>50
dV to half-life (V) from 1000 cd/m^2	<1	<1	<1	<1

Table 9.4 Properties of typical PHOLED materials in 2005[25]

	Red RD15	Green GD33	New Blue	White
CIE X	0.67	0.31	0.16	0.37
CIE Y	0.33	0.64	0.37	0.38
Efficiency (cd/A) (@ 100 cd/m^2)	12	40	22	32
Lifetime (k hours) (@ 100 cd/m^2)	100 (@ 500 cd/m^2)	20 (@ 1000 cd/m^2)	15 (@ 200 cd/m^2)	–

Table 9.5 Properties of typical P-OLED materials in 2005[24]

	Red	Green	Blue	White
CIE X	0.68	0.43	0.16	0.28
CIE Y	0.32	0.55	0.20	0.29
Voltage (V) (@ 100 cd/m^2)	3.8	3.4	5.8	5.7
Efficiency (cd/A) (@ 100 cd/m^2)	2.2	7.7	5.5	4.0
Eff'y (lm/W) (@ 100 cd/m^2)	1.8	7.0	3.0	2.1
Lifetime (k hours) (@ 100 cd/m^2)	310	255	100	65
Power factor	1.7	1.5	2	1.8
dV to half-life (V)	1.3	1.1	1.0	1.2

Table 9.6 Properties of typical PIN OLED materials in 2005[26]

	Red top (PHOS)	Green top (PHOS)	Blue top (PHOS)	White – (PHOS)
CIE X	0.68	0.31	0.14	–
CIE Y	0.32	0.66	0.23	–
Voltage (V) (@ 1000 cd/m^2)	3.0	2.5	3.4	–
Efficiency (cd/A) (@ 1000 cd/m^2)	15.5	95	9	–
Eff'y (lm/W) (@ 1000 cd/m^2)	11	90	8.4	25
Lifetime (k hours) (@ 500 cd/m^2)	100	50	6	20

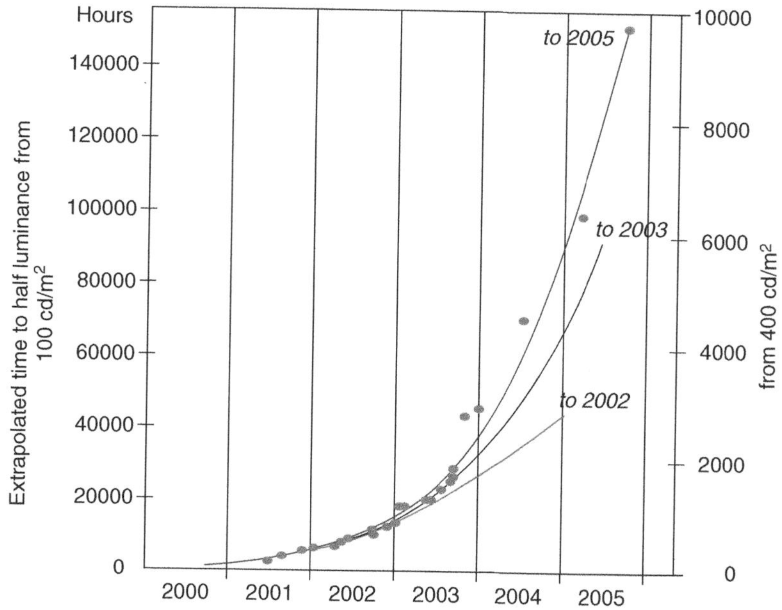

Figure 9.10 Improvement in parameter blue lifetime with time. Reprinted courtesy of Cambridge Display Technology

9.3 Device Construction and Manufacture

There are many and subtle variants on the possible structure of both molecular and polymer devices. Simple versions of each are now presented. The basic configuration for both SMOLED and P-OLED is a multi-layer thin film sandwich. Layer thicknesses tend to be of the order of a few tens of nanometers. The presence of organic layers in addition to the emitting layer itself assists with lowering the drive voltage by boosting charge injection at the interface between electrode and organics, moving the emission sites away from the electrode thus boosting efficiency (as emission is inhibited close to an electrode that attenuates the optical electric field) and balancing the proportion of holes and electrons entering the OLED layer (a preponderance of one over the other can lead to charge build up at the interface with consequential effects such as space charge limited current).

Table 9.7 offers a summary comparison of SMOLED and P-OLED materials and devices.

Table 9.7 General comparison between mainstream SMOLED and P-OLED

Conventional SMOLED	Polymer OLED
Shadow mask patterning – more precise	Inkjet printing – very large substrates, more straightforward
High-purity materials easier to prepare	Lower operating voltage
Vacuum deposition less likely to allow water/oxygen contamination	Lower fabrication cost
	Manufacturing is more compatible with flexible substrates
	Better high-temperature stability

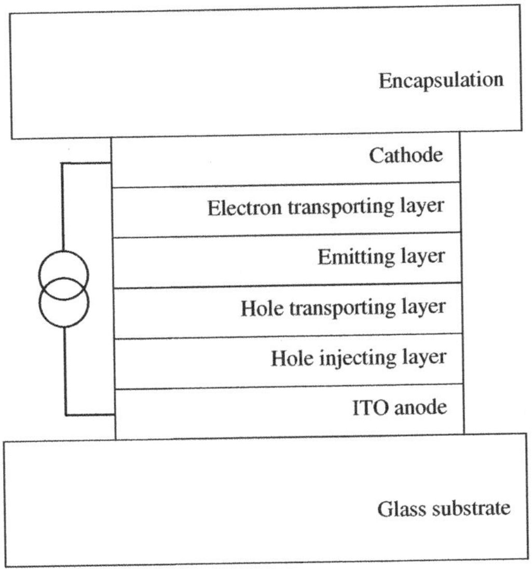

Figure 9.11 Simplified cross-section through conventional SMOLED device

The anode and cathode must be considered with great care: the materials, the method and conditions of deposition, the layer thicknesses etc. all play a part in determining the final efficiency of the device. A general comparison between mainstream SMOLED and P-OLED materials is given in Table 9.7.

9.3.1 SMOLED

Figure 9.11 shows a simplified cross-section through a conventional large-area, bottom-emitting SMOLED device. The cathode and organic materials quickly degrade in the presence of oxygen and moisture, making it essential to protect against atmosphere. Encapsulation layers deposited or bonded to the substrate provide device level protection. Packaging in a hermetically sealed can containing getter material, to absorb contaminants, reinforces protection. An effective seal against atmospheric contamination plays an important role in OLED lifetime.

The metallic cathode provides high reflectivity in a bottom-emitting device, along with low work function electronic injection. Typically, both functions are served by aluminum with a thin coat of lithium fluoride (LiF:Al) or magnesium stabilized with silver ($Mg_{0.9}Ag_{0.1}$). In a top-emitting device, the metallic cathode must be thin enough to transmit substantial light.

Efficient charge injection is required of the electron-transporting layer in contact with the cathode. A combination of host material and n-type doping provides high conductivity and charge injection. Matching the electronic energy level of the emitting layer reduces charge accumulation and voltage drop.

The light emitting layer controls light color. Energy levels should match the electron and hole boundary layers. The hole-transport layer aids in the matching of interface energy levels. Boundary layers should confine the electrons, holes, and excitons to the emitting layer.

High conductivity of the hole-injection layer enhances charge injection and is energy matched to the anode. Transparency and durability of indium tin oxide (ITO) makes it a favorite anode material. A glass substrate provides transparency, durability and strength, with established ITO deposition methods. A wide choice of glass ensures chemical inertness.

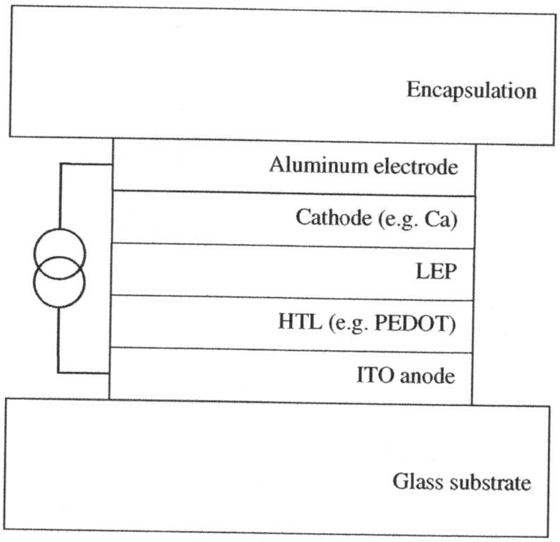

Figure 9.12 Simplified cross-section through conventional polymer OLED device

9.3.2 PHOLED

The structure of a typical PHOLED device is similar to that of a SMOLED. However, due to the fact that host emitters, such as CBP, suitable for PHOLED devices are typically hole conductors, the PHOLED structure adds a hole blocking layer (HBL) between the ETL and the emitting layer.

9.3.3 P-OLED

Figure 9.12 shows a simplified cross-section through a conventional large-area passive matrix P-OLED device. The structure has much in common with the SMOLED device. Protection against atmosphere is equally important. The cathode favors aluminum with a thin calcium coating, matching the polymeric emitter layer energy level. A layer of hole-conducting PEDOT matches the ITO anode.

The aluminum electrode reflects emitted light, which exits the device through the underlying glass substrate. This device structure is known as "down-emitting" or "bottom-emitting." Such displays are often packaged upside-down in a metal can with a glass lid so that light is emitted from the top of the packaged device.

In recent times P-OLED structures have been shown to benefit from the presence of an additional interlayer[27] between the HTL and the P-OLED. The interlayer improves P-OLED efficiency and lifetime. The efficiency is improved in two ways: (a) the interlayer acts as an exciton block thus reducing exciton quenching by the PEDOT, and (b) it acts as an electron block.

9.3.4 PIN OLED

The PIN OLED incorporates an intrinsic (ambipolar) emitting layer bounded by a p-type layer and an n-type layer as shown in Figure 9.13. The bipolar conduction of the emitting layer provides an increase in volume of the emitting region, compared with interface emission from a PN-OLED. Moreover, isolation of the emitting region extends the range of compatible substrates. Increase in drive voltage is a disadvantage of additional layers, but overcome by enhanced conductivity in doped layers.

PIN-OLED
Cathode
n-ETL
EML emitting
p-HTL
Anode

2-PIN-OLED
Cathode
n-ETL
EML(n) emitting
EML(p) emitting
p-HTL
Anode

Cathode	Aluminum	100 nm
n-doped electron transport layer	BPhen + Cs doped	50 nm
electron interlayer	BPhen	10 nm
Emission layer (n)	TAZ + 8% wt $Ir(ppy)_3$	20 nm
Emission layer (p)	TCTA + 8% wt $Ir(ppy)_3$	20 nm
Hole interlayer	Spiro-TAD	10 nm
p-doped hole transport layer	MeO-TPD + 2 mol% F_4-TCNQ	100 nm
Anode	ITO	

Green emitting 2-PIN-OLED structure

Figure 9.13 PIN-OLED and 2-PIN OLED structure. EML is intrinsic emitting layer, EML(n) is predominantly electron transport emitter and EML(p) is predominantly hole transport emitter

Further developments incorporate two distinct emission layers forming a double-emission region. The emission region on the hole side is predominantly a hole conductor, while the complementary emission layer is predominantly electron conducting. The additional degree of freedom in layer design eliminates charge accumulation at the emitters' outer interfaces, thereby maintaining efficiency at high luminance.

Figure 9.13 includes the structure of a development 2-PIN-OLED employing a double-emission layer (D-EML).[15,28] The emission layers include the phosphorescent dye $Ir(ppy)_3$, emitting green light at 100 lm/W. Essentially the same structure provides phosphorescent red or fluorescent blue light with appropriate emitter layers. High conductivity of the doped layers attenuates ohmic losses and enhances charge injection, giving operating voltage <3 V. Further development replaces the Cs n-dopant with an organic n-dopant NDN-1, improving the manufacturing process and device lifetime.[29]

9.3.5 Encapsulation and Packaging

Encapsulation is extremely important in maintaining the shelf life and the performance of all OLED displays. The organic light emitting materials are susceptible to water and air causing degradation and a consequent reduction in performance. Furthermore, the reactive electrode metals (such as calcium) must also be protected. Conventional bottom-emitting OLED displays often have intermediate deposited encapsulation layers and may also be packaged in hermetically sealed packages containing desiccants. OLED microdisplays may have packaging and encapsulation constraints in comparison with conventional OLED displays.

9.3.6 Display Efficiency

Passive matrix OLED displays typically consist of a glass substrate on which lies a stripe-patterned transparent conducting anode usually of ITO. Following deposition and patterning of the OEL material and associated organic layers, a reflective electrode such as aluminum is deposited and patterned in orthogonal stripes to form the cathode. Thus light is emitted through the substrate – the so-called bottom-emitting configuration. As with most PM displays, in this configuration, the stripes in both the anode and the cathode are separated by narrow gaps. A pixel is formed at the intersection of any pair of anode and cathode stripes.

Modeling reported by Baynes and Smith[30] has shown that, at high pixel counts, components of power consumption due to the parasitic resistance and capacitance of the stripe electrodes come to dominate the overall power consumption of the display. This provides a strong incentive for the use of active matrix addressing at high pixel counts.

In active matrix displays, the active circuits and associated interconnects are opaque and thus, in the down-emitting configuration, limit the proportion of the pixel that can emit light – the emissive aperture ratio (EAR) or emissive fill-factor (F_E).

The role of pixel aperture ratio or fill factor in determining the optical efficiency in modulating displays is covered in detail in Chapter 2. In emissive microdisplays the emissive aperture ratio of the pixel is also an important parameter. The average luminance of a pixelated display, $L_{display}$, is given by

$$L_{display} = F_E L_{pad} \tag{9.1}$$

where L_{pad} is the luminance of the emissive or active part of the pixel (the pixel pad), and F_E is the emissive aperture ratio.

For example, if the specified luminance of a pixelated display is 100 cd/m^2 and $F_E = 50\%$ then the required luminance of the pixel pad is 200 cd/m^2. A higher F_E means either that a given display luminance can be achieved at a lower drive signal (voltage or current) leading to lower power and lower stress on the material (i.e. longer lifetime) or that, for a given drive signal and material stress, a brighter display can be achieved. F_E is also one of the conversion factors required in order to relate the performance of test cells and development structures to the performance of pixelated display products.

Note that in OLED microdisplays, due to the very thin layer structure of an OLED device, there is no loss of periphery equivalent to that caused by the fringing-field effect in LCOS.

The effective quantum efficiency of an OLED device, η_{eff}, is conventionally given by

$$\eta_{eff} = \gamma q r_{st} \eta_{ext} \tag{9.2}$$

where

- γ is the ratio of excitons produced to charge carriers injected
- q is the photoluminescent efficiency (the efficiency of radiative decay of singlet excitons)
- r_{st} is the fraction of excitons that can radiatively decay (theoretically up to 25% for singlets and up to 100% for phosphorescent emission)
- η_{ext} is the optical outcoupling efficiency – consideration of the multi-layer thin film structure of the OLED device is important in determining the amount of generated light that escapes from the display.

Then the equivalent efficiency of a pixelated OLED display is reduced in proportion to the emissive aperture ratio such that

$$\eta_{eff} = F_E \gamma q r_{st} \eta_{ext}. \tag{9.3}$$

Equation (9.3) explicitly shows the different parameters that must be controlled or increased in order to make a more efficient OLED display. γ and q are material parameters. r_{st} is dependent upon the choice of fluorescent or phosphorescent OLED. η_{ext} is dependent upon good design of the thin film structure through which light is transmitted including transparent encapsulation layers. In a top-emitting microdisplay, F_E is determined by the pixel floor plan and, in particular, the size of the pixel electrode. Other factors not shown explicitly include the efficiency of charge injection and charge balance. Any electron or hole that traverses the device and fails to form an exciton in the emitting region contributes to the losses. An unequal transport of electrons and holes to the emitting region implies loss.

Total internal reflection limits the outcoupling efficiency, η_{ext}. Snell's law ($n_e \sin\theta_i = \sin\theta_r$) gives the relation between the incident angle (θ_i) and refracted angle (θ_r) for an air interface with material of refractive index n_e. Total internal reflection occurs for $\sin\theta_i > 1/n_e$ beyond the critical angle $\theta_c = \sin^{-1}(1/n_e)$. Assuming just total internal reflection and no transmission loss, outcoupling efficiency can be written for Lambertian emission ($\cos\theta$ dependence)

$$\eta_{ext} = \frac{\int_0^{\theta_c} \sin\theta\cos\theta \, d\theta}{\int_0^{\pi/2} \sin\theta\cos\theta \, d\theta} = 1/n_c^2. \tag{9.4}$$

Taking a typical value of $n_e = 1.7$ for the emitting layer of an OLED (intervening layers before the air interface do not influence the result), implies $\theta_i < 36$ degrees, giving $\eta_{ext} < 35\%$. Alternatively, assuming emission is independent of angle

$$\eta_{ext} < 1 - \left(1 - 1/n_e^2\right)^{-0.5}. \tag{9.5}$$

Substituting $n_e = 1.7$ into (9.5) gives $\eta_{ext} < 19\%$ which represents a severe loss. Typical observed values of η_{ext} are around 20%. Careful use of surface structures such as a microlens arrays can substantially enhance the value of η_{ext}.

In a microdisplay system the image is magnified by optics with *f*-number more than 1, spanning an external angular range less than ±30 degrees; this corresponds to $\theta_i < 17$ degrees. Therefore, total internal reflection is not a significant factor for a near-eye system, but surface texturing may improve the source luminance. Luminance is conserved in magnification (apart from transmission losses) making source luminance the dominant parameter. Low *f*-number magnification provides a larger pupil or eyebox for viewing comfort (see Chapter 11).

9.4 Device Characteristics

9.4.1 DC Characteristics

The DC operating characteristics of a typical historical polymer OLED device are shown in Figure 9.14. Operated in forward bias starting at 0 V there is first a region where little or no current flows. The voltage reaches a point (typically a few volts) at which the flow of current and the emission of light become significant – this is called the threshold voltage. Beyond the threshold voltage, both current flow and light output rise roughly as the square of the increase in voltage. (This dependence is consistent with space-charge being the limiting factor in diode current flow.) Useful levels of brightness for a display are typically achieved within no more than a few volts above threshold. The low level of the threshold voltage and the steep slope of the luminance curve are important factors in the compatibility of OLED materials for use in CMOS-based microdisplays. Additional characteristics are shown for a yellow material in Figure 9.15 and a white material in Figure 9.16.

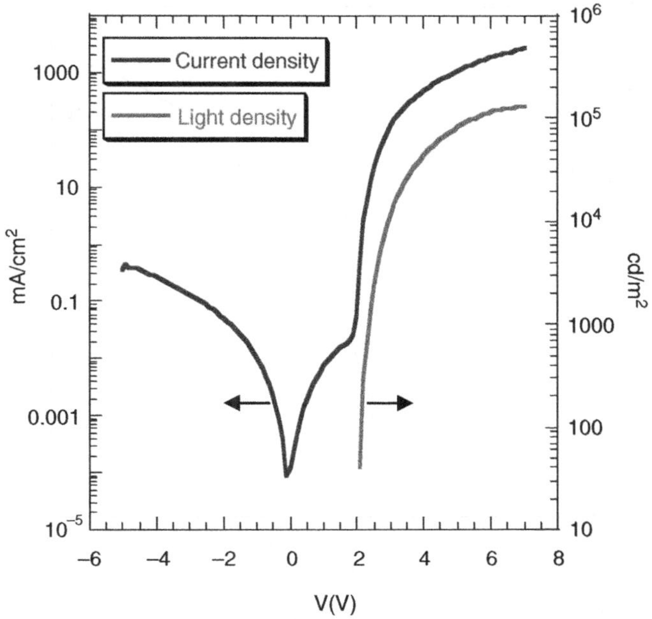

Figure 9.14 Current density and associated light density for polymer OLED display. Reprinted courtesy of Cambridge Display Technology

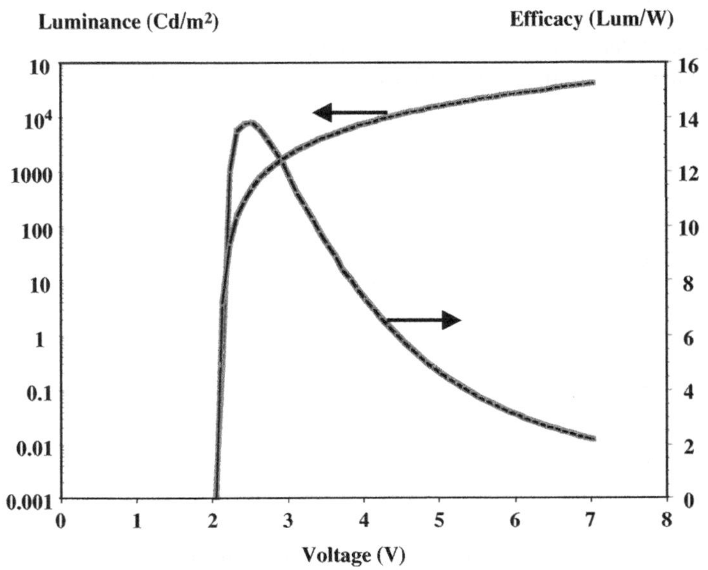

Figure 9.15 Luminance and efficacy versus voltage for monochrome yellow P-OLED in 2001. Reprinted courtesy of Cambridge Display Technology

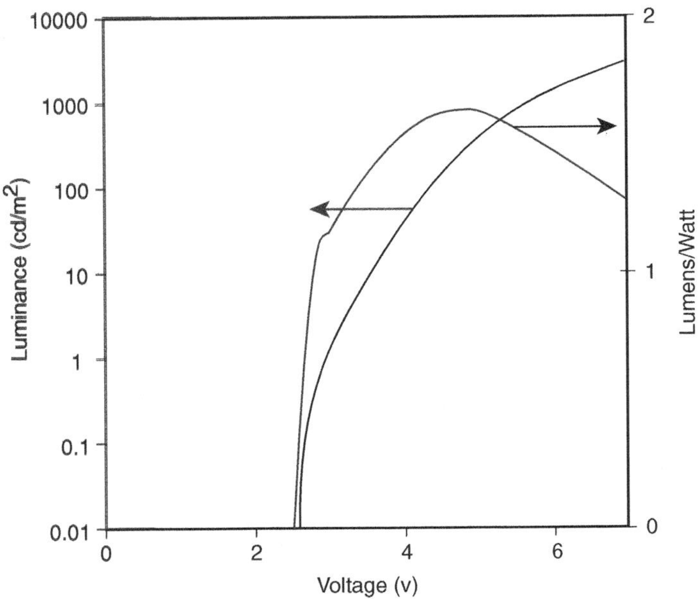

Figure 9.16 Characteristics of white P-OLED in 2001. Reprinted courtesy of Cambridge Display Technology

It is also the case that light output rises and falls linearly with drive current as shown in Figure 9.17.[31] Thus, as explained in Chapter 2, current drive is the most common method of active matrix drive for OLED displays.

9.4.2 Switching Time

The switching time of fluorescent SMOLEDs and P-OLEDs is very short in comparison with most LC materials. Switching times of the order of nanoseconds[32] were already reported in the early years. In

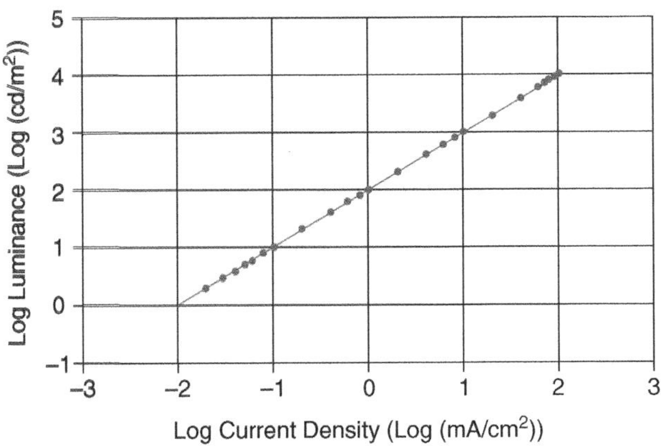

Figure 9.17 Example plot of luminance as a function of current density for OLED device. Reprinted courtesy of Society for Information Display

stark contrast to liquid crystals the switching speed is not only very fast but shows little variation with temperature over the operating temperature range of typical displays. In most practical cases, including displays, the switching time of the output light is limited by the capability of the pixel drive circuit – its *RC* time-constant.

In any case, fast switching speed gives designers flexibility to choose, for example, conventional analog-amplitude, pulse-width or pulse-coded modulation to achieve grayscale and removes any concerns over such LC phenomena as motion-blur and disappearing cursors.

9.4.3 Aging

Aging of devices is perceived by users to be an issue for all OLED displays. A detailed explanation of the mechanisms involved in aging is not appropriate in this context but a description of the potential consequences is. Baynes and Smith[30] describe three symptoms of age-related loss of performance due to degradation of some of the functional parts of the device, (i) an increase in the voltage required to achieve a set drive-current density, (ii) a decrease in the level of light output for a set drive-current density, and (iii) (for some systems) color shift.

Figure 9.18 shows an example luminance decay curve for a blue P-OLED device.[33] Such curves approximate the shape of a simple or stretched exponential depending upon the material system. The single parameter typically quoted to characterize aging is lifetime or half-life t_{50} (defined as the time taken for the display brightness to fall to half of its initial value at a constant drive current; quoted initial values typically range from 100 cd/m^2 to more than 1000 cd/m^2). Typical material lifetimes today are expressed in thousands or tens of thousands of hours a shown in Tables 9.3 to 9.6. It must be noted, however, that maintaining a constant current drive may require a significant increase in the driving voltage. This may become the lifetime-limiting factor in some systems, particularly OLED microdisplay systems in which the available drive voltage is limited to that allowed by the CMOS backplane. Material suppliers often offer, as part of a material specification, the parameter dV_{50} (often shortened to dV) – the increase in magnitude of drive voltage required to maintain a constant current

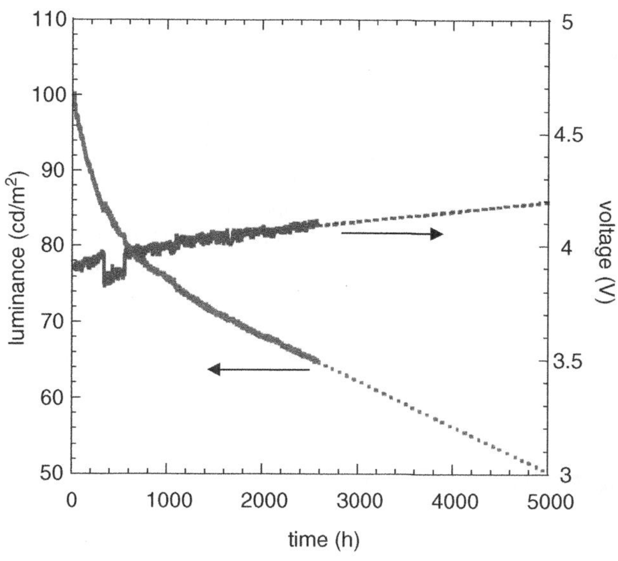

Figure 9.18 Lifetime characteristic for blue P-OLED material from 2003. Reprinted courtesy of Cambridge Display Technology

through an OLED device from time zero at a given luminance (often 100 cd/m^2) to t_{50}, the time at which the luminance reached half its initial value.

Differential aging, in which different areas of the display age at different rates due to a bias in the content of the images being displayed in the long term, can be more noticeable than global aging or dimming. For example, if a pattern of black and white stripes is shown for a long period the OLED in the white stripe areas will age more rapidly than those under the black stripes. An example of this in a practical display is a permanent fixed icon. Differential aging of pixels due to differential use of the pixels can lead to the presence of a persistent ghost image (a faint contrast-reversed copy of the previously displayed image). This differential aging is a direct consequence of the reduction of luminance with use.

Furthermore the differing lifetimes of the primary colors, and in particular the short lifetime of blue materials, can lead to a noticeable shift in color balance with time for systems that use three (red, green and blue) OLED materials to achieve color. In the case of microdisplays, today's devices use white emitters plus patterned color filters so differential aging of colors is not a significant issue. Also, this is the kind of effect that could in future conceivably be significantly reduced by incorporating suitable monitoring and adjusting capability into the CMOS circuitry in the backplane.

9.5 Color

The emission color of an OLED depends on material properties and doping that determine energy levels and transitions. Figure 9.19 shows an example of the color gamut achieved by P-OLED emission. There are, in principle, many possible means of creating color on a pixel-by-pixel basis in an emissive microdisplay (such as an OLED microdisplay). Only those based upon segmentation of the pixel into individual color elements placed side-by-side (sub-pixelation) or one on top of the other (stacked) are described below. Burrows *et al.*[5] describe such structures in more detail for SMOLED. Field-sequential-color, scrolling color and three-microdisplay systems are described in Chapters 2 and 10.

9.5.1 Patterned RGB Color Filter

Here, white light from three neighboring sub-pixels (where the pixelation is defined in the device electrodes) is passed through three precisely aligned overlying areas of red, green and blue optical filter material aligned to the electrodes. This technique is used in almost all color LCDs. It is also prevalent in the OLED microdisplays described later in this chapter. The efficiency of this system depends upon the match or overlap between the emission spectrum of the OLED and the transmission spectrum of the filters and may approximate to around 20%. A three-component OLED system with three emission peaks matched to the color filter transmission peaks is optimal. However, some white-emitting OLEDs are two-component systems that do not match well to color filters thus significantly reducing overall optical efficiency. Figure 9.20 shows an example spectrum of a two-component emitter and a three component emitter.

9.5.2 Patterned RGBW Color Filter

Patterned RGBW color filter is a variation on the previous patterned RGB color filter configuration. In the RGBW case each pixel contains four sub-pixels – red, green, blue and white. The additional sub-pixel emits unfiltered white light. At low pixel light levels the white sub-pixel remains off allowing

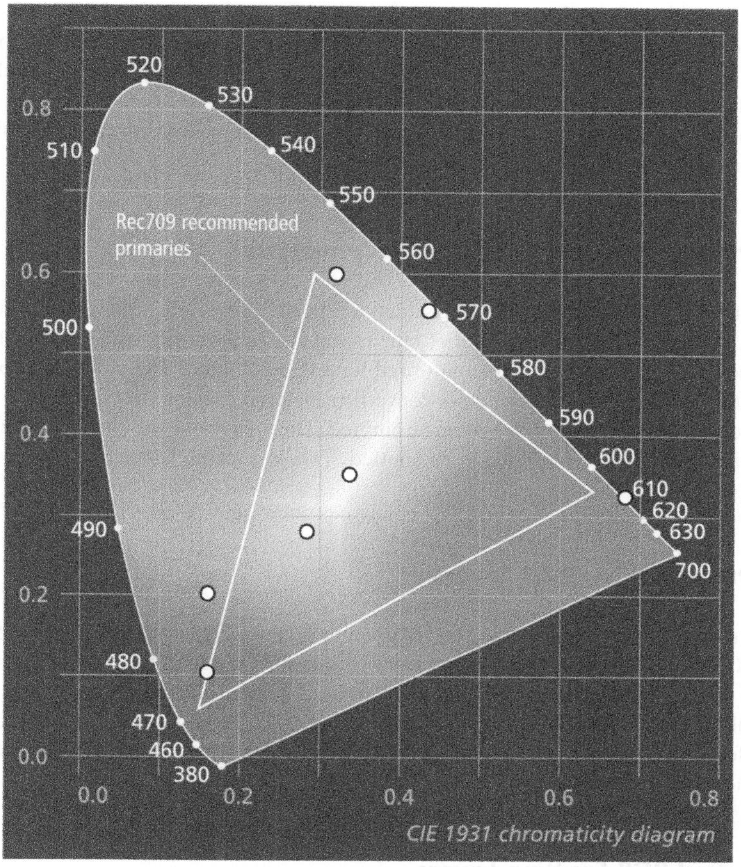

Figure 9.19 CIE color coordinates of P-OLED materials. Reprinted courtesy of Cambridge Display Technology

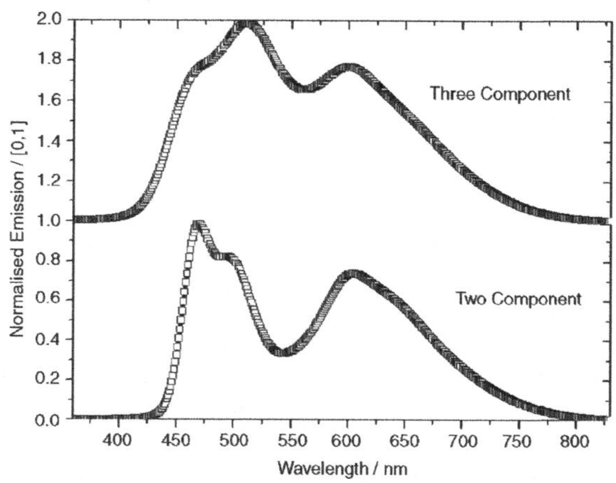

Figure 9.20 Spectrum of two-component white emitter. Reprinted courtesy of MicroEmissive Displays

strong color saturation, albeit slightly sub-optimal due to the reduced effective fill factor. At high pixel light levels the white sub-pixel dominates control of the luminance whilst the RGB sub-pixels supply the color hue. Bright colors are not saturated but high luminance levels are achieved at relatively high efficiency as the luminance is dominated by unfiltered white light. The overall effect can be a significant saving in power consumption. In microdisplays, the use of four sub-pixels may necessitate a larger pixel pitch and consequently larger and more expensive display.

9.5.3 Patterned Red and Green Color Converters

In this case, blue light from three sub-pixels (defined in the device electrodes) is emitted. The red sub-pixel electrode is overlaid with a layer of blue to red color converting material causing blue light to be absorbed and a corresponding amount of red light to be emitted. The green sub-pixel electrode is overlaid with a layer of blue to green color converting material causing the blue light to be absorbed and a corresponding amount of green light to be emitted. The area above the blue sub-pixel is not filled causing the blue light to pass through and be emitted. This is a method sometimes known as stacked RGB emitting layers.

9.5.4 Patterned RGB OLED Layers

In this case, red-emitting, green-emitting and blue-emitting material is deposited onto the appropriate sub-pixel (defined in the device electrode layers) within each. This gives the most efficient generation and emission of light. However compromises may have to be made in the device structure, as it may not be possible to simultaneously optimize the device (e.g. the cathode) for all three materials. Furthermore differential aging of the three materials may lead to a significant color shift in output light throughout the lifetime of the display.

9.5.5 Stacked RGB Emitting Layers

Here light is generated from red, green and blue OLEDS, stacked on top of each other, resulting in a complex vertical structure that has the advantage of producing all color hues from the footprint equivalent of one sub-pixel. The upper OLEDs have to be transparent to the emission of the lower OLEDs. Whilst such structures have been proposed and even demonstrated,[36] manufacturing issues are likely to inhibit near-term commercial implementation of the stacked method for microdisplays.

9.5.6 Color Summary

A schematic representation of several of the approaches to color is contained in Figure 9.21. The pros and cons of the approaches are summarized in Table 9.8.

Recall that organic emissive microdisplays are currently used only in near-to-eye applications where size, weight and cost are key factors. To date, the high level of system complexity seems to have ruled out the use of field-sequential-color and the multi-microdisplay approach. The difficulty of manufacture at the small dimensions required in microdisplays has thus far prevented the development of RGB patterned OLED and stacked OLED configurations. The combination of blue OLED and patterned color converters has been investigated[37] while color products currently on the market all use white-emitting material combined with RGB color filters. There are, to date, no reports of the use of the RGBW technique in microdisplays but no reasons in principle why it could not be implemented.

Table 9.8 Pros and cons of various methods of achieving color for microdisplays

Method	Pros	Cons
RGB color filter	Mature technology Manufacturability	Inefficient
RGBW color filter	Mature technology Manufacturability Improved efficiency	Increased pixel footprint
Color converter	Improved efficiency over color filter	Dependency on blue lifetime
RGB OLED	Most efficient Lowest power	High-resolution patterning of OLED material Difficulty of manufacture
Stacked color	No spatial separation of RGB sub-pixels	Difficulty of manufacture Efficiency of underlying layers

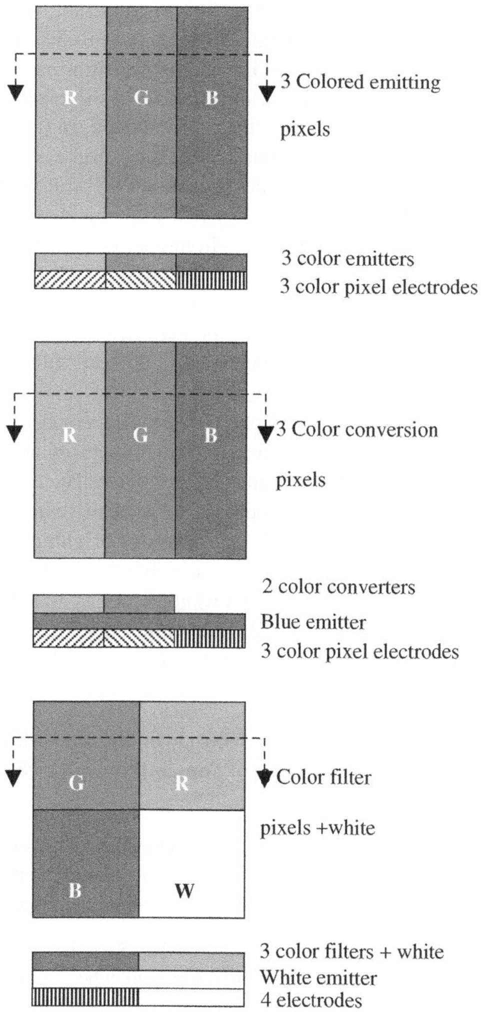

Figure 9.21 Various methods of creating color

9.6 OLED Microdisplays

9.6.1 General Properties of OLED Microdisplays

This section mentions some of the main attractions of OLED microdisplays to the user when compared to other microdisplays.

Luminance levels sufficient for electronic viewfinders and enclosed HMDs are achieved at relatively low drive voltage. This allows the use of "standard" transistors in the CMOS backplane (rather than high-voltage transistors). Luminance levels sufficient for open or semi-transparent HMDs are also possible, albeit with a higher voltage being required.

OLED microdisplays appear to exhibit low power at any given brightness when compared with other display technologies in general and other microdisplay technologies in particular. This is of great appeal for portable, battery-powered products as it allows prolonged battery life and/or the use of smaller and lighter batteries. The inherent low-power nature of the OLED microdisplays is enhanced by the inherent pattern-dependent power consumption of emissive displays. That is to say, in a modulating display the "optical" power consumption (of the backlight) is constant, irrespective of the pattern on the display and corresponds to all pixels being fully ON. In an emissive display, the "optical" power consumption is at a maximum when all pixels are ON at maximum luminance. When, for example, only half of the pixels are ON at maximum luminance or when all of the pixels are ON at half-maximum luminance, the "optical" power consumption is significantly redued. In the case of, for example, video applications, in which the average pixel use is around 20%, the power saving can be significant.

Very high contrast ratio is possible provided the electronic OFF state of the OLED pixel is below the threshold for light emission. The use of the display in a well-shielded EVF or well-enclosed HMD minimizes the contrast-reducing effect of ambient light and so better preserves the inherent contrast ratio of the display in use.

The viewing optics associated with the pseudo-Lambertian unpolarized emitted light is simpler, less bulky and less expensive than those associated with the generation and subsequent viewing of polarized light from transmissive and particularly reflective LC devices; furthermore there is no optical component requirement associated with illumination.

Currently, the main drawbacks of OLED microdisplays from a user perspective are the limited availability in terms of range and volume of products, the perceived relative immaturity of the technology and concerns that relate to OLED displays in general such as lifetime, reliability etc. The existence of a growing range of commercial products containing OLED displays should suffice to begin to dispel the latter concerns.

A practical restriction imposed by the relatively low levels of brightness available from contemporary OLED microdisplays (in comparison with those available from LCOS) is that OLED microdisplays have been, to date, used in only near-to-eye (NTE) applications and systems.

The principal differences between CMOS-based OLED microdisplays and conventional direct-view OLED displays, such as passive matrix and TFT-based active matrix OLED displays that use HTPS, LTPS or a-Si, are as follows.

- The opaque substrate and reflective aluminium electrode on the top surface of the CMOS lead naturally to the implementation of a "top-emitting" or "up-emitting" device structure in contrast to the more conventional bottom-emitting configuration.
- The small pixel-pitch of the microdisplay leads to necessary adjustments in the manufacturing process. For example, full-color microdisplays using polymer materials cannot be made using inkjet printing as this technique falls short of achieving both the necessary feature size and drop placement accuracy.[27]
- The narrow gap between the metal pixel electrodes in the CMOS (of the order of 1 μm) means that lateral conduction must be much more tightly controlled. In explanation, when a pixel is turned ON, current flows vertically through the OLED structure. Lateral conduction would allow a proportion of the pixel current to leak laterally into adjacent pixels. This could spuriously reduce the luminance value of the driven pixel whilst, at the same time, spuriously increasing the luminance value of an adjacent OFF pixel.

It is often erroneously assumed that the heterogeneous integration of materials onto a CMOS silicon substrate involves integrating onto a crystalline silicon surface. Modern CMOS circuits are implemented using interconnects comprising multiple layers of metal/oxide sandwich above the surface of the silicon substrate. In these cases the crystalline substrate is completely covered and deeply buried. Etching through to expose the crystalline silicon would involve process customization and would require the active circuits to be packed less densely, which is unattractive. Thus, for OLED microdisplays, the actual integration problem involves interfacing to a metal surface, typically sputtered or evaporated aluminum or aluminum alloy (e.g. AlSi or AlSiCu).

9.6.2 Small-Molecule OLED (SMOLED) on Silicon Devices

In the early 1990s Kim and co-workers made a series of reports[38–40] describing the use of polymer OLEDs integrated on silicon substrates containing bipolar active circuits and photo receivers; the application was photonic switching and optical interconnect rather than displays.

The Optical Sciences Center at the University of Arizona is known to have been active in the area of OLED integrated on CMOS. Mathine *et al.*,[41] for example, describe an 8 × 8 array of 90 μm pixels.

Bulovic *et al.*[42] used a Mg:Ag electrode on a silicon substrate to build an OLED device (with a novel protective cap layer) thus allowing the possibility of directly integrating the OLED material with MOS active matrix drive circuits. Zhou *et al.*[43] used p-type silicon in place of ITO as the device anode in order to avoid problems from indium diffusing into the organic light emitter.

In the late 1990s Sanford and co-workers at IBM developed a series of technology demonstrators, including a fixed-pattern monochrome green display of XGA resolution with 4 μm pixel pitch in 1998, a VGA (640 × 3 × 480) color display with a 15 μm pixel pitch (3 stripes of 15 μm × 5 μm) in 1999, and a VGA monochrome display[44,45] with a 34.3 μm pixel pitch.

At around the same time Stewart *et al.*[46] in 1998 published details of a VGA OLED display using polysilicon backplane technology before some members of that group moved on to develop OLED on crystalline silicon (CMOS) microdisplays.[47,48]

In recent years a first device has made substantial commercial progress. The SVGA+[iii] microdisplay reported by Ghosh in 2002[49] and Prache[50] amongst others uses molecular OLED materials to produce a low-voltage, high-brightness, high-resolution solution for professional, military and other demanding applications. Some attributes of this device are summarized in Table 9.9. Lin *et al.*[51] briefly describe the CMOS backplane whilst Levy *et al.*[52] describe some aspects of both the electronics and the optics of the device in more detail. Details of the power consumption in particular are reproduced in Table 9.10.

Table 9.9 Main properties of SVGA+ OLED microdisplay[50]

Pixel count	852 × 600 × RGB
Pixel pitch (μm)	15
Screen diagonal (mm)	16
Nominal brightness (cd/m^2)	100
Power consumption (mW)	Less than 300
Interface	Analog RGB
Color method	White OLED with RGB color filter / Blue OLED with RG color converter
Other features	–
Grey levels	24-bit
Pixel pattern	RGB stripe on square grid
CMOS technology	0.35 μm 4-metal 3.3/4.0 V

[iii]SVGA+ in this instance refers to 852 × 600 pixels. This can be operated as SVGA+ (852 × 600), in conventional SVGA 4:3 aspect ratio at 800 × 600 or in widescreen WVGA 16:9 aspect ratio at 852 × 480.

Table 9.10 Power consumption of SVGA+ microdisplay[50]

Pattern	I_{DD} (mA) @ $V_{DD} = 3.3\,V$	I_{AN} (mA) @ $V_{AN} = 4.0\,V$	I_{CATH} (mA) @ $V_{CATH} = -4.5\,V$	Power (mW)
White	43.9	34.3	32.9	430
SMPTE master	43.01	19	17.7	297
MS Word	43.8	28	26	373

Reprinted courtesy of Society for Information Display

The CMOS backplane, designed by Cadence Design Systems, is discussed in Chapter 2. The device structure is shown in schematic cross-section in Figure 9.22. It can be seen that the diagram leaves unanswered the question of whether the microdisplay uses color filters or color converters. It is believed that white emitter and RGB color filters are currently favored. Figure 9.23 suggests a detailed layer structure for a white-emitting device, and Figure 9.24 shows a photograph of an image displayed on the SVGA+ microdisplay. Ghosh[49] describes the process steps involved in device fabrication. These are shown diagrammatically in Figure 9.25.

In 2004, Low *et al.*[53] published a study of the detailed characteristics of the device, paying particular attention to performance over a range of temperature. A further study by Barre *et al.*,[54] published in mid 2005, describes some aspects of the operational performance of OLED on CMOS devices including the temperature dependence of color coordinates and gamma, the micro- and macro- uniformity of luminance and some lifetime characteristics. In both of the above studies, the authors were particularly interested in the use of the display for military applications.

In early 2006, consumer headsets are commercially available using the SVGA or a closely related device. The model X800 3DVisor is a consumer headset that uses two displays to produce stereoscopic 3-D moving pictures whilst the model Z800 adds a head-tracking capability.

The use of multiple SVGA displays tiled together by means of tapered fiber-optic bundles to achieve effective resolutions of UXGA and 2UXGA has recently been publicized.[55] The UXGA device is shown

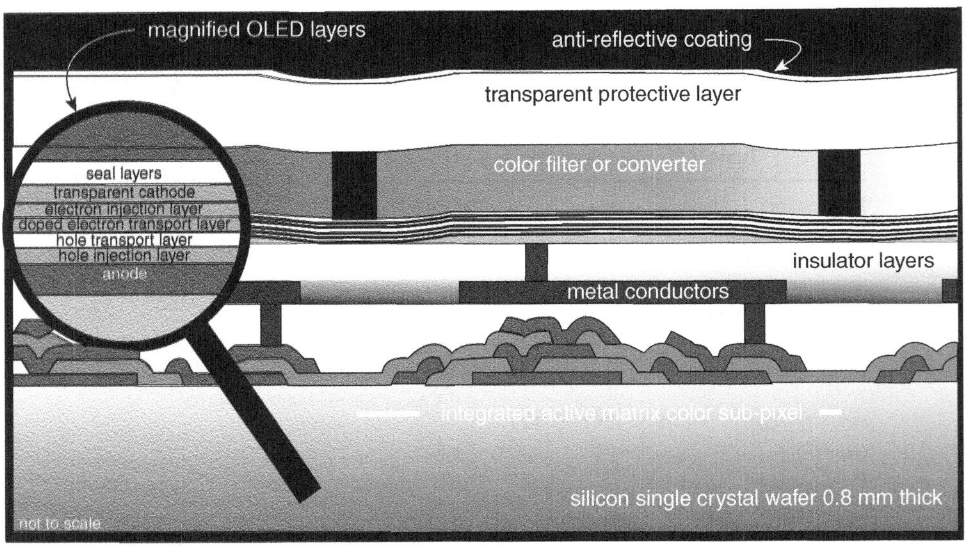

Figure 9.22 Schematic cross-section of part of SMOLED microdisplay. Reprinted courtesy of Society for Information Display

Transparent cathode=low work function metal	
ETL	Alq_3
EL	DPV + red fluorescent dye
HTL	NPB
HIL	CuPc
Reflecting anode=high work function metal	

Figure 9.23 Schematic cross-section of microdisplay white emitting OLED

Figure 9.24 Grayscale version of color photograph of SVGA+ color SMOLED microdisplay. Reprinted courtesy of Society for Information Display

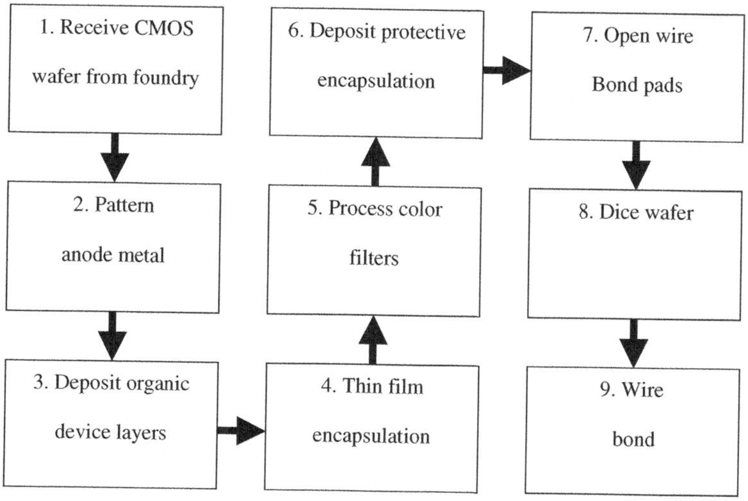

Figure 9.25 Example fabrication process steps for SMOLED microdisplay

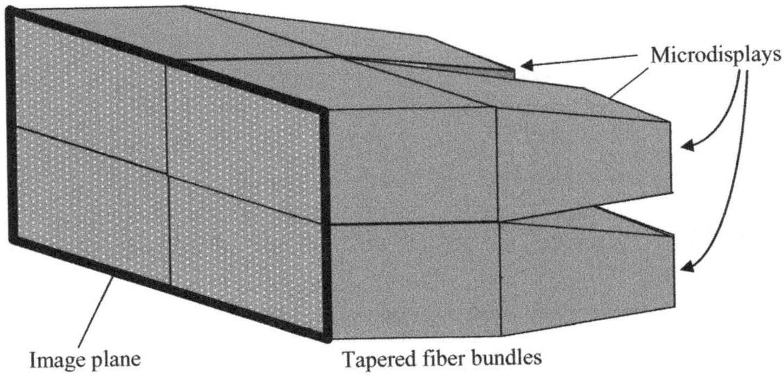

Figure 9.26 Schematic of use of fiber bundles to tile microdisplays

in Figure 9.26. The additional cost, weight and volume of this tiling solution may restrict it to specialist use. A tiling method employing Fresnel lenses applied to a wide-angle virtual-reality head-mounted display is discussed in Chapter 11.

9.6.3 Polymer OLED Microdisplays

Polymers have long been used in integrated optics[56] so that the development of light emitting polymers integrated on CMOS is no surprise.

In 1994, Parker *et al.*[57] described a device with the structure Si/MEH-PPV/Ca. In the same year Baigent *et al.*,[58] in an attempt to show a device with no reactive metal electrodes, reported on a polymer light-emitting diode in which the bottom cathode was of aluminum lying directly on a doped silicon substrate. The device structure was Si/Al/CN-PPV/PPV/ITO. In 1997 Yu *et al.*[59] described a polymer light-emitting electrochemical cell (PLEC) that had the potential to be "hybridized with integrated circuits on a silicon wafer." Heinrich *et al.*[60] reported on the compatibility of polymer OLED with CMOS in 1997 and again in 2000.[61] They used p-type silicon as a hole-injecting lower electrode and a very thin sputtered aluminum layer as the cathode. Light emission occurred upwards through the semi-transparent aluminum. The use of p-type silicon as the lower electrode is in contrast to the microdisplays described in this chapter all of which use the top-level metal of the CMOS as the basis for the substrate electrode. In a CMOS active matrix backplane, the Heinrich approach (like that of Parker *et al.*,[57] would require the OLED to sit side-by-side with the circuit and interconnect layers of the CMOS thus reducing the proportion of the pixel area available to emit light (rather like a conventional down-emitting active matrix OLED display) and would involve significant light leakage into the substrate with the accompanying circuit performance reduction and/or disruption. All of the OLED microdisplay devices in this chapter achieve emission from a very large proportion of the pixel area by placing the reflective electrode above, and covering, the pixel circuit. In this case the inter-electrode gap is minimal and determined by the metal-to-metal minimum spacing as defined by the CMOS design rules.

In 2000, Smith[34] reported briefly on the results of a development that produced a monochrome microdisplay emitting green light as shown in Figure 9.27. This was subsequently reported in more detail by Abraham *et al.*[62] in 2001, including the use of a white LEP. Both Smith[34] and Abraham *et al.*[62] comment on the early stage of development at that time and the consequent poor cosmetic quality of image then achieved; Abraham and co-workers go on to detail the required improvements. They report that the CMOS backplane used to drive the LEP is the same one reported by Blalock *et al.*[63] as a backplane for a ferroelectric LC microdisplay. The backplane of Blalock and co-workers is described in more detail in Chapter 2. This leads us to the conclusion that grayscale was achieved by pulse width modulation and the LEP was driven in constant voltage mode. To the best knowledge of the current

Figure 9.27 Photograph from 2001 of monochrome LEP on CMOS microdisplay. Reprinted courtesy of Cambridge Display Technology

authors, no subsequent progress has been reported leading us to the conclusion that the development effort ceased prior to product commercialization.

In 2003, Bodammer *et al.*[64] reported a QVGA monochrome active matrix backplane used in conjunction with a green emitter and a white emitter. The same backplane has been used to demonstrate a 160 × 120 color microdisplay[65] as shown in Figure 9.28. These brief reports carry no details of the

Figure 9.28 Grayscale photograph of color QQVGA (160 × RGBG × 120) P-OLED microdisplay. Reprinted courtesy of MicroEmissive Displays Ltd.

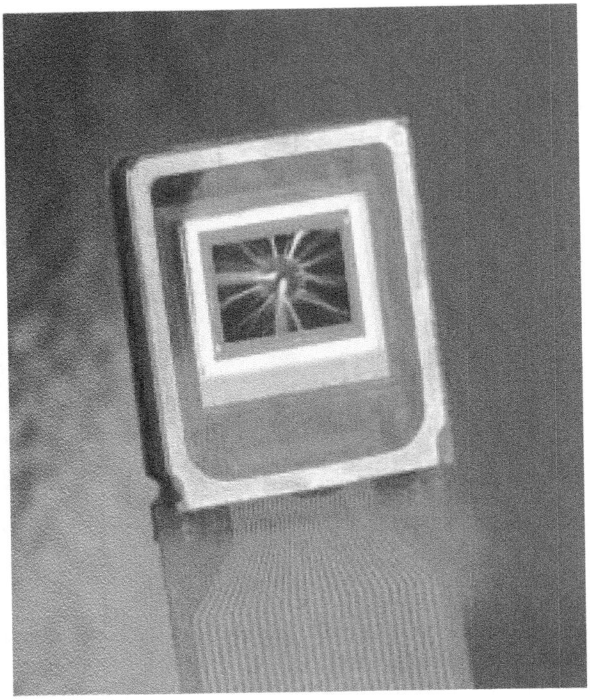

Figure 9.29 Grayscale photograph of color QVGA P-OLED microdisplay image. Reprinted courtesy of Micro-Emissive Displays Ltd.

device structure, pixel circuit or addressing scheme so the detailed specifics of their approach remain outside the public domain.

At the time of writing, a first P-OLED based microdisplay product is commercially available. The ME3203 QVGA microdisplay,[66] first reported by Underwood *et al.*[67] in 2004, is shown in Figure 9.29; it makes use of the low drive voltage and high efficiency of polymer OLED to provide a low-cost, low-power electronic solution for electronic viewfinders and consumer HMDs. Some attributes of this device are summarized in Table 9.11. Note that ME3203 has a digital interface interface that can be

Table 9.11 Summary properties of QVGA polymer OLED microdisplay

pixel count	320 × 240 × RGB
Pixel pitch (μm)	18
Screen diagonal (mm)	7.2
Nominal luminance (cd/m^2)	80
Power consumption at nominal luminance (mW)	Approx 35
Interface	BT656, RGB serial
Color method	White LEP with RGB color filter
Other features	On-board frame store Low-power still-image mode
Color depth	18-bit (6-bit per color)
Pixel pattern	RGB delta on square grid
CMOS technology	0.30 μm 4-metal 3.3 V

configured to accept RGB serial and CCIR-656 format data; it also has an on-board frame-store. Thus it appears to exhibit a very advanced level of electronic integration.

9.7 Other Emissive Microdisplays

9.7.1 Inorganic Electroluminescent Microdisplays

Active matrix inorganic electroluminescent (AMEL) microdisplays were developed during the 1990s.[68] These used phosphor materials[69] such as ZnS:Tb, ZnS:Mn and SrS:Ce to emit amber, green and blue light respectively.

Some of the unusual aspects of the design and manufacture of these displays included the use of features such as: silicon on insulator substrate; DMOS transistors capable of handling the very high drive voltages (up to 200 V in total comprising 140 V AC bias and 60 V switching signal);[70] refractory metal interconnect on the substrate in order to withstand the high temperatures (450 °C) generated; and long times required to process the EL stack that was deposited using atomic layer epitaxy (ALE).

Later displays used digital drive circuitry and pulse coded modulation to achieve 6-bit grayscale. The use of PCM also allowed the displays to operate in symbology-mode (1-bit grayscale) at significantly higher luminance. A luminance of 3400 cd/m^2 was reported[71] in a VGA device with a pixel fill-factor of 84%, achieved using 22 μm square pixels on a 24 μm pitch.

The displays produced were monochrome but a color display was proposed. A schematic cross-section of the structure of a color display is shown in Figure 9.30. The use of four square sub-pixels in a 2 × 2 RBBG array implies that, with suitable peripheral circuitry, a single backplane could be used as the active matrix for a monochrome display of $2m \times 2n$ pixels and a color display of $m \times n$ pixels. The use of color filters implies a white phosphor.

High maximum luminance is a noticeable positive feature of the AMEL technology, making it suitable for open or see-through head-mounted applications whilst the primary disadvantage is the high drive voltage requirement. The AMEL microdisplay program has been abandoned, possibly in view of the current and impending development of OLED microdisplays.

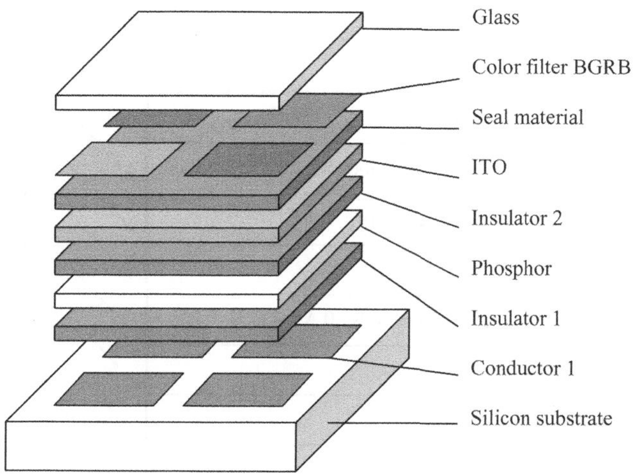

Figure 9.30 Schematic representation of quad color inorganic EL pixel. Reprinted courtesy of Society for Information Display

9.7.2 Gallium Nitride Microemitter Arrays

GaN-based (also called III-nitride) wide band-gap semiconductors are capable of emission in the blue and ultraviolet regions of the spectrum. Since the late 1990s micro-arrays[72–75] of such LEDs have been developed. One potential application of such arrays is microdisplays.[76,77]

Figure 9.31 shows a representative schematic cross-section through a III-nitride light-emitting structure such as described in detail by, for example, Jiang and Lin in 2001.[78] The layout of several pixels of a specific 128×96 array[79] is shown in Figure 9.32. In this array, pixels on a column share a common n-type region whilst pixels on a row are electrically connected by metal lines running across the array. The IV and spectral characteristics are illustrated in Figure 9.33 whilst an image is shown on such a display in Figure 9.34.

The proposed attractions of these devices include those associated with emissive microdisplays in general plus the potential for very high brightness. Color may be possible by control of the quantum well width and composition to produce different emission wavelengths, or by the down-conversion of emitted blue light to red and green or emitted UV to RGB. Active matrix addressing may be achieved by flip-chip bonding to a CMOS active matrix backplane. A transparent sapphire substrate allows light out with the potential for the integration of, for example, microlenses. At the time of writing, this technology is at an early stage of development and much of the above remains speculative. There are no reports of commercialization but the above potential advantages, if realized, might point to applications in micro-projection. The cost of inorganic LED arrays is likely to restrict them to special applications, with lower cost OLED microdisplays being likely to dominate in cost-sensitive markets such as the near-eye market.

9.7.3 Porous Silicon Microdisplays

Crystalline silicon is an indirect bandgap semiconductor that does not efficiently emit light in the visible region of the spectrum. Porous silicon,[80] on the other hand, can be configured as a nanocrystalline structure in which the dimensions of the structure control the optical properties. At appropriate dimen-

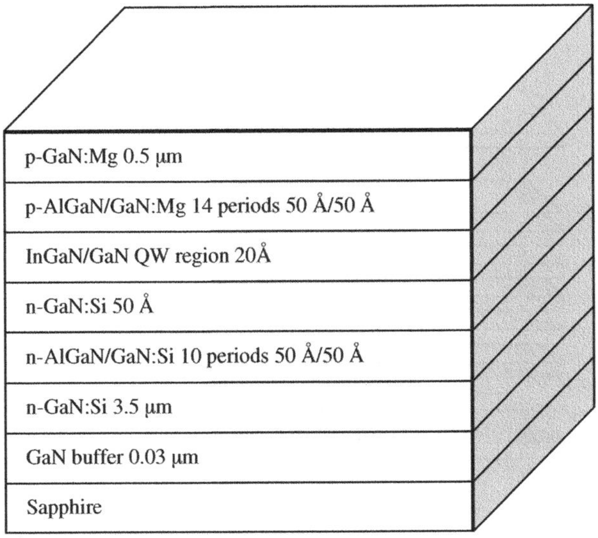

Figure 9.31 Representative layer structure of MQW device

(a)

(b)

Figure 9.32 Schematic diagram of micro-LED array. ©2004 IEEE: reprinted with permission

sions, say a few nanometers, strong visible electroluminescence and photoluminescence is possible. The energy band theory of solids is based on lattice periodicity, implying dimensions large compared with that periodicity. At nano-scale, band theory breaks down, relaxing the constraints on radiative transitions.

The earliest optical studies of porous silicon were performed by Uhlir[81] and Turner[82] in the early 1950s. In 1990, Canham[83] reported room temperature photoluminescence. Then a light emitting diode (LED) based upon reverse-biased porous silicon was demonstrated by Richter *et al.*[84] in 1991. A number of improvements in efficiency and stability have been reported over the years including that of Gelloz *et al.*[85] who used reverse-biased oxidized porous silicon. Smirnov and co-workers have recently reported on the fabrication[86] and characterization[87] of microdisplays using nanoporous silicon (NPS). A schematic cross-section of the NPS LED structure of a single pixel is shown in Figure 9.35 and a representative photoluminescence spectrum is shown in Figure 9.36. It is worthy of note that, in the configuration of Figure 9.35, the porous silicon layer is formed in contact with the n+ silicon layer in the substrate. This means that the emissive region is restricted in that it must lie alongside the active

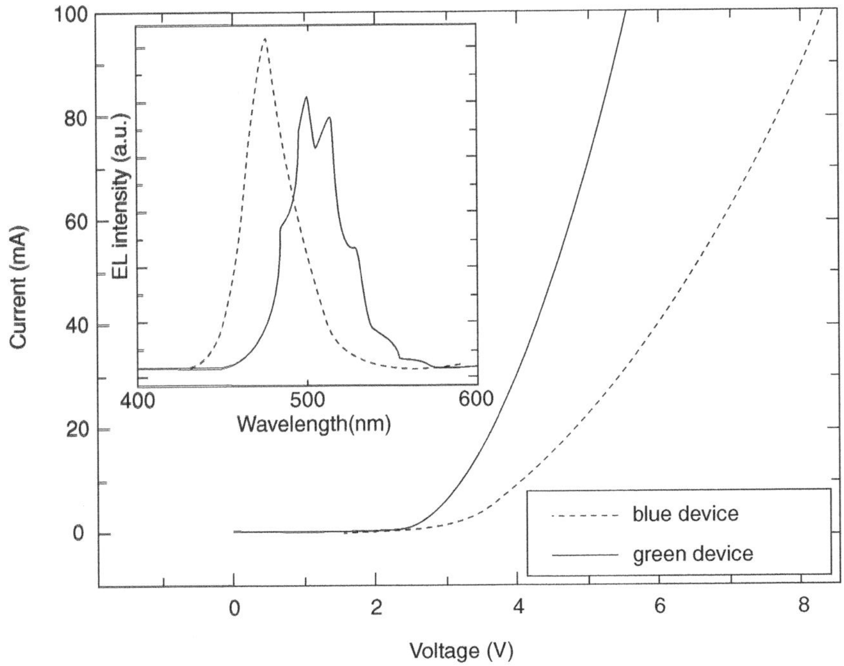

Figure 9.33 Representative IV and spectral characteristics of micro-LED array. ©2004 IEEE: reprinted with permission

electronic devices and metal interconnect. This probably restricts the number of metal interconnect layers that can sensibly be used and certainly significantly restricts the emissive aperture ratio in a manner analogous to that of transmissive LC microdisplays. So the emissive aperture ratio of these porous silicon microdisplays is unlikely to match that of OLED microdisplays in which the emissive area lies over the active circuit elements and metal interconnect.

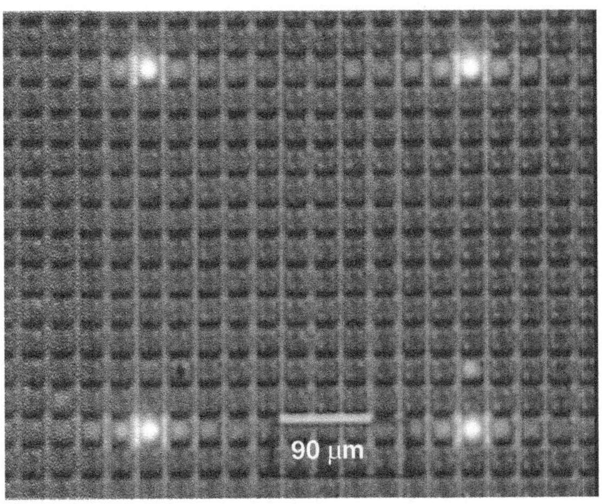

Figure 9.34 Photograph of working micro-LED array. ©2004 IEEE: reprinted with permission

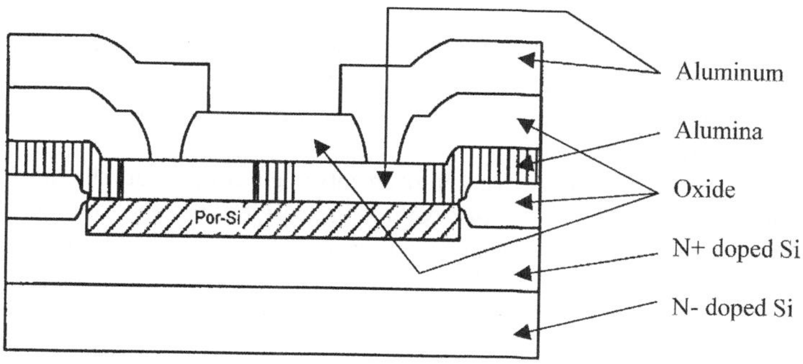

Figure 9.35 Schematic cross-section of nanoporous silicon LED structure (adapted from reference 87). Reprinted courtesy of Society for Information Display

9.8 Summary

There is no doubt that, whilst microdisplay technologies such as LCOS and DLP are today relatively mature, OLED microdisplay technology (like OLED display technology in general) is still in the early stages of development and has achieved only a small part of its potential. Fast progress is to be expected over the coming years.

The perceived advantages of the OLED microdisplay technology are highly appropriate to the near-eye market. Lower power is a key factor for any portable or personal electronic product as it leads to some combination of increased battery life and reduced battery size and weight. This is particularly relevant for head-mounted displays, especially consumer units and wireless HMDs which require an on-board battery. Integration of electronics onto the CMOS backplane is common to all CMOS-based microdisplay technologies and is an advantage over miniature TFT-based microdisplays. It lowers electronic component count as well as power. Integration of the light source onto the surface of the microdisplay lowers component count and power as the generation of light by the OLED is more efficient than that of a backlit LCD.

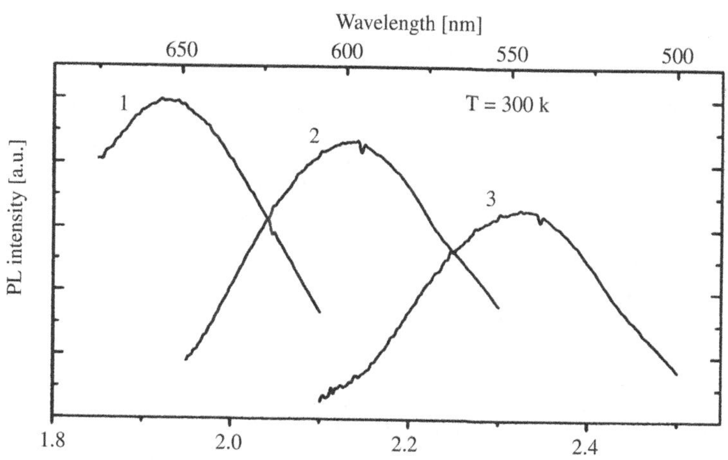

Figure 9.36 Photoluminescence spectra of NPS samples: (1) no post-anodizing treatment, (2) 20 minutes in electrolyte, (3) 30 minutes in electrolyte. Reprinted courtesy of Society for Information Display

These advantages offer clear incentive to quickly and fully resolve the perceived current issues, such as comparatively limited lifetime and reliability, associated with an early stage technology. To draw a historical comparison, in the early days of LCD and LCOS technologies, issues such as unproven reliability and low lifetime expectations were prevalent but were soon resolved. The same is likely of OLED microdisplays in the near future.

In the course of time, material and device improvements in general, the development of methods of patterning red, green and blue materials at the required dimensions (down to 5 μm pitch) and possibly the use of phosphorescent materials will lead to significant increases in the brightness and efficiency of OLED microdisplays. Even before then, the potential of the technology is such that it is likely to be widely deployed.

9.9 Bibliography

Braun, D. Semiconductor polymer LEDs. *Materials Today* **5**, 32–9 (2002).
Bulovic, V. and Forrest, S. R. Polymeric and molecular organic light-emitting devices: a comparison. *Semiconductors and Semimetals* **65**, 1–26 (2000).
Burroughes, J. H. Progress in light-emitting polymer technology. *SID Seminar Lecture Notes*, M-6/1-71 (2003).
Forrest, S. Fundamentals of organic light emitting devices. *SID Short Course* S-2 (2002).
Forsythe, E. W. and Shi, J. Fundamentals of OLED displays. *SID Short Course* S-2 (2003).
Holton, W. C. Light emitting polymers: increasing promise. *Solid State Technology*, May, 163–9 (1997).
Howard, W. E. OLED Primer: www.emagin.com/oledpri.htm
Le Barny, P. Introduction to organic electroluminescence (from basic principles to applications). *IDRC Seminar Lecture Notes*, 111–34 (2002).
Rajeswaran, G. OLED manufacturing technology. *SID Seminar Lecture Notes*, M-7/1-80 (2003).
Tang, C. W. Organic light-emitting diodes (OLEDs). *SID Seminar Lecture Notes*, M-5/1-87 (2003).
Vaeth, K. Organic and polymeric light emitting diode technology. *IDRC Seminar Lecture Notes*, 135–78 (2002).

9.10 References

1 Burroughes, J.H., Jones, C. A. and Friend, R. H. (1988) Novel semiconductor device physics in polymer diodes and transistors. *Nature* **335**, 137–41.
2 Burrows, P. E. and Wilson, E. G. (1990) The inchworm memory: a new molecular electronic device. *J. Mol. Electron.* **6**, 209–20.
3 Forrest, S. R. (1989) Organic-on-inorganic semiconductor heterojunctions: building blocks for the next generation of optoelectronic devices? *IEEE Circuits Devices* **5**, 33–41.
4 Underwood, I. and Gourlay, J. (2003) Light emitting polymer on CMOS: a new photonic technology? *Proc SPIE*, **5181**, 110–19.
5 Burrows, P. E., Gu, G., Bulovic, V. *et al.* (1997) Achieving full-color organic light-emitting devices for lightweight flat-panel displays. *IEEE Trans. Electron Dev.* **44**, 1188–203.
6 Bulovic, V., Burrows, P. E. and Forrest, S. R. (2000) Molecular organic light-emitting devices: Electroluminescence 1. In G. Mueller (ed.) *Semiconductors and Semimetals.* Academic Press, 255–306.
7 Malliaras, G. and Friend, R. H. (2005) An organic electronics primer. *Physics Today*, May, 53–8.
8 Duggal, A. R. *et al.* (2005) Large-area white OLEDs. *SID Digest* **36**, 28–31.
9 Becker, H. *et al.* (2004) Novel host materials for efficient and stable phosphorescent OLED device. *SID Digest* **35**, 796–9.
10 Adachi, C., Baldo, M. A., Thompson, M. E. and Forrest, S. R. (2001). Nearly 100% internal phosphorescence efficiency in an organic light-emitting device. *J Appl. Phys.* **90**, 5048.

11 Ma, D. G., Lupton, J. M., Samue, I. D. W. *et al.* (2002) Bright electroluminescence from a conjugated dendrimer. *Applied Physics Letters* **81**, 2285–7.
12 Pope, M., Kallmann, H. P. and Magnante, P. (1963) Electroluminescence in organic crystals. *J. Chem. Phys.* **38**, 2042.
13 Tang, C. W. and Van Slyke, S. A. (1987) Organic electroluminescent diodes. *Applied Physics Letters* **51**, 913–15.
14 Tang, C. W. (1997) An overview of organic electroluminescent materials and devices. *J. SID* **5**, 11–15.
15 Wellmann, P. *et al.* (2005) High-efficiency p–i–n organic light-emitting diodes with long lifetime. *J. SID* **13**, 393–7.
16 Burroughes, J. H., Bradley, D. D. C., Brown, A. R. *et al.* (1990) Light-emitting diodes based on conjugated polymers. *Nature* **347**, 539–41.
17 Albertalli, D. (2005) Gen 7 FPD inkjet equipment. *SID Digest* **35**, 1200–3.
18 Van Dijken, A. *et al.* (2005) High-efficiency phosphorescent polymer LEDs. *SID Digest* **35**, 1866–9.
19 Halim, M., Pillow, J. N. G., Samuel, I. D. W and Burn, P. L. (1999) Conjugated dendrimers for light-emitting diodes:effect of generation. *Advanced Materials* **11**, 371.
20 Lo, S. C., Male, N. A. H., Markham, J. P. J. (2002) A green phosphorescent dendrimer for light-emitting diodes. *Advanced Materials* **14**, 975.
21 Lo, S. C., Anthopoulos, T.D., Namdas, E. B. *et al.* (2005) Encapsulated cores: host-free organic light-emitting diodes based on solution-processible electrophosphorescent dendrimers. *Advanced Materials* **17**, 1945.
22 Anthopoulos, T. D. *et al.* (2005) Dendrimers: efficient solution-processed phosphorescent OLED. *IDRC* (Eurodisplay), paper 4–7.
23 Namdas, E. B. *et al.* (2005) Dendrimers: efficient solution-processed phosphorescent OLED materials. *SID Digest* **36**, 1862–5.
24 Burroughes, J. H. and Van Slyke, S. (2005) An overview of polymer and small molecule devices and materials. Pre-conference Seminar 1, Intertech conference "OLEDs 2005," San Diego, CA, November.
25 Mahon, J. (2005) Recent advances in phosphorescent OLED technology. *Digest of Intertech OLEDs 2005*, San Diego, CA, November.
26 Sorin, G. (2005) Key challenges for OLED industry success. *Digest of Intertech OLEDs 2005*, San Diego, CA, November.
27 Burroughes, J. H. and Van Slyke, S. (2005) An overview of polymer and small molecule devices and materials. Pre-conference Seminar 1, Intertech conference "OLEDs 2005," San Diego, CA, November.
28 He, G., Pfeiffer, M., Leo, K. *et al.* (2004) High-efficiency and low-voltage p–i–n electrophosphorescent organic light-emitting diodes with double-emission layers. *Applied Physics Letters* **85**, 3911–3.
29 Werner, A. (2005) Charge transport layers doped with molecular donors and acceptors: high power efficiency and long lifetimes for display and lighting applications. *Intl Display Workshop*, Japan, 581–4.
30 Baynes, N. and Smith, E. (2006) Organic electroluminescent displays. In *Handbook of Optoelectronics*. CRC Press.
31 Van Slyke, S. A. (2004) OLED manufacturing. *SID Seminar Lecture Notes* M-5, 1–44.
32 Braun, D. *et al.* (1992) Nanosecond transient electroluminescence from polymer light emitting diodes. *Applied Physics Letters* **61**, 3092–4.
33 Smith, E. (2003) Optical out-coupling in light emitting polymer devices. *Physics Congress 2003*, Abstract Book. Institute of Physics.
34 Smith, E. (2001) LEP technology for microdisplays. *Digest of CUPID Microdisplay Conference*, Edinburgh, Scotland, 13–14 September.
35 Ali, T. A., Jones, G. W. and Howard, W. E. (2004) Dual-doped high T_g white organic light emitting devices on silicon. *SID Digest* **35**, 1012–15.
36 Shen, Z., Burrows, P. E., Bulovic, V. *et al.* (1997) Three color organic light-emitting devices. *Science* **276**, 2009–11.
37 Graupner, W., Heller, C. M., Ghosh, A. P. and Howard, W. E. High resolution color organic light emitting diode microdisplay fabrication method. *Proc SPIE*, **4207**, 11–19.

38 Kim, H. H., Swartz, R.G., Woodward, T.K. *et al.* (1993) Prospects for silicon monolithic opto-electronics with polymer light emitting diode. *Digest of LEOS 6th Annual Meeting*, 202–3.
39 Kim, H. H., Swartz, R. G., Ota, Y. *et al.* (1994) Prospects for silicon monolithic opto-electronics with polymer light emitting diodes. *IEEE J. Lightwave Technol.* **12**, 2114–21.
40 Kim, H. H. (1995) Organic light emitting diodes and silicon receivers for monolithic silicon opto-electronic circuits. *Digest of LEOS 8th Annual Meeting*, 343–4.
41 Mathine, D. L., Woo, H. S., He, W. *et al.* (1999) Organic LEDs heterogeneously integrated with CMOS circuitry. *Digest of LEOS 12th Annual Meeting*, 216–17.
42 Bulovic, V., Tian, P., Burrows, P.E. *et al.* (1997) A surface-emitting vacuum-deposited organic light emitting device. *Applied Physics Letters* **70**, 2954–6.
43 Zhou, X., He, J., Liao, S. *et al.* (1999) Enhanced hole injection in a bilayer vacuum-deposited organic light-emitting device using a p-type doped silicon anode. *Applied Physics Letters* **74**, 609–11.
44 Sanford, J. L. and Schlig, E. S. (2001) Direct view active matrix VGA OLED-on-crystalline-silicon display. *SID Intl Symposium Digest XXXII*, 376–9.
45 Sanford, J. L., Schlig, E. S., Prache, O. *et al.* (2002) *Proc SPIE*, **4464**, 11–22.
46 Stewart, M., Howell, R. S., Pires, L. *et al.* (1998) Polysilicon VGA active matrix OLED displays: technology and performance. *Technical Digest of IEDM*, 871–4.
47 Prache, O. (2002) Active matrix molecular OLED microdisplays. *Displays* **22**, 49–56.
48 Jones, G. W. (2001) Active matrix OLED microdisplays. *SID Digest* **32**, 134–7.
49 Ghosh, A. P. (2002) Full-color OLED on silicon microdisplay. *Proc SPIE*, **4464**, 1–10.
50 Prache, O. (2002) Full color SVGA+ OLED-on-silicon microdisplay. *J. SID* **10**, 133–8.
51 Lin, H., Naviasky, E., Ebner, J. *et al.* (2002) An 852×600 pixel OLED-on-silicon color microdisplay using CMOS subthreshold-voltage-scaling current driver. *Digest of ISSCC*.
52 Levy, G. B., Evans, W., Ebner, J. *et al.* (2002) An 852×600 pixel OLED-on-silicon color microdisplay using CMOS subthreshold-voltage-scaling current drivers. *IEEE J. Solid State Circuits* **37**, 1879–89.
53 Low, K., Jones, S.K., Prache, O. and Fellowes, D.A. (2004) Active matrix OLED for rugged HMD and viewfinder applications. *Proc SPIE*, **5442**, 280–90.
54 Barre, F., Chiquard, A., Faure, S. *et al.* (2006) OLED study for military applications. *Proc SPIE*, 5961,09-1–09-11.
55 *UXGA-QT Preliminary Data Sheet*; available from eMagin Corp.: www.emagin.com
56 Eldada, L. (2002) Polymer integrated optics: promise vs. practicality. *Proc SPIE*, **4642**, 11–20.
57 Parker, I. D. and Kim, H. H. (1994) Fabrication of polymer light-emitting diodes using doped silicon electrodes. *Applied Physics Letters* **64**, 1774–6.
58 Baigent, D. R., Marks, R.N., Greenham, N. C. *et al.* (1994) Conjugated polymer light-emitting diodes on silicon substrates. *Applied Physics Letters* **65**, 2636.
59 Yu, G., Pei, Q. and Heeger, A.J. (1997) Planar light-emitting devices fabricated with luminescent electrochemical polyblends. *Applied Physics Letters* **70**, 934–6.
60 Heinrich, L. M. H., Muller, J., Hilleringmann, U. *et al.* (1997) CMOS compatible organic light-emitting diodes. *IEEE Trans. Electron Dev.* **44**, 1249–52.
61 Heinrich, L. M. H., Horstmann, J. T. and Goser, K. (2000) PLEDs for integration with MOS circuits. *Proc. ESSDERC 2000*, **30**, 596–9.
62 Abraham, H. E. *et al.* (2001) XGA resolution full video microdisplay using light emitting polymers on a silicon active matrix circuit. *Proc SPIE*, **4105**, 37–52.
63 Blalock, T. N., Gaddis, N. B., Nishimura, K. A. and Knotts, T.A. (2000) 8-bit/color 1024×768 microdisplay with analog in-pixel pulse width modulation and retinal averaging offset correction. *IEEE Symposium VLSI Circuits Digest Technical Papers*, 20–27.
64 Bodammer, G., Buckley, A., Burns, D. *et al.* (2003) Emissive microdisplay using light emitting polymer. *Society for Information Display 2003 International Symposium Digest of Technical Papers* **34**, 742–745.
65 Bodammer, G., Buckley, A., Burns, D. *et al.* (2002) Emissive color microdisplay using light emitting polymer. *Digest of International Displays Research Conference*, 2002.
66 ME3203 product brief, available at www.microemissive.com
67 Underwood, I., Wright, J., Bodammer, G. *et al.* (2004) QVGA color microdisplay using light emitting polymer on CMOS. *Society for Information Display 2004 International Symposium Digest of Technical Papers* **35**, 293–5.

68 Dolny, G., Ipri, A., Hsueh, F-L. *et al.* (1993) High density active matrix electroluminescent display using single-crystal silicon-on-insulator high-voltage IC technology. *Technical Digest of the International Electron Devices Meeting* (IEDM).
69 Aguilera, M., Arbuthnot, L. and Keyser, T. (1996) First generation VGA electroluminescent display for head-mounted applications. *Proc SPIE*, **2735**, 92–8.
70 Aguilera, M. and Aitchison, B. (1996) Fabricating high resolution AMEL flat panel displays. *Solid State Technology*, November, 109–16.
71 Arbuthnot, L., Aitchison, B., Carlsen, C. *et al.* (1999) High luminance active matrix electroluminescent (AMEL) display for see-through symbology applications. *Proc SPIE*, **3689**, 260–8.
72 Jin, S. X., Li, J., Li, J. Z. *et al.* (2000) GaN microdisk light emitting diodes. *Applied Physics Letters* **76**, 631–3.
73 Choi, H. W., Jeon, C-W. and Dawson, M. D. (2003) InGaN micro-ring light emitting diodes. *IEEE Phot. Technol. Lett.* **15**, 510–12.
74 Choi, H. W., Jeon, C-W. and Dawson, M. D. (2003) Fabrication and performance of parallel-addressed InGaN micro-LED arrays. *IEEE Phot. Technol. Lett.* **15**, 513.
75 Jeon, C-W., Choi, H. W., Gu, E. and Dawson, M. D. (2004) High-density matrix-addressable AlInGaN-based 368-nm microarray light-emitting diodes. *IEEE Phot. Technol. Lett.* **16**, 2421–3.
76 Jiang, H. X., Jin, S. X., Li, J. *et al.* (2001) III-nitride blue microdisplays. *Applied Physics Letters* **78**, 1303–5.
77 Jeon, C-W., Choi, H. W. and Dawson, M. D. (2003) Fabrication of matrix-addressable InGaN-based microdisplays of high array density. *IEEE Phot. Technol. Lett.* **15**, 1516–18.
78 Jiang, H. X. and Lin, J. Y. (2001) Microdisplays based on III-nitride wide band-gap semiconductors put the future in our hands. *OE Magazine*, July.
79 Choi, H. W., Jeon, C. W. and Dawson, M. D. (2004) High-resolution 128 × 96 microdisplay. *IEEE Electron Device Letters* **25**, 277–9.
80 Feng, Z. C. and Tsu, R. (eds) (1994) *Porous Silicon*. World Scientific Publishing.
81 Uhlir, A. (1956) Electrolytic shaping of germanium and silicon. *Bell System Tech. J.* **35**, 333.
82 Turner, D. R. (1958) Electropolishing silicon in hydrofluoric acid solutions. *J. Electrochem. Soc.* **105**, 402–8.
83 Canham, L. T. (1990) Silicon quantum wire array fabrication by electrochemical and chemical dissolution of wafers. *Applied Physics Letters* **57**, 1046–8.
84 Richter, A., Steiner, P., Kozlowski, F. and Lang, W. (1991) Current induced light emission from a porous silicon device. *IEEE Electron Device Letters* **12**, 691–3.
85 Gelloz, B. and Koshida, N. (2000) Electroluminescence with high and stable quantum efficiency and low threshold voltage from anodically oxidized thin porous silicon diode. *Applied Physics Letters* **88**, 4319–24.
86 Lazarouk, S., Smirnov, A. and Labunov, V. (2003) Fabrication of a full inorganic EL microdisplay integrated with addressing circuitry onto a silicon chip. *Proc. Intl Conf. IDMC*, Taipei, Taiwan, 3–8.
87 Smirnov, A., Labunov, V. and Lazarouk, S. (2005) Si-based LED microdisplay: design and manufacturing process. *Proc IDRC* **25**, 474–6.

10

Projection Displays

10.1 Background

Projection displays are a large topic that forms the subject of a volume in this series.[1] We review the principles of projection in relation to microdisplays, with emphasis on reflecting devices that have the potential to win substantial market share. Polysilicon addressed transmission LCDs created the micro-display projector market, and remain strong competitors. Their transmission design sacrifices aperture ratio, but has the advantage of routing input and output light along different paths. In reflective devices the aperture ratio approaches 100%, but separation of output from input light requires beam-splitting components, or off-axis systems. Expert polarization engineering is crucial to the design of high-contrast LCD projectors, and the subject of a recent volume in the SID series.[2] The upgrade of consumer television to HDTV has expanded market opportunities for rear-projection systems based on microdisplays. The microdisplays and supporting projector components are improving rapidly to meet that challenge.[3]

10.2 Throughput

10.2.1 Lumen Scale

The optical flux in lumens delivered to the projection screen characterizes optical throughput. Screen luminance is also determined by screen area and screen gain. A Lambertian screen scatters light through angle θ, where scattered luminous intensity follows a $\cos\theta$ law. The projected screen area in the direction θ follows the same law, making the screen luminance independent of viewing angle. A screen fabricated to provide more luminance in a desired direction by sacrificing luminance in other directions is said to have screen gain relative to the unity gain Lambertian screen. For example, a unity gain (Lambertian) screen of area $1\,m^2$, with unit reflectivity, illuminated by 500 lumens has an illuminance of $500\,lm/m^2$ (500 lux), and a hemispherical average of $\cos\theta$ gives screen luminance

Introduction to Microdisplays D. Armitage, I. Underwood and S-T. Wu

$500/\pi = 159\,\text{lm/m}^2/\text{sr}$ (159 candela/m^2), independent of viewing angle. Increasing the screen gain to 3 concentrates luminance to $477\,\text{cd/m}^2$ in the favored viewing direction (on-axis), at the expense of other viewing directions, while screen illuminance remains 500 lux. Describing the projector output in lumen flux eliminates the image size and screen details, enabling a direct comparison between projectors.

A projected image will always appear better in low background light, due to higher contrast and color saturation. The higher the projector output lumens, the more tolerance to ambient light. Increased lumen output generally comes with increased cost, greater fan noise, and lower lamp life. Large screen projectors, such as cinema projectors, demand highest output ~10,000 lm. Pocket projectors can function with ~10 lm. HDTV rear projection requires >100 lm, with benchmark currently 500 lm.

10.2.2 Etendue Limit

High output lumen flux is desirable in a projector. However, raising the lumen source in a projector may not increase the output, due to optical collimation limits imposed by etendue (E) conservation:[1]

$$E = n^2 \int\int \cos\theta \, dA d\Omega \tag{10.1}$$

where n is refractive index, propagation angle θ is relative to the normal direction of area element dA, and the integral extends over the solid angle Ω. In an optical system, the etendue increases, or at best remains constant, as light propagates through the system. Thermodynamics imposes this entropy-like property on etendue, ensuring the highest luminance is at the source, and conservation of etendue implies conservation of luminance in a lossless system.

An arc lamp source has an effective emitting volume and surface area determined by the arc dimensions; radiation is independent of angle, but restricted by the opaque electrodes. The lamp etendue E_L is calculated from Equation (10.1), and given the output lumen flux L, the lamp luminance can be expressed L/E_L. The lamp collection and collimating optics increase optical propagation area and decrease angular divergence, while minimizing etendue expansion. The projector optical system etendue E_S is usually inadequate to accommodate the source etendue, implying:

$$\textit{Projector lumen throughput} < LE_S/E_L = (\text{source luminance})(\text{system etendue})\,. \tag{10.2}$$

This places an upper bound on throughput in etendue limited systems. High source luminance requires a low etendue source, achieved by reducing the arc gap, while generating sufficient lumen flux.

Applying (10.1) to the output of a microdisplay with area A, collected by a projection lens with f-number $1/2\sin\theta_o$, gives the output etendue:

$$E_o = \pi A \sin^2\theta_o = \frac{\pi A}{4F^2}\,. \tag{10.3}$$

E_o is normally the limiting etendue of the system ($E_S = E_o$), as indicated in Figure 10.1. An SXGA (1280×1024) microdisplay with $9\,\mu\text{m}$ pixel pitch has an area $A = 106\,\text{mm}^2$, and assuming $F = 2.4$, $E_o = 14.5\,\text{mm}^2\text{sr}$.

For a given lamp, the flux collected from the lamp is a function $L(E)$ of the collection etendue, which can be determined experimentally. A formula expressing the experimental data is given by Equation 10.6 (section 10.4) for the Philips UHP lamp, where $L(14.5) = 3406\,\text{lm}$, when $P = 100\,\text{W}$ and arc-gap $= 1.25\,\text{mm}$. Losses such as polarization, absorption, reflection, and color correction reduce the projector output to order 500 lm, depending on design details. The cost benefit of reduced area results in microdisplay projectors generally operating under "etendue limited" conditions, with high-luminance sources.

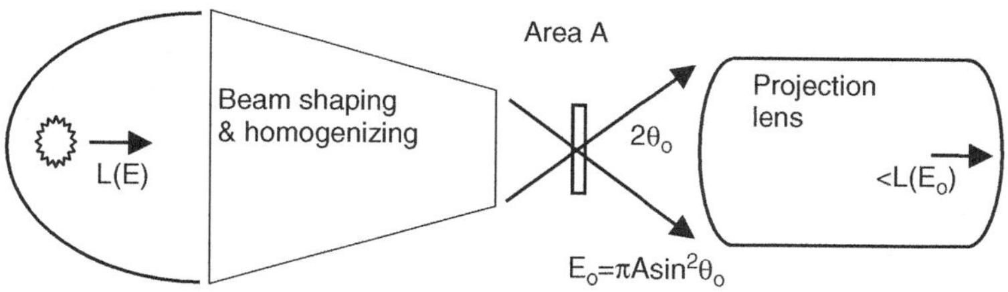

Figure 10.1 Lamp flux $L(E)$ is restricted to $L(E_o)$ by limiting etendue E_o

Tuned dielectric mirrors can superimpose etendues of different wavelength. There is no etendue penalty in splitting white light into primary colors for modulation, followed by recombination for projection. Etendues of the same wavelength cannot be superimposed. Compounding a number of sources, such as same-color LEDs, cannot increase luminance.

Arrangements of polarizing beam-splitters and rotators convert unpolarized light to polarized form, but the restriction on etendue overlap requires a doubling of the etendue with polarizing conversion. The only exception is a reflecting polarizer recycling the reflected polarization through the source to achieve conversion without an etendue penalty, but the source absorbs some of the recycled light.

10.3 Laser and LED Sources

10.3.1 Lasers

Lasers are the optimum light source in the long term. They are compact, low etendue, polarized, high color saturation and provide wide color gamut, with potential for high efficacy (lumen/watt) and long life.[4,5] Prototype displays have demonstrated that speckle issues can be resolved.[6,7] Steady progress in laser diodes has provided sufficient red power and lifetime, while similar developments in green and blue should follow. There is consensus that improved performance and lower cost will eventually bring visible laser diodes to the mass display industry, but the time scale may be several years or substantially longer.

Microlasers are compact solid-state lasers pumped by infrared diode lasers, and produce enough light for a projector.[6] The diode pump laser is inexpensive and long-lived. However, the manufacturing costs of microlasers need to be much lower if they are to compete in the projector market.

Recent developments by Novalux establish a path to large-scale manufacturing at competitive pricing.[8] Novalux is developing novel lasers based on frequency-doubled VECSELs (vertical extended-cavity surface-emitting lasers), also known as NECSELs (Novalux extended-cavity surface-emitting lasers). The small etendue favors combination of an array of vertically emitting lasers, distributing the thermal dissipation and attenuating speckle. Thousands of infrared lasers are fabricated on a four-inch diameter wafer and can be tested before separation into arrays or individual lasers. Addition of a periodically poled lithium niobate crystal achieves frequency doubling of the IR light into the visible. Predicted service life >20,000 hours implies a light source matching the lifetime cycle of a consumer projector. Several microdisplay projector companies are developing systems based on NECSEL lasers.

Electrical on/off switching of lasers simplifies the structure of a color sequential projection system. The overall simplicity and effectiveness of the system is an incentive for rapid development. Lasers provide polarized light, simplifying the design of liquid crystal systems and raising interest in color sequential liquid crystal projectors.

The laser power required depends on the laser wavelengths, lumen output, and desired white (x,y) color coordinates, as illustrated by the following calculation.[1] Table 10.1 gives the laser wavelengths, with color matching functions from tabulated values. Assume 500 screen lumens, with white correlated

Table 10.1 Lower limit of laser power for tabulated conditions

Color	Red	Green	Blue	8200 K white
λ (nm)	630	530	450	sum
CIE CMF $\bar{X}$	0.64240	0.16550	0.33620	Color coordinate $x_w = 0.3137$
CIE CMF $\bar{Y}$	0.26500	0.86200	0.03800	Color coordinate $y_w = 0.3290$
CIE CMF $\bar{Z}$	0.00005	0.04216	1.77211	$z_w = 1 - x_w - y_w = 0.3573$
Flux (lumen)	$L_R = 119.7$	$L_G = 366.4$	$L_B = 13.9$	Sum = 500
Power (watt)	$P_R = 0.661$	$P_G = 0.622$	$P_B = 0.536$	Sum = 1.819

color temperature 8200 K having color coordinates ($x_w = 0.293$, $y_w = 0.303$), typical for a projector, and generating the following tristimulus X, Y, Z values:

$$Y = 500, \quad X = \frac{Yx_w}{y_w} = 483.5, \quad Z = \frac{Y(1 - x_w - y_w)}{y_w} = 666.7. \tag{10.4}$$

The lumen flux of each color L_R, L_G, L_B is given by

$$\begin{bmatrix} L_R \\ L_G \\ L_B \end{bmatrix} = \begin{bmatrix} \frac{\bar{X}_R}{\bar{Y}_R} & \frac{\bar{X}_G}{\bar{Y}_G} & \frac{\bar{X}_B}{\bar{Y}_B} \\ 1 & 1 & 1 \\ \frac{\bar{Z}_R}{\bar{Y}_R} & \frac{\bar{Z}_G}{\bar{Y}_G} & \frac{\bar{Z}_B}{\bar{Y}_B} \end{bmatrix}^{-1} \begin{bmatrix} X \\ Y \\ Z \end{bmatrix}. \tag{10.5}$$

Ideal laser output powers P_R, P_G, P_B, calculated from lumens values are:

$$P_R = \frac{L_R}{683\bar{Y}_R}, \quad P_G = \frac{L_G}{683\bar{Y}_G}, \quad P_B = \frac{L_B}{683\bar{Y}_B}. \tag{10.6}$$

The ideal laser power is a lower limit that assumes no loss between laser output and that delivered to the screen. A demonstration projector powered by lasers achieved 33% throughput efficiency, including beam shaping and speckle reduction losses.[6] Such a system would require a threefold increase in power values listed in Table 10.1. Projectors employing NECEL lasers demonstrate much higher throughput efficiency. Microdisplay projection is tolerant to poor beam quality, 1000 times diffraction limits. Competing scanning laser systems require high beam quality in at least one dimension.

10.3.2 Light Emitting Diodes

The performance and cost of LEDs is improving at a faster rate compared to laser diodes, due to simpler structure and demands from general lighting. However, LEDs are approximate Lambertian emitters; consequently, system etendue limits the throughput. Packaged LEDs have provided the light source for small projectors (pocket projectors) that display projected images comparable in size and luminance to a laptop.[9] Such projectors may employ field-sequential-color, exploiting LED pulse modulation. Predicted LED output rises considerably over the next few years, as efficiency is improved and higher current achieved.[10,11]

Etendue considerations restrict the LED die emitting area to ~5 mm^2, which must dissipate ~10 W power to provide adequate luminance for the consumer rear-projection market. Predicted service life of 50,000 hours degrades with increasing junction temperature, imposing severe heat sinking requirements. To achieve adequate luminance, projector design should emphasize etendue conservation and accommodate maximum source etendue.

The highest performance requires an array of bare LED die-mounted on an efficient heat sink, and coupled to a tapered waveguide preserving etendue.[12] Such an experimental system is consistent with 280 screen lumens from a projector employing three 0.78-inch diagonal LCOS microdisplays with *f*/1.8 projection lens. Polarization conversion with a retro-reflecting polarizer preserves etendue. Anticipated improvements in performance expect to boost output to a commercially acceptable value approaching 400 screen lumens.[12] A substantial improvement in lumen throughput is achieved by sub-wavelength photonic lattice structures in the LED surface that enhance the optical output in a smaller angular aperture.[13] Further developments in LED performance, particularly the green LED, plus the photonic lattice innovation, are opening the RPTV market to LED projectors. Demonstration of several LED powered microdisplay rear-projectors has been well received.

The light from a number of same-wavelength LEDs cannot be added in projection without etendue expansion; however, the color wavebands of typical R, G, and B primaries are broad enough to accommodate more than 1 LED/band provided they can be efficiently combined with appropriate dielectric mirrors.

10.4 Arc Lamps

10.4.1 Lamp Structure

Highest power projectors use xenon arc lamps, operating at 6 kW or higher, standard cinema equipment. Their visible spectrum is similar to sunlight and color temperature remains constant over lifetime; small arc dimensions ~1 mm provide low source etendue. Modest lifetime and efficacy 18–40 lm/W are substantial drawbacks, and cold-state pressure of 10 atmospheres is an explosion hazard during lamp replacement. Erosion of the tungsten electrodes limits lifetime, causing the gap to increase and the arc to become unstable. Experimental studies on operating lamps below rated power showed extended life.[14]

Standard mercury arc lamps are high luminance and long-lived, but unsuitable for displays because of red spectral deficiency. Metal halide arc lamps introduce metal halide species into a mercury discharge to enhance emission in desired regions of the spectrum, resulting in efficacy as high as 100lm/W. Bromine is included in the burner to recycle tungsten sputtered from the electrodes, preventing it from obscuring the burner envelope. Short arcs approaching 1 mm are essential to meet the etendue demands of microdisplays. Efforts to develop a short-arc metal halide lamp that would meet the consumer goal of 10,000-hour lifetime have not succeeded. The addition of metal halides to improve the spectrum and efficacy of a mercury discharge sacrifices the very high luminance of a pure mercury discharge. Microdisplay projectors demand long-lived source luminance beyond 1 Gcd/mm^2, which is unlikely to be achieved in a metal halide lamp.[15]

The search for an efficient long-life short-arc lamp led Philips to develop mercury lamps with very high pressure approaching 300 atmospheres. Raising the operating pressure increases the continuum radiation, improving the color balance and enhancing the efficacy to 60lm/W; Figure 10.2 illustrates the spectrum. The discharge voltage also increases with pressure, allowing a reduction in current and longer life due to lower electrode erosion. The Philips UHP lamp is an ultra-high-pressure mercury lamp, with a small quantity of bromine and oxygen included to scavenge and recycle tungsten evaporated from the electrodes.[15] Ultra-high-pressure mercury discharge lamps now dominate the low-power projector market, due to lifetime beyond 10,000 hours and high luminance. When switched off the lamp cools and the mercury condenses allowing the pressure to approach atmosphere. There is only a risk of explosion during operation, and small burner volume limits the energy released.

A number of important advances followed the introduction of the UHP lamp. Erosion of the electrodes creates a rough electrode structure susceptible to arc jumping instability. Arc fluctuations cause screen luminance variations perceived as flicker. A current pulse designed to raise the local electrode temperature before current reversal stabilizes the arc.[16] Lamp switch-on requires 20 kV to initiate arc formation by field emission at the electrodes. A photoelectric initiation method reduces the arc ignition voltage to 5 kV, enabling space and cost savings in the power supply.[16] A reflector deposited on one

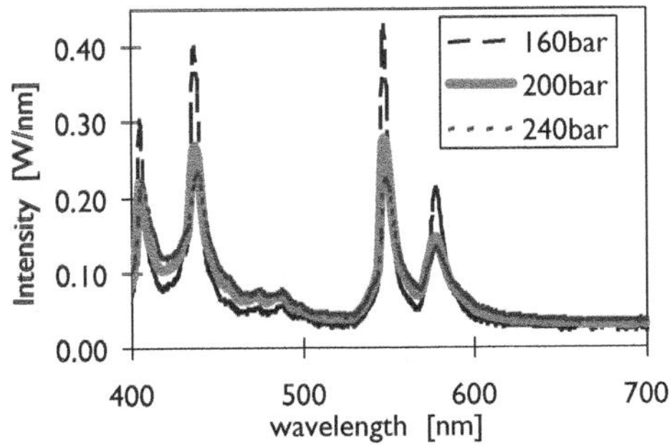

Figure 10.2 Spectrum of UHP lamp. Reprinted courtesy of Philips

hemisphere of the burner reduces the etendue by retro reflection through the arc plasma, as shown in Figure 10.3. The lamp luminance increases 55%, and the color efficiency improves because transmission through the arc plasma is greater in the continuum region of the spectrum.[16] The lower source etendue allows a more compact optical collector. Further developments progress towards compact, efficient, and inexpensive lamp/power supplies for microdisplay projectors.

10.4.2 Lamp Output

Intense study of the UHP lamp over the past few years has provided a comprehensive formula that predicts the lumen output for given operating conditions, enabling an estimate of throughput for a given projector design. The standard UHP lamp output flux approximates:[17]

$$L = 40.5 \frac{P}{\frac{V_{\mathrm{elec}}}{adp} + 1} \tan^{-1}\left(\frac{E}{3.8d^2 + 0.9d + 0.8} \right) \qquad (10.7)$$

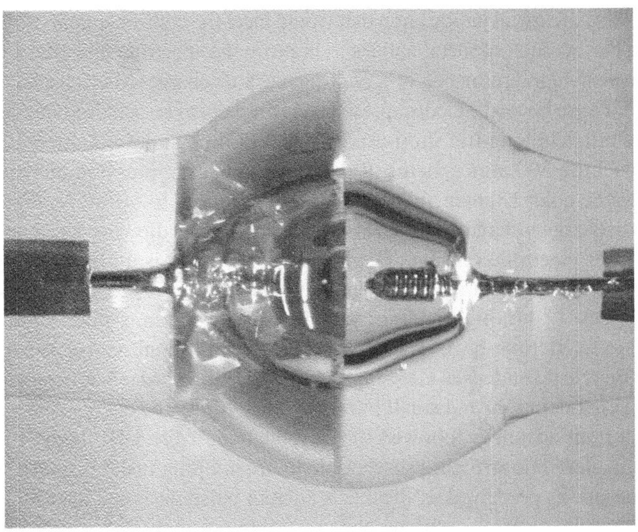

Figure 10.3 Dielectric mirror covering half of burner. Reprinted courtesy of Philips

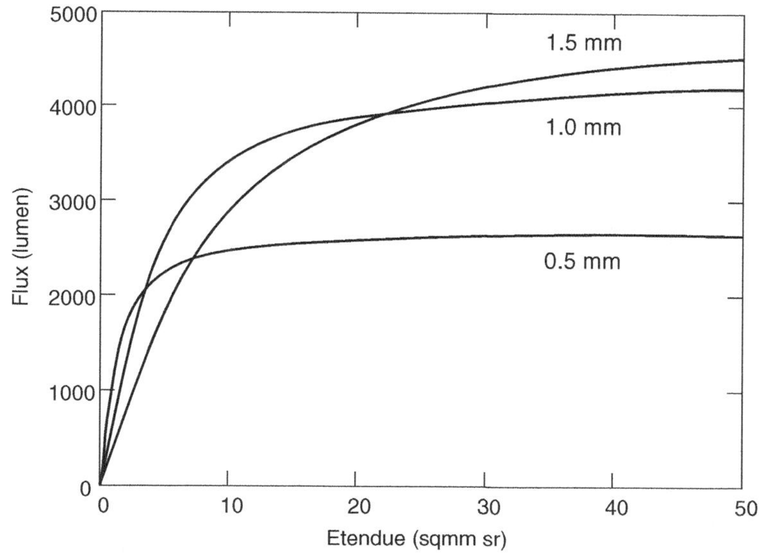

Figure 10.4 Collected flux against etendue for arc gaps: 0.5, 1.0 and 1.5 mm, pressure 200 bar, power 100 W. Highest flux corresponds to smallest gap at low etendue.

where L is flux (lumen), P lamp power (watt), E etendue ($mm^2 sr$), d arc gap (mm), p pressure (bar), $a = 0.26$ (V/mmbar), and electrodes volt drop $V_{elec} = 18$ V. The arc-to-electrode voltage drop represents power absorbed by the electrodes rather than arc plasma, and increases in significance at smaller arc gaps. Raising the pressure increases the arc voltage, compensating reduction in arc gap, giving rise to the dp product in Equation (10.7). The constant 40.5 is adjusted to match the flux output of a 100 W, 1.3 mm arc as a function of etendue. Figure 10.4 shows a plot of (10.7), giving the flux available to a system restricted by etendue value E. Small etendue favors small arc gap, but for a given E value, the output flux is maximized for an appropriate choice of d.

The xenon and metal halide lamps scale to high powers of order 10 kW, but UHP mercury lamps normally operate over a limited range of power 50–275 W, due to temperature restrictions on the burner wall. An upper temperature is set by crystallization of the quartz containment, which is true for all lamps. The UHP lamp also has a lower limit on burner wall temperature to maintain the operating pressure. The modest temperature range (<230 K) of the burner wall determines the allowed power level. Moreover, the burner volume is restricted to limit convection that would induce excessive temperature gradients. A recent development, directed cooling of the lamp, extends power limit ~800 W, providing cinema level lumen output.[17]

Figure 10.5 shows the advantage of a lamp, with a half-mirror burner, at low etendue. The standard lamp provides more flux at sufficient etendue, because of the loss in burner retro reflection. Difficulties in fabricating a dielectric mirror to withstand operating temperature and temperature cycling may inhibit production.

Electrode limited lifetime is overcome by microwave powered lamps, and such electrode-less lamps also eliminate glass to metal lamp seals.[18] They can accommodate a wider range of fill materials such as sulfur, as well as mercury and metal halides. Difficulty in reducing the lamp etendue and the ascendancy of UHP lamps has resulted in a decline of interest in electrode-less lamps.

10.5 Polarizing Optics

Liquid crystal devices are polarization modulators, requiring the support of polarizing optics. Arc lamps produce unpolarized light, which involves loss in converting to polarized form. Microdisplays such as

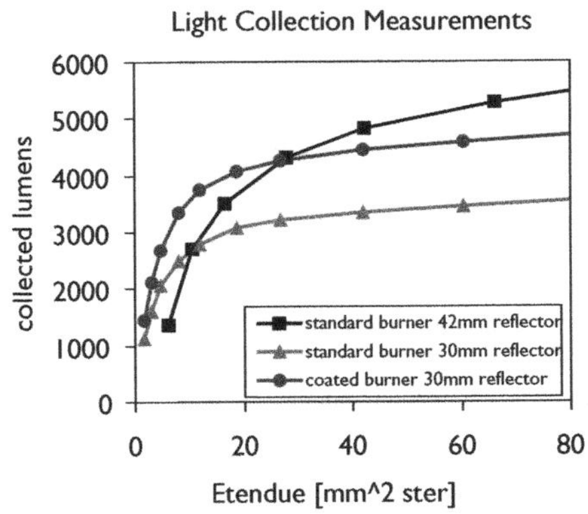

Figure 10.5 Advantage of half-mirror burner at low etendue. Reprinted courtesy of Philips

the DMD that accept unpolarized light have a throughput advantage. Liquid crystal devices based on scattering (PDLC) do not require polarized light, but have difficulty in achieving high throughput and high contrast. Diffractive LC devices are less sensitive to polarization, but diffraction introduces an expansion in etendue, with consequent sacrifice in throughput. The polarizing optical components used in projectors are described, stressing recent developments.

10.5.1 Absorbing Polarizer

Absorbing polarizers contain dyes oriented in a polymer film to absorb light polarized in the oriented direction, while transmitting light in the orthogonal polarization. The transmission parallel $T(\lambda)_{\parallel}$ and perpendicular $T(\lambda)_{\perp}$ to the polarizer transmission axis characterize the polarizer as a function of wavelength:

$$\mathit{Extinction\ ratio} = \frac{T(\lambda)_{\parallel}}{T(\lambda)_{\perp}}. \tag{10.8}$$

High extinction ratio enables high contrast ratio, but the highest extinction ratio requires some sacrifice in $T(\lambda)_{\parallel}$, and consequent throughput loss. Extinction ratio of order 1000 can be achieved with $T(\lambda)_{\parallel} = 88\%$. The polarization axis is orthogonal to the dye orientation, and is not sensitive to the incidence angle of the optical beam.

Absorption of light by the polarizer raises the temperature in and around the polarizer, preventing high-power microdisplay applications. The material has poor thermal conductivity and must be air-cooled. Absorption polarizers are compact and inexpensive, finding favor in transmission microdisplays of medium or lower power.

10.5.2 Polarizing Beam-splitter

Calcite polarizing beam-splitters (PBS) are expensive. The PBS found in reflective microdisplay projectors are usually of the MacNielle type, which depend on the optical properties of dielectric layers. The polarized state of an incident beam determines p-vector and s-vector orthogonal components at a dielectric interface. The p-component is in the incident plane of the beam, and the s-component orthogonal to that plane. The transmitted and reflected p and s components have differing dependence

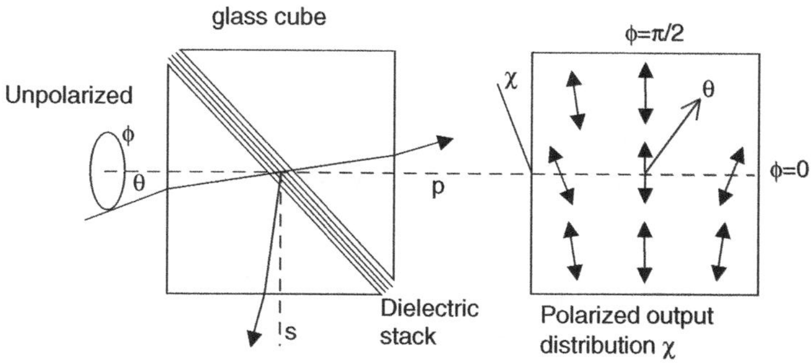

Figure 10.6 MacNielle polarizing beam-splitter showing the rotation χ of skew ray polarization in the output plane

on incident angle, enabling separation of s and p components. At the Brewster, or polarizing angle, the reflected p component is zero. The MacNielle polarizer amplifies the interface effect by a dielectric stack designed to optimize the performance over wavelength and angular range. The dielectric stack is located on an internal diagonal plane of a glass cube, centering the incident angle at 45 degrees, where the input and output beam directions are orthogonal, as shown in Figure 10.6.

Propagation directions of the transmitted and reflected rays follow the standard laws of refraction and reflection. The transmitted ray ($T_p + T_s$) is largely polarized T_p, with a small T_s, whereas the reflected ray ($R_p + R_s$) is mainly polarized R_s with small R_p. Transmission CR $= T_p/T_s$, and reflection CR $= R_s/R_p$. The input angle of a ray in air is written (θ,ϕ), consistent with stack plane normal direction $(\pi/4,\pi/2)$. The angular relation of ray and stack plane determine the polarization orientations of s and p. The s-polarization vector is given by the vector product of the ray (in medium of refraction n) and surface normal: $(\pi/4,\pi/2)\times(\theta/n,\phi)$; however, there is a small change in polarization for an oblique ray exiting the cube surface. Depolarizing effects should be considered at all dielectric interfaces in the system.[19] It is sufficient to consider the limit of small θ, where the rotation (χ) of s and p vectors becomes:

$$\theta \to 0, \ \chi = \frac{\theta}{n}\cos\phi \, . \tag{10.9}$$

High-index ($n \sim 1.7$) PBS materials minimize the sensitivity to incidence angle. Integrating χ^2 over a uniformly filled circular aperture gives an upper limit on extinction, or contrast ratio:

$$\theta \to 0, \ \mathrm{CR} < \frac{\pi\theta^2}{\int_0^{\theta}\int_0^{2\pi}\frac{\theta^2}{n^2}\theta\cos^2\phi\, d\theta\, d\phi} = \frac{4n^2}{\theta^2} \, . \tag{10.10}$$

The arrangement of a PBS aligned with a quarter-wave retarder and mirror, as shown in Figure 10.7, eliminates the round-trip skew ray depolarization known as the compound angle effect.[19–21] The first-pass depolarization $-\chi$ would double at the second-pass reflection from the dielectric stack, since the mirror reflection preserves linear polarization. However, the double-pass quarter-wave retarder acts as a half-wave retarder, rotating the depolarization $-\chi$ into depolarization $+\chi$, which matches the polarization plane at the output reflection. The CR given by Equation (10.10) with twice the angle, $CR < (n/\theta)^2$, is the value before correction by the retarder. Numerical apertures $0.1 < \sin\theta < 0.25$ apply to projection systems, where it is essential to correct the compound angle effect to achieve high CR. With correction, CR and throughput are limited by the dielectric stack details. An alternate compensation method employing an o-plate retarder was recently demonstrated.[22]

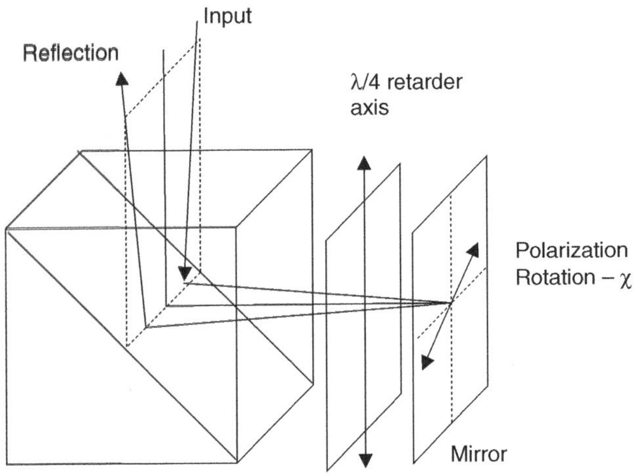

Figure 10.7 Compound angle effects. The reflected ray polarization $-\chi$ is corrected by the retarder to $+\chi$, matching the polarization plane at the final reflection (refraction angles suppressed for clarity)

The MacNielle PBS employs inorganic dielectric layers such as SiO_2 and TiO_2, which have proved highly stable at elevated temperature and high light intensity. The coating design must compromise between high transmission and high reflection over the chosen waveband and angular range, resulting in some sacrifice of throughput. Restricting the design waveband to a single color enhances the performance over the broadband case. The glass index is matched to the polarizing angle of the dielectric layers, and a compromise between high transmission and low stress birefringence is made. Light absorbed by the beam-splitter will enhance thermal gradients, raising the stress level.

The 3M PBS is a recent development in polarizing beam-splitters that replaces the dielectric stack of inorganic coatings with polymer films.[23] It is similar in appearance to the MacNielle PBS, but exploits birefringence, rather than Brewster angle effects. Film birefringence sets a unique optic axis independent of incident angle, making the s and p vectors redundant descriptors. Cartesian coordinates identify the polarization components as indicated in Figure 10.8, giving rise to the term "Cartesian PBS." The alternating polymer film stack is uniform for a unique polarization plane p_{yz}, allowing such light to pass without reflection. The orthogonal polarization p_{xz} encounters a strongly reflecting quarter-wave polymer stack.

Excellent wide-band and wide-angle performance is achieved with film layer counts of order 800, but a mismatch in glass index with the significantly thick polymer stack causes appreciable astigmatism.

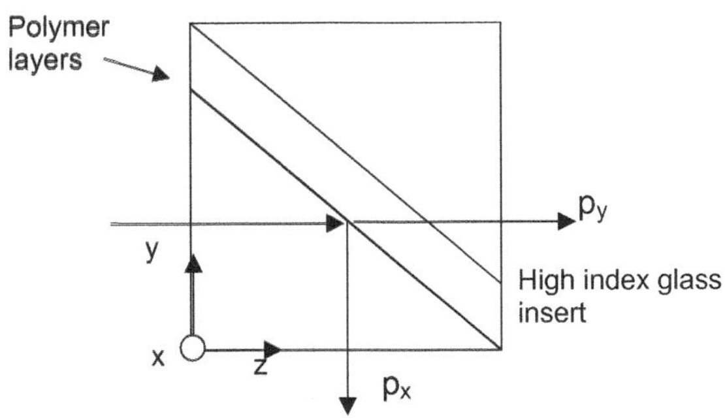

Figure 10.8 3M PBS employing birefringent polymer film stack

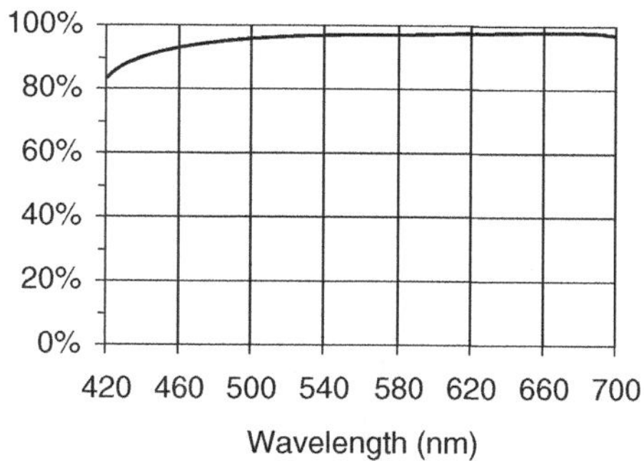

Figure 10.9 Transmission of 3M PBS at *f*2. Reprinted courtesy of 3M

Low stress-birefringence favors high-index flint glass, subject to replacement when suitable glass with $n \sim 1.55$ becomes available. Insertion of a high-index glass plate as shown in Figure 10.8 corrects the astigmatism in the present structure.

Polarized recycling systems for flat panel displays employ stretched polymer film stacks, illustrating the large manufacturing base for polymer film structures. The new polymer PBS is conservatively rated beyond 20,000 hours provided UV and temperature levels are restricted to levels acceptable to typical liquid crystal devices. A quarter-wave retarder to eliminate residual compound angle effects improves the performance. Figures 10.9 and 10.10 illustrate the performance at *f*/2, where contrast in a double-pass reflective microdisplay configuration is shown. The 3M PBS will have a substantial impact on reflective microdisplay projectors when it becomes widely available.[24]

10.5.3 Wire Grid Polarizers

Wire grid polarizers (WGP) are effective when the grid structure is small compared with wavelength, and have been used for many years in infrared optics. Advances in lithography have enabled the fabrication of WGPs for visible light, a recent innovation.[25–28] The WGP shown in Figure 10.11 has aluminum rib width 69 nm, thickness 100–200 nm, and period 144 nm, supported by a glass substrate. The polarization component parallel ($R_{\parallel}$) to the grid direction is reflected, while the orthogonal component ($T_{\perp}$) is transmitted. The spatial orientation of the grid sets the polarization properties for all rays, freeing the WGP of the skew ray depolarization found in MacNielle-type polarizers.

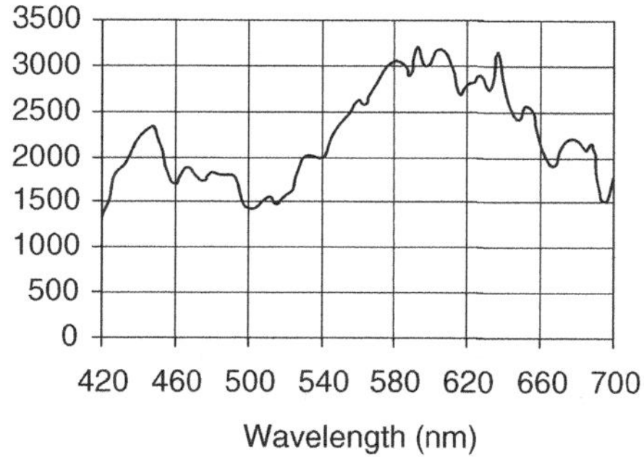

Figure 10.10 Contrast ratio of 3M PBS at *f*2. Reprinted courtesy of 3M

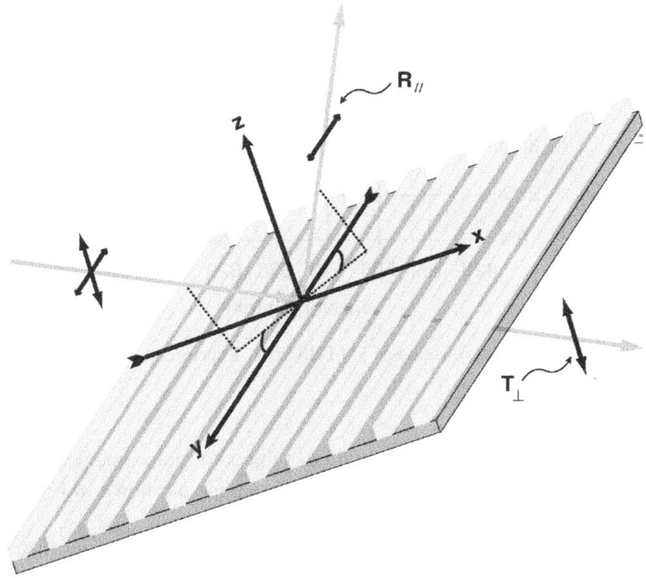

Figure 10.11 Wire grid polarizer: aluminum grid on glass substrate. Reflected polarization $R_{\parallel}$ is parallel to wire grid and transmitted polarization $T_{\perp}$ orthogonal to grid. Reprinted courtesy of Moxtek

Advantages of the WGP include wide angular range, low stress birefringence, modest absorption, and low cost. It has been shown to withstand continuous power density of at least 7 Mlx, although the temperature rises to approximately 100 °C. Modification to the structure enables a range of polarizers from high transmission to high contrast. It can also be optimized for normal incidence, or 45° incidence to form a polarizing beam-splitter on optical flat glass.

Figures 10.12–10.15 show the performance of a WGP beam-splitter. The transmission CR is much higher than the reflection CR. The WGP can substitute for absorbing polarizers in a standard transmis-

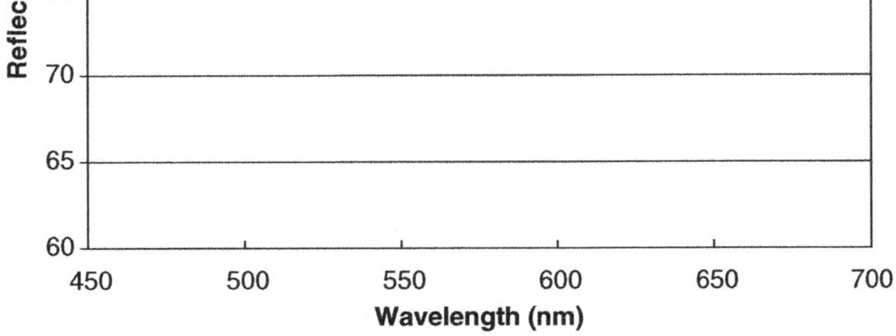

Figure 10.12 Reflectivity $R_{\parallel}$ of wire grid polarizing beamsplitter at 45-degree incidence. Reprinted courtesy of Moxtek

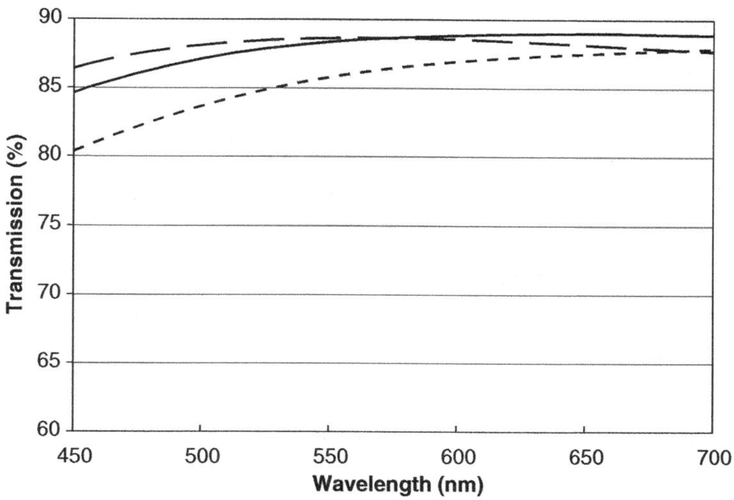

Figure 10.13 Transmission $T_{\perp}$ of wire grid polarizing beam-splitter at 35 (dashed), 45 (solid) and 55 (dotted) degrees of incidence. Reprinted courtesy of Moxtek

sion microdisplay projector, provided the reflected light is properly routed. In addition, WGP can substitute for a polarizing beam-splitter in a reflective microdisplay projector, with the help of a "clean up" dichroic polarizer to boost CR.[28,29] Optically flat glass in the WGP minimizes distortion and maintains MTF in the image path of the microdisplay. The WGPs are in service as pre-polarizers and beam-splitters for projectors and expected to make further gains over traditional polarizers.

10.5.4 Polarization Conversion

An unpolarized light source is polarized by simply inserting any of the polarizers previously described, but half the light is discarded. The rejected component of polarization can be manipulated

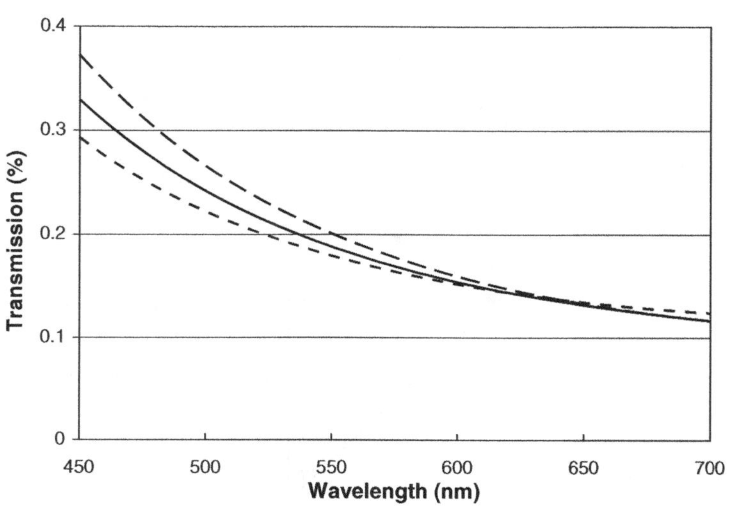

Figure 10.14 Transmission $T_{\parallel}$ of polarization parallel to wire grid polarizing beamsplitter at 35 (dashed), 45 (solid) and 55 (dotted) degrees of incidence. Reprinted courtesy of Moxtek

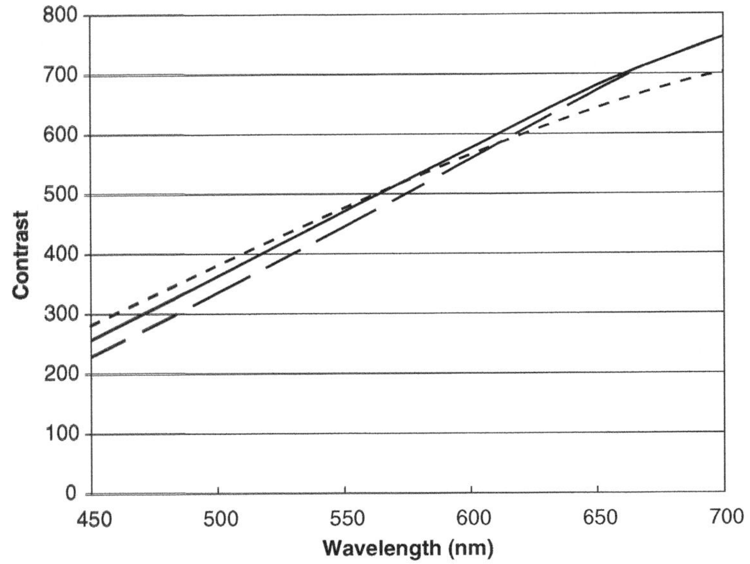

Figure 10.15 Transmission contrast ratio $T_{\perp}/T_{\parallel}$ of wire grid polarizing beamsplitter at 35 (dashed), 45 (solid) and 55 (dotted) degrees of incidence. Reprinted courtesy of Moxtek

by polarizing rotation devices to reinforce the accepted component as shown in Figure 10.16. The PBS separates unpolarized light into *s* and *p* components, and the *s* component is converted to *p*-polarization by an appropriately oriented $\lambda/2$ retarding plate. The light area is doubled, consistent with a doubling of etendue.

Polarization conversion systems (PCS) are usually combined with beam homogenizing systems as illustrated in Figure 10.17.[30] In this case, the area remains the same, but the optical divergence doubles. Direct illumination of the mirrors is blocked, since such light would emerge with the wrong polarization.[1] An integrator rod homogenizer incorporating PCS has been demonstrated.[31]

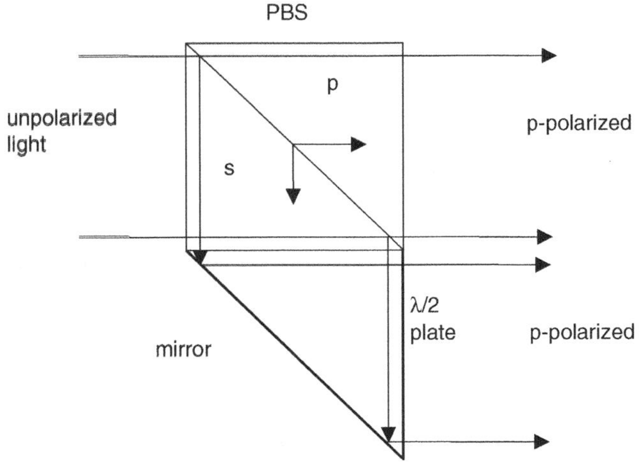

Figure 10.16 Polarization conversion

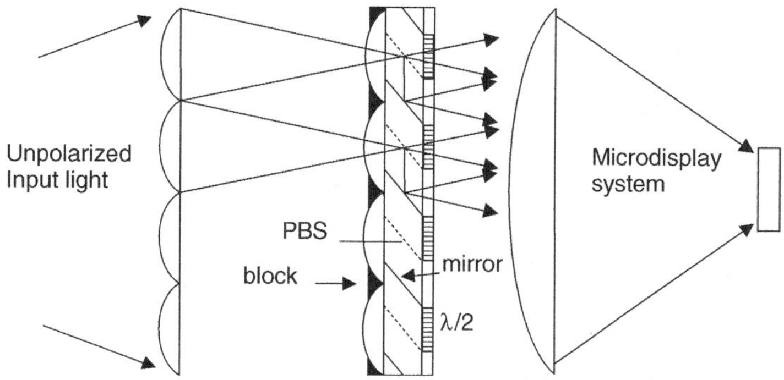

Figure 10.17 Polarization conversion and beam homogenizing

Polarization conversion efficiency approaches 80%, but the doubling of etendue sacrifices throughput. Details of the lamp and system etendue will determine if a particular PCS provides significant enhancement of the throughput. The example quoted in section 10.1 gave $L(14.5) = 3406$ lm, and without polarization conversion would provide less than $3406/2 = 1703$ polarized lumens. With polarization conversion the source etendue is doubled, which is equivalent to halving the system etendue to give $L(7.3) = 2384$ lm, and 80% conversion efficiency provides 1907 polarized lumens. Conversion provides a small increase (12%) in polarized lumens. However, if the mirrored-burner plot of Figure 10.5 is used, $L(7.3) = 3250$ giving 2600 polarized lumens, a substantial enhancement of 33% over the $L(14.5)/2 = 1950$ value.

10.5.5 Optical Compensation

The dark state of an LC device depends on residual birefringence effects, which largely determine the CR. Optical compensation introduces auxiliary retarder plates to compensate the device residual polarization shift, improving the dark state while leaving the bright state essentially unchanged, thereby raising the CR. Extensive optical compensation of direct-view LCDs is required to extend the range of viewing angles. Magnification intrinsic to microdisplays averages the dark state over the angular aperture of the viewing optics. The effective viewing angle is less than 15 degrees (*f*/1.93 projection optics) in all azimuth directions, averaging out grayscale shifts with viewing angle. The design of optical compensation for microdisplays needs to take into account the LC configuration and projection *f*-number.

The zero pretilt vertically aligned nematic cell approaches infinite CR for on-axis rays, but off-axis rays at *f*/2 limit the CR to about 1000. Compensation with an equal and opposite retardation requires a negative C-plate. Tilting of the C-plate can also compensate the nematic pretilt needed to suppress reverse-tilt disclinations.[32] Uniformity requirements of retardation compensation are stringent.[33] Compensation can also be achieved by slight rotation of the quarter-wave plate incorporated in reflective systems employing MacNielle beam-splitters. Twisted nematic cells are partly compensated by simple retardation plates.[34] However, general compensation is achieved by twisted birefingence structures.[19]

10.6 Color Management

Superposed primary color images provide the full color image in projection systems. The superposition may be in time sequence, where the frame rate is sufficient for color fusion in the viewer's eye. Alternatively, the primaries appear together, passing through the system in parallel. Color management addresses the choice of color options and color routing through the system, and considers the various compromises to achieve a particular goal.

In color temperature, one might expect the D65 standard representing the appearance of color in daylight to be favored. However, given an option to vary color temperature the typical viewer prefers a higher color temperature. The eye has a logarithmic response to luminance, which determines gray level settings in the projector. Given a gray level adjustment (gamma) control, the average viewer will distort the gray levels to give a more pleasing image. These adjustments to color and gray level also vary with the image characteristics, e.g. facial or landscape. The manufacturers of consumer displays manipulate the video signal to present the most appealing image to their customers. Electronic look-up tables perform such fine-tuning of the image, transforming video signal into addressing voltages. Color temperature and gamma settings are subsidiary to the general aims of color management.

10.6.1 Field-Sequential-Color

Given an arc lamp providing white light, a rapidly rotating color wheel, synchronized to the microdisplay addressing electronics, generates a sequence of primary colors. Figure 10.18 illustrates a simplified DMD system.[35] The number of dichroic color filter segments on the wheel is a tradeoff in wheel speed. Electrically switched liquid crystal shutters provide an alternative to the mechanical color wheel.[36]

A standard video signal provides color and luminance for red, green, and blue at 60 frame/s, demanding a minimum color sequential DMD frame rate 180 frames/s. Color fusion occurs at that frame rate, but color breakup due to eye motion is intolerable. The time sequential primary colors must superimpose precisely on the viewer's retina to provide the intended color image. Sufficient head or eye motion will shift the primary colors in edge details of the image, creating white edges in the image. Tripling the frame rate to 540 frames/s provides adequate protection against color breakup. Pulse width modulation simulates grayscale in the DMD, requiring eight sub-frames (bit-frames) per primary for 24-bit color. Different color bit-frames can be interleaved to simulate the required color frame rate needed to suppress color breakup. Achieving sufficient frame rate is the first hurdle that a color sequential system must overcome to provide an acceptable image. Efficiency is the second hurdle.

The color wheel transmits 1/3 of the light at best, which is a severe handicap to throughput efficiency. There is throughput advantage in increasing the field time for red exposure to compensate for red deficiency in UHP lamps, rather than just attenuating green and blue.[37,38] Including a white section on

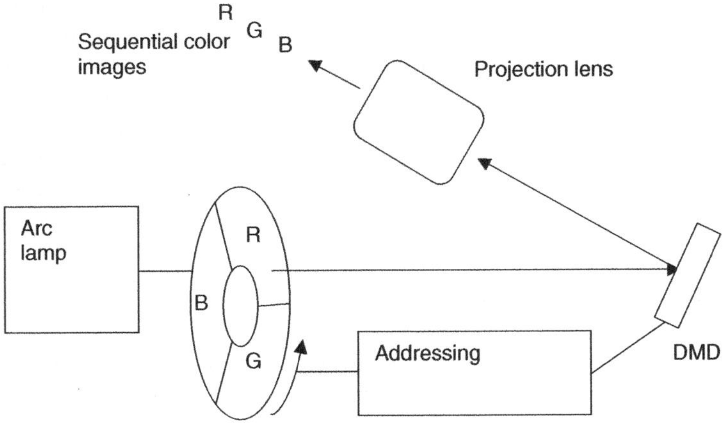

Figure 10.18 Color sequential system with red, green, blue color wheel synchronized to microdisplay addressing electronics

the color wheel raises the efficiency, but sacrifices color saturation. In consumer applications, restricting the lamp power to 100 W achieves maximum lifetime, and reduced heating and fan noise are also valued. Inadequate projector efficiency implies higher lamp power, forfeiting the advantages of a low powered lamp.

A single-microdisplay system eliminates the cost of additional microdisplays, supporting components, and the cost of achieving precise alignment over the projector life. These advantages are strong motivations for color sequential systems and account for the success of the DMD in consumer RPTV. Liquid crystal microdisplays have difficulty in achieving sufficient frame rate together with adequate throughput efficiency. The switching speed of the pulse width modulated FLC falls short of requirements. Work continues on the development of high-speed nematic liquid crystal microdisplays, and 450 frames/s have been demonstrated, together with reasonable system throughput.[37,38]

10.6.2 Color Scrolling Systems

Color scrolling introduces considerable complication in the color sequential system, to achieve higher throughput efficiency.[39,40] The LCOS microdisplay system shown in Figure 10.19 illustrates the method. Three rotating prisms, synchronized to the addressing electronics, form a scrolling RGB band pattern on the microdisplay. As a color moves off the bottom of the scroll it reappears at the top, so the light from each band is always projected. Separating the colors in this way expands the etendue by a factor three; therefore, the area of the microdisplay increases by the same factor for given *f*-number and throughput. Red band area may be increased to compensate for red deficiency, similar to the field-time adjustment mentioned earlier. The small dark gaps between color scrolling bands allow some time for the liquid crystal to respond to the new color addressing voltage.

Color break-up is suppressed by a 150 Hz color frame rate, i.e. total 450 frames/s. The presence of all three color-bands on the screen appears to raise the tolerance to color break-up. A nematic liquid

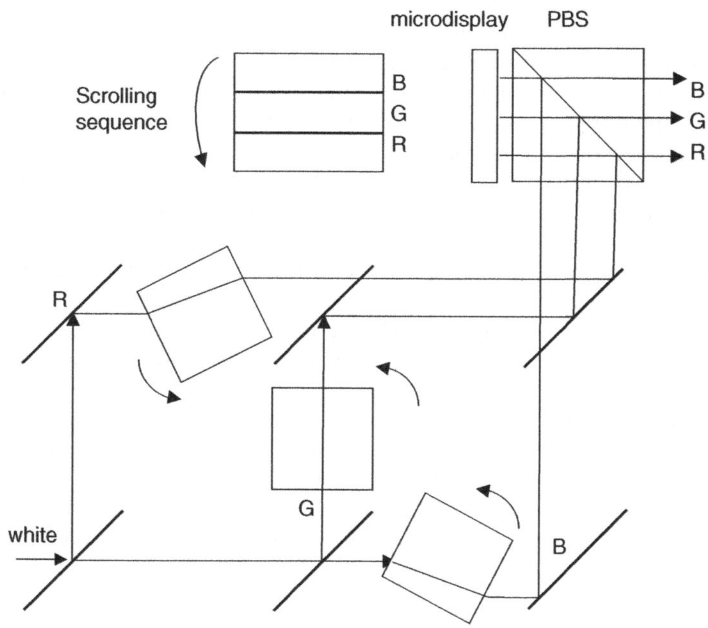

Figure 10.19 Rotating prism color scrolling system

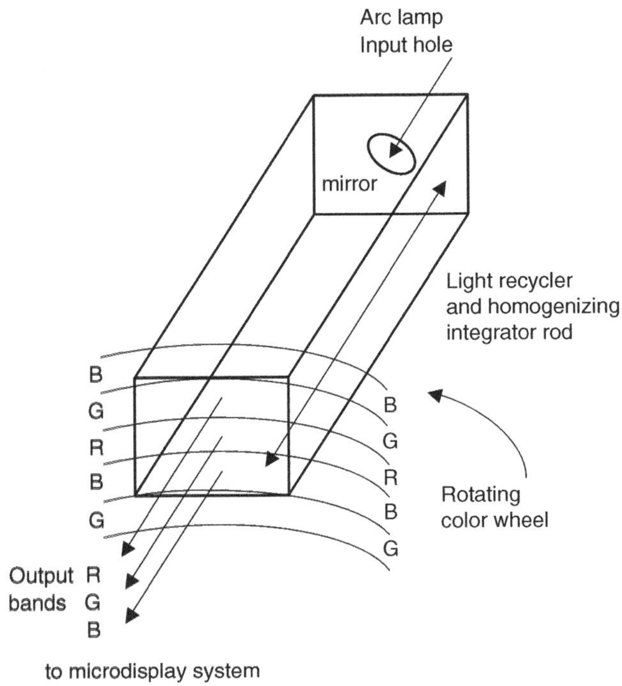

Figure 10.20 Scrolling color wheel with reflected light recovery

crystal thickness of 1 μm is necessary to achieve the required frame rate. The throughput is comparable to a three-microdisplay system with the same total area as the single scrolling microdisplay. The color convergence advantage of color sequential systems applies; moreover, the larger pixel size of the scrolling system reduces the significance of fringe field effects in the liquid crystal.

The color wheel system shown in Figure 10.20 achieves color scrolling. Color patterned in the form of an Archimedes spiral (radius proportional to rotation angle) assures constant speed of the color bands in the radial direction from the microdisplay viewpoint. A glass rod integrator homogenizes the input light to the microdisplay aspect ratio. The color wheel filters transmit red, green, and blue, while reflecting the complementary colors. The reflected colors are recycled through the integrator to illuminate the output filters again. The system is called "sequential color recapture" (SCR).[41] The efficiency of color recapture is determined from reflection losses and the area of the input light aperture. Reducing the aperture increases color recapture, but reduces the integrator input etendue. The aperture is optimized for maximum throughput efficiency. The performance of a demonstration projector was consistent with an estimated SCR gain of 1.67, but more work is required to confirm this value.

The mechanical precision required to achieve adequate performance handicaps SCR.[41] The integrator-wheel interface gap should be as small as possible, to minimize light leakage. Qualitative arguments indicate that the total runout of the rotating wheel should be less than 0.1 mm. Excess wheel wobble causes optical modulation at the wheel rotation rate, which may require wheel speeds beyond 3000 rpm to suppress flicker. Reducing the wheel diameter introduces curvature in the color bands presented to the microdisplay, but does not rule out development in that direction to improve the runout tolerance and compactness. The ideal efficiency is inferior to the rotating prism method; however, when losses are taken into account the efficiencies are comparable.[41] SCR is a simple modification to the non-scrolling system of Figure 10.18, with little added loss, except the etendue compromise. The rotating prism system introduces significant additional component loss; however, loss is reduced in a light guide system.[42]

Other scrolling systems of interest are a rotating lens structure employing a spiral lens wheel,[43] and a rotating drum containing color filters.[44]

10.6.3 RGB Pixel Systems

Direct-view LCDs use red, green, and blue pixels defined by pixel dyes. Early projectors based on such RGB pixel color suffered low efficiency, and gave way to more efficient systems employing multiple microdisplays or sequential color. However, a niche market for small inexpensive projectors has revived some interest in color pixels.[45] Color pixels have the advantage of built-in color convergence and modest response speed requirement, prompting efforts to eliminate the dye loss. If readout light is separated into R,G,B primaries, and the red light directed to designated red pixels and the green light to the designated green pixels, etc., the ideal efficiency is the same as color scrolling, and color separation imposes the same etendue penalty.

Figure 10.21 shows a holographic color filter (HCF) designed to separate white light into primary colors, routing them to designated pixels of an LCOS microdisplay.[46,47] The HCF is optically recorded on a photopolymer. The holographic cylindrical microlens array extends over the width of the microdisplay area. Oblique *s*-polarized read light diffracts to the appropriate pixels, where it is modulated in polarization and reflected back to the HCF. The *s*-polarized component diffracts towards the read light source, while the *p*-polarized component passes to the projection lens. The efficiency of the HCF

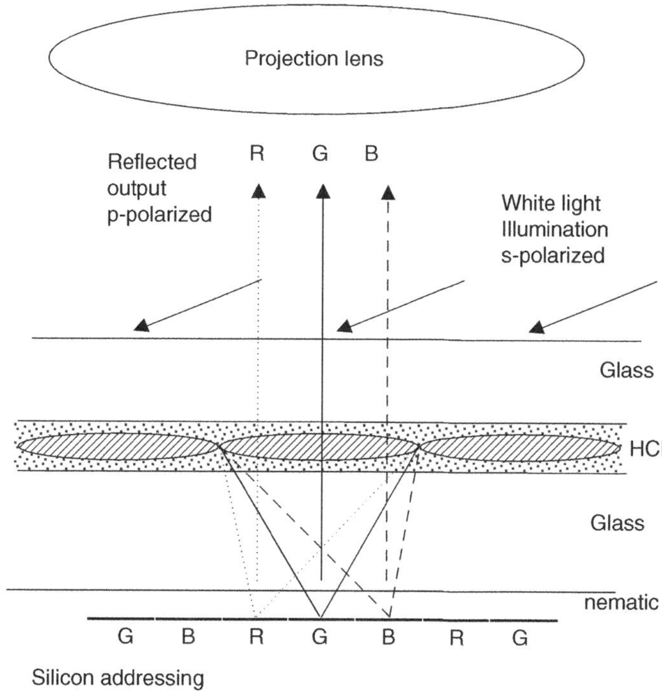

Figure 10.21 Hologram color filter separates and routes s-polarized color to designated pixel, and p-polarized output color to projection lens. Pixel and HCL exaggerated to show focusing of HCL

is improved by optimizing the input angle of each primary color forming the white read light, which imposes a further etendue penalty. The HCF must be precisely aligned with the microdisplay.

The demonstration rear-projection single-panel TV developed using the HCF has pixel pitch 7.6 μm in the diffracted direction shown in Figure 10.21, with 14.8 μm pixel pitch in the orthogonal direction. The HCF lens pitch triples the pixel pitch at 22.8 μm. The LCOS diagonal is 1.22 inches for SXGA resolution at 16:9 aspect ratio. The HCF efficiency of 40% is the weak link in the system that limits the throughput efficiency.

Transmission systems have been developed that employ conventional microlenses and angularly separated primary colors.[1] They would benefit from a HCF, if the efficiency could be raised.

10.6.4 Parallel Color

Parallel color routes each primary color to a microdisplay dedicated to that color; then the three primaries unite in a color-combining prism, prior to projection. Figure 10.22 illustrates the standard architecture for transmission microdisplays. The microdisplays are complete with dichroic polarizers, confining the polarization optics to the microdisplay package. Transmission nematic cells invariably use the 90-degree twisted geometry, providing on/off CR ~ 400, which is boosted CR ~ 800 with compensation retarder at *f*/2.7.[48,49] Color convergence requires the three microdisplays align to within a fraction of a pixel, and the alignment is secure over the projector life. Such alignment is common to all parallel color systems and becomes more stringent at reduced pixel size. Provision of surplus pixels facilitates electronic alignment by image translation, limiting the error to less than 0.5 pixel, providing the angular orientation of the microdisplays is precise.

The outstanding advantage of transmission systems is separate optical paths for the input and output light, as illustrated in Figure 10.22, enabling simple, inexpensive, high-performance systems. The major

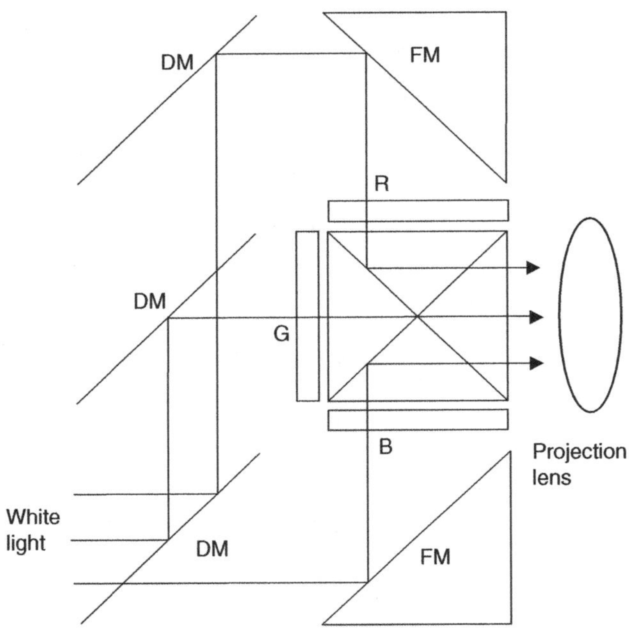

Figure 10.22 Simplified transmission architecture. Dichroic mirrors (DM) and folding mirrors (FM) route read light through red, green, and blue microdisplays, superimposed by a dichroic prism

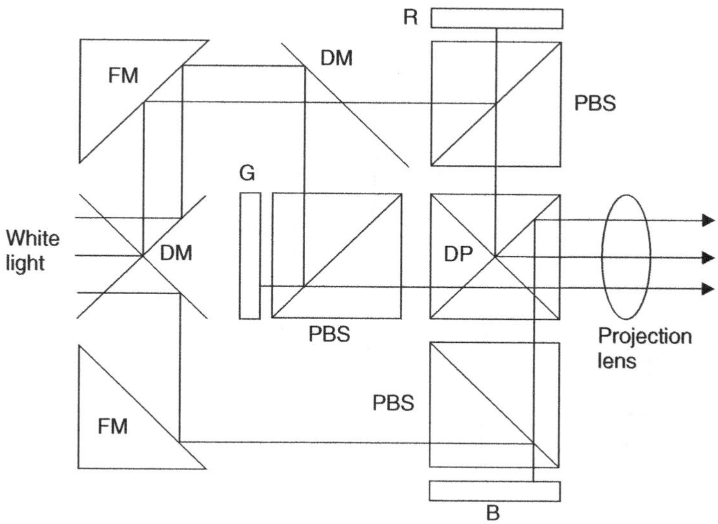

Figure 10.23 Simplified reflective architecture. Dichroic mirrors (DM), folding mirrors (FM) and polarizing beam-splitters (PBS) route read light to red, green, and blue LCOS microdisplays, superimposed by a dichroic prism (DP). Quarter-wave retarders to correct compound angle depolarization are not shown

handicap is low aperture ratio, which becomes increasingly severe with pixel contraction, e.g. 60% for a 12μm pixel in current production. Focusing the light into the pixel with a microlens array boosts the effective aperture ratio to approach 90%.[50] The focusing action reduces the effective area, with a corresponding expansion of optical divergence to conserve etendue, implying *f*-number reduction to preserve throughput. A 1280 × 720 microdisplay at 12μm pixel pitch has 0.7 inches diameter. Polarization conversion limits the loss and power dissipation in the microdisplay input polarizer, but the output polarizer must absorb and dissipate the channel output optical power. Such power dissipation limits the upper lumen output of transmission projectors using dichroic polarizers. The output polarizer is isolated from the microdisplay in the highest power systems, to improve cooling and stabilize the microdisplay temperature.

Figure 10.23 shows a similar basic arrangement for reflective microdisplays.[51] A system of dichroic mirrors and folding mirrors distributes the polarized read light to red, green, and blue LCOS microdisplays, followed by recombination in a dichroic prism. Each microdisplay has its own PBS, enabling the PBS to be optimized for that color band, and minimizing the stress birefringent path length. Quarter-wave retarders (not shown), inserted between each microdisplay and PBS, correct for skew-ray depolarization described in section 10.5.2. Small rotations of the retarders correct for residual birefringence, to boost contrast ratio.

LCOS microdisplays exploit the high thermal conductivity of the silicon substrate to heat sink the light absorbed from the readout beam and electronic power dissipation. Efficient heat sinking is consistent with small devices and high lumen throughput. Multi-reflections between the counter electrode and the highly reflective pixels influence the CR and optical uniformity, requiring precise index matching at the counter-electrode. Index matching requires the input polarization to be along one of the principal axes of the liquid crystal, which restricts the choice of nematic configuration.[39] Illumination with sharp spectral peaks exacerbates optical interference effects.

There are many variations of projector architecture using reflective displays, and developments continue as improved polarization components become available. They offer savings in cost, space, and various trades of throughput against CR. The color-quad® system employs four polarizing beam-splitters, and four retarder-stack filters, which encode color polarization to assist in routing.[52,53] The color-corner system uses a single PBS, polarization color encoding and dichroic color separation.[54] The

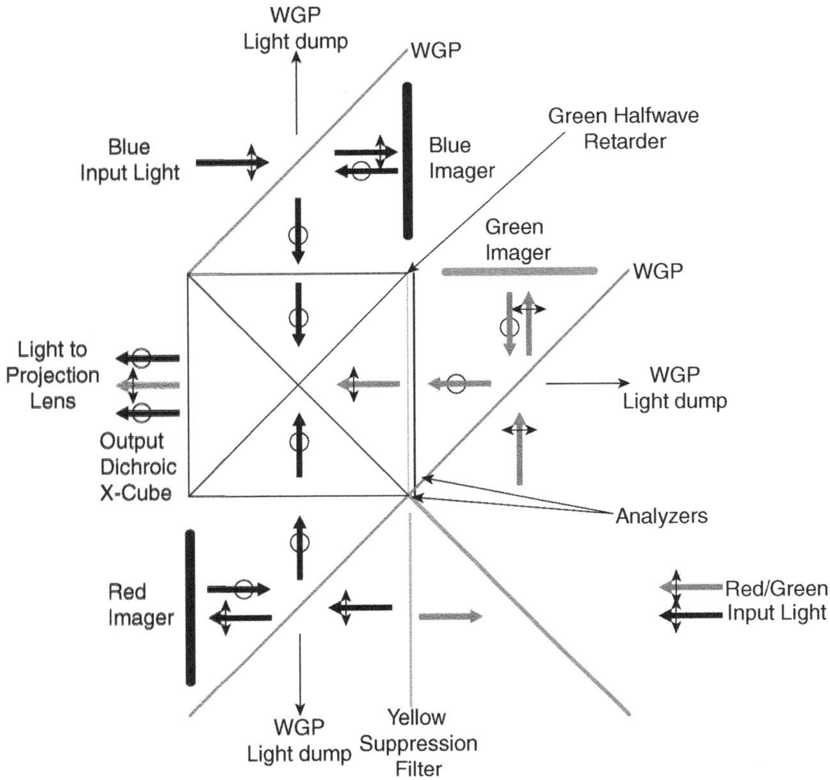

Figure 10.24 Reflective system employing wire grid polarizers, with cleanup dichroic analyzers. Halfwave retarder rotates green polarization to match X-cube. Reprinted courtesy of Society for Information Display

Philips color separation prism works in conjunction with a PBS.[55] A wire-grid polarizing design is gaining support.[29] Finally, the off-axis system eliminates PBS completely.[56]

A WGP version of Figure 10.23 appears in Figure 10.24, where the polarized optical path is essentially air, eliminating stress birefringence associated with glass.[29] Cleanup dichroic analyzers are essential to high contrast ratio, but dichroic power absorption is low since the reflecting WGP carries most of the polarizing load. A yellow blocking filter is included to accommodate UHP illumination. Linear polarized orthogonal directions follow "(—)" and "O" notation, signifying a 90° rotation of green output polarization by the half-wave retarder, optimizing the dichroic X-cube performance. The WGP enjoys the highest CR, while sacrificing throughput relative to PBS systems.

Figure 10.25 illustrates the color-quad® system, which employs stacked birefringent filters of stretched polycarbonate sheets to rotate the polarization of one color relative to its complement. The input polarized white light is split into orthogonal green and magenta polarizations by the input G/M filter. PBS1 and PBS2 direct the green light to the green microdisplay, while magenta is split into red and blue polarizations by the R/B filter and routed by PBS4 to appropriate microdisplays. A second R/B filter equalizes the red and blue polarizations so they are routed to the output by PBS3, together with green. The final G/M filter equalizes the polarization of all three primaries, and is followed by a clean-up polarizer. The two R/B filters are identical, but the G/M filters are mismatched to define the color gamut of the system and allow notching of excessive yellow and cyan light.

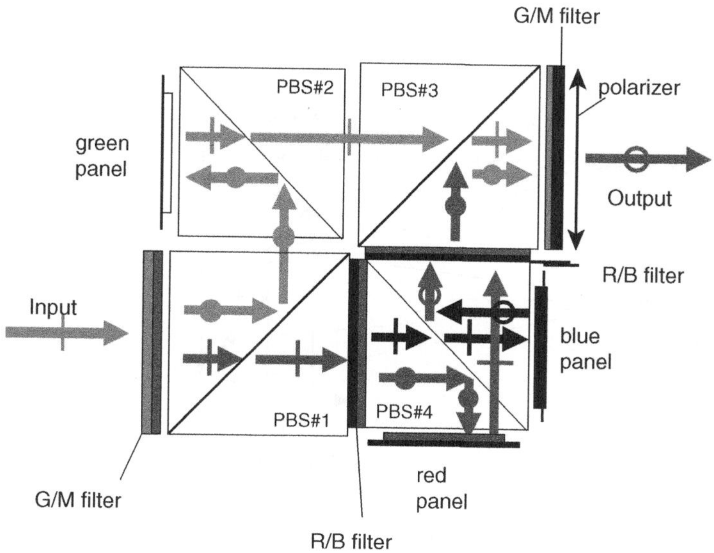

Figure 10.25 Color Quad® architecture. Reprinted courtesy of ColorLink Inc.

The DMD has the advantage of operating with unpolarized light. Angular modulation of the readout light determines the on/off states. The switch in angle can act directly to turn the readout light away from the projection aperture, as indicated in Figure 10.18. Alternatively, angle modulation can exploit total internal reflection beyond a critical incidence angle at a dielectric interface, illustrated in Figure 10.26 for a parallel color system.[57] The use of three microdisplays not only increases the throughput efficiency, but the slower frame rate enables expanded bit depth in grayscale resolution.

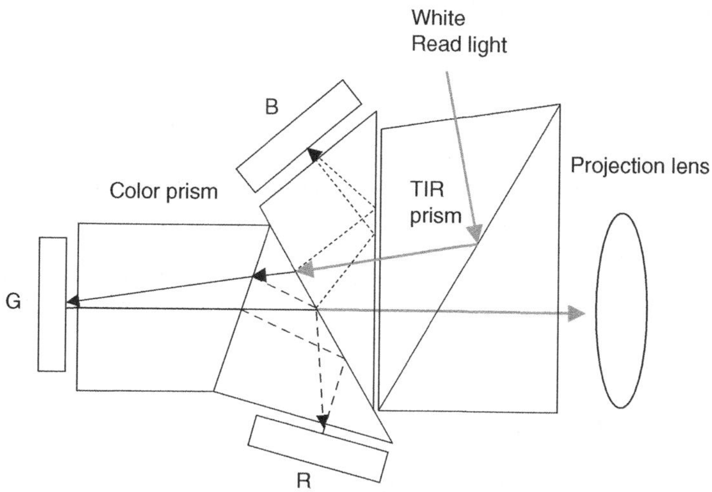

Figure 10.26 DMD parallel color system with three DMDs, R,G,B, and total internal reflecting prism separating input and output light

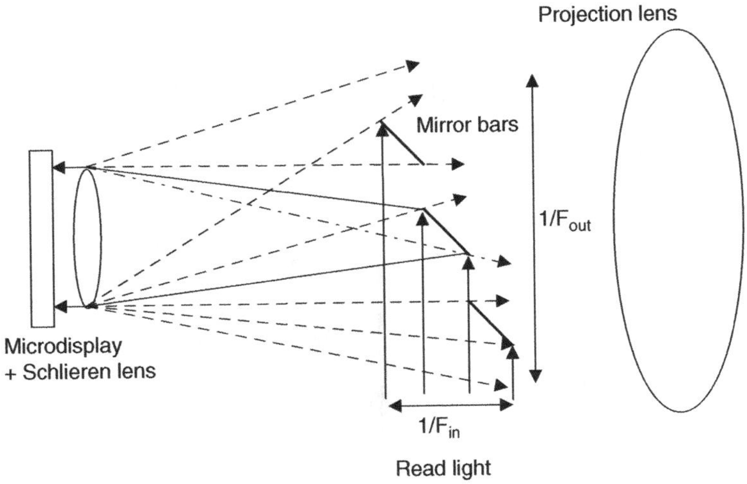

Figure 10.27 Schlieren projection system. Mirror bars in focal plane of Schlieren lens, diffraction shown dotted. Only center mirror bar shown illuminated for simplicity

10.7 Schlieren Projector

Schlieren projection applies to a microdisplay phase modulator illuminated with unpolarized light. In Figure 10.27, the mirror bars at the focal plane of the Schlieren lens are illuminated with *f*-number F_{in}, where only the light cone of the center bar is shown. Output diffracted light passes through the mirror bar gaps to the projection lens. The mirror bar positioning blocks the zero and even-order diffractions, while transmitting the odd orders. The ideal throughput is greater than 90%, but the output *f*-number F_{out} has essentially halved in the diffraction direction, where the mirror bars block half of the projection aperture. Decrease in effective etendue is the weak link in diffraction methods and lowers the throughput in comparison with the DMD at the same projection *f*-number.[58]

The output diffraction luminance is dependent on the phase modulation amplitude at the microdisplay as described in Chapter 8. Absorbing strips may be placed on the boundaries of the mirror bars to accommodate Schlieren focal errors, thereby raising the CR at expense of throughput. Absence of polarization losses and high-power capability, e.g. Eidophor projection, are Schlieren strengths. Liquid crystal phase modulators are at a disadvantage due to polarization dependence.

A Schlieren system based on optical scattering rather than diffraction is described in Chapter 6. The liquid crystal dispersion providing the scattering effect cannot be driven to a completely clear state, leaving a residual haze. High contrast favors a design that scatters the readout light outside the projection lens for the dark state. High throughput requires low *f*-number (large aperture) projection, but compromises the CR, requiring a design tradeoff between throughput and CR. A projection system based on scattering liquid crystal dispersions was developed to prototype level, but abandoned due to improved performance of non-scattering systems.[59]

10.8 Laser Scanning

A laser scanning system employing the grating light valve (GLV) is illustrated in Figure 10.28 for a single color. Full color requires color sequential operation, or three GLVs, superposed with dichroic mirrors. The laser light is formed into a narrow line focused on the GLV via the Schlieren mirror, and the diffracted output projected onto the screen through the horizontal scanning mirror. The fast addressing

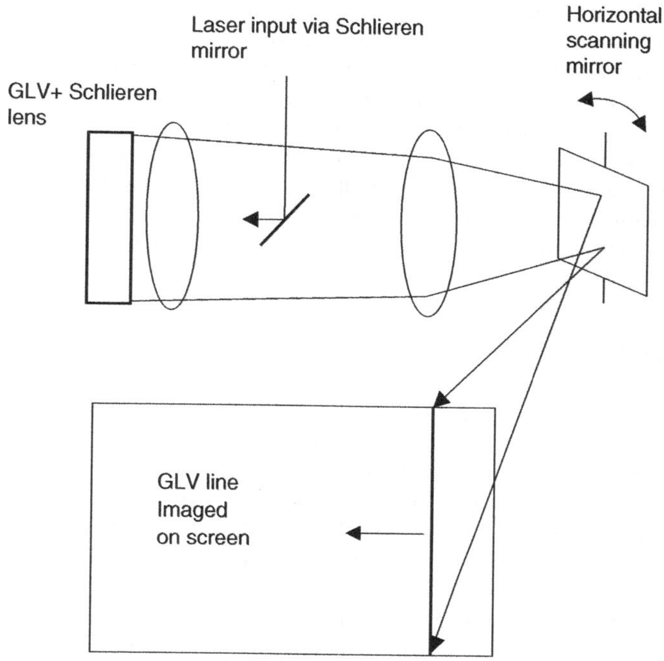

Figure 10.28 Laser scanning system using the grating light valve

speed of the GLV forms vertical lines of the display, which traverse the screen under the control of the slow horizontal scanning mirror. The laser beam width need only be narrow in one dimension to provide a high-resolution image. A prototype system achieved 1920×1080 resolution, with $CR > 200$.[60] Methods to attenuate speckle effects are discussed.[61] The latest results employing the blazed grating described in Chapter 8 gives $CR = 10{,}000$ in a 5000 lumen system.

With advances in laser diodes and microlasers, power levels and beam quality will improve to make a variety of scanning systems competitive with microdisplay systems.

10.9 Projector Performance

The overall performance of a projector is characterized by a set of performance metrics prescribed by the American National Standards Institute (ANSI).[1,62] We are essentially interested in how the microdisplay affects the projector performance, and assume an ideal projection system.

10.9.1 Flicker

Fluctuations in display luminance appear as an annoying flicker. With increase of flicker frequency, apparent flicker attenuates, approaching threshold perception at a critical flicker frequency (CFF), as discussed in Chapter 1. More stringent flicker criteria can be applied, depending on the overall display performance, including addressing stability and speckle effects.[63]

Binary-code pulse modulated grayscale used in the DMD and FLC microdisplays gives rise to a flicker effect associated with fluctuations in lamp illumination over its AC drive cycle. It is minimized by a bit-splitting method that distributes the gray level pulses more evenly over the lamp cycle, and synchronizing the lamp to the addressing cycle, preventing beating of frame and lamp frequencies.[64] A more even distribution of gray level pulses over the frame time also reduces image artifacts associated with PWM and eye movement. A false gray level registers if the eye shifts before recording the full pulse count on the retina.

Liquid crystal devices operate with alternating voltage, odd frames having opposite voltage polarity to even frames. Lack of perfect symmetry in operation generates flicker at half the frame rate. A video rate of 60 frames/s gives rise to a 30 Hz flicker, which is approximately sinusoidal when limited by the nematic LC viscous response. The 30 Hz flicker modulation must be less than 0.5% at luminance beyond 300 cd/m^2 to escape perception. LCOS microdisplays are susceptible to flicker that arises from the aluminum/ITO electrode asymmetry.[65] Such devices are operated at 120 frames/s to raise the flicker frequency to 60 Hz, where flicker is more tolerable.

10.9.2 Grayscale

In DMD cinema, triple microdisplay projection achieves required luminance level, but 14-bit depth to match the performance of film exceeds the LSB time available per frame. The problem is solved by a "hybrid frame rate" that is slower for less significant bits, exploiting the fall in flicker perception with lower screen luminance; frame rate varies from 24 frames/s for LSB to >96 frames/s for highest level bits.[66]

A single DMD operated in color sequential mode has difficulty in satisfying the gray level demand, due to limited response time. Spatial dithering exchanges time resolution for space resolution, adding extra bits to the effective gray level. Time dithering adds an extra bit of gray by only showing the bit on alternate frames. However, spatial and time dithering are perceived as image noise and should be applied sparingly. Alternatively, the systematic introduction of image noise related to our visual system can minimize bit depth artifacts, and is not restricted to PWM systems.[67] Inadequate bit depth gives rise to color temperature variations at low luminance. The addition of a dark green band to the color wheel provides adjustments in color and bit depth.

A recent development alleviates bit-depth issues by modulating the source illumination with a variable aperture, reducing luminance during dark scenes, where limited bit depth is more apparent.[68] Lowering the source luminance is equivalent to reducing grayscale steps, and applies to any digital-based modulation system, e.g. analog liquid crystal displays that derive their analog signal from digital-to-analog conversion (DAC). A 10-bit DAC is normally required to generate the proper gray levels, but cheaper 8-bit DACs suffice with variable aperture, or visual system noise processing.

10.9.3 Lumen Output

Total light flux output could be measured by the standard integrating sphere method, which would require expensive equipment. The ANSI standard IT7.215 takes a simpler approach, specifying measurement of illuminance at nine representative screen points, averaging the values to get the average illuminance, then multiplying by screen area to get the ANSI lumen flux. Location of the illuminance points are shown in Figure 10.29. It is assumed that on full white the projector is adjusted to the proper color temperature.

Polarization requirements, *f*-number limit, and efficiency of the microdisplay have a direct influence on projector throughput. DMD and LCOS employing aluminum reflectivity have comparable throughput efficiency, but LCOS suffers a polarization loss, while DMD is restricted to *f*/2.4. LCOS with dielectric enhancement of reflectivity is more competitive. The latest generation of LC transmission microdisplays maintains a competitive market presence.

10.9.4 Contrast Ratio

There are two descriptions of contrast ratio. The simplest is the ratio of center luminance with full white screen, divided by center luminance with full black screen. It is known as the *sequential CR* = (full

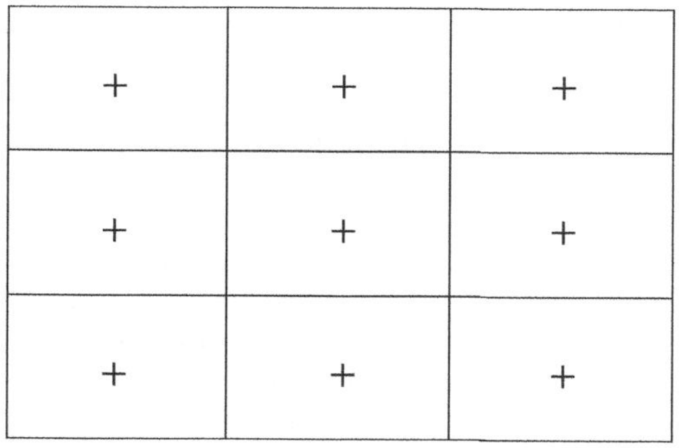

Figure 10.29 ANSI lumens point distribution over the projection screen

white)/(full black), and is the highest CR that is quoted for a projector. It identifies with the CR of the microdisplay and readout optics at the particular *f*-number of the projection system, in the absence of projection lens scattering.

The ANSI (92) standard employs the light and dark extremes in a checkerboard pattern shown in Figure 10.30, to measure the contrast ratio. The luminance at the center of each rectangle, averaged over the eight white, or eight black values, gives the ANSI contrast as the ratio of the two averages. Here, scattering effects in the projection optics are included in the measurement. The ANSI contrast ratio is always lower than the sequential CR, and the best indication of image quality.

The on-axis CR of the microdisplay, with standard retardation compensation, presents an upper limit to sequential CR. Random errors in the optical system will lower CR, as will off-axis behavior. The overall projector contrast ratio CR_p can be expressed as

$$\frac{1}{CR_p} = \frac{1}{CR_{sys}} + \frac{1}{CR_{md}} \tag{10.11}$$

where CR_{sys} is the system CR without the microdisplays, and CR_{md} is the microdisplay CR.

Projectors with sequential CR ~ 10,000 are reported, which has become a selling point. However, there is little to be gained in image quality for CR beyond 1000, particularly with an aperture-modulated source.

Figure 10.30 ANSI contrast ratio pattern on screen

10.9.5 Color Uniformity

Deviation in spatial uniformity (grayscale uniformity) is first observed as deviations in color uniformity, because the eye is more sensitive to color shift than luminance change. Digital precision determines the color uniformity in the DMD, and the issues previously raised at low illumination levels and time dependence of backlight apply.

All the liquid crystal microdisplays are prone to spatial nonuniformity due to cell gap variation. Variation in pretilt and contaminant distribution are possible, but generally insignificant compared with cell gap uniformity. A shift in local cell gap implies a shift in modulation level for that particular region, which can be calculated or measured for a particular cell design. In multi-microdisplay projectors, random variation in cell gaps gives rise to random color modulations, characterized as nonuniform color. Gap uniformity $<1\%$ ensures adequate color uniformity, but has a substantial impact on manufacturing yield. Greater variation in cell gap is accommodated by a compensating adjustment of the local pixel voltage, which is derived from a look-up table addressed by the video signal.[50] Color correction of the overall system follows from making the adjustment at projector level.

In color sequential projectors employing a single microdisplay, color uniformity is not guaranteed. Nonlinear response to wavelength and cell gap allows gap variations to influence color uniformity. However, gap tolerance is much higher.

10.9.6 Resolution

All electronically addressed microdisplays have a resolution that is determined by pixel count. However, fringe-field effects can compromise the resolution of liquid crystal devices as shown in Chapters 5 and 6.[69] The fringe field effect reduces the contrast at pixel level, implying some degradation of system MTF. In viewing conditions putting the pixel beyond visual acuity, fringing becomes imperceptible, mitigating fringe effects on image quality. The projection lens MTF is generally less than 50% at limiting pixel resolution, yet the image quality of a top-class projector is highly rated.

10.10 References

1 Stupp, E. H. and Brennesholtz, M. S. (1999) *Projection Displays*. John Wiley. ISBN 0 471 98253 9
2 Robinson, M. G., Chen, J. and Sharp, G. D. (2005) *Polarization Engineering for LCD Projection*. John Wiley. ISBN-13 978-0-470-87105-8(HB) and ISBN-10 0-470-87105-9(HB)
3 McLaughlin, C. W. (2002), Progress in Projection and Large-Area Displays. *Proc. IEEE* **90**, 521–32.
4 Melcher, R. L. (1999) Potential impact of laser illumination on light valve based projection displays. *Proc. SPIE*, **3626**, 62–72.
5 Glenn, W. E. *et al.* (1997) Laser-based electronic displays. *Proc. SPIE*, **3013**, 21–6.
6 Hargis, D. (1998) Diode pumped microlasers for display applications. *Proc. SPIE*, **3285**, 115–25.
7 Hargis, D. *et al.* (1999) Solid-state laser-based displays. *SID Digest* **30**, 986–9.
8 Jansen M. *et al.* (2006) Visible laser and laser array sources for projection displays. *Proc. SPIE*, **6135**, 61350T-1-6.
9 M. H. Keuper, G. Harbers, S. Paolini, (2004) RGB LED Illuminator for Pocket-Sized Projectors. *SID Tech. Digest* **35**, 943–5.
10 G. Harbers, S. Bierhuizen, (2006) Performance of High Power Light Emitting Diodes in Projection Applications. *SID Tech. Digest* **37**, 2007–10.
11 J. M. Teijido, *et al.* (2006) Compact Three Panel LED Projector Engine for Portable Applications. *Tech. Digest* **37**, 2011–4.
12 A. Conner, *et al.* (2005) Performance of LED as the Illumination Source for Microdisplay Rear Projection TV, *Int., Display Workshop*, 1907–10 and LEDs for Microdisplay Illumination. *SID Tech. Digest* **37**, 1819–22, (2006).

13 C. Hoepfner, (2006) PhlatLight™ Photonic Lattice LEDs for RPTV Light Engines. *SID Tech. Digest* **37**, 1808–11.
14 Roberts, R. *et al.* (1996) Long-lived sealed-beam xenon short-arc lamps. *SID Digest* **27**, 771–3.
15 Moench, H. and Ritz, A. (2003) Higher-output more-compact UHP lamp systems. *J. SID* **11**, 167–75.
16 Moench, H. *et al.* (2002) UHP lamps for projection. *J. SID* **10**, 87–94.
17 Moench, H. *et al.* (2003) UHP lamps with increased efficiency. *SID Digest* **34**, 758–61 and Jens Pollmann–Retsch, (2006) Compact Power Lights for High-end Projection Applications. *SID Tech. Digest* 1716–9.
18 Kipling, K. *et al.* (1999) High-brightness electrodeless lamp for projection displays. *SID Digest* **30**, 1080–3.
19 Rosenbluth, A. E. *et al.* (2000) Correction of contrast in projection systems by means of phase-controlled prism coatings and band-shifted twist compensators. *Proc. SPIE*, **3954**, 63–90.
20 Miyatake, Y. (1990) Polarizing beamsplitter apparatus and lightvalve image projection system. European Patent EP 389,240.
21 Sterling, R. D. and Bleha, W. P. (1997) ILA Projectors for advanced applications. *Intl Display Workshop*, Japan, 809–11.
22 Robinson, M. G. *et al.* (2003) Wide field of view compensating scheme for cube polarizing beam splitters. *SID Digest* **34**, 874–7.
23 Eckhardt, S. *et al.* (2003) 3M PBS for high performance LCOS optical engine. *Proc. SPIE*, **5002**, 106–10.
24 Bruzzone, C. L. *et al.* (2003) High performance LCoS optical engine using cartesian polarizer technology. *SID Digest* **34**, 126–9.
25 Pentico, C. *et al.* (2001) New, high performance, durable polarizers for projection displays. *SID Digest* **32**, 1287–9.
26 Hansen, D. *et al.* (2002) The display applications and physics of the ProFlux™ wire grid polarizer. *SID Digest* **33**, 730–3.
27 Ho, G. H. *et al.* (2002) The mechanical-optical properties of wire-grid type polarizer in projection display system. *SID Digest* **33**, 648–51.
28 Yu, X. J. and Kwok, H. S. (2003) Application and optimized design of wire-grid polarizers. *SID Digest* **34**, 878–81.
29 Pentico, C. *et al.* (2003) Ultra high contrast color management system for projection displays. *SID Digest* **34**, 130–3.
30 Itoh, Y. *et al.* (1997) Ultra-high-efficiency LC projector using a polarized light illuminating system. *SID Digest* **28**, 993–6.
31 Duelli, M. (2002) Integrator rod with polarization recycling functionality. *SID Digest* **33**, 1078–80.
32 Armitage, D. (2000) Contrast ratio of vertically aligned nematic cell. *Proc. SPIE*, **3954**, 197–205.
33 Duelli, M. *et al.* (2005) High performance contrast enhancing films for VAN-mode LCoS panels. *SID Digest* **36**, 892–5.
34 Gandhi, J. *et al.* (1999) Performance enhancement of reflective CMOS TN displays in projection applications using compensated films. *SID Digest* **30**, 990–3.
35 Shikama, S. (2002) Optical system of ultra-thin rear projector equipped with refractive-reflective projection optics. *SID Digest* **33**, 1250–3.
36 Bachels, T. *et al.* (2001) Advanced electronic color switch for time-sequential projectors. *SID Digest* **32**, 1080–3.
37 Jepson, M. L. (2003) Single panel LCOS lumen throughput. *SID Digest* **34**, 122–5.
38 Jepsen, M. L. *et al.* (2002) High resolution LCOS microdisplay for single-, double- or triple-panel projection systems. *Displays* **23**, 109–14.
39 Shimizu, J. A. (2001) Scrolling color LCOS for HDTV rear projection. *SID Digest* **32**, 1072–5.
40 Janssen, P. *et al.* (2002) Design aspects of a scrolling color LCoS display. *Displays* **23**, 99–108.
41 Dewald, D. S. *et al.* (2001) Sequential color recapture and dynamic filtering: a method of scrolling color. *SID Digest* **32**, 1076–9.
42 Shimizu, J. A. and Janssen, P. J. (2005) A light guide engine for single-panel LCoS. *SID Digest* **36**, 1810–13.
43 Cho, K-H. *et al.* (2003) Case Study: A New Simple and Inexpensive Single-Panel LCos System for HDTV Projection Systems. *Information Display* **19/10**, 26–30.

44 Brennesholtz, M. S. *et al.* (2005) A single panel lcos engine with a rotating drum and a wide color gamut. *SID Digest* **36**, 1814–17.
45 Huang, H. C. *et al.* (2005) Color filter liquid-crystal-on-silicon microdisplays. *SID Digest* **36**, 880–3.
46 Yamazaki, T. *et al.* (2000) The single-panel D-ILA hologram device for ILA™. *Intl Display Workshop*, Japan, 1077–80.
47 Negishi, I. (2000) The ILA projection TV using single D-ILA™ hologram device. *Intl Display Workshop*, Japan, 1081.
48 Okamoto, N. (2001) Developments in p-Si TFT LCD projectors. *SID Digest* **32**, 1176–9.
49 Karasawa, J. *et al.* (2001) 50-inch HD rear projection monitor. *Intl Display Workshop*, Japan, 1275–8.
50 Aruga, S. (2002) High temperature poly-Si TFT LCD for projection. *Intl Display Manufacturing Conference*, 551–6.
51 Melcher, R. L. *et al.* (1998) High-information-content projection display based on reflective LC on silicon light valves. *SID Digest* **29**, 25–32.
52 Sharp, G. D. *et al.* (2002) LCoS projection color management using retarder stack technology. *Displays* **23**, 139–44.
53 Sharp, G. D. *et al.* (2002) Skew-ray compensated retarder-stack filters for LCOS projection. *SID Digest* **33**, 1338–41.
54 Heine, C. and Linz-Dittrich, S. (2001) Color management system for 3-panel LCOS-projectors. *SID Digest* **32**, 1189–91.
55 Greenberg, M. R. and Bryars, B. J. (2000) Skew ray compensated color separation prism for projection display. *SID Digest* **31**, 88–91.
56 Bone, M. (2001) Method to improve contrast in an off-axis projection engine. *SID Digest* **32**, 1180–3.
57 Hornbeck, L. (1997) Digital light processing™ for high-brightness, high-resolution applications. *Proc. SPIE*, **3013**, 27–40.
58 Armitage, D. (1997) Diffractive projection displays. *Proc. SPIE*, **3013**, 112–25.
59 Nagae, Y. *et al.* (1995) Compact liquid-crystal projectors with high optical efficiency. *SID Digest* **27**, 223–6.
60 Amm, D. T. and Corrigan, R. W. (1999) Optical performance of the grating light valve technology. *Proc. SPIE*, **3634**, 71–8.
61 Trisnadi, J. I. (2002) Speckle contrast reduction in laser projection displays. *Proc. SPIE*, **4657**, 131–7.
62 Keller, P. A. (1997) *Electronic Display Measurements*. John Wiley.
63 Goldenberg, J. F. *et al.* (1997) Rear screens for light-valve projection systems. *Proc. SPIE*, **3013**, 49–59.
64 Watanabe, S. *et al.* (2002) Advantages of 2nd generation home entertainment DLP™ technology. *Intl Display Workshop*, Japan, 1245–8.
65 Lu, M. and Yang, K. H. (2000) Reflective nematic LC devices for LCOS applications. *Intl Display Research Conference*, 1–7.
66 Hewlett, G. and Pettitt, G. (2000) DLP Cinema™ projection: a hybrid frame rate technique for flicker-free performance. *Microdisplay 2000*, applications paper 1.
67 Daly, S. and Feng, X-F. (2004) Bit-depth-extension: overcoming LCD-driver limitations by using models of the equivalent input noise of the visual system. *SID Digest* **35**, 1470–3.
68 Toyooka, T. *et al.* (2004) Illumination control system for adaptive dynamic range control. *SID Digest* **35**, 174–7.
69 Katayama, T. *et al.* (2001) D-ILA™ device for top-end projection display (QXGA). *SID Digest* **32**, 976–9.

11

Near-to-Eye Systems

11.1 Background

Near-to-eye (NTE) is a general term for display systems that present their image to a single eye or pair of eyes. Head-mounted displays and camera electronic viewfinders are established examples. Potential applications include mobile phones and other portable equipment. NTE displays have an obvious advantage in size and weight, moreover in cramped conditions may be the only display option. The small amount of output light needed at the eye pupil is an important asset, allowing the use of LED illumination, or emissive displays that fall well below projection requirements. The advantages of NTE should be weighed against the unnatural viewing conditions, which are prone to induce eyestrain and fatigue.

The need for magnification is inherent to microdisplays and NTE poses interesting optical design problems. It is a challenge to provide good viewing conditions at high magnification and wide field of view. To minimize size and weight requires folded optical paths and unorthodox magnification schemes. Head-mounted systems are particularly sensitive to size and weight. A recent book on head-mounted displays,[1] chapters on head-mounted displays,[2,3] and an article[4] are recommended further reading.

11.2 Magnification

11.2.1 Virtual Image

We examine some conditions imposed by optical physics on magnification in general, before considering practical NTE systems. Figure 11.1 shows the simplest magnifier using a positive lens to present a virtual image to the viewer. In practice, the magnifier is a composite of optical elements designed to reduce aberrations. Representation by a simple lens is a first-order optics approximation, sufficient for our purpose.

Introduction to Microdisplays D. Armitage, I. Underwood and S-T. Wu

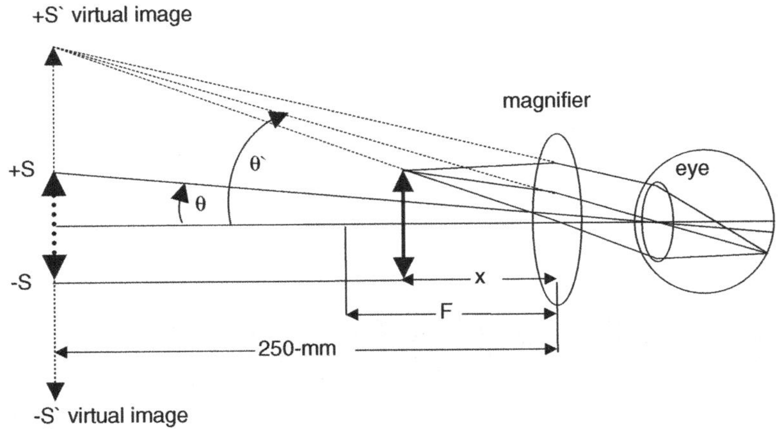

Figure 11.1 Simple single-stage magnification

The distance of most distinct vision is generally taken as 250 mm, where an object of height S subtends an angle θ. The object is then placed at position x within the focal length F of the magnifier to form virtual image S', subtending angle θ'. Applying the lens formula $[1/x+1/(-250)=1/F]$ gives x and hence the magnification M.

$$M = \frac{S'}{S} = \frac{250}{x} = 1 + \frac{250}{F} \quad x \rightarrow F,\ M = \frac{\theta'}{\theta} = \frac{250}{F}. \tag{11.1}$$

At $x = F$ the virtual image is at infinity and the magnification is expressed as an angular ratio. We draw the conclusion that high magnification requires a short-focus lens, and for substantial magnification, there is little difference in magnification between a virtual image at infinity and one at 250 mm. A virtual image distance of 2 m is a comfortable choice for NTE displays, and magnification is essentially $250/F$. Figure 11.1 assumes unit refractive index in the back focal region; a higher index (n) will increase the focal length to nF, while the magnification remains the same provided an air gap is preserved. Extending the back focal length creates space for the illumination beam-splitter required by a reflecting microdisplay.

11.2.2 Eyebox

Eye positioning is an important consideration. Figure 11.2 illustrates a pupil forming system, where all the output rays are contained in a minimum diameter P when the object rays are restricted to field angle ϕ. The viewer's eye positioned at P allows the maximum lateral motion without image clipping (vignetting). The shaded diamond area contains some rays from all points of the object, indicating the eye space where the complete image is viewable. Image clipping occurs when the viewer's eye moves outside the shaded region, progressively eliminating rays from the object boundary. The shaded region or output pupil diameter is the eyebox, and the distance of the viewer's eye from the magnifier is the relief distance R. A large eyebox provides comfortable viewing, and relief distance accommodates spectacles. A focus adjustment on the magnifier could eliminate viewer spectacles (except astigmatism).

Clearly, the eyebox will increase if the field angle of the object fills the magnifier aperture as shown in Figure 11.3. The magnifier aperture is the output pupil, and such systems are described as non-pupil

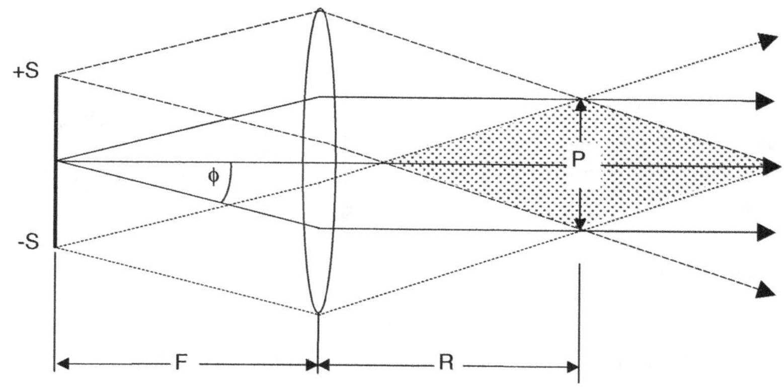

Figure 11.2 Simple pupil forming magnifier with eyebox shaded

forming. The geometry of Figure 11.3 gives the relation for eyebox

$$P_r = D - 2r\tan\theta' = D - 2r\tan\left[\frac{FOV}{2}\right]. \tag{11.2}$$

Large eyebox P_r favors large D, small FOV or magnification, and small eye relief r. An overall increase in the scale of Figure 11.3 obviously increases the eyebox, but is a tradeoff against increased weight and volume. Large magnification always sacrifices eyebox.

The typical eye pupil varies from 2 to 7 mm in diameter depending on light level and is less than 4 mm at NTE light levels. The eye rotates about an axis located approximately 10 mm behind the pupil. Rotation of the eye in scanning an image field of 35 degrees produces a lateral displacement of 6 mm. The eyebox must exceed the pupil diameter and displacement, to provide adequate viewing. Assuming 4 mm eye pupil with 6 mm displacement gives 10 mm total, but an additional 2 mm provides for viewing comfort, particularly where viewing stability is a factor as in portable equipment, indicating a minimum 12 mm eyebox diameter.[4]

The magnifier diameter should be substantially larger than the microdisplay active diameter, for adequate luminance uniformity in the image. The magnifier *f*-number ($f_\#$) is given by $f_\# = F/D$.

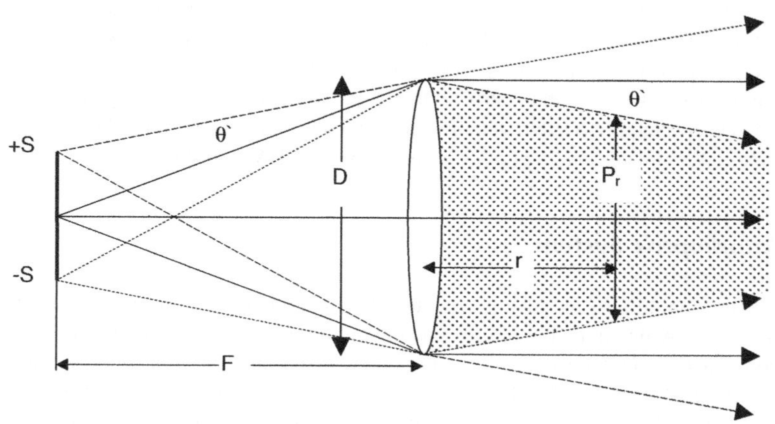

Figure 11.3 Non-pupil forming system with eyebox shaded

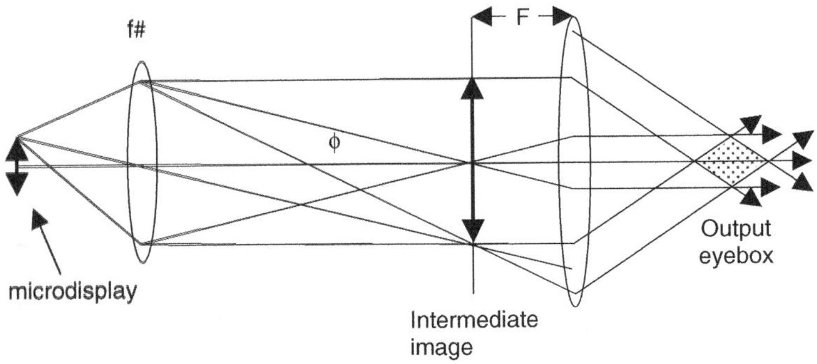

Figure 11.4 Two-stage magnification with output eyebox shaded

Complexity and expense of the magnifier lens increases rapidly below about $f/2$, as the aberrations and distortions become increasingly difficult to control. Magnification is determined by the focal length, where $M = 25$ implies focal length $F = 250/25 = 10\,\text{mm}$, and $f/2$ gives diameter $D = 5\,\text{mm}$. A microdisplay substantially less than 5 mm diameter could be magnified $\times 25$, but the output pupil would also be less than 5 mm.

11.2.3 Compound Magnification

Two-stage systems, similar to the microscope, achieve greater magnification. Figure 11.4 shows a primitive arrangement, where the objective lens $f_{\#}$ forms an intermediate image of the microdisplay, and the eyepiece lens provides a magnified virtual image of that image. The exit pupil is the image formed by the eyepiece of the objective lens, and since the microdisplay is outside the focal length of the objective lens, the geometry of Figure 10.4 indicates the output eyebox

$$P = 2F \tan\phi < \frac{250}{(f_{\#})M} \tag{11.3}$$

where M is the overall magnification; and if $M = 25$, with $f_{\#} = 2$, then $P < 5\,\text{mm}$, comparable with the single-stage case, and consistent with etendue conservation. A screen introduced at the focus of the intermediate image to scatter or diffract light through a wider angle expands the eyebox. The screen must not have visible structure nor compromise the resolution.[5] The reduction in magnification per stage allows greater focal length and aperture diameter, but compromises the eyebox, unless a regenerative screen is used. Such two-stage magnification finds little application in NTE displays, but may be incorporated in the image-relaying optics of more complex HMD systems.

11.2.4 Catadioptric System

A concave mirror magnifier has advantages such as freedom from chromatic distortion and less aberration at wide viewing angle, but requires a beam-splitter as indicated in Figure 11.5. Combining reflective and refractive elements to form a catadioptric system is important in canceling aberrations and flattening the field. A polarizing beam-splitter, in conjunction with a quarter-wave retarder, enhances the efficiency. Head-mounted displays invariably favor catadioptric systems for wide viewing angle.

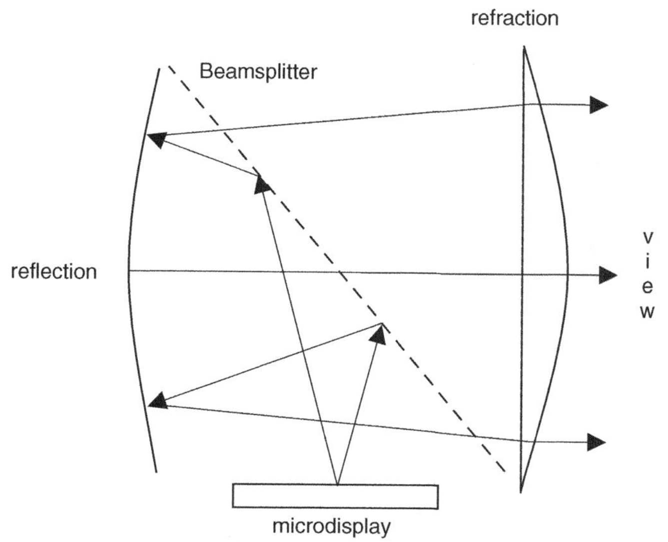

Figure 11.5 Magnifying mirror and refractive element (catadioptric) system with beam-splitter

11.3 Field of View

The field of view (FOV) of the optical system is an important parameter, influenced by the pixel count and eye resolution. Under good viewing conditions, eye resolution approximates 1 min of arc, but approaches 2 min per resolution element in many NTE applications when the modulation transfer function of the magnifier is taken into account. For comparison, an SXGA (1280×1024) display with 15-inch diagonal viewed from a distance of 20 inches has FOV $= 41°$ and resolution of 1.5 min/pixel. An SVGA (800×600) microdisplay at 2 min/pixel requires a diagonal FOV of $1000 \times 2/60 = 33°$ to barely resolve the pixels. Lower magnification giving a smaller FOV would sacrifice resolution, while a larger FOV will amplify the pixel structure. Moreover, a large FOV imposes more eye scanning, exacerbating fatigue and eyestrain. It is generally found that a FOV in excess of about 33 degrees is not tolerated in head-mounted applications subject to prolonged viewing of information type images.[4,6] Experiments simulating QVGA (240×320) microdisplays showed that at this resolution the display was uncomfortable to view at 17 degrees FOV, where the pixel extent is 3.1 minutes. At both SVGA and QVGA resolution, the tolerable FOV increased a few degrees for scenic images compared with data images. The QVGA results suggest a tradeoff between image quality and expansiveness.[6]

The relation between FOV $2\theta'$, eyebox P_r, eye relief r, and optical diameter D of the magnifier is given by rearranging (11.2) to give

$$D = P_r + 2r \tan\theta' . \qquad (11.4)$$

Choosing $P_r = 12$ mm, the required diameter is plotted against FOV for various eye relief in Figure 11.6. For comfortable 25 mm eye relief, providing room for spectacles, the magnifier aperture is 28 mm to give 35 degrees FOV.

From Figure 11.1, for microdisplay active diagonal d millimeters:

$$FOV = 2\theta' \rightarrow \frac{d}{F} = \frac{d}{D(f_\#)}, \quad f_\# \rightarrow \frac{d}{2\theta' D} < \frac{1}{2\theta'} . \qquad (11.5)$$

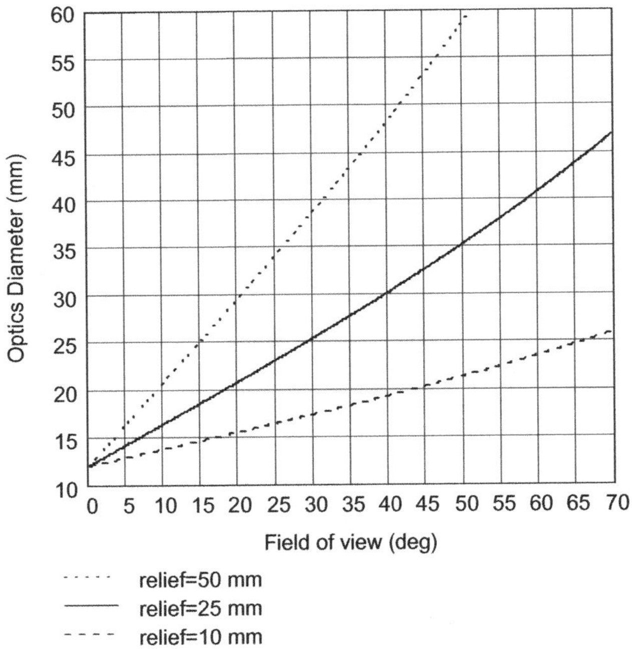

Figure 11.6 Output optical aperture against field of view for various eye relief and 12 mm eyebox

For adequate image luminance uniformity $d < D$, and (11.5) shows that the *f*-number is inversely proportional to FOV, implying less than *f*/1.6 for 35 degrees FOV. The *f*-number is increased by making *d* comparable to *D*. Large FOV, large eyebox, and small *d* demand low *f*-number.

11.4 Microdisplay Factors

The design of the magnifier takes into account the readout optics of the microdisplay. Rear-illuminated transmission LCDs and emissive displays present their output to the magnifier in similar form. Reflective microdisplays introduce the readout light on the same side as the magnifier and must accommodate beam-splitting optics to separate illumination from image light, as show in Figure 11.7. The beam-splitter restricts the minimum focal length of the magnifier.

Weight and mechanical stability issues eliminate three-microdisplay parallel color from NTE systems. Field-sequential-color is attractive because the single microdisplay minimizes the weight and stability problems, while LED illumination simplifies the mechanics of sequential color. LEDs provide a further advantage in pulse modulated systems by controlling the minimum bit duration by LED electrical switching, rather than pixel switching, extending the bit count to 24-bit color images.

Color breakup in field-sequential-color systems generally imposes a color-field rate of 450 frames/s or more. Eye and head movement prevent precise superposition of successive color fields on the retina, and color breakup gives rise to color fringing at sharp edges of an image. Head movement relative to the image is severely restricted in NTE displays, making them less sensitive to color breakup. A narrow field of view is a further advantage in reducing sensitivity to breakup. Static images are more tolerable to breakup compared to video images that encourage more eye movement. In general, a color frame rate tolerable in a NTE display would not be acceptable in a projector. In NTE systems the frame

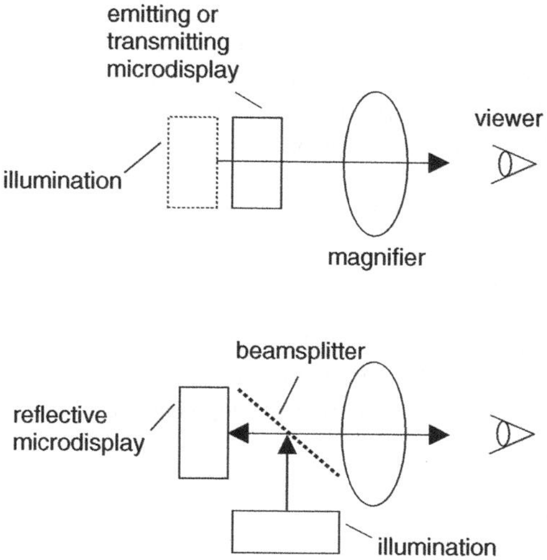

Figure 11.7 Comparison of emitting and transmitting microdisplays with reflective microdisplays

rate is boosted by trading throughput efficiency for speed, since throughput efficiency is not a critical design aspect, provided overall power consumption is not excessive. Electronic addressing power also increases with frame rate, and in optimizing the overall system, color breakup may be compromised to extend battery life. The doubling of optical path in reflecting nematic liquid crystal devices allows a thinner cell, providing a four-fold response speed advantage over transmitting devices. Reflecting devices also have an important advantage in providing greater fill factor, making the pixelation less obtrusive.[7,8]

Color sub-pixels are an extension of the direct-view method into the micro domain, and a popular option in transmission microdisplays. Sub-pixels increase the image pixel size and triple the total number of addressed pixels, expanding the overall size of the microdisplay. However, the data rate for RGB sub-pixels is generally less than that suppressing color breakup in a color field sequential system. Low aperture ratio and losses in the dye colors lead to poor throughput efficiency of ~3%. Pixel size is limited by the decline in aperture ratio and fringe field effects. The latter effect is more severe in sub-pixel colored microdisplays, where fringing is present in uniform color fields. Low aperture ratio has a strong effect on pixel visibility (screen window effect).

The contrast ratio of liquid crystal devices degrades with off-axis angle. Figures 11.3 and 11.4 show that the off-axis angle at the microdisplay is determined by the collection *f*-number. Off-axis compensation would improve the CR of wide-angle systems.[9,10] Methods and instruments to measure the performance of virtual displays are being developed, including a camera with optics that simulates the optics of the eye.[11]

Ferroelectric liquid crystal (FLC) microdisplays are appealing in NTE systems. LED illumination simplifies pulse width modulated gray levels. The need for DC balance in the addressing cycle may sacrifice 50% optical throughput, but the throughput efficiency is still higher than a high-speed, high-contrast, nematic microdisplay. Contrast ratio is less sensitive to off-axis angle. Color bit-frames can be interleaved to suppress color breakup.[12,13]

OLED microdisplays are competitive in NTE applications.[14–16] They employ colored sub-pixels, with consequent expansion in area and addressed elements. However, small depth and current drive inherent to OLEDs eliminates field fringing, allowing the minimum pixel size to be determined by lithographic limits and luminance factors. Lambertian emission automatically fills the magnifier aperture, providing maximum eyebox. Light is only generated according to image demands, providing power saving compared with devices that modulate a continuous light source. OLED response speed is essentially the response time of current change through the device, and has no impact on frame rate. OLEDs are newcomers to the display world and need to establish performance and lifetime in a practical working environment. Early success is anticipated in microdisplays, where addressing and lifetime issues are easier to resolve.

The restrictive angular range of the DMD at ±12 degrees is a disadvantage in NTE applications, making the magnifier >*f*/2.4; lower *f*- is possible with some loss of contrast ratio. The DMD has a throughput advantage in projection displays because it operates with unpolarized light and achieves adequate frame rate for field-sequential-color, justifying higher cost. Throughput advantage is not critical in NTE applications, and LED readout enables adequate speed response in competing LCDs. Moreover, the DMD pixel pitch at 13.8 μm is substantially higher than competing microdisplays, and electronic addressing power is higher. There are no reported developments in NTE application of the DMD.

11.5 Magnifiers

Some general characteristics of magnifiers are of interest in relation to NTE displays. In an ideal lossless system, the luminance of a magnified image is the same as the object (microdisplay). That is a result that follows from etendue conservation: as the image area increases, the angular divergence decreases in proportion, and luminance = lumen/m^2/sr remains constant. In practical systems, reflection and transmission loss in the optical elements are significant, but outweighed by beam-splitter losses in folded optical path systems.

A wide FOV and large eyebox favors a low *f*-number at the microdisplay. For non-emissive microdisplays, low *f* illumination is readily achieved by Lambertian emission of LEDs and diffuser plates. For example, at *f*/2 a microdisplay area 1 cm^2 illuminated with a flux of 1 millilumen has a luminance = flux/etendue = 51 cd/m^2. Clearly, LEDs can easily supply image luminance levels much higher than 100 cd/m^2. Experiments to determine comfortable luminance levels in microdisplays gave approximately 700 cd/m^2 in ordinary room light and 1500 cd/m^2 in daylight. An increase in luminance of about 10% was noted when changing from an information image to a picture image, and a further increase of 10% for a video image.[6] However, the result depends on how well shielded the eye is from ambient light, and 100 cd/m^2 is often considered adequate.

Full-color LED illuminators designed for LCDs have the red, green and blue diodes mounted in the same package. Collection optics and polarizing components deliver the LED light to the microdisplay at the design *f*-number. Adequate image luminance can be achieved with low system throughput efficiency of order 1%. More light is generated by increasing the LED current, but may compromise LED lifetime. Battery life will favor efforts to enhance throughput efficiency.

The microdisplay active diameter and resolution are incorporated in the magnifier design, along with illumination requirements and desired field of view. The virtual image is usually focused at about 2 m distance to provide relaxed viewing. A gradual 50% luminance variation from center to image boundary is not noticeable. A few percent image distortion is tolerable. Color fringing (lateral color) should be less than 2 arc minutes, or within two pixels. Field curvature requires the eye to change focus in moving from the center towards the image boundary. A design restricting field curvature to ±0.4 diopters prevents eye fatigue provoked by focus shift in scanning the image.

The sinusoidal modulation transfer function (MTF) quantifies the magnifier resolution. The MTF approaches 100% at low spatial frequency and must fall to zero with increasing spatial frequency (ν) as the diffraction limit $\nu_0 = (1/f_{\#}\lambda)$ is approached. The physical constraints on microdisplay magnifiers keep them well below diffraction limit. Spatial frequency is usually expressed in line-pair/mm = lp/mm = cycle/mm. Pixel pitch of 10 μm corresponds to spatial frequency 50 lp/mm, where magnifier MTF is generally below 40%. The magnifier is a compromise between performance, weight, size, style, and cost. Manufacturing volume is important in reducing the cost of molded optical components.

Perception of the full information content of the image, such as fine text, reveals the pixel structure; magnification and MTF are chosen to resolve the pixel. Usually pixel structure is a distraction and can be attenuated by lowering magnification and MTF. Defocusing the image is an effective way of lowering MTF. Image quality is preserved by reducing magnification rather than MTF. If the MTF remained at 100% to the highest spatial frequency of interest, the spatial image quality would be determined by the pixel resolution, which could be set at the limit of human vision. However, a moderate resolution microdisplay providing a wide viewing angle would reveal a distracting pixel structure, which could be hidden by a fall-off in MTF, or more sophisticated depixelation schemes.

11.6 Camera Viewfinder

The optical viewfinder of a camera directs some of the image light to an eyepiece presenting a view of the scene to the photographer. The photographer readily accepts replacement of the optical viewfinder by an electronic viewfinder (EVF), to the extent that the viewing action is similar. The image recording CCD array drives the viewfinder, ensuring focus and image location. The viewfinder also displays the camera settings and stored images, which remain easy to see in all lighting environments. A less obvious advantage is the reduction in size of a high-zoom camera with a microdisplay design, compared with an optical viewfinder.[17–19] The camera battery must supply power for focus, flash, image recording, viewfinder, and display. Power requirement is clearly an important factor in viewfinder design.

Popular viewfinders have QVGA resolution (240×320), with field sequential or RGB pixel color. Typical FOV is ~20 degrees with resolution ~3 min/pixel. A small direct-view display supplements the viewfinder, where two or more people can view the pictures. Figure 11.8 shows a viewfinder employing a nematic liquid crystal transmission microdisplay with RGB color pixels. Resolution is QVGA, with 5 μm RGB pixel pitch generating 6.1 mm diagonal active area; viewfinder magnification is $\times 15$, with 22 degrees FOV, and 16 mm eye relief, and focus adjustment -3 to $+1$ diopter.[20] The microdisplay employing transferred-silicon technology was adapted from a military helmet display program.[10]

A reflecting QVGA FLC-on-silicon-backplane (FLCOS) viewfinder with 12 μm pixel pitch (4.8 mm diagonal) displays field-sequential 24-bit color video at 50–100 frames/s, without noticeable flicker or color breakup.[12,21] LED illumination provides a 440 cd/m^2 output image at CR > 100 in a 7 mm diameter pupil, with 17 mm eye relief, and -3 to $+1$ diopter focal adjustment. Total power consumption at

Figure 11.8 Transmission nematic liquid crystal microdisplay viewfinder, $20 \times 18 \times 25$ mm, 6.6 gm: CyberEVF™ 230 K. Reprinted courtesy of Kopin Corp.

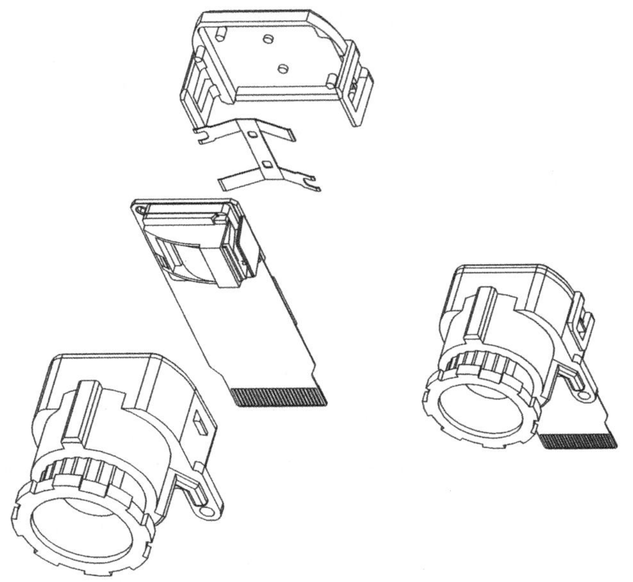

Figure 11.9 Exploded view of reflective microdisplay viewfinder, showing magnifier, microdisplay with illumination unit, and enclosure. Assembled depth is 30 mm and total weight 5.33 gm. Reprinted courtesy of Displaytech Inc.

60 Hz frame rate is less than 135 mW, including LED power. The ×15 magnifier and display assembly are shown in Figure 11.9.

Quick response and smooth motion are critical in live viewfinder use, where the EVF frame rate should match to the camera image processor for optimum performance. In playback mode, the frame rate can be increased to 75 Hz or higher to provide the best image view. For sophisticated cameras, a high-resolution SXGA (1280 × 1024) viewfinder employing FLCOS color sequential with 1000:1 dynamic range at 120 frames/s is available.[22] High-resolution viewfinders will find favor in consumer applications, providing cost and power are sufficiently low.

Two companies are developing viewfinders using OLED microdisplays (details presented in OLED chapter). OLEDs should compete for market share in the near future.[14–16]

11.7 Head-Mounted Displays

11.7.1 General Considerations

Mounting the display on the viewer's head is the simplest way of delivering the image for continuous viewing over substantial periods. Head movement does not shift the optical output pupil relative to the eye, providing a more relaxed viewing experience. However, we are accustomed to our view changing with head movement; consequently, isolation of the image from head movement is a source of stress for the long-term viewer. In sophisticated virtual reality viewers, the image changes with head movement as the head is tracked and coupled to the image generator.

Head mounting frees the hands for keyboard operation, or any mechanical adjustments prompted by the displayed information. The ability to view complex information such as circuit diagrams, or mechanical assembles, while inspecting or manipulating hardware is an advantage that can enhance productivity. Such commercial applications may require see-through images that superimpose the displayed image on normal viewing.

The term "total immersion" describes NTE displays that completely isolate the viewer from any visual contact with the real world. Total immersion is essential for a "virtual reality" experience, although the disorienting effects associated with poor design can be unpleasant. A "partial immersion" system allows sufficient visual contact with the outside world to maintain orientation and avoid discomfort such as motion sickness. Video entertainment headsets allow sufficient vision outside the eyepiece to prevent disorientation, while not distracting from the displayed image. Industrial NTE displays may be of the monocular form to allow the free eye to act as guidance, or of "see-through" design for guidance or comparison purpose, also known as "augmented reality."

In binocular vision, both eyes fixate on the same point and the brain fuses the two closely related images into a single image. Any disturbance to fusional lock of the images results in discomfort and fatigue. Stereopsis is the ability to form depth perception from the slightly different viewpoints of the two images. In a head-mounted display, images separately delivered to each eye must converge at their focal distance, typically 2 m, to provide fusional lock. Compactness is served by assigning a microdisplay to each eye; moreover, the microdisplays can be driven to provide a stereoscopic effect.

All head-mounted displays gain in comfort from a reduction in weight and volume of the display. Design of a head-mounted display is a compromise between weight, volume, and performance.[23] Electronics and power supply for the display attach to a lower region of the body, but optical magnification components need to be close to the eye. Design of the optical magnifier is critical in head-mounted displays. The balance and mounting structure are optimized, as well as the weight and volume. A simple linear extension of the optical train, as used in viewfinder design, would protrude as stalks in a HMD. Maintaining the position of such an unbalanced structure would require additional structural support with associated increase in volume and weight. If the optical path is folded, the opto-mechanical system moves closer to the eye, making it easier to support. When the HMD is reduced in size and weight comparable to ordinary spectacles, similar support by a nose bridge gives rise to the term "face-mounted" display.

Figure 11.10 shows a stereoscopic headset employing a color sequential SVGA reflective LC microdisplay, with specifications: 26 degrees FOV, 1.6 min/pixel, 17×6 mm pupil, 25 mm eye relief, 24-bit

Figure 11.10 SVGA reflective LC microdisplay 3D headset. Reprinted courtesy of i-O Display Systems

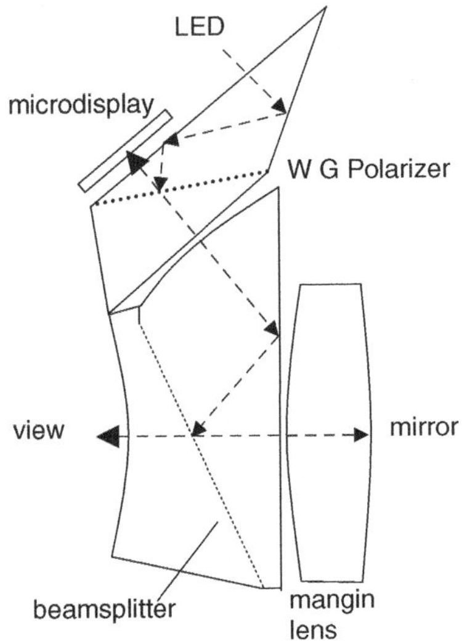

Figure 11.11 Folded path eyepiece and illumination for reflective microdisplay. Reprinted courtesy of Society for Information Display

color, total weight 198 gm. A folded catadioptric optical system similar to Figure 11.5 is used, modified for reflective readout. The large horizontal eyebox accommodates a wide range of inter-pupil distance.

11.7.2 TIR Prism

Total internal reflection (TIR) provides an advantage in prism systems. Figure 11.11 illustrates a compact eyepiece and illuminator for a reflective microdisplay.[24] The illuminator appears separately in Figure 11.12. The LED light source is an RGB package suitable for color sequential readout, and is efficiently collected (15%) by a nonimaging reflector design. Duplicate LED packages and reflectors are tiled to generate a 12 × 6 mm exit pupil. The large horizontal pupil facilitates a binocular display, without the added mechanical complexity of inter-pupil distance adjustment. The large exit pupil and

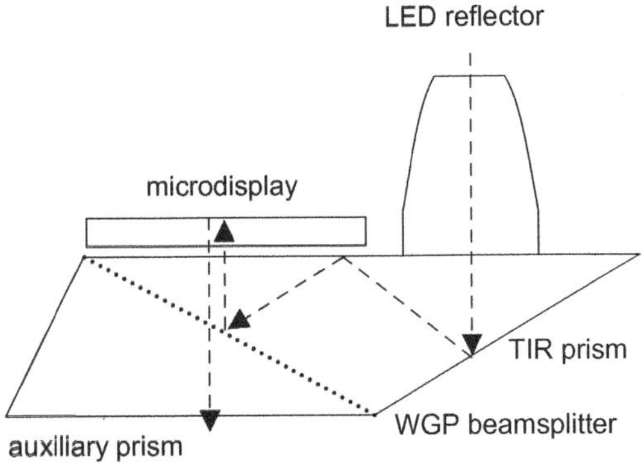

Figure 11.12 Illumination of a reflective microdisplay. Reprinted courtesy of Society for Information Display

wide FOV requires very large numerical aperture at the display, *f*/1.3 horizontal and *f*/2.6 vertical. The illumination, prepolarized by a dichroic polarizer, is directed by TIR and a wire-grid polarizing beam-splitter to the microdisplay. Depolarization at the TIR surface is limited by a phase-optimized dielectric coating. The auxiliary prism removes astigmatism.

The eyepiece illustrated in Figure 11.11 consists of a three-element magnifier with folded light path controlled by TIR and a beam-splitting surface. Polarized image light exiting the auxiliary prism is constrained by a field lens molded into the entrance of the following prism. TIR at the second prism surface followed by reflection at the beam-splitting surface directs light to the mangin lens. A mangin lens has a reflecting surface introducing catadioptric advantages to the design and a further folding of the optical path. The output light traverses the beam-splitter and a final prism with an aspheric divergent output surface. The divergent surface extends back-focal length to accommodate the illumination prism, and includes a molded diffractive kinoform for lateral color correction. All of the powered optical surfaces are on-axis and rotationally symmetric, simplifying the design, verification process, and cost for small production runs. In comparison, the design of anamorphic prism optics is complex and expensive to develop.[25] If the Mangin reflector is modified to a half-mirror, the system becomes see-through with only 7% distortion.

The eyepiece designed for an SVGA (800 × 600) reflecting nematic microdisplay with active area diagonal 12 mm has an effective focal length 18 mm, diagonal FOV 36 degrees, 2.2 min/pixel, eyebox 12 × 6 mm, eye relief 25 mm, average luminance 140 cd/m^2 (200 cd/m^2 with anticipated improvement in front-light and eyepiece efficiency). The luminance falls 50% from center to edge, and CR = 80 at the center. Monocular dimensions are 46.5 × 38 × 25.5 mm, and weight 35 gm. Total LED power is 220 mW. Optical characteristics: distortion 4%, lateral color 2 pixels, field curvature 0.2 diopter, MTF 40% at 40 lp/mm. The MTF easily resolves the 12 μm pixel periodicity, which is equivalent to 42 lp/mm. A dichroic polarizer inserted between the auxiliary prism and field lens improves contrast and attenuates ghost imaging due to light returned from the eyepiece beam-splitter. Figure 11.13 illustrates a binocular face mounted version of the eyepiece.

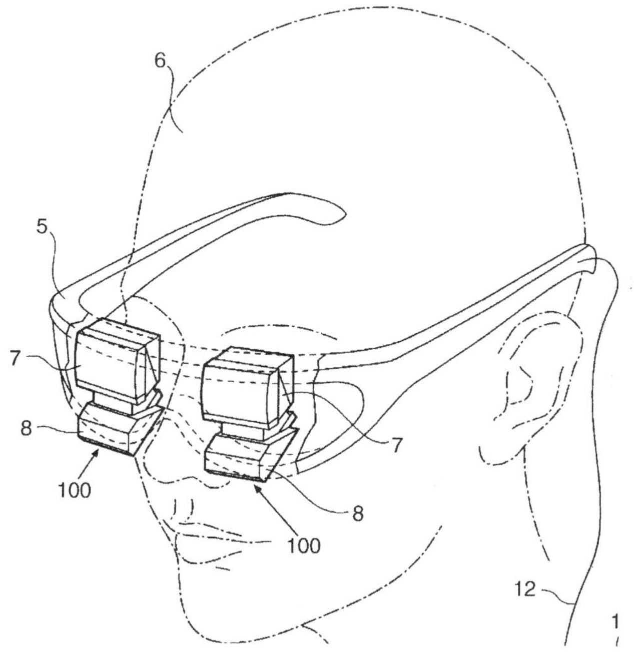

Figure 11.13 Structure of face-mounted display. Reprinted courtesy of Syntax-Brillian Corp.

Light weight and low cost favors molded plastic optics, but excessive stress birefringence distorts polarization, forcing a half-mirror at the second beam-splitter position. Optical glass would allow a polarizing beam-splitter (plus quarter-wave retarder) with a fourfold increase in eyepiece efficiency ($18 \times 4 = 72\%$), raising the luminance to approximately $800\,\mathrm{cd/m^2}$.

11.8 Free-Surface Prisms

Compact magnifier design favors off-axis optical systems, but reduction in symmetry creates severe aberrations that defy correction by rotationally symmetric optical elements. An optical design theory for asymmetric optical surfaces that correct off-axis aberrations was developed to design a compact eyepiece.[25] The fabricated eyepiece takes the form of a prism shown in Figure 11.14. Three surfaces of the prism function as optical elements, which can in general be molded to any surface form, giving rise to the term free-form surface (FFS) prism (or anamorphic prism). The surfaces maintain relative position as determined by the prism body, which also enables TIR further enhancing compactness. A 34 degrees FOV is achieved with a total optical system thickness of 15 mm, and monocular weight

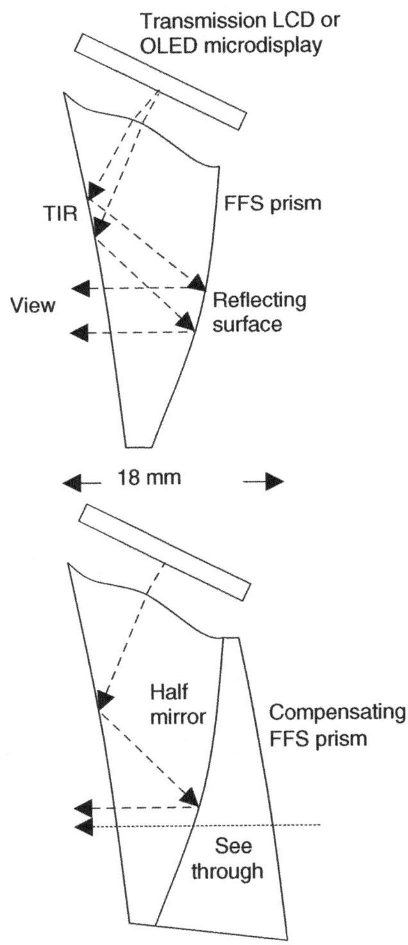

Figure 11.14 Free-form-surface prism eyepiece in immersion and see-through version. Reprinted with permission of SPIE

Figure 11.15 Transmission LC microdisplay headset: PUD-J5A. Reprinted courtesy of Sony Corp.

80 gm. An augmented reality version produced an VGA resolution, 25 mm active diagonal, 60 degrees FOV, 4.5 min/pixel, pupil 12 mm, eye relief 20 mm, and total thickness 18 mm.[26]

Several head-mount displays employing FFS prism eyepieces are available commercially. The headset illustrated in Chapter 1, Figure 1.2, employs dual 0.6-inch diagonal transmission displays with RGB color filter pixels. A headset designed for computer games, shown in Figure 11.15, uses transmission LC microdisplays providing resolution $300 \times 200 \times 3$ RGB pixels over a 30 degrees FOV. A head tracker provides 360-degree viewing directions. The display unit weighs 340 gm and ergonomic design enables prolonged viewing experience.

A monocular headset employing an SVGA OLED microdisplay[16] with anamorphic prism optics is illustrated in Figure 11.16. Specifications are: 15.6 mm display diagonal, 37 degrees FOV, 2.2 min/pixel, 12×6 mm pupil, 25 mm eye relief, 24-bit color, 100 cd/m^2 luminance, 100:1 contrast, 140 gm weight, 0.6 W total system power.

The smallest weight and volume are achieved when the FFS prism is designed as an eyepiece (no intermediate image) magnifier. However, at sufficiently high magnification, aberrations can be reduced by incorporating relay optics providing an intermediate image within a prism structure.[27]

The FFS prism eyepiece has been applied to transmission LCDs and OLEDs, but is more of a challenge for reflective displays.[28] Insertion of a glass illumination prism extends the back-focal length by virtue of its refractive index, and provides a polarizing beam-splitting surface, as shown in Figure 11.17. The LED illuminator includes a light guide disperser and polarizer. A pseudo on-axis configuration achieves high resolution and large exit pupil: microdisplay diagonal 12 mm, FOV 40 degrees, 63 lp/mm at 20% MTF would resolve XGA (1024×768) 1.9 min/pixel, pupil 12 mm, eye relief 20 mm, but field curvature at ± 0.6 diopter is excessive.[28]

A reflective microdisplay design employing a holographic lens with color sequential LED readout gives the best see-through efficiency, as illustrated in Figure 11.18. The holographic optical element

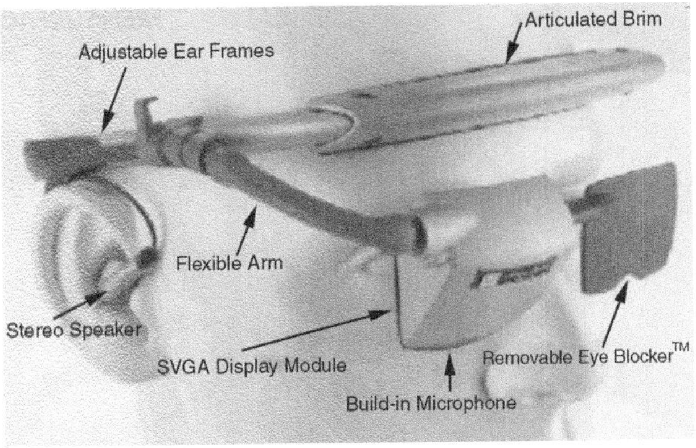

Figure 11.16 SVGA OLED monocular display. Reprinted courtesy of eMagin Corp.

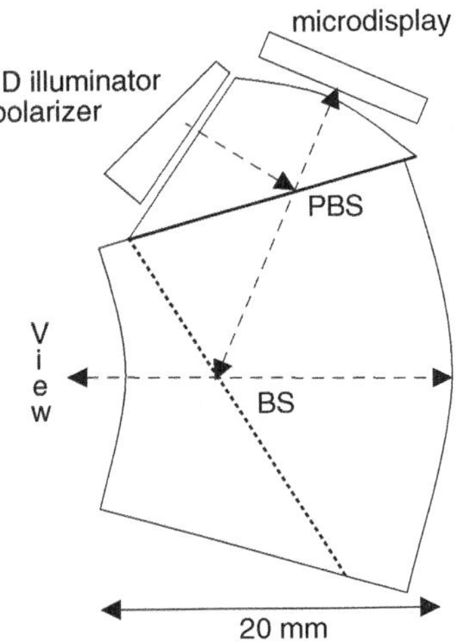

Figure 11.17 Prism eyepiece for reflective microdisplay. PBS = polarizing beamsplitter, BS = beam-splitter. Reprinted courtesy of Society for Information Display

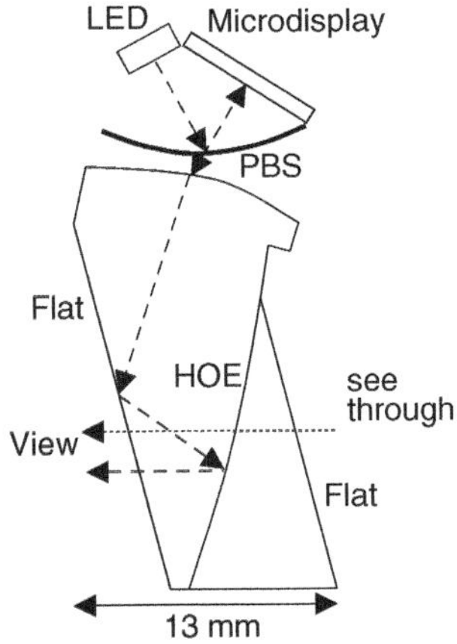

Figure 11.18 See-through prism eyepiece with holographic lens. PBS = polarizing beamsplitter, HOE = holographic element. Reprinted courtesy of Society for Information Display

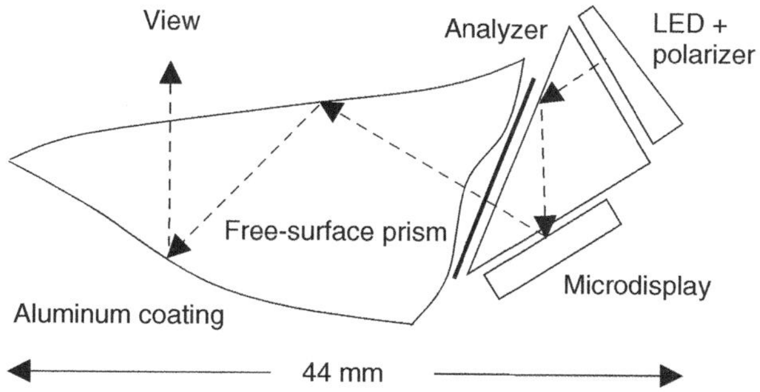

Figure 11.19 Prism eyepiece for viewfinder or cellular phone. Reprinted courtesy of Society for Information Display

(HOE) focuses the well-defined red, green and blue readout wavelengths, while the broadband ambient light passes with little loss. Experimental results:[28] eyepiece efficiency for green LED 8.3%, see-through efficiency 78%, CR 100:1, display diagonal 4.8 mm, FOV 20 degrees, resolution 35 lp/mm at 20% MTF, 8 mm pupil, 25 mm eye relief.

FFS prism optics also applies to camera viewfinders and portable equipment, as illustrated in Figure 11.19.[28] The efficiency of the eyepiece approaches 37% with unpolarized light input. Results: display diagonal 12 mm, FOV 33 degrees, and resolution 40 lp/mm at 20% MTF, pupil 12 mm, eye relief 35 mm.

11.9 Eyewear-based Displays

Ordinary spectacles and sunglasses have evolved for continuous wear, such that any inconvenience is hardly noticed in normal circumstance. "Eyewear-based" displays exploit this established platform to support a microdisplay.[29–32] Figure 11.20 illustrates a relay optics system embedded in an eyeglass, fed by a temple-mounted microdisplay. The displayed image is superimposed on the external view by a beam-splitter, which incorporates a quarter-wave retarder in the case of a polarizing beam-splitter. Small-scale optics limits the FOV, but the inherent small eye relief ~15 mm supports adequate pupil diameter. Figure 11.21 shows an QVGA (320 × 240), color sequential transmission LCD, 6.4 mm diameter, giving FOV 13 degrees, 2 min/pixel, weight 53 gm including eyeglasses.

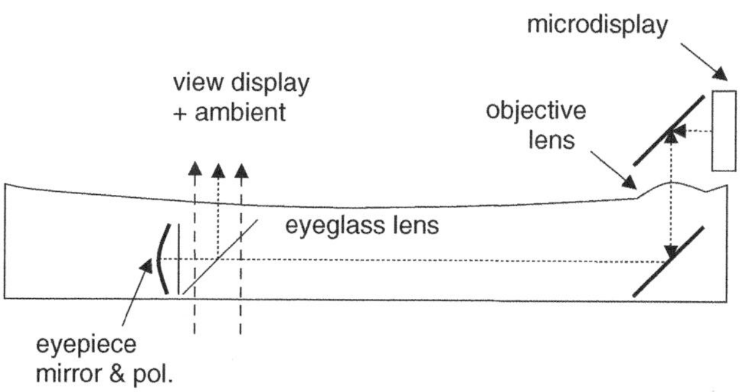

Figure 11.20 Relay optics embedded in eyeglass lens. Reprinted courtesy of MicroOptical Corp.

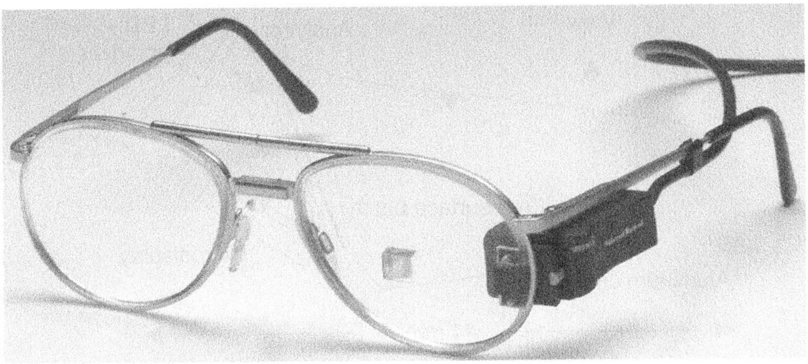

Figure 11.21 Eyeglasses with embedded relay optics and temple mounted microdisplay. Reprinted courtesy of MicroOptical Corp.

A VGA clip-on light pipe system is available as shown in Figure 11.22. Relay optics are mounted in the light pipe positioned in front of the eyeglass, and supported by the spectacle frame. The viewer can see through the light pipe, which is blackened on the upper and lower surface to enhance the display contrast against ambient light. A focal adjustment sets the virtual image 2–15 feet in front of the viewer. The greater FOV for VGA resolution required an upgrade of the system to doublet eyepiece and objective optics. The microdisplay is supported at the temple. Specification: VGA (640×480), 24-bit color, 20 degrees FOV, 1.5 min/pixel, 35 gm weight.

Attaching clip-on microdisplays on both sides of the spectacle frame to form a binocular display provides stereo image capability. An optimum design for a stereo display based on light pipe viewers is under development.[32] The light pipe approach gives one of the most compact and lightweight systems.

11.10 Light-Guide Systems

Conveying the image to the eye by a solid light-guiding material enables remote mounting of the microdisplay away from the visual field. Minimizing the guide in at least one dimension serves compactness, but requires innovative optical techniques to expand the output pupil and field of view.

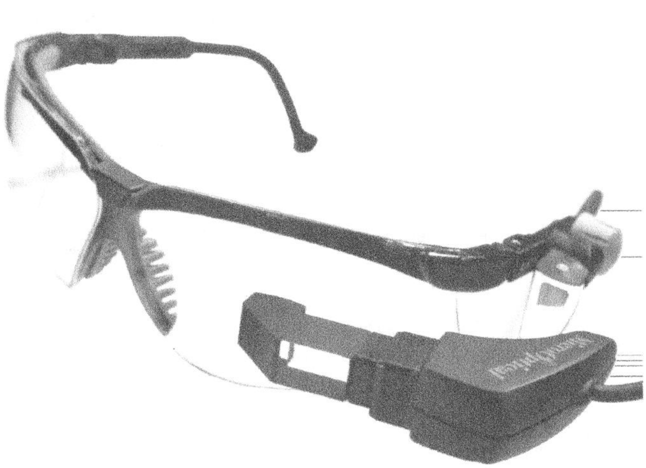

Figure 11.22 Clip-on light pipe relay optics and microdisplay. Reprinted courtesy of MicroOptical Corp.

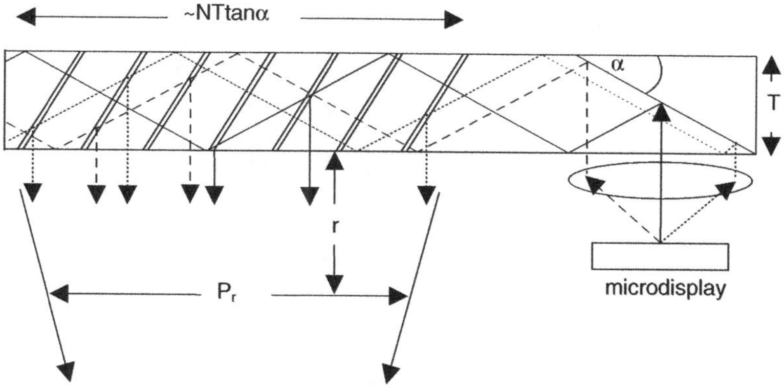

Figure 11.23 Light-guide optical element with $N = 7$ partially reflecting output surfaces. Reprinted courtesy of Society for Information Display

11.10.1 Reflective Array

A novel light-guide optical element (LOE), provides a large output pupil or eyebox (P_r), at substantial eye relief (r), for modest input numerical aperture.[33–35] Figure 11.23 illustrates the concept. The guide has thickness T, where a mirror tilted through angle α directs collimated input light along the guide. Total internal reflection from upper and lower surfaces guides the light until encountering an array of partially reflecting surfaces that deflect into the output pupil. Design of the array minimizes reflection at near normal incidence, while controlling the reflectivity at shallow angles according to the sequence of surface elements (1 to N). The angular selectivity of the reflecting array eliminates multiple images, and distribution of sequential reflectivity gives uniform intensity over P_r.

The maximum field of view (FOV_{max}) approximates to

$$FOV_{max} \approx \tan^{-1}\left(\frac{NT\tan\alpha - P_r}{r}\right). \tag{11.6}$$

The partially reflecting output surfaces expand etendue to provide a large eyebox by translating the angular spectrum, while preserving angle. The viewer's eye focuses the angular spectrum into a retinal image, identifying image points with angular direction. The average eye has an angular resolution of 0.4 mrad, implying all reflecting surfaces in the LOE should maintain a cumulative precision <0.4 mrad to provide ideal resolution. In practice, a precision compatible with the resolution of the microdisplay pixel is sufficient.

Dichroic coatings form the reflecting elements, tuned for the required angular discrimination, and wavelength independence. Appropriate increase of reflectivity with progression along the train maintains uniform intensity over the output pupil. Wavelength independence facilitates color imaging. Figure 11.24 shows on-axis rays, which completely fill the output pupil without overlap, but off-axis rays have gaps or overlap in angular space; the effect on luminance uniformity is mitigated by angular dependence of reflectivity, and eye pupil ~4 mm that accepts rays from more than one reflector.[34] Coating performance influences contrast ratio and image quality, as does tilt angle α. The LOE should be inexpensive to manufacture in large volumes using an injection molded optics approach.[33–35] Pupil expansion in two dimensions favors orthogonal reflecting arrays as indicated in Figure 11.24.

Figure 11.25 illustrates LOE near-to-eye display products, showing how the microdisplay can be spaced substantial distance from the output pupil, facilitating an extremely compact viewing system. Other applications include portable handheld devices such as mobile phones that would benefit from a superior display, and head-up displays.

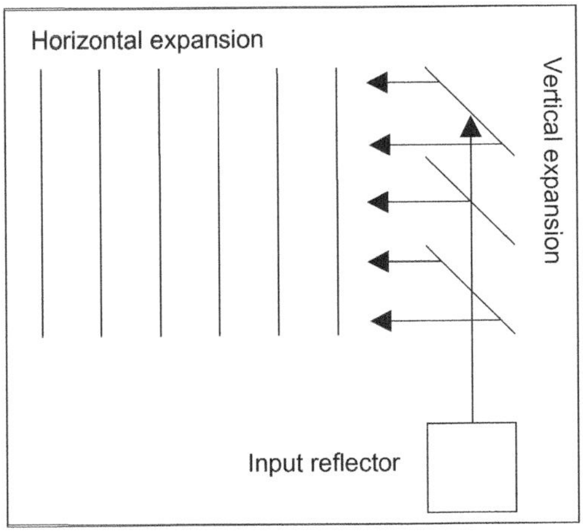

Figure 11.24 LOE designed for vertical and horizontal pupil expansion. Reprinted courtesy of Society for Information Display

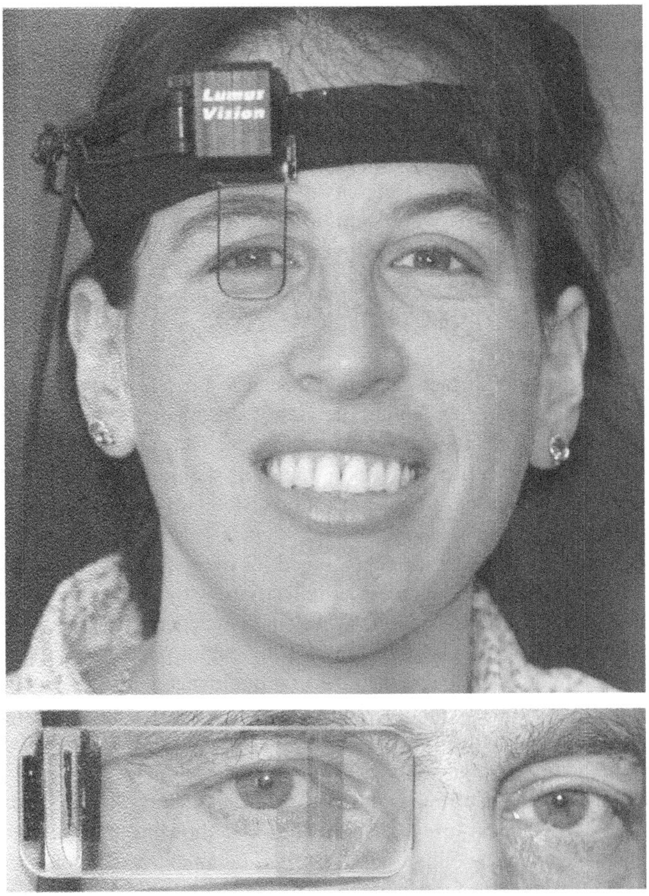

Figure 11.25 LOE near-to-eye display products. Reprinted courtesy of Lumus Ltd

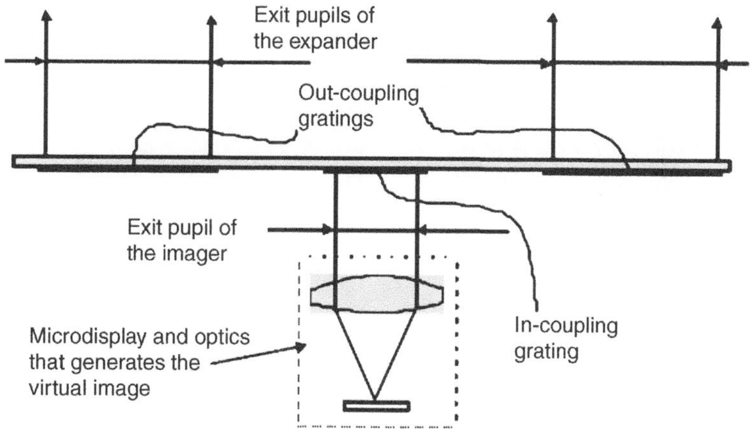

Figure 11.26 Diffractive optical exit pupil expander. Reprinted courtesy of Society for Information Display

A design example with $r = 28\,\text{mm}$, $P_r = 12\,\text{mm}$, $\alpha = 30°$, $T = 4\,\text{mm}$, produced FOV $= 26 \times 19.5$ degrees. The entire field of view presented a sharp color image at $1500\,\text{cd/m}^2$ luminance, without noticeable chromatic effects. The complete system, including microdisplay, collimation lens, and mechanical enclosure weighs less than 60 gm. The LOE provides a natural see-though image for augmented reality applications, while total immersion follows from simple blocking of external light.

11.10.2 Diffractive Array

The exit pupil expansion (EPE) described in the previous section can be achieved by diffractive optic elements, as indicated in Figure 11.26.[36] High-resolution lithography facilitates gratings fabrication with periods below visible wavelength. Figure 11.27 shows the input image propagating along an optical waveguide by total internal reflection until encountering the out-coupling gratings. The gratings are designed to optimize the first-order diffraction, suppressing higher orders. The angular range and wavelength range of the input image complicate the grating structures; moreover, two-dimensional EPE imposes further complexity on the gratings.

The diffractive system has not reached prototype demonstration at present, but the ready availability of high-resolution lithography and sophisticated diffraction theory should encourage development. The problem of uniformity in the expanded pupil can be overcome by stacking waveguide structures separated by low-index material. Sophisticated EPE devices that enable lightweight structures minimizing visual discomfort may be the key to mass marketing of NTE displays.

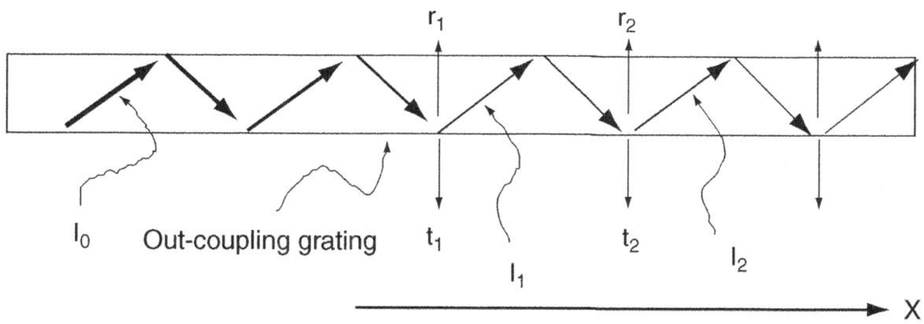

Figure 11.27 Waveguiding of image to out coupling gratings. Reprinted courtesy of Society for Information Display

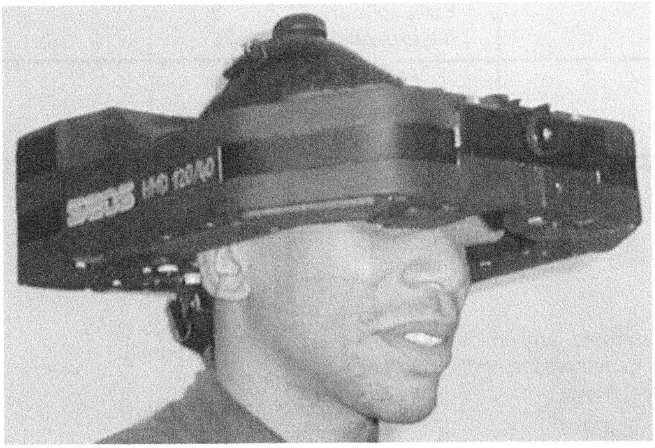

Figure 11.28 Wide-angle field of view HMD. Reprinted courtesy of SOES Displays

11.11 Wide Field of View

11.11.1 Single Microdisplay

Virtual reality or simulation displays require a 120-degree FOV to maintain the illusion of being part of the scene rather than viewing it on a screen. Wide FOV comes with the burden of weight, volume, and balance. A recent approach to minimize weight employs polymer foam as a structural material to support the relay optics and eyepiece.[5] The HMD shown in Figure 11.28 weighs less than 1 kg, including head tracker. Linking the view to head-position is essential for the virtual reality illusion. The HMD has a center of gravity coincident with the head, and fit designed to minimize attachment stress.

The relay optics shown in Figure 11.29 includes four aspheric elements and folds the optical path around the side of the head. The image is relayed to a screen behind the eyepiece, where optical scattering from the screen fills the large exit pupil of the eyepiece. The exit pupil has a soft edge, as the

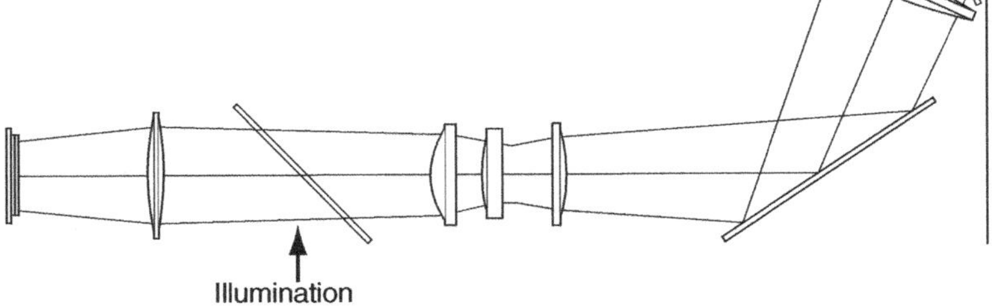

Figure 11.29 Relay optics and eyepiece achieving 80×67 degrees FOV per eye. Reprinted courtesy of SOES Displays

luminance and resolution decay outside the design eye-pupil. Expansion of the angular field by the screen allows the *f*-number of the relay optics to increase, lowering the weight. Conventional screen materials have a visible structure when magnified by the eyepiece, and sacrifice resolution. A screen material developed for this application has no visible structure, but some loss of resolution, which has the virtue of attenuating the pixel structure. A catadioptric eyepiece was developed to achieve large FOV and eyebox in compact form. It derives from an on-axis mirror collimator invented by LaRussa, employing reflections between a spherical and flat beam-splitter.[3] Lacroix improved the performance by introduction of a polarizing beam-splitter, realized by a wire-grid polarizer in the present design.

An SXGA (1280×1024) color sequential FLCOS microdisplay feeds the system at 60 frames/s, with interleaved color fields at high enough rate to suppress color break-up.[22] Performance: 80×67 degrees FOV/eye, 3.8 min/pixel, total FOV 120×67 with 40-degree stereo overlap, 24-bit color, 12 mm eye relief, ±4-diopter focus adjustment.

11.11.2 Tiled Microdisplays

An alternative approach to wide-angle viewing tiles a number of microdisplays through an array of lenses.[37,38] Tiling is a way around the FOV/resolution tradeoff, and achieves desired field curvature without distortion. Figure 11.30 illustrates the equatorial plane view, with four microdisplays and four lenses distributed circularly about the center of eye rotation. Each display positioned at the focus of a lens presents its magnified virtual image to the eye. Each eye has four stacked rows of microdisplays and lenses arranged spherically about the center of eye rotation, to give 16 displays/eye as indicated in Figure 11.31. The montage of images has sufficient overlap to present a continuous image as the eye scans the field of view.

The image sources are SVGA OLED-on-silicon microdisplays with active area diagonal 15 mm, providing 24-bit color at 60 Hz.[16] Emissive sources simplify an already complicated design. A molded Fresnel lens array, with 18 mm eye relief, images the microdisplays onto the pupil space. The circuit board supporting each microdisplay interferes with the microdisplay relative positioning, necessitating an increase in the back-focal length by insertion of glass blocks. Here is an example where modification of the microdisplay structure could reduce the size and weight of the HMD. Performance: 150-degree horizontal FOV, 70-degree binocular overlap, 100-degree vertical FOV view, 3 min/pixel. With an additional column of microdisplays for each eye, the horizontal FOV could be increased to 220 degrees. Mechanical adjustments allow horizontal positioning of the optics for each eye, and along the primary visual axis together. Figure 11.32 illustrates the head attachment of the display. A comparison with Figure 11.21 shows the burden imposed by wide FOV HMD, and the opportunities for microdisplay and optics R&D.

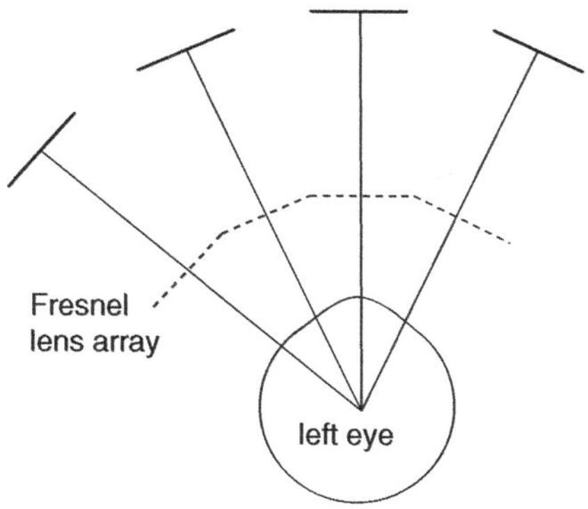

Figure 11.30 HMD with tiled microdisplays: polar view of equatorial (horizontal) plane. Reprinted courtesy of Johns Hopkins University

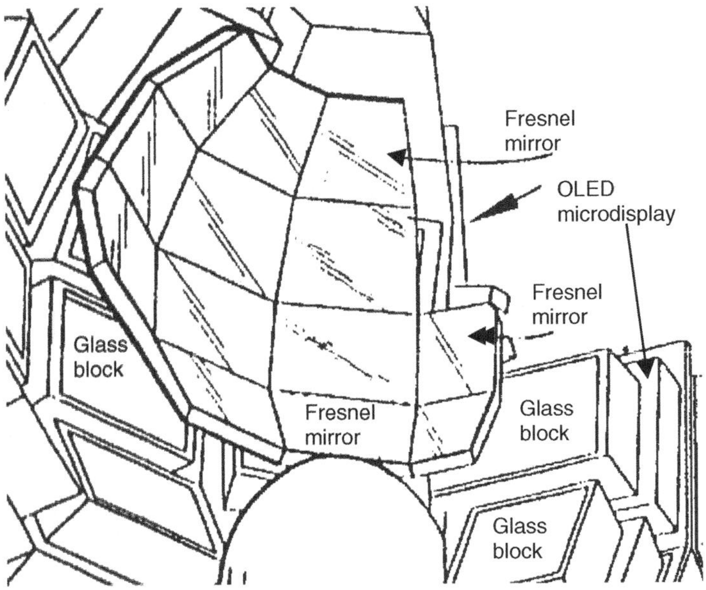

Figure 11.31 Molded Fresnel lens array and tiled microdisplays (for left eye). Reprinted courtesy of Johns Hopkins University

The system demands substantial computing power. Head tracking determines the image fed to each microdisplay, and eye tracking determines the local resolution required. A master computer handles the head and eye tracking data, distributing the information to 32 slave computers, each generating the image data for a particular microdisplay. The slave computers receive image alignment data from

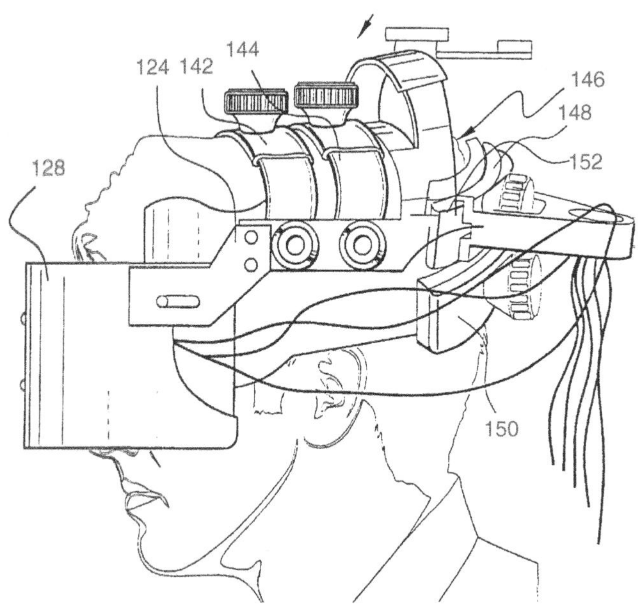

Figure 11.32 Head mounting structure for total immersion display. Reprinted courtesy of Johns Hopkins University

a calibration process that achieves pixel-level precision. The master computer also distributes hue and brightness data to achieve uniformity across the display montage. Video frame rate is in the range 20–60 Hz.

Prolonged wearing of a total immersion HMD can cause eyestrain, nausea and motion sickness, disorientation etc., collectively referred to as "simulator sickness," which may continue for several hours after removal of the HMD. Failure of a simulator to mimic normal visual experience provokes simulator sickness. The tiled HMD attenuates the dominant causes of simulator sickness by providing: sufficient FOV at adequate resolution, low image latency in response to head and eye movement, and freeze-frame override for rapid movement.[38]

11.12 Portable Equipment

Small portable equipment such as cell phones, hand-held computers, digital assistants, etc., are battery powered and have special display requirements.[39] Low power consumption is important and display area restricted. It appears to be a fertile area for NTE displays.[40,41] However, direct-view displays dominate, although the information is limited by the small available display area, requiring scrolling methods to scan a map, spreadsheet, or web page. A NTE display with SVGA resolution would provide the advantage of a monitor-like view, with head-mount optics easily adapted for hand-held equipment. A longer eye relief of about 35 mm is favored, with eyebox in the region of 12 mm; the anamorphic prism optics illustrated in Figure 11.19 is an example.

Figure 11.33 shows a NTE display mounted on a prototype hand-held web browser, presenting a 21-inch diagonal monitor-like image, at VGA resolution. It accommodated a variety of inputs,

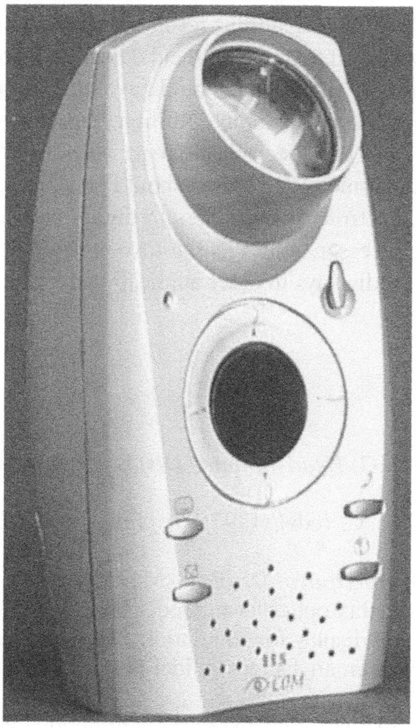

Figure 11.33 Virtual display mounted in prototype hand-held web browser, 4.6 × 2.5 × 1.4 inches, weight 6 oz. Reprinted courtesy of Icuity Corp.

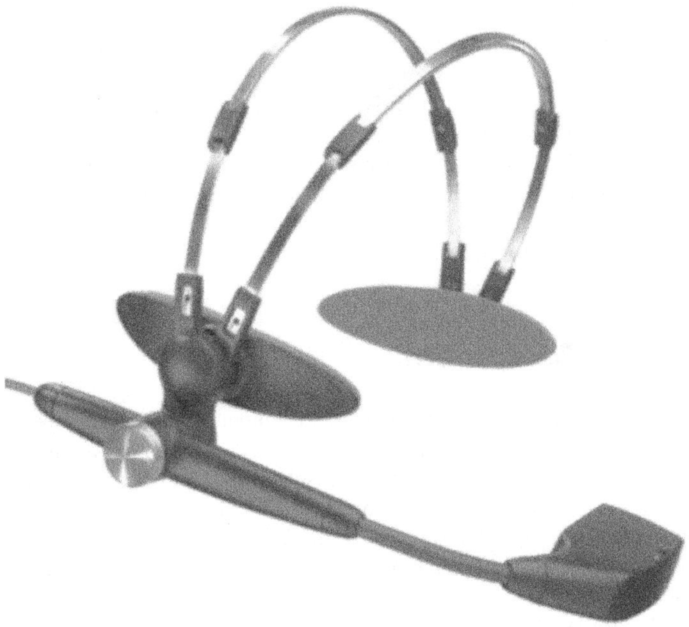

Figure 11.34 Head-mounted display providing 26 degrees FOV and VGA resolution: M920-CF. Reprinted courtesy of Icuiti Corp.

including cellular phone. First shown at a Japanese exhibition in late 2001, it did not go into production. The HMD shown in Figure 11.34 has similar resolution, with more comfortable viewing during extended periods, and can be connected to any hand-held device that provides a display outlet.

NTE displays mounted on portable equipment should have more success when viewed for brief periods, and scrolling is not acceptable. For example, military applications may make such demands. HMDs are particularly attractive to the spectacle-wearing viewer, providing a platform for a small clip-on display; otherwise, the head-mounting frame should be designed to fold into a small volume. The biggest market for NTE is electronic viewfinders in digital cameras and camcorders, where the ergonomics are similar to its predecessor. As EVFs improve in resolution and FOV, viewers may grow to appreciate the value of the NTE displays in other applications.

11.13 References

1 Melzer, J. E. and Moffitt, K. (1997) *Head Mounted Displays: Designing for the User.* McGraw-Hill. ISBN 0-07-041819-5
2 MacDonald, L. W. and Low, A. C. (eds) (1997) *Display Systems Design and Applications.* John Wiley. ISBN 0 471 95870 0
3 Karim, M. A. (ed.) (1992) *Electro-Optical Displays.* Marcel Dekker.
4 Hildebrand, A. P. (1999) High-quality optics for microdisplays. *Information Display,* April/May, 22–6.
5 Coates, N. (2003) Head mounted display for simulation and visualization. *SID Digest* **34**, 184–7.
6 McLaughlin, C. *et al.* (2003) Human factors guidelines for binocular near-to-eye displays. *SID Digest* **34**, 280–3.
7 Schott, D. J. (1999) Reflective LCOS light valves and their application to virtual displays. *Proc. Intl Display Research Conference*, 485–8.

8 Bergstrom, N. *et al.* (2000) Ergonomic wearable personal display. *SID Digest* **31**, 1138–41.
9 Bartlett, C. T. and Freeman, J. (2002) Flat panel displays in the helmet mounted display. *Proc. SPIE*, **4711**, 33–47.
10 Woodard, O. C. *et al.* (2000) Developing the 1280 by 1024 AMLCD for the RAH-66 Comanche. *Proc. SPIE*, **4021**, 203–13.
11 Levola, T. and Viinikanoja, J. (2002) Using peripheral view in virtual display characterization. *Proc. Intl Display Research Conference*, 111–14.
12 O'Callaghan, M. J. and Handschy, M. A. (2001) Ferroelectric liquid crystal SLMs: from prototypes to production. *Proc. SPIE*, **4457**, 31–42.
13 Banas, D. *et al.* (1997) Miniature FLC/CMOS color sequential display systems. *J. SID* **5**, 17–31.
14 Bodammer, G. *et al.* (2003) Emissive microdisplay using light emitting polymer. *SID Digest* **34**, 742–5.
15 Prache, O. (2002) Full-color SVGA+ OLED-on-silicon microdisplay, *J. SID* **10**, 133–8.
16 Ghosh, A. P. (2002) Full color OLED-on-silicon microdisplay. *Proc. SPIE*, **4464**, 1–10.
17 Zavracky, P. M. and Sebastian, F. (2000) Microdisplays for the camcorder viewfinder market. *SID Microdisplay Tech. Digest.*
18 O'Donnell, P. and Doi, A. (2001) Microdisplays in digital cameras. *Information Display* **17/12**, 16–19.
19 Reinhard, S. and Handschy, M. (2002) Effect of pixel pitch on microdisplays for electronic viewfinders. *SID Microdisplay Tech. Digest*, 41.
20 CyberEVF™ 230K, Kopin Corp., Westborough, MA 01581.
21 LightView™ QVGA Display Engine, QDE-0076-SE1, Displaytech Inc., Longmont, CO 80503.
22 Whittley, T. *et al.* (2002) Applications of FLCOS microdisplays. *SID Microdisplay Tech. Digest*, 26–9.
23 Chen, C-H. *et al.* (1999) Overview of head mounted display technology and applications. *Intl Display Workshops*, Japan, 319–22.
24 Gleckman, P. and Schuck, M. (2001) Optical characteristics of high performance LCoS virtual display. *SID Digest* **32**, 62–6.
25 Hoshi, H. *et al.* (1996) Off-axial HMD optical system consisting of aspherical surfaces without rotational symmetry. *Proc. SPIE*, **2653**, 234–42.
26 Yamazaki, S. *et al.* (1999) A thin, wide field-of-view HMD with free-form-surface prism and applications. *Proc. SPIE*, **3639**, 453–62.
27 Tekeyama, T. (2002) Image display apparatus. US Patent 6,342,871.
28 Mukaw, H. *et al.* (2000) Novel virtual image optics for reflective microdisplays. *Proc. Intl Display Research Conference*, 96–9.
29 Spitzer, M. B. *et al.* (1998) Optical approaches to the incorporation of displays within eyeglasses. *SID Digest* **29**, 1007–9.
30 Spitzer, M. B. *et al.* (2000) Eye-based displays for personal electronics. *Proc. SPIE*, **4021**, 27–31.
31 Spitzer, M. B. *et al.* (2001) Eyeware platforms for miniature displays. *SID Digest* **32**, 258–61.
32 Spitzer, M. B. *et al.* (2003) Wearable stereo eyewear display. *SID Digest* **34**, 261–3.
33 Amitai, Y. (2004) Extremely compact high-performance HMDs based on substrate-guided optical element. *SID Digest* **35**, 310–13.
34 Amitai, Y. (2003) Substrate-guided optical beam expander. US Patent Application 2003/0165017.
35 Amitai, Y. (2005) A two-dimensional aperture expander for ultra-compact, high-performance head-worn displays. *SID Digest* **36**, 360–3.
36 Levola, T. (2005) Diffractive optics for virtual reality displays. *Proc. Intl Display Research Conference*, 542–5.
37 Massof, R. W. *et al.* (2003) Full-field high-resolution binocular HMD. *SID Digest* **34**, 1145–7.
38 Massof, R. W. *et al.* (2003) Head mounted display with full field of view and high magnification. US Patent 6,529,331.
39 Akins, R. (1999) Displays for hand-held portable electronic product. *J. SID* **7**, 273–6.
40 Hildebrand, A. P. and Kintz, G. J. (1999) Twice folded compound magnified virtual image electronic display. US Patent 5,905,478.
41 Hakkinen, J. (2004) A virtual display for mobile use. *SID Digest* **35**, 1586–9.

Table of Symbols

Symbol	Meaning
α	Nematic pretilt angle
α	Guide propagation angle
β	Angle between polarization axis and input LC director
β	Maximum deflection angle
χ	Polarization rotation of skew ray
$\Delta\varepsilon$	Dielectric anisotropy $\varepsilon_{\parallel}-\varepsilon_{\perp}$
ΔL	Luminance difference
Δn	Birefringence n_e-n_o
$\varepsilon_{\perp}$	Nematic perpendicular relative permittivity
$\varepsilon_{\parallel}$	Nematic parallel relative permittivity
ε_0	Permittivity of free space
$\varepsilon_{\mathrm{LC}}$	Liquid crystal relative permittivity
$\varepsilon_{\mathrm{Si}}$	Relative permittivity of crystalline silicon
ϕ	Nematic twist angle
ϕ	Mechanical deflection angle
Φ_{B}	Bulk or substrate potential
Φ_{MS}	Work function difference between gate and channel
ϕ_{s}	Stable deflection limit
γ	Grayscale gamma
γ	Body effect parameter
Γ	Phase retardation
γ	Ratio of excitons produced to injected charge carriers
γ_1	Nematic rotational viscosity
η	Optical throughput efficiency
$\eta(\lambda)$	Pixel mirror reflectivity
η_{eff}	OLED pixelated efficiency
η_{ext}	Optical outcoupling efficiency

Introduction to Microdisplays D. Armitage, I. Underwood and S-T. Wu

Symbol	Meaning
μ	Carrier mobility
μ	Ion mobility
ν	Spatial frequency
ν_o	Diffraction limit
θ	Planarization angle
θ	FLC tilt angle
(θ,ϕ)	Skew ray polar coordinates
$2\theta'$	Field of view
θ_i	Angle of incidence
θ_n	Diffraction angle order n
θ_p	Nematic pretilt angle
θ_r	Angle of reflection
λ	Channel length modulation factor
λ	Wavelength
λ_o	Center band wavelength
σ	Bulk ionic conductivity
τ_0	Nematic reorientation time constant
τ_{dwell}	Row dwell time
τ_{field}	Field address time
a	Lamp constant 0.26 V/mmbar
A	Wafer area
A_C	Critical wafer area
$[A_e,B_e]$	Cauchy dispersion constants for n_e
$[A_o,B_o]$	Cauchy dispersion constants for n_o
b	Mirror length
C	Contrast
C	Cost per good die
C_0	Gate dielectric capacitance/area
C_G	Gate oxide capacitance
C_{GD}	Capacitance gate to drain
C_{LC}	Liquid crystal capacitance of pixel
C_M	Mirror and light block capacitance
C_{PAD}	Pixel total capacitance
C_{PC}	Photoconductor capacitance
CR	Contrast ratio
CR_{md}	Microdisplay contrast ratio
CR_p	Overall projector contrast ratio
CR_{sys}	Projection system contrast ratio
C_S	Pixel storage capacitance
d	Liquid crystal cell gap
d	Lamp arc length
D	Lens diameter
D_0	Defect density
D_{avrg}	Average data rate
DE	Diffraction efficiency
dV	Increase in voltage to maintain current (at t_{50})
dV_{50}	Voltage increase to maintain current at t_{50}
E	Electric field

Symbol	Meaning
E	Etendue
E_o	Output etendue
E_S	System etendue
F	Frame rate
F	f/number
F	Focal length
F_B	Backplane fill factor
F_E	Emissive fill factor
f_{field}	Field update rate
F_P	Pixel fill factor (aperture)
G	Number of good dice
H_R	Pixel reflection efficiency
I_{D0}	Process parameter $\sim$ 20 nA
I_{DS}	Drain to source current
$I_{DS,SAT}$	Drain to source saturation current
k	Boltzmann constant
k	Elastic constant
$K_{\parallel}$	Nematic splay elastic constant
K_{33}	Nematic bend elastic constant
L	Luminance
L	Channel length
L	Output lumen flux
L_B	Blue lumens
L_G	Green lumens
L_R	Red lumens
m	Flicker modulation
M	Number of pixel rows
M	Magnification
MTF	Modulation transfer function
n	Carrier concentration
n	Process parameter $\sim$ 2
n	Free ion concentration
n	Refractive index
N	Number of pixel columns
N	Number of dice
N	Guide reflecting elements
N_A	Average dopant density in the channel
n_e	Extraordinary optic index
n_e	Emitting layer refractive index
n_o	Ordinary optic index
n_p	Polymer refractive index
p	Chiral pitch
p	Mirror array pitch
p	Grating period
p	Lamp pressure
p	Polarization component in plane of incidence
P	Lamp power
P	Exit pupil (eye box) diameter

Symbol	Meaning
P_B	Blue radiant power
P_G	Green radiant power
P_R	Red radiant power
P_s	Spontaneous polarization
q	Electron charge
q	Ion charge
q	Photoluminescent efficiency
Q_{SCL}	Limiting space charge
Q_{SS}	Surface charge/area
r	Radius of usable wafer area
R	Surface planar relaxation distance
R	Surface reflectivity
R	eye relief distance
$R_{\perp}$	Normalized reflectance
$R_{\parallel}$	Reflectance parallel to wire grid
r_{st}	Fraction of excitons that can radiatively decay
s	Pixel aperture ratio
s	Polarization component orthogonal to incidence plane
S	Contrast sensitivity function
S	Nematic order parameter
t	Time
T	Total torque
T	Light guide thickness
$T_{\perp}$	Normalized transmission of TN cell
$T_{\perp}$	Transmittance perpendicular to wire grid
t_0	Binary-0 time
t_1	Binary-1 time
t_{50}	Lifetime to luminance fall 50%
T_{decay}	Optical decay time (90-10%)
T_e	Electrostatic torque
t_F	Addressing frame period
t_r	Ion transit time
T_{rise}	Optical rise time (10-90%)
u	Spatial frequency
V	Liquid crystal cell voltage
V_0	Binary-0 pixel voltage
V_1	Pixel odd frame voltage
V_1	Binary-1 pixel voltage
V_2	Pixel even frame voltage
V_a	Binary addressing voltage
V_A	Common electrode (ITO) modulation voltage
$V_{\bar{a}}$	Complementary binary addressing voltage
V_b	Bias Voltage
V_C	Common electrode (ITO) potential
V_{DD}	Supply voltage
V_{DS}	Drain to source voltage
V_{elect}	Lamp arc-to-electrode voltage drop
V_g	Ghost voltage

Symbol	Meaning
V_{GS}	Gate to source voltage
V_{LC}	Liquid crystal voltage
V_{PAD}	Pixel voltage
V_{RMS}	Root mean square voltage
V_s	Stable voltage limit
V_{SB}	Substrate bias (source to substrate) voltage
V_t	Thermal potential kT/q
V_T	Threshold voltage
V_{th}	Nematic threshold voltage
V_{TO}	Threshold voltage at $V_{SB}=0$
w	Flicker frequency
W	Channel width
(x_w, y_w)	White point color coordinates
Y	Dice yield

Index

Introduction to Microdisplays D. Armitage, I. Underwood and S-T. Wu

Printed and bound by CPI Group (UK) Ltd, Croydon, CR0 4YY
16/09/2026
14963473-0004